STUDENT'S SOLUTIONS MANUAL

JUDITH A. PENNA

Indiana University Purdue University Indianapolis

ELEMENTARY ALGEBRA: CONCEPTS AND APPLICATIONS

SEVENTH EDITION

Marvin L. Bittinger

Indiana University Purdue University Indianapolis

David J. Ellenbogen

Community College of Vermont

Boston San Francisco New York
London Toronto Sydney Tokyo Singapore Madrid
Mexico City Munich Paris Cape Town Hong Kong Montreal

Reproduced by Pearson Addison-Wesley from electronic files supplied by the author.

Publishing as Pearson Addison-Wesley, 75 Arlington Street, Boston, MA 02116.

ISBN 0-321-26954-3

4 5 6 BB 08 07 06

Contents

Chapter 1

Introduction to Algebraic Expressions

Exercise Set 1.1

1. $4x+7$ does not contain an equals sign, so it is an expression.

3. $2x-5=9$ contains an equals sign, so it is an equation.

5. $38=2t$ contains an equals sign, so it is an equation.

7. $4a-5b$ does not contain an equals sign, so it is an expression.

9. $2x-3y=8$ contains an equals sign, so it is an equation.

11. $7-4rt$ does not contain an equals sign, so it is an expression.

13. Substitute 9 for a and multiply.

$$4a = 4\cdot 9 = 36$$

15. Substitute 7 for x and multiply.

$$8\cdot 7 = 56$$

17. $\dfrac{a}{b}=\dfrac{45}{9}=5$

19. $\dfrac{x+y}{4}=\dfrac{2+14}{4}=\dfrac{16}{4}=4$

21. $\dfrac{15+20}{7}=\dfrac{35}{7}=5$

23. $\dfrac{m-n}{2}=\dfrac{20-8}{2}=\dfrac{12}{2}=6$

25.
$$\begin{aligned} bh &= (6\text{ ft})(4\text{ ft})\\ &= (6)(4)(\text{ft})(\text{ft})\\ &= 24\text{ ft}^2,\text{ or 24 square feet}\end{aligned}$$

27.
$$\begin{aligned} A &= \frac{1}{2}bh\\ &= \frac{1}{2}(5\text{ cm})(6\text{ cm})\\ &= \frac{1}{2}(5)(6)(\text{cm})(\text{cm})\\ &= \frac{5}{2}\cdot 6\text{ cm}^2\\ &= 15\text{ cm}^2,\text{ or 15 square centimeters}\end{aligned}$$

29. $\dfrac{h}{a}=\dfrac{7}{19}\approx 0.368$

31. Let r represent Ron's age. Then we have $r+5$, or $5+r$.

33. $b+6$, or $6+b$

35. $c-9$

37. $6+q$, or $q+6$

39. Let s represent Phil's speed. Then we have $9s$, or $s\cdot 9$.

41. $y-x$

43. $x\div w$, or $\dfrac{x}{w}$

45. $n-m$

47. Let l and h represent the box's length and height, respectively. Then we have $l+h$, or $h+l$.

49. $9\cdot 2m$, or $2m\cdot 9$

51. Let y represent "some number." Then we have $\dfrac{1}{4}y$, or $\dfrac{y}{4}$, or $y/4$, or $y\div 4$.

53. Let x represent the number of women attending. Then we have 64% of x, or $0.64x$.

55.
$$\begin{array}{r|l} \multicolumn{2}{c}{x+17=32} & \text{Writing the equation}\\ \hline 15+17 & 32 & \text{Substituting 15 for } x\\ \multicolumn{2}{c}{=}\\ 32 \ ? & 32 & 32=32 \text{ is TRUE.}\end{array}$$

Since the left-hand and right-hand sides are the same, 15 is a solution.

57.
$$\begin{array}{r|l} \multicolumn{2}{c}{a-28=75} & \text{Writing the equation}\\ \hline 93-28 & 75 & \text{Substituting 93 for } a\\ \multicolumn{2}{c}{=}\\ 65 \ ? & 75 & 65=75 \text{ is FALSE.}\end{array}$$

Since the left-hand and right-hand sides are not the same, 93 is not a solution.

59.
$$\begin{array}{r|l} \multicolumn{2}{c}{\frac{t}{7}=9}\\ \hline \frac{63}{7} & 9\\ \multicolumn{2}{c}{=}\\ 9 \ ? & 9 \quad 9=9 \text{ is TRUE.}\end{array}$$

Since the left-hand and right-hand sides are the same, 63 is a solution.

61.
$$\begin{array}{r|l} \multicolumn{2}{c}{\frac{108}{x}=36}\\ \hline \frac{108}{3} & 36\\ \multicolumn{2}{c}{=}\\ 36 \ ? & 36 \quad 36=36 \text{ is TRUE.}\end{array}$$

Since the left-hand and right-hand sides are the same, 3 is a solution.

63. Let x represent the number.

What number added to 73 is 201?

Translating: $x \quad + \quad 73 \ = \ 201$

$x+73=201$

65. Let y represent the number.

Rewording: 42 times what number is 2352?

Translating: $42 \cdot y = 2352$

$42y = 2352$

67. Let s represent the number of unoccupied squares.

Rewording: The number of unoccupied squares added to 19 is 64.

Translating: $s + 19 = 64$

$s + 19 = 64$

69. Let x represent the total amount of waste generated, in millions of tons.

Rewording: 27% of the total amount of waste is 56 million tons.

Translating: $27\% \cdot x = 56$

$27\% \cdot x = 56$, or $0.27x = 56$

71. The sum of two numbers m and n is $m+n$, and twice the sum is $2(m+n)$. Choice (f) is the correct answer.

73. Two more than a number t is $t+2$. If this expression is equal to 5, we have the equation $t+2=5$. Choice (d) is the correct answer.

75. The sum of a number t and 5 is $t+5$, and 3 times the sum is $3(t+5)$. Choice (g) is the correct answer.

77. The product of two numbers a and b is ab, and 1 less than this product is $ab-1$. If this expression is equal to 49, we have the equation $ab-1=49$. Choice (e) is the correct answer.

79. *Writing Exercise*

81. *Writing Exercise*

83. Area of sign: $A = \frac{1}{2}(3\text{ ft})(2.5\text{ ft}) = 3.75\text{ ft}^2$

Cost of sign: $\$90(3.75) = \$337.50$

85. When x is twice y, then y is one-half x, so $y = \frac{12}{2} = 6$.

$$\frac{x-y}{3} = \frac{12-6}{3} = \frac{6}{3} = 2$$

87. When a is twice b, then b is one-half a, so $b = \frac{16}{2} = 8$.

$$\frac{a+b}{4} = \frac{16+8}{4} = \frac{24}{4} = 6$$

89. The next whole number is one more than $w+3$:

$$w+3+1 = w+4$$

91. $l+w+l+w$, or $2l+2w$

93. If t is Molly's race time, then Joe's race time is $t+3$ and Ellie's race time is

$$t+3+5 = t+8.$$

Exercise Set 1.2

1. Commutative

3. Associative

5. Distributive

7. Associative

9. Commutative

11. $x+7$ Changing the order

13. $c+ab$

15. $3y+9x$

17. $5(1+a)$

19. $a \cdot 2$ Changing the order

21. ts

23. $5+ba$

25. $(a+1)5$

27. $a+(5+b)$

29. $(r+t)+7$

31. $ab+(c+d)$

33. $8(xy)$

35. $(2a)b$

37. $(3 \cdot 2)(a+b)$

39. a) $r+(t+6) = (t+6)+r$ Using the commutative law

$= (6+t)+r$ Using the commutative law again

b) $r+(t+6) = (t+6)+r$ Using the commutative law

$= t+(6+r)$ Using the associative law

Answers may vary.

41. a) $(17a)b = b(17a)$ Using the commutative law

$= b(a17)$ Using the commutative law again

b) $(17a)b = (a17)b$ Using the commutative law

$= a(17b)$ Using the associative law

Answers may vary.

43. $(5 + x) + 2$

$= (x + 5) + 2$ Commutative law

$= x + (5 + 2)$ Associative law

$= x + 7$ Simplifying

45. $(m \cdot 3)7 = m(3 \cdot 7)$ Associative law

$= m \cdot 21$ Simplifying

$= 21m$ Commutative law

47. $4(a + 3) = 4 \cdot a + 4 \cdot 3 = 4a + 12$

49. $6(1 + x) = 6 \cdot 1 + 6 \cdot x = 6 + 6x$

51. $3(x + 1) = 3 \cdot x + 3 \cdot 1 = 3x + 3$

53. $8(3 + y) = 8 \cdot 3 + 8 \cdot y = 24 + 8y$

55. $9(2x + 6) = 9 \cdot 2x + 9 \cdot 6 = 18x + 54$

57. $5(r + 2 + 3t) = 5 \cdot r + 5 \cdot 2 + 5 \cdot 3t = 5r + 10 + 15t$

59. $(a + b)2 = a(2) + b(2) = 2a + 2b$

61. $(x + y + 2)5 = x(5) + y(5) + 2(5) = 5x + 5y + 10$

63. $x + xyz + 19$

The terms are separated by plus signs. They are x, xyz, and 19.

65. $2a + \frac{a}{b} + 5b$

The terms are separated by plus signs. They are $2a$, $\frac{a}{b}$, and $5b$.

67. $2a + 2b = 2(a + b)$ The common factor is 2.

Check: $2(a + b) = 2 \cdot a + 2 \cdot b = 2a + 2b$

69. $7 + 7y = 7 \cdot 1 + 7 \cdot y$ The common factor is 7.

$= 7(1 + y)$ Using the distributive law

Check: $7(1 + y) = 7 \cdot 1 + 7 \cdot y = 7 + 7y$

71. $18x + 3 = 3 \cdot 6x + 3 \cdot 1 = 3(6x + 1)$

Check: $3(6x + 1) = 3 \cdot 6x + 3 \cdot 1 = 18x + 3$

73. $5x + 10 + 15y = 5 \cdot x + 5 \cdot 2 + 5 \cdot 3y = 5(x + 2 + 3y)$

Check: $5(x+2+3y) = 5 \cdot x + 5 \cdot 2 + 5 \cdot 3y = 5x + 10 + 15y$

75. $12x + 9 = 3 \cdot 4x + 3 \cdot 3 = 3(4x + 3)$

Check: $3(4x + 3) = 3 \cdot 4x + 3 \cdot 3 = 12x + 9$

77. $3a + 9b = 3 \cdot a + 3 \cdot 3b = 3(a + 3b)$

Check: $3(a + 3b) = 3 \cdot a + 3 \cdot 3b = 3a + 9b$

79. $44x + 11y + 22z = 11 \cdot 4x + 11 \cdot y + 11 \cdot 2z =$
$11(4x + y + 2z)$

Check: $11(4x + y + 2z) = 11 \cdot 4x + 11 \cdot y + 11 \cdot 2z =$
$44x + 11y + 22z$

81. $st = s \cdot t$

The factors are s and t.

83. $3(x + y) = 3 \cdot (x + y)$

The factors are 3 and $(x + y)$.

85. The factors of $7 \cdot a$ are 7 and a.

87. $(a - b)(x - y) = (a - b) \cdot (x - y)$

The factors are $(a - b)$ and $(x - y)$.

89. *Writing Exercise*

91. Let k represent Kara's salary. Then we have $2k$.

93. *Writing Exercise*

95. The expressions are equivalent by the distributive law.

$8 + 4(a + b) = 8 + 4a + 4b = 4(2 + a + b)$

97. The expressions are equivalent by the commutative law of multiplication and the distributive law.

$(rt + st)5 = 5(rt + st) = 5 \cdot t(r + s) = 5t(r + s)$

99. The expressions are not equivalent.

Let $x = 1$ and $y = 0$. Then we have:

$30 \cdot 0 + 1 \cdot 15 = 0 + 15 = 15$, but

$5[2(1 + 3 \cdot 0)] = 5[2(1)] = 5 \cdot 2 = 10$.

101. *Writing Exercise*

Exercise Set 1.3

1. 9 is composite because it has more than two different factors. They are 1, 3, and 9.

3. 31 is prime because it has only two different factors, 31 and 1.

5. 25 is composite because it has more than two different factors. They are 1, 5, and 25.

7. 2 is prime because it has only two different factors, 2 and 1.

9. The terms "prime" and "composite" apply only to natural numbers. Since 0 is not a natural number, it is neither prime nor composite.

11. Since $35 = 5 \cdot 7$, choice (b) is correct.

13. Since 65 is an odd number and has more than two different factors, choice (d) is correct.

15. We write two factorizations of 50. There are other factorizations as well.

$2 \cdot 25,\ 5 \cdot 10$

List all of the factors of 50:

1, 2, 5, 10, 25, 50

17. We write two factorizations of 42. There are other factorizations as well.

$2 \cdot 21,\ 6 \cdot 7$

List all of the factors of 42:

1, 2, 3, 6, 7, 14, 21, 42

19. $26 = 2 \cdot 13$

21. We begin factoring 30 in any way that we can and continue factoring until each factor is prime.

$30 = 2 \cdot 15 = 2 \cdot 3 \cdot 5$

23. We begin by factoring 27 in any way that we can and continue factoring until each factor is prime.

$27 = 3 \cdot 9 = 3 \cdot 3 \cdot 3$

25. We begin by factoring 18 in any way that we can and continue factoring until each factor is prime.

$18 = 2 \cdot 9 = 2 \cdot 3 \cdot 3$

27. We begin by factoring 40 in any way that we can and continue factoring until each factor is prime.

$40 = 4 \cdot 10 = 2 \cdot 2 \cdot 2 \cdot 5$

29. 43 has exactly two different factors, 43 and 1. Thus, 43 is prime.

31. $210 = 2 \cdot 105 = 2 \cdot 3 \cdot 35 = 2 \cdot 3 \cdot 5 \cdot 7$

33. $115 = 5 \cdot 23$

35. $\frac{14}{21} = \frac{7 \cdot 2}{7 \cdot 3}$ Factoring numerator and denominator

$= \frac{7}{7} \cdot \frac{2}{3}$ Rewriting as a product of two fractions

$= 1 \cdot \frac{2}{3}$ $\quad \frac{7}{7} = 1$

$= \frac{2}{3}$ Using the identity property of 1

37. $\frac{16}{56} = \frac{2 \cdot 8}{7 \cdot 8} = \frac{2}{7} \cdot \frac{8}{8} = \frac{2}{7} \cdot 1 = \frac{2}{7}$

39. $\frac{6}{48} = \frac{1 \cdot 6}{8 \cdot 6}$ Factoring and using the identity property of 1 to write 6 as $1 \cdot 6$

$= \frac{1}{8} \cdot \frac{6}{6}$

$= \frac{1}{8} \cdot 1 = \frac{1}{8}$

41. $\frac{49}{7} = \frac{7 \cdot 7}{1 \cdot 7} = \frac{7}{1} \cdot \frac{7}{7} = \frac{7}{1} \cdot 1 = 7$

43. $\frac{19}{76} = \frac{1 \cdot 19}{4 \cdot 19}$ Factoring and using the identity property of 1 to write 19 as $1 \cdot 19$

$= \frac{1 \cdot \cancel{19}}{4 \cdot \cancel{19}}$ Removing a factor equal to 1: $\frac{19}{19} = 1$

$= \frac{1}{4}$

45. $\frac{150}{25} = \frac{6 \cdot 25}{1 \cdot 25}$ Factoring and using the identity property of 1 to write 25 as $1 \cdot 25$

$= \frac{6 \cdot \cancel{25}}{1 \cdot \cancel{25}}$ Removing a factor equal to 1: $\frac{25}{25} = 1$

$= \frac{6}{1}$

$= 6$ Simplifying

47. $\frac{42}{50} = \frac{2 \cdot 21}{2 \cdot 25}$ Factoring the numerator and the denominator

$= \frac{\cancel{2} \cdot 21}{\cancel{2} \cdot 25}$ Removing a factor equal to 1: $\frac{2}{2} = 1$

$= \frac{21}{25}$

49. $\frac{120}{82} = \frac{2 \cdot 60}{2 \cdot 41}$ Factoring

$= \frac{\cancel{2} \cdot 60}{\cancel{2} \cdot 41}$ Removing a factor equal to 1: $\frac{2}{2} = 1$

$= \frac{60}{41}$

51. $\frac{210}{98} = \frac{2 \cdot 7 \cdot 15}{2 \cdot 7 \cdot 7}$ Factoring

$= \frac{\cancel{2} \cdot \cancel{7} \cdot 15}{\cancel{2} \cdot \cancel{7} \cdot 7}$ Removing a factor equal to 1: $\frac{2 \cdot 7}{2 \cdot 7} = 1$

$= \frac{15}{7}$

53. $\frac{1}{2} \cdot \frac{3}{7} = \frac{1 \cdot 3}{2 \cdot 7}$ Multiplying numerators and denominators

$= \frac{3}{14}$

55. $\frac{9}{2} \cdot \frac{3}{4} = \frac{9 \cdot 3}{2 \cdot 4} = \frac{27}{8}$

57. $\frac{1}{8} + \frac{3}{8} = \frac{1+3}{8}$ Adding numerators; keeping the common denominator

$= \frac{4}{8}$

$= \frac{1 \cdot \cancel{4}}{2 \cdot \cancel{4}} = \frac{1}{2}$ Simplifying

59. $\frac{4}{9}+\frac{13}{18}=\frac{4}{9}\cdot\frac{2}{2}+\frac{13}{18}$ Using 18 as the common denominator

$=\frac{8}{18}+\frac{13}{18}$

$=\frac{21}{18}$

$=\frac{7\cdot\cancel{3}}{6\cdot\cancel{3}}=\frac{7}{6}$ Simplifying

61. $\frac{3}{a}\cdot\frac{b}{7}=\frac{3b}{7a}$ Multiplying numerators and denominators

63. $\frac{4}{a}+\frac{3}{a}=\frac{7}{a}$ Adding numerators; keeping the common denominator

65. $\frac{3}{10}+\frac{8}{15}=\frac{3}{10}\cdot\frac{3}{3}+\frac{8}{15}\cdot\frac{2}{2}$ Using 30 as the common denominator

$=\frac{9}{30}+\frac{16}{30}$

$=\frac{25}{30}$

$=\frac{5\cdot\cancel{5}}{6\cdot\cancel{5}}=\frac{5}{6}$ Simplifying

67. $\frac{9}{7}-\frac{2}{7}=\frac{7}{7}=1$

69. $\frac{13}{18}-\frac{4}{9}=\frac{13}{18}-\frac{4}{9}\cdot\frac{2}{2}$ Using 18 as the common denominator

$=\frac{13}{18}-\frac{8}{18}$

$=\frac{5}{18}$

71. Note that $\frac{20}{30}=\frac{2}{3}$. Thus, $\frac{20}{30}-\frac{2}{3}=0$.
We can also do this exercise by finding a common denominator:
$\frac{20}{30}-\frac{2}{3}=\frac{20}{30}-\frac{20}{30}=0$

73. $\frac{7}{6}\div\frac{3}{5}=\frac{7}{6}\cdot\frac{5}{3}$ Multiplying by the reciprocal of the divisor

$=\frac{35}{18}$

75. $\frac{8}{9}\div\frac{4}{15}=\frac{8}{9}\cdot\frac{15}{4}=\frac{2\cdot\cancel{4}\cdot\cancel{3}\cdot5}{\cancel{3}\cdot3\cdot\cancel{4}}=\frac{10}{3}$

77. $12\div\frac{3}{7}=\frac{12}{1}\cdot\frac{7}{3}=\frac{4\cdot\cancel{3}\cdot7}{1\cdot\cancel{3}}=28$

79. Note that we have a number divided by itself. Thus, the result is 1. We can also do this exercise as follows:
$\frac{7}{13}\div\frac{7}{13}=\frac{7}{13}\cdot\frac{13}{7}=\frac{7\cdot13}{7\cdot13}=1$

81. $\frac{\frac{2}{7}}{\frac{5}{3}}=\frac{2}{7}\div\frac{5}{3}=\frac{2}{7}\cdot\frac{3}{5}=\frac{2\cdot3}{7\cdot5}=\frac{6}{35}$

83. $\frac{9}{\frac{1}{2}}=9\div\frac{1}{2}=\frac{9}{1}\cdot\frac{2}{1}=\frac{9\cdot2}{1\cdot1}=18$

85. *Writing Exercise*

87. $5(x+3)=5(3+x)$ Commutative law of addition
Answers may vary.

89. *Writing Exercise*

91.

Product	56	63	36	72	140	96	168
Factor	7	7	2	36	14	8	8
Factor	8	9	18	2	10	12	21
Sum	15	16	20	38	24	20	29

93. $\frac{16\cdot9\cdot4}{15\cdot8\cdot12}=\frac{\cancel{4}\cdot\cancel{4}\cdot\cancel{3}\cdot\cancel{3}\cdot\cancel{2}\cdot2}{\cancel{3}\cdot5\cdot\cancel{2}\cdot\cancel{4}\cdot\cancel{3}\cdot\cancel{4}}=\frac{2}{5}$

95. $\frac{27pqrs}{9prst}=\frac{3\cdot\cancel{9}\cdot\cancel{p}\cdot q\cdot\cancel{r}\cdot\cancel{s}}{\cancel{9}\cdot\cancel{p}\cdot\cancel{r}\cdot\cancel{s}\cdot t}=\frac{3q}{t}$

97. $\frac{15\cdot4xy\cdot9}{6\cdot25x\cdot15y}=\frac{\cancel{15}\cdot\cancel{2}\cdot2\cdot\cancel{x}\cdot\cancel{y}\cdot\cancel{3}\cdot3}{\cancel{2}\cdot\cancel{3}\cdot25\cdot\cancel{x}\cdot\cancel{15}\cdot\cancel{y}}=\frac{6}{25}$

99. $\frac{\frac{27ab}{15mn}}{\frac{18bc}{25np}}=\frac{27ab}{15mn}\div\frac{18bc}{25np}=\frac{27ab}{15mn}\cdot\frac{25np}{18bc}=$

$\frac{27ab\cdot25np}{15mn\cdot18bc}=\frac{\cancel{3}\cdot\cancel{9}\cdot a\cdot\cancel{b}\cdot\cancel{5}\cdot5\cdot\cancel{n}\cdot p}{\cancel{3}\cdot\cancel{5}\cdot m\cdot\cancel{n}\cdot2\cdot\cancel{9}\cdot\cancel{b}\cdot c}=\frac{5ap}{2mc}$

101. $\frac{5\frac{3}{4}rs}{4\frac{1}{2}st}=\frac{\frac{23}{4}rs}{\frac{9}{2}st}=\frac{\frac{23rs}{4}}{\frac{9st}{2}}=\frac{23rs}{4}\div\frac{9st}{2}=$

$\frac{23rs}{4}\cdot\frac{2}{9st}=\frac{23rs\cdot2}{4\cdot9st}=\frac{23\cdot r\cdot\cancel{s}\cdot\cancel{2}}{\cancel{2}\cdot2\cdot9\cdot\cancel{s}\cdot t}=\frac{23r}{18t}$

103. $A=lw=\left(\frac{4}{5}\text{ m}\right)\left(\frac{7}{9}\text{ m}\right)$

$=\left(\frac{4}{5}\right)\left(\frac{7}{9}\right)(\text{m})(\text{m})$

$=\frac{28}{45}\text{ m}^2$, or $\frac{28}{45}$ square meters

105. $P=4s=4\left(3\frac{5}{9}\text{ m}\right)=4\cdot\frac{32}{9}\text{ m}=\frac{128}{9}\text{ m}$, or $14\frac{2}{9}\text{ m}$

107. There are 12 edges, each with length $2\frac{3}{10}$ cm. We multiply to find the total length of the edges.

$12\cdot2\frac{3}{10}\text{ cm}=12\cdot\frac{23}{10}\text{ cm}$

$=\frac{12\cdot23}{10}\text{ cm}$

$=\frac{\cancel{2}\cdot6\cdot23}{\cancel{2}\cdot5}\text{ cm}$

$=\frac{138}{5}\text{ cm}$, or $27\frac{3}{5}$ cm

Exercise Set 1.4

1. Since $\frac{4}{7} = 0.\overline{571428}$, the correct choice is "repeating."

3. The set of integers consists of all whole numbers along with their opposites, so the correct choice is "integer."

5. A "rational number" has the form described.

7. A "natural number" can be thought of as a counting number.

9. The real number -1349 corresponds to 1349 ft below sea level. The real number 29,035 corresponds to 29,035 ft above sea level

11. The real number 950,000,000 corresponds to 950 million°F. The real number -460 corresponds to 460°F below zero.

13. The real number 2 corresponds to two over par. The real number -6 corresponds to six under par.

15. The real number 750 corresponds to a \$750 deposit, and the real number -125 corresponds to a \$125 withdrawal.

17. The Jets are 34 pins behind, so the real number -34 corresponds to the situation from the Jets' viewpoint. The Strikers are 34 pins ahead, so the real number 34 corresponds to the situation from the Strikers' point of view.

19. Since $\frac{10}{3} = 3\frac{1}{3}$, its graph is $\frac{1}{3}$ of a unit to the right of 3.

$\frac{10}{3}$

$-5\ -4\ -3\ -2\ -1\ 0\ 1\ 2\ 3\ 4\ 5$

21. The graph of -4.3 is $\frac{3}{10}$ of a unit to the left of -4.

-4.3

$-5\ -4\ -3\ -2\ -1\ 0\ 1\ 2\ 3\ 4\ 5$

23. $-5\ -4\ -3\ -2\ -1\ 0\ 1\ 2\ 3\ 4\ 5$

25. $\frac{7}{8}$ means $7 \div 8$, so we divide.

$$\begin{array}{r} 0.8\,7\,5 \\ 8\,\overline{)\,7.0\,0\,0} \\ 6\,4 \\ \hline 6\,0 \\ 5\,6 \\ \hline 4\,0 \\ 4\,0 \\ \hline 0 \end{array}$$

We have $\frac{7}{8} = 0.875$.

27. We first find decimal notation for $\frac{3}{4}$. Since $\frac{3}{4}$ means $3 \div 4$, we divide.

$$\begin{array}{r} 0.7\,5 \\ 4\,\overline{)\,3.0\,0} \\ 2\,8 \\ \hline 2\,0 \\ 2\,0 \\ \hline 0 \end{array}$$

Thus, $\frac{3}{4} = 0.75$, so $-\frac{3}{4} = -0.75$.

29. $\frac{7}{6}$ means $7 \div 6$, so we divide.

$$\begin{array}{r} 1.1\,6\,6 \\ 6\,\overline{)\,7.0\,0\,0} \\ 6 \\ \hline 1\,0 \\ 6 \\ \hline 4\,0 \\ 3\,6 \\ \hline 4\,0 \\ 3\,6 \\ \hline 4 \end{array}$$

We have $\frac{7}{6} = 1.1\overline{6}$.

31. $\frac{2}{3}$ means $2 \div 3$, so we divide.

$$\begin{array}{r} 0.6\,6\,6\,\ldots \\ 3\,\overline{)\,2.0\,0\,0} \\ 1\,8 \\ \hline 2\,0 \\ 1\,8 \\ \hline 2\,0 \\ 1\,8 \\ \hline 2 \end{array}$$

We have $\frac{2}{3} = 0.\overline{6}$.

33. We first find decimal notation for $\frac{1}{2}$. Since $\frac{1}{2}$ means $1 \div 2$, we divide.

$$\begin{array}{r} 0.5 \\ 2\,\overline{)\,1.0} \\ 1\,0 \\ \hline 0 \end{array}$$

Thus, $\frac{1}{2} = 0.5$, so $-\frac{1}{2} = -0.5$.

35. Since the denominator is 100, we know that $\frac{13}{100} = 0.13$. We could also divide 13 by 100 to find this result.

37. Since -9 is to the left of 4, we have $-9 < 4$.

39. Since 7 is to the right of 0, we have $7 > 0$.

41. Since -6 is to the left of 6, we have $-6 < 6$.

43. Since -8 is to the left of -5, we have $-8 < -5$.

45. Since -5 is to the right of -11, we have $-5 > -11$.

47. Since -12.5 is to the left of -9.4, we have $-12.5 < -9.4$.

49. We convert to decimal notation.
$\frac{5}{12} = 0.41\overline{6}$ and $\frac{11}{25} = 0.44$. Thus, $\frac{5}{12} < \frac{11}{25}$.

51. $-7 > x$ has the same meaning as $x < -7$.

53. $-10 \leq y$ has the same meaning as $y \geq -10$.

55. $-3 \geq -11$ is true, since $-3 > -11$ is true.

57. $0 \geq 8$ is false, since neither $0 > 8$ nor $0 = 8$ is true.

59. $-8 \leq -8$ is true because $-8 = -8$ is true.

61. $|-58| = 58$ since -58 is 58 units from 0.

63. $|17| = 17$ since 17 is 17 units from 0.

65. $|5.6| = 5.6$ since 5.6 is 5.6 units from 0.

67. $|329| = 329$ since 329 is 329 units from 0.

69. $\left|-\frac{9}{7}\right| = \frac{9}{7}$ since $-\frac{9}{7}$ is $\frac{9}{7}$ units from 0.

71. $|0| = 0$ since 0 is 0 units from itself.

73. $|x| = |-8| = 8$

75. $-83, -4.7, 0, \frac{5}{9}, 8.31, 62$

77. $-83, 0, 62$

79. All are real numbers.

81. *Writing Exercise*

83. $3xy = 3 \cdot 2 \cdot 7 = 42$

85. *Writing Exercise*

87. *Writing Exercise*

89. List the numbers as they occur on the number line, from left to right: $-23, -17, 0, 4$

91. Converting to decimal notation, we can write
$\frac{4}{5}, \frac{4}{3}, \frac{4}{8}, \frac{4}{6}, \frac{4}{9}, \frac{4}{2}, -\frac{4}{3}$ as
$0.8, 1.3\overline{3}, 0.5, 0.6\overline{6}, 0.4\overline{4}, 2, -1.3\overline{3}$, respectively. List the numbers (in fractional form) as they occur on the number line, from left to right:
$-\frac{4}{3}, \frac{4}{9}, \frac{4}{8}, \frac{4}{6}, \frac{4}{5}, \frac{4}{3}, \frac{4}{2}$

93. $|4| = 4$ and $|-7| = 7$, so $|4| < |-7|$.

95. $|23| = 23$ and $|-23| = 23$, so $|23| = |-23|$.

97. $|x| = 7$

x represents a number whose distance from 0 is 7. Thus, $x = 7$ or $x = -7$.

99. $2 < |x| < 5$

x represents an integer whose distance from 0 is greater than 2 and also less than 5. Thus, $x = -4, -3, 3, 4$

101. $0.9\overline{9} = 3(0.3\overline{3}) = 3 \cdot \frac{1}{3} = \frac{3}{3}$

103. $7.7\overline{7} = 70(0.1\overline{1}) = 70 \cdot \frac{1}{9} = \frac{70}{9}$

(See Exercise 100.)

105. *Writing Exercise*

Exercise Set 1.5

1. Choice (f), $-3n$, has the same variable factor as $8n$.

3. Choice (e), 9, is a constant as is 43.

5. Choice (b), $5x$, has the same variable factor as $-2x$.

7. Start at 5. Move 8 units to the left.

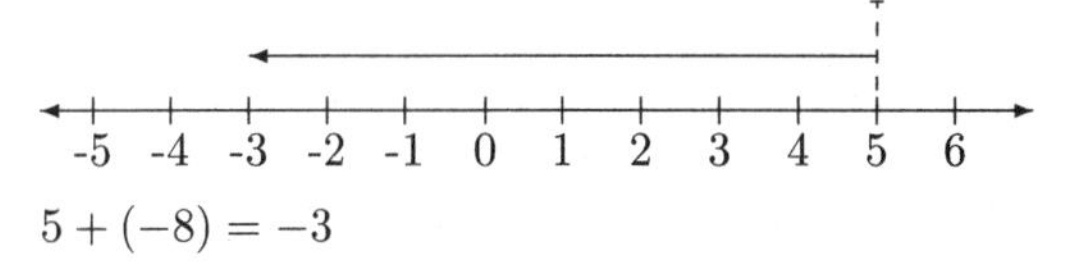

$5 + (-8) = -3$

9. Start at -5. Move 9 units to the right.

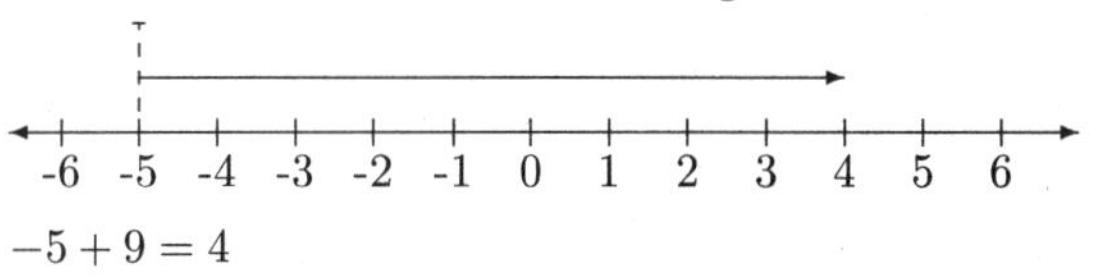

$-5 + 9 = 4$

11. Start at 8. Move 8 units to the left.

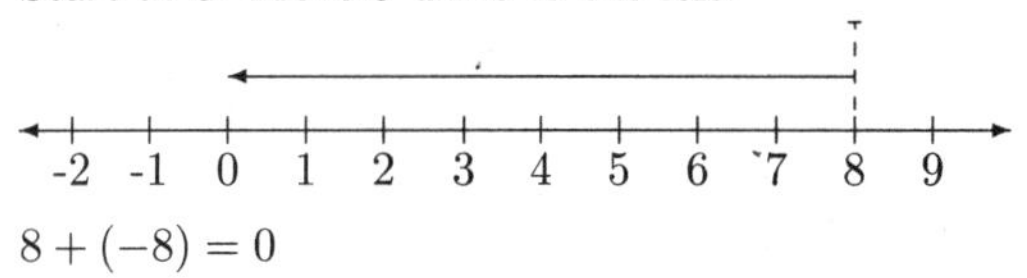

$8 + (-8) = 0$

13. Start at -3. Move 5 units to the left.

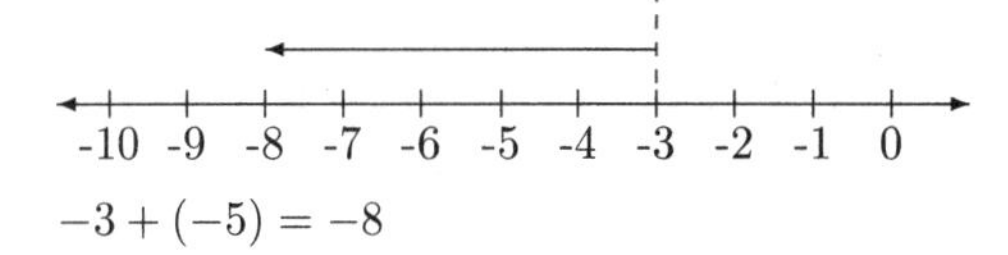

$-3 + (-5) = -8$

15. $-27+0$ One number is 0. The answer is the other number. $-27 + 0 = -27$

17. $0+(-8)$ One number is 0. The answer is the other number. $0 + (-8) = -8$

19. $12 + (-12)$ The numbers have the same absolute value. The sum is 0. $12 + (-12) = 0$

21. $-24 + (-17)$ Two negatives. Add the absolute values, getting 41. Make the answer negative. $-24 + (-17) = -41$

23. $-13 + 13$ The numbers have the same absolute value. The sum is 0. $-13 + 13 = 0$

25. $18 + (-11)$ The absolute values are 18 and 11. The difference is $18 - 11$, or 7. The positive number has the larger absolute value, so the answer is positive. $18 + (-11) = 7$

27. $10 + (-12)$ The absolute values are 10 and 12. The difference is $12 - 10$, or 2. The negative number has the larger absolute value, so the answer is negative. $10 + (-12) = -2$

29. $-3+14$ The absolute values are 3 and 14. The difference is $14-3$, or 11. The positive number has the larger absolute value, so the answer is positive. $-3+14=11$

31. $-14+(-19)$ Two negatives. Add the absolute values, getting 33. Make the answer negative.
$-14+(-19)=-33$

33. $19+(-19)$ The numbers have the same absolute value. The sum is 0. $19+(-19)=0$

35. $23+(-5)$ The absolute values are 23 and 5. The difference is $23-5$ or 18. The positive number has the larger absolute value, so the answer is positive. $23+(-5)=18$

37. $-31+(-14)$ Two negatives. Add the absolute values, getting 45. Make the answer negative.
$-31+(-14)=-45$

39. $40+(-40)$ The numbers have the same absolute value. The sum is 0. $40+(-40)=0$

41. $85+(-65)$ The absolute values are 85 and 65. The difference is $85-65$, or 20. The positive number has the larger absolute value, so the answer is positive. $85+(-65)=20$

43. $-3.6+1.9$ The absolute values are 3.6 and 1.9. The difference is $3.6-1.9$, or 1.7. The negative number has the larger absolute value, so the answer is negative. $-3.6+1.9=-1.7$

45. $-5.4+(-3.7)$ Two negatives. Add the absolute values, getting 9.1. Make the answer negative. $-5.4+(-3.7)=-9.1$

47. $\frac{-3}{5}+\frac{4}{5}$ The absolute values are $\frac{3}{5}$ and $\frac{4}{5}$. The difference is $\frac{4}{5}-\frac{3}{5}$, or $\frac{1}{5}$. The positive number has the larger absolute value, so the answer is positive.
$\frac{-3}{5}+\frac{4}{5}=\frac{1}{5}$

49. $\frac{-4}{7}+\frac{-2}{7}$ Two negatives. Add the absolute values, getting $\frac{6}{7}$. Make the answer negative.
$\frac{-4}{7}+\frac{-2}{7}=\frac{-6}{7}$

51. $-\frac{2}{5}+\frac{1}{3}$ The absolute values are $\frac{2}{5}$ and $\frac{1}{3}$. The difference is $\frac{6}{15}-\frac{5}{15}$, or $\frac{1}{15}$. The negative number has the larger absolute value, so the answer is negative.
$-\frac{2}{5}+\frac{1}{3}=-\frac{1}{15}$

53. $\frac{-4}{9}+\frac{2}{3}$ The absolute values are $\frac{4}{9}$ and $\frac{2}{3}$. The difference is $\frac{6}{9}-\frac{4}{9}$, or $\frac{2}{9}$. The positive number has the larger absolute value, so the answer is positive.
$\frac{-4}{9}+\frac{2}{3}=\frac{2}{9}$

55. $35+(-14)+(-19)+(-5)$
$=35+[(-14)+(-19)+(-5)]$ Using the associative law of addition
$=35+(-38)$ Adding the negatives
$=-3$ Adding a positive and a negative

57. $-4.9+8.5+4.9+(-8.5)$
Note that we have two pairs of numbers with different signs and the same absolute value: -4.9 and 4.9, 8.5 and -8.5. The sum of each pair is 0, so the result is $0+0$, or 0.

59. Rewording: First increase plus decrease plus second increase is change in price.

Translating: $5¢ + (-3¢) + 7¢ =$ change in price

Since $5+(-3)+7$
$=2+7$
$=9$,
the price rose 9¢ during the given period.

61. Rewording: July bill plus payment plus August charges is new balance.

Translating: $-82 + 50 + (-63) =$ new balance

Since $-82+50+(-63)=-32+(-63)$
$=-95$,
Maya's new balance was \$95.

63. Rewording: First try yardage plus second try yardage plus third try yardage is total gain or loss.

Translating: $13 + (-12) + 21 =$ total gain or loss

Since $13+(-12)+21$
$=1+21$
$=22$,
the total gain was 22 yd.

65. Rewording: Original balance plus first payment plus

Translating: -470 $+$ 45 $+$

new charges plus second payment is new balance.

-160 $+$ 500 $=$ new balance

Since $-470+45+(-160)+500$

$= [-470+(-160)]+(45+500)$

$= -630+545$

$= -85,$

Lyle owes \$85 on his credit card.

67. Rewording: Elevation of base plus total height is elevation of peak.

Translating: $-19,684$ $+$ $33,480$ $=$ elevation of peak.

Since $-19,684+33,480=13,796$, the elevation of the peak is 13,796 ft above sea level.

69. $8a+6a=(8+6)a$ Using the distributive law

$=14a$

71. $-3x+12x=(-3+12)x$ Using the distributive law

$=9x$

73. $5t+8t=(5+8)t=13t$

75. $7m+(-9m)=[7+(-9)]m=-2m$

77. $-5a+(-2a)=[-5+(-2)]a=-7a$

79. $-3+8x+4+(-10x)$

$=-3+4+8x+(-10x)$ Using the commutative law of addition

$=(-3+4)+[8+(-10)]x$ Using the distributive law

$=1-2x$ Adding

81. Perimeter $=8+5x+9+7x$

$=8+9+5x+7x$

$=(8+9)+(5+7)x$

$=17+12x$

83. Perimeter $=3t+3r+7+5t+9+4r$

$=3t+5t+3r+4r+7+9$

$=(3+5)t+(3+4)r+(7+9)$

$=8t+7r+16$

85. Perimeter $=9+6n+7+8n+4n$

$=9+7+6n+8n+4n$

$=(9+7)+(6+8+4)n$

$=16+18n$

87. *Writing Exercise*

89. $7(3z+y+2)=7\cdot 3z+7\cdot y+7\cdot 2=21z+7y+14$

91. *Writing Exercise*

93. Starting with the final value, we "undo" the rise and drop in value by adding their opposites. The result is the original value.

Rewording: Final value plus opposite of rise plus

Translating: 64.38 $+$ (-2.38) $+$

opposite of drop is original value.

3.25 $=$ original value.

Since $64.38+(-2.38)+3.25=62.00+3.25$

$=65.25,$

the stock's original value was \$65.25.

95. $4x+\_\_+(-9x)+(-2y)$

$=4x+(-9x)+\_\_+(-2y)$

$=[4+(-9)]x+\_\_+(-2y)$

$=-5x+\_\_+(-2y)$

This expression is equivalent to $-5x-7y$, so the missing term is the term which yields $-7y$ when added to $-2y$. Since $-5y+(-2y)=-7y$, the missing term is $-5y$.

97. $3m+2n+\_\_+(-2m)$

$=2n+\_\_+(-2m)+3m$

$=2n+\_\_+(-2+3)m$

$=2n+\_\_+m$

This expression is equivalent to $2n+(-6m)$, so the missing term is the term which yields $-6m$ when added to m. Since $-7m+m=-6m$, the missing term is $-7m$.

99. Note that, in order for the sum to be 0, the two missing terms must be the opposites of the given terms. Thus, the missing terms are $-7t$ and -23.

101. $-3+(-3)+2+(-2)+1=-5$

Since the total is 5 under par after the five rounds and $-5=-1+(-1)+(-1)+(-1)+(-1)$, the golfer was 1 under par on average.

Exercise Set 1.6

1. $-x$ is read "the opposite of x," so choice (d) is correct.

3. $12 - (-x)$ is read "twelve minus the opposite of x," so choice (f) is correct.

5. $x - (-12)$ is read "x minus negative twelve," so choice (a) is correct.

7. $-x - x$ is read "the opposite of x minus x," so choice (b) is correct.

9. $4 - 10$ is read "four minus ten."

11. $2 - (-9)$ is read "two minus negative nine."

13. $9 - (-t)$ is read "nine minus the opposite of t."

15. $-x - y$ is read "the opposite of x minus y."

17. $-3 - (-n)$ is read "negative three minus the opposite of n."

19. The opposite of 39 is -39 because $39 + (-39) = 0$.

21. The opposite of -9 is 9 because $-9 + 9 = 0$.

23. The opposite of -3.14 is 3.14 because $-3.14 + 3.14 = 0$.

25. If $x = 23$, then $-x = -(23) = -23$. (The opposite of 23 is -23.)

27. If $x = -\frac{14}{3}$, then $-x = -\left(-\frac{14}{3}\right) = \frac{14}{3}$.

$\left(\text{The opposite of } -\frac{14}{3} \text{ is } \frac{14}{3}.\right)$

29. If $x = 0.101$, then $-x = -(0.101) = -0.101$.

(The opposite of 0.101 is -0.101.)

31. If $x = -72$, then $-(-x) = -(-72) = 72$

(The opposite of the opposite of 72 is 72.)

33. If $x = -\frac{2}{5}$, then $-(-x) = -\left[-\left(-\frac{2}{5}\right)\right] = -\frac{2}{5}$.

$\left(\text{The opposite of the opposite of } -\frac{2}{5} \text{ is } -\frac{2}{5}.\right)$

35. When we change the sign of -1 we obtain 1.

37. When we change the sign of 7 we obtain -7.

39. $6 - 8 = 6 + (-8) = -2$

41. $0 - 5 = 0 + (-5) = -5$

43. $3 - 9 = 3 + (-9) = -6$

45. $0 - 10 = 0 + (-10) = -10$

47. $-9 - (-3) = -9 + 3 = -6$

49. Note that we are subtracting a number from itself. The result is 0. We could also do this exercise as follows:

$$-8 - (-8) = -8 + 8 = 0$$

51. $14 - 19 = 14 + (-19) = -5$

53. $30 - 40 = 30 + (-40) = -10$

55. $-7 - (-9) = -7 + 9 = 2$

57. $-9 - (-9) = -9 + 9 = 0$

(See Exercise 49.)

59. $5 - 5 = 5 + (-5) = 0$

(See Exercise 49.)

61. $4 - (-4) = 4 + 4 = 8$

63. $-7 - 4 = -7 + (-4) = -11$

65. $6 - (-10) = 6 + 10 = 16$

67. $-4 - 15 = -4 + (-15) = -19$

69. $-6 - (-5) = -6 + 5 = -1$

71. $5 - (-12) = 5 + 12 = 17$

73. $0 - 5 = 0 + (-5) = -5$

75. $-5 - (-2) = -5 + 2 = -3$

77. $-7 - 14 = -7 + (-14) = -21$

79. $0 - (-5) = 0 + 5 = 5$

81. $-8 - 0 = -8 + 0 = -8$

83. $3 - (-7) = 3 + 7 = 10$

85. $2 - 25 = 2 + (-25) = -23$

87. $-42 - 26 = -42 + (-26) = -68$

89. $-51 - 7 = -51 + (-7) = -58$

91. $3.2 - 8.7 = 3.2 + (-8.7) = -5.5$

93. $0.072 - 1 = 0.072 + (-1) = -0.928$

95. $\frac{2}{11} - \frac{9}{11} = \frac{2}{11} + \left(-\frac{9}{11}\right) = -\frac{7}{11}$

97. $\frac{-1}{5} - \frac{3}{5} = \frac{-1}{5} + \left(\frac{-3}{5}\right) = \frac{-4}{5}$, or $-\frac{4}{5}$

99. $-\frac{4}{17} - \left(-\frac{9}{17}\right) = -\frac{4}{17} + \frac{9}{17} = \frac{5}{17}$

101. We subtract the smaller number from the larger.

Translate: $3.8 - (-5.2)$

Simplify: $3.8 - (-5.2) = 3.8 + 5.2 = 9$

103. We subtract the smaller number from the larger.

Translate: $114 - (-79)$

Simplify: $114 - (-79) = 114 + 79 = 193$

105. $-21 - 37 = -21 + (-37) = -58$

107. $9 - (-25) = 9 + 25 = 34$

109. $25 - (-12) - 7 - (-2) + 9 = 25 + 12 + (-7) + 2 + 9 = 41$

111. $-31 + (-28) - (-14) - 17 = (-31) + (-28) + 14 + (-17) = -62$

113. $-34-28+(-33)-44=(-34)+(-28)+(-33)+(-44)=-139$

115. $-93+(-84)-(-93)-(-84)$

Note that we are subtracting -93 from -93 and -84 from -84. Thus, the result will be 0. We could also do this exercise as follows:

$$-93+(-84)-(-93)-(-84)=-93+(-84)+93+84=0$$

117. $-7x-4y=-7x+(-4y)$, so the terms are $-7x$ and $-4y$.

119. $9-5t-3st=9+(-5t)+(-3st)$, so the terms are 9, $-5t$, and $-3st$.

121. $4x-7x$

$= 4x+(-7x)$ Adding the opposite

$= (4+(-7))x$ Using the distributive law

$= -3x$

123. $7a-12a+4$

$= 7a+(-12a)+4$ Adding the opposite

$= (7+(-12))a+4$ Using the distributive law

$= -5a+4$

125. $-8n-9+n$

$= -8n+(-9)+n$ Adding the opposite

$= -8n+n+(-9)$ Using the commutative law of addition

$= -7n-9$ Adding like terms

127. $3x+5-9x$

$= 3x+5+(-9x)$

$= 3x+(-9x)+5$

$= -6x+5$

129. $2-6t-9-2t$

$= 2+(-6t)+(-9)+(-2t)$

$= 2+(-9)+(-6t)+(-2t)$

$= -7-8t$

131. $5y+(-3x)-9x+1-2y+8$

$= 5y+(-3x)+(-9x)+1+(-2y)+8$

$= 5y+(-2y)+(-3x)+(-9x)+1+8$

$= 3y-12x+9$

133. $13x-(-2x)+45-(-21)-7x$

$= 13x+2x+45+21+(-7x)$

$= 13x+2x+(-7x)+45+21$

$= 8x+66$

135. We subtract the lower temperature from the higher temperature:

$25-(-125)=25+125=150$

The temperature range is 150°C.

137. We subtract the lower elevation from the higher elevation:

$$29,035-(-1349)=29,035+1349=30,384$$

The difference in elevation is 30,384 ft.

139. We subtract the lower elevation from the higher elevation:

$$-40-(-156)=-40+156=116$$

Lake Assal is 116 m lower than the Valdes Peninsula.

141. *Writing Exercise*

143. Area $= lw =$ (36 ft)(12 ft)=432 ft^2

145. *Writing Exercise*

147. If the clock reads 8:00 A.M. on the day following the blackout when the actual time is 3:00 P.M., then the clock is 7 hr behind the actual time. This indicates that the power outage lasted 7 hr, so power was restored 7 hr after 4:00 P.M., or at 11:00 P.M. on August 14.

149. False. For example, let $m=-3$ and $n=-5$. Then $-3>-5$, but $-3+(-5)=-8 \not> 0$.

151. True. For example, for $m=4$ and $n=-4$, $4=-(-4)$ and $4+(-4)=0$; for $m=-3$ and $n=3$, $-3=-3$ and $-3+3=0$.

153. [(-)] [9] [-] [(-)] [7] [ENTER]

Exercise Set 1.7

1. The product of two reciprocals is 1.

3. The sum of a pair of additive inverses is 0.

5. The number 0 has no reciprocal.

7. The number 1 is the multiplicative identity.

9. A nonzero number divided by itself is 1.

11. $-3\cdot 8=-24$ Think: $3\cdot 8=24$, make the answer negative.

13. $-8\cdot 7=-56$ Think: $8\cdot 7=56$, make the answer negative.

15. $8\cdot(-3)=-24$

17. $-9\cdot 8=-72$

19. $-6\cdot(-7)=42$ Multiplying absolute values; the answer is positive.

21. $-5\cdot(-9)=45$ Multiplying absolute values; the answer is positive.

23. $-19\cdot(-10)=190$

25. $-12\cdot 12=-144$

27. $-25\cdot(-48)=1200$

29. $-3.5\cdot(-28)=98$

31. $6\cdot(-13)=-78$

33. $-7\cdot(-3.1)=21.7$

35. $\frac{2}{3}\cdot\left(-\frac{3}{5}\right)=-\left(\frac{2\cdot 3}{3\cdot 5}\right)=-\left(\frac{2}{5}\cdot\frac{3}{3}\right)=-\frac{2}{5}$

37. $-\frac{3}{8}\cdot\left(-\frac{2}{9}\right)=\frac{\cancel{3}\cdot\cancel{2}\cdot 1}{4\cdot\cancel{2}\cdot\cancel{3}\cdot 3}=\frac{1}{12}$

39. $(-5.3)(2.1)=-11.13$

41. $-\frac{5}{9}\cdot\frac{3}{4}=-\frac{5\cdot\cancel{3}}{\cancel{3}\cdot 3\cdot 4}=-\frac{5}{12}$

43. $3\cdot(-7)\cdot(-2)\cdot 6$

$=-21\cdot(-12)$ Multiplying the first two numbers and the last two numbers

$=252$

45. 0, The product of 0 and any real number is 0.

47. $-\frac{1}{3}\cdot\frac{1}{4}\cdot\left(-\frac{3}{7}\right)=-\frac{1}{12}\cdot\left(-\frac{3}{7}\right)=\frac{3}{12\cdot 7}=$

$\frac{\cancel{3}\cdot 1}{\cancel{3}\cdot 4\cdot 7}=\frac{1}{28}$

49. $-2\cdot(-5)\cdot(-3)\cdot(-5)=10\cdot 15=150$

51. 0, The product of 0 and any real number is 0.

53. $(-8)(-9)(-10)=72(-10)=-720$

55. $(-6)(-7)(-8)(-9)(-10)=42\cdot 72\cdot(-10)=3024\cdot(-10)=-30,240$

57. $14\div(-2)=-7$ Check: $-7\cdot(-2)=14$

59. $\frac{36}{-9}=-4$ $-4\cdot(-9)=36$

61. $\frac{-56}{8}=-7$ Check: $-7\cdot 8=-56$

63. $\frac{-48}{-12}=4$ Check: $4(-12)=-48$

65. $\frac{-72}{9}=-8$ Check: $-8\cdot 9=-72$

67. $-100\div(-50)=2$ Check: $2(-50)=-100$

69. $-108\div 9=-12$ Check: $-12\cdot 9=-108$

71. $\frac{400}{-50}=-8$ Check: $-8\cdot(-50)=400$

73. Undefined

75. $-4.8\div 1.2=-4$ Check: $-4(1.2)=-4.8$

77. $\frac{0}{-9}=0$

79. $\frac{9.7(-2.8)0}{4.3}$

Since the numerator has a factor of 0, the product in the numerator is 0. The denominator is nonzero, so the quotient is 0.

81. $\frac{-8}{3}=\frac{8}{-3}$ and $\frac{-8}{3}=-\frac{8}{3}$

83. $\frac{29}{-35}=\frac{-29}{35}$ and $\frac{29}{-35}=-\frac{29}{35}$

85. $-\frac{7}{3}=\frac{-7}{3}$ and $-\frac{7}{3}=\frac{7}{-3}$

87. $\frac{-x}{2}=\frac{x}{-2}$ and $\frac{-x}{2}=-\frac{x}{2}$

89. The reciprocal of $\frac{4}{-5}$ is $\frac{-5}{4}$ $\left(\text{or equivalently, } -\frac{5}{4}\right)$ because $\frac{4}{-5}\cdot\frac{-5}{4}=1$.

91. The reciprocal of $-\frac{47}{13}$ is $-\frac{13}{47}$ because $-\frac{47}{13}\cdot\left(-\frac{13}{47}\right)=1$.

93. The reciprocal of -10 is $\frac{1}{-10}$ $\left(\text{or equivalently, } -\frac{1}{10}\right)$ because $-10\left(\frac{1}{-10}\right)=1$.

95. The reciprocal of 4.3 is $\frac{1}{4.3}$ because $4.3\left(\frac{1}{4.3}\right)=1$.

Since $\frac{1}{4.3}=\frac{1}{4.3}\cdot\frac{10}{10}=\frac{10}{43}$, the reciprocal can also be expressed as $\frac{10}{43}$.

97. The reciprocal of $\frac{-9}{4}$ is $\frac{4}{-9}$ $\left(\text{or equivalently, } -\frac{4}{9}\right)$ because $\frac{-9}{4}\cdot\frac{4}{-9}=1$.

99. The reciprocal of 0 does not exist. (There is no number n for which $0\cdot n=1$.)

101. $\left(\frac{-7}{4}\right)\left(-\frac{3}{5}\right)$

$=\left(-\frac{7}{4}\right)\left(-\frac{3}{5}\right)$ Rewriting $\frac{-7}{4}$ as $-\frac{7}{4}$

$=\frac{21}{20}$

103. $\left(\frac{-6}{5}\right)\left(\frac{2}{-11}\right)$

$=\left(\frac{-6}{5}\right)\left(\frac{-2}{11}\right)$ Rewriting $\frac{2}{-11}$ as $\frac{-2}{11}$

$=\frac{12}{55}$

105. $\frac{-3}{8}+\frac{-5}{8}=\frac{-8}{8}=-1$

107. $\left(\frac{-9}{5}\right)\left(\frac{5}{-9}\right)$

Note that this is the product of reciprocals. Thus, the result is 1.

109. $\left(-\frac{3}{11}\right)+\left(-\frac{6}{11}\right)=-\frac{9}{11}$

111. $\frac{7}{8}\div\left(-\frac{1}{2}\right)=\frac{7}{8}\cdot\left(-\frac{2}{1}\right)=-\frac{14}{8}=-\frac{7\cdot\cancel{2}}{\cancel{2}\cdot 4\cdot 1}=-\frac{7}{4}$

113. $\frac{9}{5}\cdot\frac{-20}{3}=\frac{9}{5}\left(-\frac{20}{3}\right)=-\frac{180}{15}=-\frac{\cancel{3}\cdot 3\cdot 4\cdot\cancel{5}}{\cancel{5}\cdot\cancel{3}\cdot 1}=-12$

115. $\left(-\frac{18}{7}\right)+\left(-\frac{3}{7}\right)=-\frac{21}{7}=-3$

117. $-\frac{5}{9} \div \left(-\frac{5}{9}\right)$

Note that we have a number divided by itself. Thus, the result is 1.

119. $-44.1 \div (-6.3) = 7$ Do the long division. The answer is positive.

121. $\frac{5}{9} - \frac{7}{9} = -\frac{2}{9}$

123. $\frac{-3}{10} + \frac{2}{5} = \frac{-3}{10} + \frac{2}{5} \cdot \frac{2}{2} = \frac{-3}{10} + \frac{4}{10} = \frac{1}{10}$

125. $\frac{7}{10} \div \left(\frac{-3}{5}\right) = \frac{7}{10} \div \left(-\frac{3}{5}\right) = \frac{7}{10} \cdot \left(-\frac{5}{3}\right) = -\frac{35}{30} =$
$-\frac{7 \cdot \cancel{5}}{2 \cdot \cancel{5} \cdot 3} = -\frac{7}{6}$

127. $\frac{5}{7} - \frac{1}{-7} = \frac{5}{7} - \left(-\frac{1}{7}\right) = \frac{5}{7} + \frac{1}{7} = \frac{6}{7}$

129. $\frac{-4}{15} + \frac{2}{-3} = \frac{-4}{15} + \frac{-2}{3} = \frac{-4}{15} + \frac{-2}{3} \cdot \frac{5}{5} = \frac{-4}{15} + \frac{-10}{15} =$
$\frac{-14}{15}$, or $-\frac{14}{15}$

131. *Writing Exercise*

133. $\frac{264}{468} = \frac{\cancel{2} \cdot \cancel{2} \cdot 2 \cdot \cancel{3} \cdot 11}{\cancel{2} \cdot \cancel{2} \cdot \cancel{3} \cdot 3 \cdot 13} = \frac{22}{39}$

135. *Writing Exercise*

137. Consider the sum $2 + 3$. Its reciprocal is $\frac{1}{2+3}$, or $\frac{1}{5}$, but $\frac{1}{2} + \frac{1}{3} = \frac{5}{6}$.

139. When n is negative, $-n$ is positive, so $\frac{m}{-n}$ is the quotient of a negative and a positive number and, thus, is negative.

141. When n is negative, $-n$ is positive, so $\frac{-n}{m}$ is the quotient of a positive and a negative number and, thus, is negative. When m is negative, $-m$ is positive, so $-m \cdot \left(\frac{-n}{m}\right)$ is the product of a positive and a negative number and, thus, is negative.

143. $m + n$ is the sum of two negative numbers, so it is negative; $\frac{m}{n}$ is the quotient of two negative numbers, so it is positive. Then $(m + n) \cdot \frac{m}{n}$ is the product of a negative and a positive number and, thus, is negative.

145. a) m and n have different signs;

b) either m or n is zero;

c) m and n have the same sign

147. *Writing Exercise*

Exercise Set 1.8

1. a) $4 + 8 \div 2 \cdot 2$

There are no grouping symbols or exponential expressions, so we multiply and divide from left to right. This means that we divide first.

b) $7 - 9 + 15$

There are no grouping symbols, exponential expressions, multiplications, or divisions, so we add and subtract from left to right. This means that we subtract first.

c) $5 - 2(3 + 4)$

We perform the operation in the parentheses first. This means that we add first.

d) $6 + 7 \cdot 3$

There are no grouping symbols or exponential expressions, so we multiply and divide from left to right. This means that we multiply first.

e) $18 - 2[4 + (3 - 2)]$

We perform the operation in the innermost grouping symbols first. This means that we perform the subtraction in the parentheses first.

f) $\frac{5 - 6 \cdot 7}{2}$

Since the denominator does not need to be simplified, we consider the numerator. There are no grouping symbols or exponential expressions, so we multiply and divide from left to right. This means that we multiply first.

3. $\underbrace{2 \cdot 2 \cdot 2}_{\text{3 factors}} = 2^3$

5. $\underbrace{x \cdot x \cdot x \cdot x \cdot x \cdot x \cdot x}_{\text{7 factors}} = x^7$

7. $3t \cdot 3t \cdot 3t \cdot 3t \cdot 3t = (3t)^5$

9. $3^2 = 3 \cdot 3 = 9$

11. $(-4)^2 = (-4)(-4) = 16$

13. $-4^2 = -(4 \cdot 4) = -16$

15. $4^3 = 4 \cdot 4 \cdot 4 = 16 \cdot 4 = 64$

17. $(-5)^4 = (-5)(-5)(-5)(-5) = 25 \cdot 25 = 625$

19. $7^1 = 7$ (1 factor)

21. $(3t)^4 = (3t)(3t)(3t)(3t) =$
$3 \cdot 3 \cdot 3 \cdot 3 \cdot t \cdot t \cdot t \cdot t = 81t^4$

23. $(-7x)^3 = (-7x)(-7x)(-7x) =$
$(-7)(-7)(-7)(x)(x)(x) = -343x^3$

25. $5 + 3 \cdot 7 = 5 + 21$ Multiplying
$= 26$ Adding

27. $8 \cdot 7 + 6 \cdot 5 = 56 + 30$ Multiplying
$= 86$ Adding

29. $19 - 5 \cdot 3 + 3 = 19 - 15 + 3$ Multiplying
$= 4 + 3$ Subtracting and add-
$= 7$ ing from left to right

31. $9 \div 3 + 16 \div 8 = 3 + 2$ Dividing
$= 5$ Adding

33. $14 \cdot 19 \div (19 \cdot 14)$

Since $14 \cdot 19$ and $19 \cdot 14$ are equivalent, we are dividing the product $14 \cdot 19$ by itself. Thus the result is 1.

35. $3(-10)^2 - 8 \div 2^2$
$= 3(100) - 8 \div 4$ Simplifying the exponential expressions
$= 300 - 8 \div 4$ Multiplying and
$= 300 - 2$ dividing from left to right
$= 298$ Subtracting

37. $8 - (2 \cdot 3 - 9)$
$= 8 - (6 - 9)$ Multiplying inside the parentheses
$= 8 - (-3)$ Subtracting inside the parentheses
$= 8 + 3$ Removing parentheses
$= 11$ Adding

39. $(8 - 2)(3 - 9)$
$= 6(-6)$ Subtracting inside the parentheses
$= -36$ Multiplying

41. $13(-10)^2 + 45 \div (-5)$
$= 13(100) + 45 \div (-5)$ Simplifying the exponential expression
$= 1300 + 45 \div (-5)$ Multiplying and
$= 1300 - 9$ dividing from left to right
$= 1291$ Subtracting

43. $2^4 + 2^3 - 10 \div (-1)^4 = 16 + 8 - 10 \div 1 =$
$16 + 8 - 10 = 24 - 10 = 14$

45. $5 + 3(2 - 9)^2 = 5 + 3(-7)^2 = 5 + 3 \cdot 49 = 5 + 147 = 152$

47. $[2 \cdot (5 - 8)]^2 = [2 \cdot (-3)]^2 = (-6)^2 = 36$

49. $\dfrac{7 + 2}{5^2 - 4^2} = \dfrac{9}{25 - 16} = \dfrac{9}{9} = 1$

51. $8(-7) + |6(-5)| = -56 + |-30| = -56 + 30 = -26$

53. $\dfrac{(-2)^3 + 4^2}{3 - 5^2 + 3 \cdot 6} = \dfrac{-8 + 16}{3 - 25 + 3 \cdot 6} = \dfrac{8}{3 - 25 + 18} =$
$\dfrac{8}{-22 + 18} = \dfrac{8}{-4} = -2$

55. $\dfrac{27 - 2 \cdot 3^2}{8 \div 2^2 - (-2)^2} = \dfrac{27 - 2 \cdot 9}{8 \div 4 - 4} = \dfrac{27 - 18}{2 - 4} = \dfrac{9}{-2} = -\dfrac{9}{2}$

57. $9 - 4x = 9 - 4 \cdot 5$ Substituting 5 for x
$= 9 - 20$ Multiplying
$= -11$ Subtracting

59. $24 \div t^3$
$= 24 \div (-2)^3$ Substituting -2 for t
$= 24 \div (-8)$ Simplifying the exponential expression
$= -3$ Dividing

61. $45 \div 3 \cdot a = 45 \div 3 \cdot (-1)$ Substituting -1 for a
$= 15 \cdot (-1)$ Dividing
$= -15$ Multiplying

63. $5x \div 15x^2$
$= 5 \cdot 3 \div 15(3)^2$ Substituting 3 for x
$= 5 \cdot 3 \div 15 \cdot 9$ Simplifying the exponential expression
$= 15 \div 15 \cdot 9$ Multiplying and dividing
$= 1 \cdot 9$ in order from
$= 9$ left to right

65. $45 \div 3^2 x(x - 1)$
$= 45 \div 3^2 \cdot 3(3 - 1)$ Substituting 3 for x
$= 45 \div 3^2 \cdot 3(2)$ Subtracting inside the parentheses
$= 45 \div 9 \cdot 3(2)$ Evaluating the exponential expression
$= 5 \cdot 3(2)$ Dividing and
$= 15(2)$ multiplying
$= 30$ from left to right

67. $-x^2 - 5x = -(-3)^2 - 5(-3) = -9 - 5(-3) =$
$-9 + 15 = 6$

69. $\dfrac{3a - 4a^2}{a^2 - 20} = \dfrac{3 \cdot 5 - 4(5)^2}{(5)^2 - 20} = \dfrac{3 \cdot 5 - 4 \cdot 25}{25 - 20} =$
$\dfrac{15 - 100}{5} = \dfrac{-85}{5} = -17$

71. $-(9x + 1) = -9x - 1$ Removing parentheses and changing the sign of each term

73. $-[5 - 6x] = -5 + 6x$ Removing grouping symbols and changing the sign of each term

75. $-(4a - 3b + 7c) = -4a + 3b - 7c$

77. $-(3x^2 + 5x - 1) = -3x^2 - 5x + 1$

79. $8x - (6x + 7)$
$= 8x - 6x - 7$ Removing parentheses and changing the sign of each term
$= 2x - 7$ Collecting like terms

81. $2a - (5a - 9) = 2a - 5a + 9 = -3a + 9$

83. $2x + 7x - (4x + 6) = 2x + 7x - 4x - 6 = 5x - 6$

85. $9t - 5r - 2(3r + 6t) = 9t - 5r - 6r - 12t = -3t - 11r$

87. $15x - y - 5(3x - 2y + 5z)$
$= 15x - y - 15x + 10y - 25z$ Multiplying each term in parentheses by -5
$= 9y - 25z$

89. $3x^2 + 7 - (2x^2 + 5) = 3x^2 + 7 - 2x^2 - 5$
$= x^2 + 2$

91. $5t^3 + t - 3(t + 2t^3) = 5t^3 + t - 3t - 6t^3$
$= -t^3 - 2t$

93. $12a^2 - 3ab + 5b^2 - 5(-5a^2 + 4ab - 6b^2)$
$= 12a^2 - 3ab + 5b^2 + 25a^2 - 20ab + 30b^2$
$= 37a^2 - 23ab + 35b^2$

95. $-7t^3 - t^2 - 3(5t^3 - 3t)$
$= -7t^3 - t^2 - 15t^3 + 9t$
$= -22t^3 - t^2 + 9t$

97. $5(2x - 7) - [4(2x - 3) + 2]$
$= 5(2x - 7) - [8x - 12 + 2]$
$= 5(2x - 7) - [8x - 10]$
$= 10x - 35 - 8x + 10$
$= 2x - 25$

99. *Writing Exercise*

101. Let x represent "a number." Then we have $2x+9$, or $9+2x$

103. *Writing Exercise*

105. $5t - \{7t - [4r - 3(t - 7)] + 6r\} - 4r$
$= 5t - \{7t - [4r - 3t + 21] + 6r\} - 4r$
$= 5t - \{7t - 4r + 3t - 21 + 6r\} - 4r$
$= 5t - \{10t + 2r - 21\} - 4r$
$= 5t - 10t - 2r + 21 - 4r$
$= -5t - 6r + 21$

107. $\{x - [f - (f - x)] + [x - f]\} - 3x$
$= \{x - [f - f + x] + [x - f]\} - 3x$
$= \{x - [x] + [x - f] - 3x$
$= \{x - x + x - f\} - 3x$
$= x - f - 3x$
$= -2x - f$

109. *Writing Exercise*

111. True; $m - n = -n + m = -(n - m)$

113. False; let $m = 2$ and $n = 1$. Then $-2(1-2) = -2(-1) = 2$, but $-(2 \cdot 1 + 2^2) = -(2 + 4) = -6$.

115. $[x + 3(2 - 5x) \div 7 + x](x - 3)$
When $x = 3$, the factor $x - 3$ is 0, so the product is 0.

117. $4 \cdot 20^3 + 17 \cdot 20^2 + 10 \cdot 20 + 0 \cdot 2$
$= 4 \cdot 8000 + 17 \cdot 400 + 10 \cdot 20 + 0 \cdot 2$
$= 32,000 + 6800 + 200 + 0$
$= 39,000$

119. The tower is composed of cubes with sides of length x. The volume of each cube is $x \cdot x \cdot x$, or x^3. Now we count the number of cubes in the tower. The two lowest levels each contain 3×3, or 9 cubes. The next level contains one cube less than the two lowest levels, so it has $9 - 1$, or 8 cubes. The fourth level from the bottom contains one cube less than the level below it, so it has $8-1$, or 7 cubes. The fifth level from the bottom contains one cube less than the level below it, so it has $7 - 1$, or 6 cubes. Finally, the top level contains one cube less than the level below it, so it has $6-1$, or 5 cubes. All together there are $9+9+8+7+6+5$, or 44 cubes, each with volume x^3, so the volume of the tower is $44x^3$.

Chapter 2

Equations, Inequalities, and Problem Solving

Exercise Set 2.1

1. A solution is a replacement that makes an equation true.

3. The 9 in $9ab$ is a coefficient.

5. The multiplication principle is used to solve $\frac{2}{3} \cdot x = -4$.

7. $x + 6 = 23$
$x + 6 - 6 = 23 - 6$ Subtracting 6 from both sides
$x = 17$ Simplifying

Check: $x + 6 = 23$
$17 + 6 \mid 23$
$23 \stackrel{?}{=} 23$ TRUE

The solution is 17.

9. $y + 7 = -4$
$y + 7 - 7 = -4 - 7$ Subtracting 7 from both sides
$t = -11$

Check: $y + 7 = -4$
$-11 + 7 \mid -4$
$-4 \stackrel{?}{=} -4$ TRUE

The solution is -11.

11. $t + 9 = -12$
$t + 9 - 9 = -12 - 9$
$t = -21$

Check: $t + 9 = -12$
$-21 + 9 \mid -12$
$-12 \stackrel{?}{=} -12$ TRUE

The solution is -21.

13. $-6 = y + 25$
$-6 - 25 = y + 25 - 25$
$-31 = y$

Check: $-6 = y + 25$
$-6 \mid -31 + 25$
$-6 \stackrel{?}{=} -6$ TRUE

The solution is -31.

15. $x - 8 = 5$
$x - 8 + 8 = 5 + 8$
$x = 13$

Check: $x - 8 = 5$
$13 - 8 \mid 5$
$5 \stackrel{?}{=} 5$ TRUE

The solution is 13.

17. $12 = -7 + y$
$7 + 12 = 7 + (-7) + y$
$19 = y$

Check: $12 = -7 + y$
$12 \mid -7 + 19$
$12 \stackrel{?}{=} 12$ TRUE

The solution is 19.

19. $-5 + t = -9$
$5 + (-5) + t = 5 + (-9)$
$t = -4$

Check: $-5 + t = -9$
$-5 + (-4) \mid -9$
$-9 \stackrel{?}{=} -9$ TRUE

The solution is -4.

21. $r + \frac{1}{3} = \frac{8}{3}$
$r + \frac{1}{3} - \frac{1}{3} = \frac{8}{3} - \frac{1}{3}$
$r = \frac{7}{3}$

Check: $r + \frac{1}{3} = \frac{8}{3}$
$\frac{7}{3} + \frac{1}{3} \mid \frac{8}{3}$
$\frac{8}{3} \stackrel{?}{=} \frac{8}{3}$ TRUE

The solution is $\frac{7}{3}$.

23. $x + \frac{3}{5} = -\frac{7}{10}$
$x + \frac{3}{5} - \frac{3}{5} = -\frac{7}{10} - \frac{3}{5}$
$x = -\frac{7}{10} - \frac{3}{5} \cdot \frac{2}{2}$
$x = -\frac{7}{10} - \frac{6}{10}$
$x = -\frac{13}{10}$

Check:

$$\begin{array}{c|c} x+\frac{3}{5} & -\frac{7}{10} \\ \hline -\frac{13}{10}+\frac{3}{5} & -\frac{7}{10} \\ -\frac{13}{10}+\frac{6}{10} & \\ -\frac{7}{10} \stackrel{?}{=} & -\frac{7}{10} \quad \text{TRUE} \end{array}$$

The solution is $-\frac{13}{10}$.

25.

$$\begin{aligned} x-\frac{5}{6}&=\frac{7}{8} \\ x-\frac{5}{6}+\frac{5}{6}&=\frac{7}{8}+\frac{5}{6} \\ x&=\frac{7}{8}\cdot\frac{3}{3}+\frac{5}{6}\cdot\frac{4}{4} \\ x&=\frac{21}{24}+\frac{20}{24} \\ x&=\frac{41}{24} \end{aligned}$$

Check:

$$\begin{array}{c|c} x-\frac{5}{6} & \frac{7}{8} \\ \hline \frac{41}{24}-\frac{5}{6} & \frac{7}{8} \\ \frac{41}{24}-\frac{20}{24} & \frac{21}{24} \\ \frac{21}{24} \stackrel{?}{=} & \frac{21}{24} \quad \text{TRUE} \end{array}$$

The solution is $\frac{41}{24}$.

27.

$$\begin{aligned} -\frac{1}{5}+z&=-\frac{1}{4} \\ \frac{1}{5}-\frac{1}{5}+z&=\frac{1}{5}-\frac{1}{4} \\ z&=\frac{1}{5}\cdot\frac{4}{4}-\frac{1}{4}\cdot\frac{5}{5} \\ z&=\frac{4}{20}-\frac{5}{20} \\ z&=-\frac{1}{20} \end{aligned}$$

Check:

$$\begin{array}{c|c} -\frac{1}{5}+z & -\frac{1}{4} \\ \hline -\frac{1}{5}+\left(-\frac{1}{20}\right) & -\frac{1}{4} \\ -\frac{4}{20}+\left(-\frac{1}{20}\right) & -\frac{5}{20} \\ -\frac{5}{20} \stackrel{?}{=} & -\frac{5}{20} \quad \text{TRUE} \end{array}$$

The solution is $-\frac{1}{20}$.

29.

$$\begin{aligned} m+3.9&=5.4 \\ m+3.9-3.9&=5.4-3.9 \\ m&=1.5 \end{aligned}$$

Check:

$$\begin{array}{c|c} m+3.9 & 5.4 \\ \hline 1.5+3.9 & 5.4 \\ 5.4 \stackrel{?}{=} & 5.4 \quad \text{TRUE} \end{array}$$

The solution is 1.5.

31.

$$\begin{aligned} -9.7&=-4.7+y \\ 4.7+(-9.7)&=4.7+(-4.7)+y \\ -5&=y \end{aligned}$$

Check:

$$\begin{array}{c|c} -9.7 & -4.7+y \\ \hline -9.7 & -4.7+(-5) \\ -9.7 \stackrel{?}{=} & -9.7 \quad \text{TRUE} \end{array}$$

The solution is -5.

33.

$$\begin{aligned} 5x&=70 \\ \frac{5x}{5}&=\frac{70}{5} && \text{Dividing both sides by 5} \\ 1\cdot x&=14 && \text{Simplifying} \\ x&=14 && \text{Identity property of 1} \end{aligned}$$

Check:

$$\begin{array}{c|c} 5x & 70 \\ \hline 5\cdot 14 & 70 \\ 70 \stackrel{?}{=} & 70 \quad \text{TRUE} \end{array}$$

The solution is 14.

35.

$$\begin{aligned} 9t&=36 \\ \frac{9t}{9}&=\frac{36}{9} && \text{Dividing both sides by 9} \\ 1\cdot t&=4 && \text{Simplifying} \\ t&=4 && \text{Identity property of 1} \end{aligned}$$

Check:

$$\begin{array}{c|c} 9t & 36 \\ \hline 9\cdot 4 & 36 \\ 36 \stackrel{?}{=} & 36 \quad \text{TRUE} \end{array}$$

The solution is 4.

37.

$$\begin{aligned} 84&=7x \\ \frac{84}{7}&=\frac{7x}{7} && \text{Dividing both sides by 7} \\ 12&=1\cdot x \\ 12&=x \end{aligned}$$

Check:

$$\begin{array}{c|c} 84 & 7x \\ \hline 84 & 7\cdot 12 \\ 84 \stackrel{?}{=} & 84 \quad \text{TRUE} \end{array}$$

The solution is 12.

39.

$$\begin{aligned} -x&=23 \\ -1\cdot x&=23 \\ -1\cdot(-1\cdot x)&=-1\cdot 23 \\ 1\cdot x&=-23 \\ x&=-23 \end{aligned}$$

Check:
$$\begin{array}{c|c} \multicolumn{2}{c}{-x = 23} \\ \hline -(-23) & 23 \\ \multicolumn{2}{c}{23 \stackrel{?}{=} 23 \quad \text{TRUE}} \end{array}$$

The solution is -23.

41. $-t = -8$

The equation states that the opposite of t is the opposite of 8. Thus, $t = 8$. We could also do this exercise as follows.

$$\begin{aligned} -t &= -8 \\ -1(-t) &= -1(-8) \quad \text{Multiplying both sides by } -1 \\ t &= 8 \end{aligned}$$

Check:
$$\begin{array}{c|c} \multicolumn{2}{c}{-t = -8} \\ \hline -(8) & -8 \\ \multicolumn{2}{c}{-8 \stackrel{?}{=} -8 \quad \text{TRUE}} \end{array}$$

The solution is 8.

43.
$$\begin{aligned} 7x &= -49 \\ \frac{7x}{7} &= \frac{-49}{7} \\ 1 \cdot x &= -7 \\ x &= -7 \end{aligned}$$

Check:
$$\begin{array}{c|c} \multicolumn{2}{c}{7x = -49} \\ \hline 7(-7) & -49 \\ \multicolumn{2}{c}{-49 \stackrel{?}{=} -49 \quad \text{TRUE}} \end{array}$$

The solution is -7.

45.
$$\begin{aligned} -1.3a &= -10.4 \\ \frac{-1.3a}{-1.3} &= \frac{-10.4}{-1.3} \\ a &= 8 \end{aligned}$$

Check:
$$\begin{array}{c|c} \multicolumn{2}{c}{-1.3a = -10.4} \\ \hline -1.3(8) & -10.4 \\ \multicolumn{2}{c}{-10.4 \stackrel{?}{=} -10.4 \quad \text{TRUE}} \end{array}$$

The solution is 8.

47.
$$\begin{aligned} \frac{y}{-8} &= 11 \\ -\frac{1}{8} \cdot y &= 11 \\ -8\left(-\frac{1}{8}\right) \cdot y &= -8 \cdot 11 \\ y &= -88 \end{aligned}$$

Check:
$$\begin{array}{c|c} \multicolumn{2}{c}{\frac{y}{-8} = 11} \\ \hline -\frac{88}{-8} & 11 \\ \multicolumn{2}{c}{11 \stackrel{?}{=} 11 \quad \text{TRUE}} \end{array}$$

The solution is -88.

49.
$$\begin{aligned} \frac{4}{4} &= 16 \\ \frac{5}{4} \cdot \frac{4}{5}x &= \frac{5}{4} \cdot 16 \\ x &= \frac{5 \cdot \not{4} \cdot 4}{\not{4} \cdot 1} \\ y &= 20 \end{aligned}$$

Check:
$$\begin{array}{c|c} \multicolumn{2}{c}{\frac{4}{5}x = 16} \\ \hline \frac{4}{5} \cdot 20 & 16 \\ \multicolumn{2}{c}{16 \stackrel{?}{=} 16 \quad \text{TRUE}} \end{array}$$

The solution is 20.

51.
$$\begin{aligned} \frac{-x}{6} &= 9 \\ -\frac{1}{6} \cdot x &= 9 \\ -6\left(-\frac{1}{6}\right) \cdot x &= -6 \cdot 9 \\ x &= -54 \end{aligned}$$

Check:
$$\begin{array}{c|c} \multicolumn{2}{c}{\frac{-x}{6} = 9} \\ \hline \frac{-(-54)}{6} & 9 \\ \frac{54}{6} & \\ \multicolumn{2}{c}{9 \stackrel{?}{=} 9 \quad \text{TRUE}} \end{array}$$

The solution is -54.

53.
$$\begin{aligned} \frac{1}{9} &= \frac{z}{5} \\ \frac{1}{9} &= \frac{1}{5} \cdot z \\ 5 \cdot \frac{1}{9} &= 5 \cdot \frac{1}{5} \cdot z \\ \frac{5}{9} &= z \end{aligned}$$

Check:
$$\begin{array}{c|c} \multicolumn{2}{c}{\frac{1}{9} = \frac{z}{5}} \\ \hline \frac{1}{9} & \frac{5/9}{5} \\ & \frac{5}{9} \cdot \frac{1}{5} \\ \multicolumn{2}{c}{\frac{1}{9} \stackrel{?}{=} \frac{1}{9} \quad \text{TRUE}} \end{array}$$

The solution is $\frac{5}{9}$.

55. $-\frac{3}{5}r = -\frac{3}{5}$

The solution of the equation is the number that is multiplied by $-\frac{3}{5}$ to get $-\frac{3}{5}$. That number is 1. We could also do this exercise as follows:

$$-\frac{3}{5}r = \frac{3}{5}$$
$$-\frac{5}{3}\cdot\left(-\frac{3}{5}r\right) = -\frac{5}{3}\left(-\frac{3}{5}\right)$$
$$r = 1$$

Check: $-\frac{3}{5}r = -\frac{3}{5}$

$$-\frac{3}{5}\cdot 1 \quad \Big| \quad -\frac{3}{5}$$
$$-\frac{3}{5} \stackrel{?}{=} -\frac{3}{5} \quad \text{TRUE}$$

The solution is 1.

57. $$\frac{-3r}{2} = -\frac{27}{4}$$
$$-\frac{3}{2}r = -\frac{27}{4}$$
$$-\frac{2}{3}\cdot\left(-\frac{3}{2}r\right) = -\frac{2}{3}\cdot\left(-\frac{27}{4}\right)$$
$$r = \frac{\not{2}\cdot\not{3}\cdot 3\cdot 3}{\not{3}\cdot\not{2}\cdot 2}$$
$$r = \frac{9}{2}$$

Check: $\frac{-3r}{2} = -\frac{27}{4}$

$$-\frac{3}{2}\cdot\frac{9}{2} \quad \Big| \quad -\frac{27}{4}$$
$$-\frac{27}{4} \stackrel{?}{=} -\frac{27}{4} \quad \text{TRUE}$$

The solution is $\frac{9}{2}$.

59. $$4.5 + t = -3.1$$
$$4.5 + t - 4.5 = -3.1 - 4.5$$
$$t = -7.6$$

The solution is -7.6.

61. $$-8.2x = 20.5$$
$$\frac{-8.2x}{-8.2} = \frac{20.5}{-8.2}$$
$$x = -2.5$$

The solution is -2.5.

63. $$x - 4 = -19$$
$$x - 4 + 4 = -19 + 4$$
$$x = -15$$

The solution is -15.

65. $$3 + t = 21$$
$$-3 + 3 + t = -3 + 21$$
$$t = 18$$

The solution is 18.

67. $$-12x = 72$$
$$\frac{-12x}{-12} = \frac{72}{-12}$$
$$1\cdot x = -6$$
$$x = -6$$

The solution is -6.

69. $$48 = -\frac{3}{8}y$$
$$-\frac{8}{3}\cdot 48 = -\frac{8}{3}\left(-\frac{3}{8}y\right)$$
$$-\frac{8\cdot\not{3}\cdot 16}{\not{3}} = y$$
$$-128 = y$$

The solution is -128.

71. $$a - \frac{1}{6} = -\frac{2}{3}$$
$$a - \frac{1}{6} + \frac{1}{6} = -\frac{2}{3} + \frac{1}{6}$$
$$a = -\frac{4}{6} + \frac{1}{6}$$
$$a = -\frac{3}{6}$$
$$a = -\frac{1}{2}$$

The solution is $-\frac{1}{2}$.

73. $$-24 = \frac{8x}{5}$$
$$-24 = \frac{8}{5}x$$
$$\frac{5}{8}(-24) = \frac{5}{8}\cdot\frac{8}{5}x$$
$$-\frac{5\cdot\not{8}\cdot 3}{\not{8}\cdot 1} = x$$
$$-15 = x$$

The solution is -15.

75. $$-\frac{4}{3}t = -16$$
$$-\frac{3}{4}\left(-\frac{4}{3}t\right) = -\frac{3}{4}(-16)$$
$$t = \frac{3\cdot\not{4}\cdot 4}{\not{4}}$$
$$t = 12$$

The solution is 12.

77. $$-483.297 = -794.053 + t$$
$$-483.297 + 794.053 = -794.053 + t + 794.053$$
$$310.756 = t \qquad \text{Using a calculator}$$

The solution is 310.756.

79. *Writing Exercise*

81. $9-2\cdot 5^2+7$

$= 9-2\cdot 25+7$ Simplifying the exponential expression

$= 9-50+7$ Multiplying

$= -41+7$ Subtracting and

$= -34$ Adding from left to right

83. $16\div(2-3\cdot 2)+5$

$= 16\div(2-6)+5$ Simplifying inside

$= 16\div(-4)+5$ the parentheses

$= -4+5$ Dividing

$= 1$ Adding

85. *Writing Exercise*

87. $2x = x+x$

$2x = 2x$ Adding on the right side

This is an identity.

89. $5x = 0$

$$\frac{5x}{5} = \frac{0}{5}$$

$x = 0$

The solution is 0.

91. $x+8 = 3+x+7$

$x+8 = 10+x$ Adding on the right side

$x+8-x = 10+x-x$

$8 = 10$

This is a contradiction.

93. $2|x| = -14$

$$\frac{2|x|}{2} = -\frac{14}{2}$$

$|x| = -7$

Since the absolute value of a number is always nonnegative, this is a contradiction.

95. $mx = 9.4m$

$$\frac{mx}{m} = \frac{9.4m}{m}$$

$x = 9.4$

The solution is 9.4.

97. $cx+5c = 7c$

$cx+5c-5c = 7c-5c$

$cx = 2c$

$$\frac{cx}{c} = \frac{2c}{c}$$

$x = 2$

The solution is 2.

99. $7+|x| = 20$

$-7+7+|x| = -7+20$

$|x| = 13$

x represents a number whose distance from 0 is 13. Thus $x = -13$ or $x = 13$.

101. $t-3590 = 1820$

$t-3590+3590 = 1820+3590$

$t = 5410$

$t+3590 = 5410+3590$

$t+3590 = 9000$

103. To "undo" the last step, divide 22.5 by 0.3.

$22.5\div 0.3 = 75$

Now divide 75 by 0.3.

$75\div 0.3 = 250$

The answer should be 250 not 22.5.

Exercise Set 2.2

1. $3x-1 = 7$

$3x-1+1 = 7+1$ Adding 1 to both sides

$3x = 7+1$

Choice (c) is correct.

3. $6(x-1) = 2$

$6x-6 = 2$ Using the distributive law

Choice (a) is correct.

5. $4x = 3-2x$

$4x+2x = 3-2x+2x$ Adding $2x$ to both sides

$4x+2x = 3$

Choice (b) is correct.

7. $2x+9 = 25$

$2x+9-9 = 25-9$ Subtracting 9 from both sides

$2x = 16$ Simplifying

$$\frac{2x}{2} = \frac{16}{2}$$ Dividing both sides by 2

$x = 8$ Simplifying

Check:

$2x+9 = 25$	
$2\cdot 8+9$	25
$16+9$	

$25 \stackrel{?}{=} 25$ TRUE

The solution is 8.

9. $6z+4 = 46$

$6z+4-4 = 46-4$ Subtracting 4 from both sides

$6z = 42$ Simplifying

$$\frac{6z}{6} = \frac{42}{6}$$ Dividing both sides by 6

$z = 7$ Simplifying

Check:

$6z+4 = 46$	
$6\cdot 7+4$	46
$42+4$	

$46 \stackrel{?}{=} 46$ TRUE

The solution is 7.

11.
$$\begin{aligned} 7t - 8 &= 27 \\ 7t - 8 + 8 &= 27 + 8 \quad \text{Adding 8 to both sides} \\ 7t &= 35 \\ \frac{7t}{7} &= \frac{35}{7} \quad \text{Dividing both sides by 7} \\ t &= 5 \end{aligned}$$

Check: $7t - 8 = 27$
$$\begin{array}{r|l} 7 \cdot 5 - 8 & 27 \\ 35 - 8 & \\ 27 & \stackrel{?}{=} 27 \quad \text{TRUE} \end{array}$$

The solution is 5.

13.
$$\begin{aligned} 3x - 9 &= 33 \\ 3x - 9 + 9 &= 33 + 9 \\ 3x &= 42 \\ \frac{3x}{3} &= \frac{42}{3} \\ x &= 14 \end{aligned}$$

Check: $3x - 9 = 33$
$$\begin{array}{r|l} 3 \cdot 14 - 9 & 33 \\ 42 - 9 & \\ 33 & \stackrel{?}{=} 33 \quad \text{TRUE} \end{array}$$

The solution is 14.

15.
$$\begin{aligned} 8z + 2 &= -54 \\ 8z + 2 - 2 &= -54 - 2 \\ 8z &= -56 \\ \frac{8z}{8} &= \frac{-56}{8} \\ z &= -7 \end{aligned}$$

Check: $8z + 2 = -54$
$$\begin{array}{r|l} 8(-7) + 2 & -54 \\ -56 + 2 & \\ -54 & \stackrel{?}{=} -54 \quad \text{TRUE} \end{array}$$

The solution is -7.

17.
$$\begin{aligned} -91 &= 9t + 8 \\ -91 - 8 &= 9t + 8 - 8 \\ -99 &= 9t \\ \frac{-99}{9} &= \frac{9t}{9} \\ -11 &= t \end{aligned}$$

Check: $-91 = 9t + 8$
$$\begin{array}{r|l} -91 & 9(-11) + 8 \\ & -99 + 8 \\ -91 & \stackrel{?}{=} -91 \quad \text{TRUE} \end{array}$$

The solution is -11.

19.
$$\begin{aligned} 12 - 4x &= 108 \\ -12 + 12 - 4x &= -12 + 108 \\ -4x &= 96 \\ \frac{-4x}{-4} &= \frac{96}{-4} \\ x &= -24 \end{aligned}$$

Check: $12 - 4x = 108$
$$\begin{array}{r|l} 12 - 4(-24) & 108 \\ 12 + 96 & \\ 108 & \stackrel{?}{=} 108 \quad \text{TRUE} \end{array}$$

The solution is -24.

21.
$$\begin{aligned} -6z - 18 &= -132 \\ -6z - 18 + 18 &= -132 + 18 \\ -6z &= -114 \\ \frac{-6z}{-6} &= \frac{-114}{-6} \\ z &= 19 \end{aligned}$$

Check: $-6z - 18 = -132$
$$\begin{array}{r|l} -6 \cdot 19 - 18 & -132 \\ -114 - 18 & \\ -132 & \stackrel{?}{=} -132 \quad \text{TRUE} \end{array}$$

The solution is 19.

23.
$$\begin{aligned} 4x + 5x &= 10 \\ 9x &= 10 \quad \text{Combining like terms} \\ \frac{9x}{9} &= \frac{10}{9} \\ x &= \frac{10}{9} \end{aligned}$$

Check: $4x + 5x = 10$
$$\begin{array}{r|l} 4 \cdot \frac{10}{9} + 5 \cdot \frac{10}{9} & 10 \\ \frac{40}{9} + \frac{50}{9} & \\ \frac{90}{9} & \\ 10 & \stackrel{?}{=} 10 \quad \text{TRUE} \end{array}$$

The solution is $\frac{10}{9}$.

25.
$$\begin{aligned} 32 - 7x &= 11 \\ -32 + 32 - 7x &= -32 + 11 \\ -7x &= -21 \\ \frac{-7x}{-7} &= \frac{-21}{-7} \\ x &= 3 \end{aligned}$$

Check: $32 - 7x = 11$
$$\begin{array}{r|l} 32 - 7 \cdot 3 & 11 \\ 32 - 21 & \\ 11 & \stackrel{?}{=} 11 \quad \text{TRUE} \end{array}$$

The solution is 3.

27.
$$\frac{3}{5}t - 1 = 8$$
$$\frac{3}{5}t - 1 + 1 = 8 + 1$$
$$\frac{3}{5}t = 9$$
$$\frac{5}{3}\cdot\frac{3}{5}t = \frac{5}{3}\cdot 9$$
$$t = \frac{5\cdot\not{3}\cdot 3}{\not{3}\cdot 1}$$
$$t = 15$$

Check: $\frac{3}{5}t - 1 = 8$

$\frac{3}{5}\cdot 15 - 1$	8
$9 - 1$	

$8 \stackrel{?}{=} 8$ TRUE

The solution is 15.

29.
$$4 + \frac{7}{2}x = -10$$
$$-4 + 4 + \frac{7}{2}x = -4 - 10$$
$$\frac{7}{2}x = -14$$
$$\frac{2}{7}\cdot\frac{7}{2}x = \frac{2}{7}(-14)$$
$$x = -\frac{2\cdot 2\cdot\not{7}}{\not{7}\cdot 1}$$
$$x = -4$$

Check: $4 + \frac{7}{2}x = -10$

$4 + \frac{7}{2}(-4)$	-10
$4 - 14$	

$-10 \stackrel{?}{=} -10$ TRUE

The solution is -4.

31.
$$-\frac{3a}{4} - 5 = 2$$
$$-\frac{3a}{4} - 5 + 5 = 2 + 5$$
$$-\frac{3a}{4} = 7$$
$$-\frac{4}{3}\left(-\frac{3a}{4}\right) = -\frac{4}{3}\cdot 7$$
$$a = -\frac{28}{3}$$

Check: $-\frac{3a}{4} - 5 = 2$

$-\frac{3}{4}\left(-\frac{28}{3}\right) - 5$	2
$7 - 5$	

$2 \stackrel{?}{=} 2$ TRUE

The solution is $-\frac{28}{3}$.

33. $-5z - 6z = -44$

$-11z = -44$ Combining like terms

$$\frac{-11z}{-11} = \frac{-44}{-11}$$

$z = 4$

Check: $-5z - 6z = -44$

$-5\cdot 4 - 6\cdot 4$	-44
$-20 - 24$	

$-44 \stackrel{?}{=} -44$ TRUE

The solution is 4.

35. $4x - 6 = 6x$

$-6 = 6x - 4x$ Subtracting $4x$ from both sides

$-6 = 2x$ Simplifying

$\frac{-6}{2} = \frac{2x}{2}$ Dividing both sides by 2

$-3 = x$

Check: $4x - 6 = 6x$

$4(-3) - 6$	$6(-3)$
$-12 - 6$	-18

$-18 \stackrel{?}{=} -18$ TRUE

The solution is -3.

37. $5y - 2 = 28 - y$

$5y - 2 + y = 28 - y + y$ Adding y to both sides

$6y - 2 = 28$ Simplifying

$6y - 2 + 2 = 28 + 2$ Adding 2 to both sides

$6y = 30$ Simplifying

$\frac{6y}{6} = \frac{30}{6}$ Dividing both sides by 6

$y = 5$

Check: $5y - 2 = 28 - y$

$5\cdot 5 - 2$	$28 - 5$
$25 - 2$	23

$23 \stackrel{?}{=} 23$ TRUE

The solution is 5.

39. $7(2a - 1) = 21$

$14a - 7 = 21$ Using the distributive law

$14a = 21 + 7$ Adding 7

$14a = 28$

$a = 2$ Dividing by 14

Check: $7(2a - 1) = 21$

$7(2\cdot 2 - 1)$	21
$7(4 - 1)$	
$7\cdot 3$	

$21 \stackrel{?}{=} 21$ TRUE

The solution is 2.

41. We can write $8 = 8(x+1)$ as $8 \cdot 1 = 8(x+1)$. Then $1 = x+1$, or $x = 0$. The solution is 0.

43.
$$\begin{aligned} 2(3+4m)-6 &= 48 \\ 6+8m-6 &= 48 \\ 8m &= 48 \quad \text{Combining like terms} \\ m &= 6 \end{aligned}$$

Check: $2(3+4m)-6 = 48$

$$\begin{array}{r|l} 2(3+4\cdot 6)-6 & 48 \\ 2(3+24)-6 & \\ 2\cdot 27-6 & \\ 54-6 & \end{array}$$

$48 \overset{?}{=} 48$ TRUE

The solution is 6.

45.
$$\begin{aligned} 7r-(2r+8) &= 32 \\ 7r-2r-8 &= 32 \\ 5r-8 &= 32 \quad \text{Combining like terms} \\ 5r &= 32+8 \\ 5r &= 40 \\ r &= 8 \end{aligned}$$

Check: $7r-(2r+8) = 32$

$$\begin{array}{r|l} 7\cdot 8-(2\cdot 8+8) & 32 \\ 56-(16+8) & \\ 56-24 & \end{array}$$

$32 \overset{?}{=} 32$ TRUE

The solution is 8.

47.
$$\begin{aligned} 6x+3 &= 2x+3 \\ 6x-2x &= 3-3 \\ 4x &= 0 \\ \frac{4x}{4} &= \frac{0}{4} \\ x &= 0 \end{aligned}$$

Check: $6x+3 = 2x+3$

$$\begin{array}{r|l} 6\cdot 0+3 & 2\cdot 0+3 \\ 0+3 & 0+3 \end{array}$$

$3 \overset{?}{=} 3$ TRUE

The solution is 0.

49.
$$\begin{aligned} 5-2x &= 3x-7x+25 \\ 5-2x &= -4x+25 \\ 4x-2x &= 25-5 \\ 2x &= 20 \\ \frac{2x}{2} &= \frac{20}{2} \\ x &= 10 \end{aligned}$$

Check:

$5-2x = 3x-7x+25$

$$\begin{array}{r|l} 5-2\cdot 10 & 3\cdot 10-7\cdot 10+25 \\ 5-20 & 30-70+25 \\ -15 & -40+25 \end{array}$$

$-15 \overset{?}{=} -15$ TRUE

The solution is 10.

51.
$$\begin{aligned} 7+3x-6 &= 3x+5-x \\ 3x+1 &= 2x+5 \quad \text{Combining like terms on each side} \\ 3x-2x &= 5-1 \\ x &= 4 \end{aligned}$$

Check: $7+3x-6 = 3x+5-x$

$$\begin{array}{r|l} 7+3\cdot 4-6 & 3\cdot 4+5-4 \\ 7+12-6 & 12+5-4 \\ 19-6 & 17-4 \end{array}$$

$13 \overset{?}{=} 13$ TRUE

The solution is 4.

53.
$$\begin{aligned} 4y-4+y+24 &= 6y+20-4y \\ 5y+20 &= 2y+20 \\ 5y-2y &= 20-20 \\ 3y &= 0 \\ y &= 0 \end{aligned}$$

Check:

$4y-4+y+24 = 6y+20-4y$

$$\begin{array}{r|l} 4\cdot 0-4+0+24 & 6\cdot 0+20-4\cdot 0 \\ 0-4+0+24 & 0+20-0 \end{array}$$

$20 \overset{?}{=} 20$ TRUE

The solution is 0.

55.
$$\begin{aligned} 13-3(2x-1) &= 4 \\ 13-6x+3 &= 4 \\ 16-6x &= 4 \\ -6x &= 4-16 \\ -6x &= -12 \\ x &= 2 \end{aligned}$$

Check: $13-3(2x-1) = 4$

$$\begin{array}{r|l} 13-3(2\cdot 2-1) & 4 \\ 13-3(4-1) & \\ 13-3\cdot 3 & \\ 13-9 & \end{array}$$

$4 \overset{?}{=} 4$ TRUE

The solution is 2.

57.
$$\begin{aligned} 7(5x-2) &= 6(6x-1) \\ 35x-14 &= 36x-6 \\ -14+6 &= 36x-35x \\ -8 &= x \end{aligned}$$

Check:

$7(5x-2) = 6(6x-1)$

$$\begin{array}{r|l} 7(5(-8)-2) & 6(6(-8)-1) \\ 7(-40-2) & 6(-48-1) \\ 7(-42) & 6(-49) \end{array}$$

$-294 \overset{?}{=} -294$ TRUE

The solution is -8.

59.
$$\begin{aligned} 19-(2x+3) &= 2(x+3)+x \\ 19-2x-3 &= 2x+6+x \\ 16-2x &= 3x+6 \\ 16-6 &= 3x+2x \\ 10 &= 5x \\ 2 &= x \end{aligned}$$

Check:

$19-(2x+3)=2(x+3)+x$	
$19-(2\cdot 2+3)$	$2(2+3)+2$
$19-(4+3)$	$2\cdot 5+2$
$19-7$	$10+2$

$12 \stackrel{?}{=} 12$ TRUE

The solution is 2.

61. $\frac{5}{4}x+\frac{1}{4}x=2x+\frac{1}{2}+\frac{3}{4}x$

The number 4 is the least common denominator, so we multiply by 4 on both sides.

$$\begin{aligned}4\left(\frac{5}{4}x+\frac{1}{4}x\right)&=4\left(2x+\frac{1}{2}+\frac{3}{4}x\right)\\4\cdot\frac{5}{4}x+4\cdot\frac{1}{4}x&=4\cdot 2x+4\cdot\frac{1}{2}+4\cdot\frac{3}{4}x\\5x+x&=8x+2+3x\\6x&=11x+2\\6x-11x&=2\\-5x&=2\\\frac{-5x}{-5}&=\frac{2}{-5}\\x&=-\frac{2}{5}\end{aligned}$$

Check:

$\frac{5}{4}x+\frac{1}{4}x=2x+\frac{1}{2}+\frac{3}{4}x$	
$\frac{5}{4}\left(-\frac{2}{5}\right)+\frac{1}{4}\left(-\frac{2}{5}\right)$	$2\left(-\frac{2}{5}\right)+\frac{1}{2}+\frac{3}{4}\left(-\frac{2}{5}\right)$
$-\frac{1}{2}-\frac{1}{10}$	$-\frac{4}{5}+\frac{1}{2}-\frac{3}{10}$
$-\frac{5}{10}-\frac{1}{10}$	$-\frac{8}{10}+\frac{5}{10}-\frac{3}{10}$

$-\frac{6}{10} \stackrel{?}{=} -\frac{6}{10}$ TRUE

The solution is $-\frac{2}{5}$.

63. $\frac{2}{3}+\frac{1}{4}t=6$

The number 12 is the least common denominator, so we multiply by 12 on both sides.

$$\begin{aligned}12\left(\frac{2}{3}+\frac{1}{4}t\right)&=12\cdot 6\\12\cdot\frac{2}{3}+12\cdot\frac{1}{4}t&=72\\8+3t&=72\\3t&=72-8\\3t&=64\\t&=\frac{64}{3}\end{aligned}$$

Check:

$\frac{2}{3}+\frac{1}{4}t=6$	
$\frac{2}{3}+\frac{1}{4}\left(\frac{64}{3}\right)$	6
$\frac{2}{3}+\frac{16}{3}$	
$\frac{18}{3}$	

$6 \stackrel{?}{=} 6$ TRUE

The solution is $\frac{64}{3}$.

65. $\frac{2}{3}+4t=6t-\frac{2}{15}$

The number 15 is the least common denominator, so we multiply by 15 on both sides.

$$\begin{aligned}15\left(\frac{2}{3}+4t\right)&=15\left(6t-\frac{2}{15}\right)\\15\cdot\frac{2}{3}+15\cdot 4t&=15\cdot 6t-15\cdot\frac{2}{15}\\10+60t&=90t-2\\10+2&=90t-60t\\12&=30t\\\frac{12}{30}&=t\\\frac{2}{5}&=t\end{aligned}$$

Check:

$\frac{2}{3}+4t=6t-\frac{2}{15}$	
$\frac{2}{3}+4\cdot\frac{2}{5}$	$6\cdot\frac{2}{5}-\frac{2}{15}$
$\frac{2}{3}+\frac{8}{5}$	$\frac{12}{5}-\frac{2}{15}$
$\frac{10}{15}+\frac{24}{15}$	$\frac{36}{15}-\frac{2}{15}$

$\frac{34}{15} \stackrel{?}{=} \frac{34}{15}$ TRUE

The solution is $\frac{2}{5}$.

67. $\frac{1}{3}x+\frac{2}{5}=\frac{4}{15}+\frac{3}{5}x-\frac{2}{3}$

The number 15 is the least common denominator, so we multiply by 15 on both sides.

$$\begin{aligned}15\left(\frac{1}{3}x+\frac{2}{5}\right)&=15\left(\frac{4}{15}+\frac{3}{5}x-\frac{2}{3}\right)\\15\cdot\frac{1}{3}x+15\cdot\frac{2}{5}&=15\cdot\frac{4}{15}+15\cdot\frac{3}{5}x-15\cdot\frac{2}{3}\\5x+6&=4+9x-10\\5x+6&=-6+9x\\5x-9x&=-6-6\\-4x&=-12\\\frac{-4x}{-4}&=\frac{-12}{-4}\\x&=3\end{aligned}$$

Check:

$$\begin{array}{r|l}
\frac{1}{3}x+\frac{2}{5} & \frac{4}{15}+\frac{3}{5}x-\frac{2}{3} \\
\hline
\frac{1}{3}\cdot 3+\frac{2}{5} & \frac{4}{15}+\frac{3}{5}\cdot 3-\frac{2}{3} \\
1+\frac{2}{5} & \frac{4}{15}+\frac{9}{5}-\frac{2}{3} \\
\frac{5}{5}+\frac{2}{5} & \frac{4}{15}+\frac{27}{15}-\frac{10}{15} \\
\frac{7}{5} & \frac{21}{15}
\end{array}$$

$$\frac{7}{5} \stackrel{?}{=} \frac{7}{5} \qquad \text{TRUE}$$

The solution is 3.

69.

$$\begin{aligned}
2.1x+45.2 &= 3.2-8.4x \\
&\text{Greatest number of decimal places is 1} \\
10(2.1x+45.2) &= 10(3.2-8.4x) \\
&\text{Multiplying by 10 to clear decimals} \\
10(2.1x)+10(45.2) &= 10(3.2)-10(8.4x) \\
21x+452 &= 32-84x \\
21x+84x &= 32-452 \\
105x &= -420 \\
x &= \frac{-420}{105} \\
x &= -4
\end{aligned}$$

Check:

$$\begin{array}{r|l}
\multicolumn{2}{c}{2.1x+45.2 = 3.2-8.4x} \\
\hline
2.1(-4)+45.2 & 3.2-8.4(-4) \\
-8.4+45.2 & 3.2+33.6
\end{array}$$

$$36.8 \stackrel{?}{=} 36.8 \qquad \text{TRUE}$$

The solution is -4.

71.

$$\begin{aligned}
0.76+0.21t &= 0.96t-0.49 \\
&\text{Greatest number of decimal places is 2} \\
100(0.76+0.21t) &= 100(0.96t-0.49) \\
&\text{Multiplying by 100 to clear decimals} \\
100(0.76)+100(0.21t) &= 100(0.96t)-100(0.49) \\
76+21t &= 96t-49 \\
76+49 &= 96t-21t \\
125 &= 75t \\
\frac{125}{75} &= t \\
\frac{5}{3} &= t, \text{ or} \\
1.\overline{6} &= t
\end{aligned}$$

The answer checks. The solution is $\frac{5}{3}$, or $1.\overline{6}$.

73.

$$\frac{2}{5}x-\frac{3}{2}x = \frac{3}{4}x+2$$

The least common denominator is 20.

$$\begin{aligned}
20\left(\frac{2}{5}x-\frac{3}{2}x\right) &= 20\left(\frac{3}{4}x+2\right) \\
20\cdot\frac{2}{5}x-20\cdot\frac{3}{2}x &= 20\cdot\frac{3}{4}x+20\cdot 2 \\
8x-30x &= 15x+40 \\
-22x &= 15x+40 \\
-22x-15x &= 40 \\
-37x &= 40 \\
\frac{-37x}{-37} &= \frac{40}{-37} \\
x &= -\frac{40}{37}
\end{aligned}$$

Check:

$$\begin{array}{r|l}
\multicolumn{2}{c}{\frac{2}{5}x-\frac{3}{2}x = \frac{3}{4}x+2} \\
\hline
\frac{2}{5}\left(-\frac{40}{37}\right)-\frac{3}{2}\left(-\frac{40}{37}\right) & \frac{3}{4}\left(-\frac{40}{37}\right)+2 \\
-\frac{16}{37}+\frac{60}{37} & -\frac{30}{37}+\frac{74}{37}
\end{array}$$

$$\frac{44}{37} \stackrel{?}{=} \frac{44}{37} \qquad \text{TRUE}$$

The solution is $-\frac{40}{37}$.

75.

$$\begin{aligned}
\frac{1}{3}(2x-1) &= 7 \\
3\cdot\frac{1}{3}(2x-1) &= 3\cdot 7 \\
2x-1 &= 21 \\
2x &= 22 \\
x &= 11
\end{aligned}$$

Check:

$$\begin{array}{r|l}
\multicolumn{2}{c}{\frac{1}{3}(2x-1) = 7} \\
\hline
\frac{1}{3}(2\cdot 11-1) & 7 \\
\frac{1}{3}\cdot 21 &
\end{array}$$

$$7 \stackrel{?}{=} 7 \qquad \text{TRUE}$$

The solution is 11.

77.

$$\begin{aligned}
\frac{3}{4}(3t-6) &= 9 \\
\frac{4}{3}\cdot\frac{3}{4}(3t-6) &= \frac{4}{3}\cdot 9 \\
3t-6 &= 12 \\
3t &= 18 \\
t &= 6
\end{aligned}$$

Check:
$$\begin{array}{c|c} \multicolumn{2}{c}{\frac{3}{4}(3t-6) = 9} \\ \hline \frac{3}{4}(3\cdot 6-6) & 9 \\ \frac{3}{4}\cdot 12 & \\ 9 \stackrel{?}{=} 9 & \end{array} \quad \text{TRUE}$$

The solution is 6.

79.
$$\begin{aligned} \frac{1}{6}\left(\frac{3}{4}x-2\right) &= -\frac{1}{5} \\ 30\cdot\frac{1}{6}\left(\frac{3}{4}x-2\right) &= 30\left(-\frac{1}{5}\right) \\ 5\left(\frac{3}{4}x-2\right) &= -6 \\ \frac{15}{4}x-10 &= -6 \\ \frac{15}{4}x &= 4 \\ 4\cdot\frac{15}{4}x &= 4\cdot 4 \\ 15x &= 16 \\ x &= \frac{16}{15} \end{aligned}$$

Check:
$$\begin{array}{c|c} \multicolumn{2}{c}{\frac{1}{6}\left(\frac{3}{4}x-2\right) = -\frac{1}{5}} \\ \hline \frac{1}{6}\left(\frac{3}{4}\cdot\frac{16}{15}-2\right) & -\frac{1}{5} \\ \frac{1}{6}\left(\frac{4}{5}-2\right) & \\ \frac{1}{6}\left(-\frac{6}{5}\right) & \\ -\frac{1}{5} \stackrel{?}{=} -\frac{1}{5} & \end{array} \quad \text{TRUE}$$

The solution is $\frac{16}{15}$.

81.
$$\begin{aligned} 0.7(3x+6) &= 1.1-(x+2) \\ 2.1x+4.2 &= 1.1-x-2 \\ 10(2.1x+4.2) &= 10(1.1-x-2) \quad \text{Clearing decimals} \\ 21x+42 &= 11-10x-20 \\ 21x+42 &= -10x-9 \\ 21x+10x &= -9-42 \\ 31x &= -51 \\ x &= -\frac{51}{31} \end{aligned}$$

The check is left to the student. The solution is $-\frac{51}{31}$.

83.
$$\begin{aligned} a+(a-3) &= (a+2)-(a+1) \\ a+a-3 &= a+2-a-1 \\ 2a-3 &= 1 \\ 2a &= 1+3 \\ 2a &= 4 \\ a &= 2 \end{aligned}$$

Check:
$$\begin{array}{c|c} \multicolumn{2}{c}{a+(a-3) = (a+2)-(a+1)} \\ \hline 2+(2-3) & (2+2)-(2+1) \\ 2-1 & 4-3 \\ 1 \stackrel{?}{=} 1 & \end{array} \quad \text{TRUE}$$

The solution is 2.

85. *Writing Exercise*

87. $3-5a = 3-5\cdot 2 = 3-10 = -7$

89. $7x-2x = 7(-3)-2(-3) = -21+6 = -15$

91. *Writing Exercise*

93.
$$\begin{aligned} 8.43x-2.5(3.2-0.7x) &= -3.455x+9.04 \\ 8.43x-8+1.75x &= -3.455x+9.04 \\ 10.18x-8 &= -3.455x+9.04 \\ 10.18x+3.455x &= 9.04+8 \\ 13.635x &= 17.04 \\ x &= 1.\overline{2497}, \text{ or } \frac{1136}{909} \end{aligned}$$

The solution is $1.\overline{2497}$, or $\frac{1136}{909}$.

95.
$$\begin{aligned} -2[3(x-2)+4] &= 4(5-x)-2x \\ -2[3x-6+4] &= 20-4x-2x \\ -2[3x-2] &= 20-6x \\ -6x+4 &= 20-6x \\ 4 &= 20 \quad \text{Adding } 6x \text{ to both sides} \end{aligned}$$

This is contradiction.

97.
$$\begin{aligned} 3(x+4) &= 3(4+x) \\ 3x+12 &= 12+3x \\ 3x+12-12 &= 12+3x-12 \\ 3x &= 3x \end{aligned}$$

This is an identity.

99.
$$\begin{aligned} 2x(x+5)-3(x^2+2x-1) &= 9-5x-x^2 \\ 2x^2+10x-3x^2-6x+3 &= 9-5x-x^2 \\ -x^2+4x+3 &= 9-5x-x^2 \\ 4x+3 &= 9-5x \quad \text{Adding } x^2 \\ 4x+5x &= 9-3 \\ 9x &= 6 \\ x &= \frac{2}{3} \end{aligned}$$

The solution is $\frac{2}{3}$.

101.
$$\begin{aligned} 9-3x &= 2(5-2x)-(1-5x) \\ 9-3x &= 10-4x-1+5x \\ 9-3x &= 9+x \\ 9-9 &= x+3x \\ 0 &= 4x \\ 0 &= x \end{aligned}$$

The solution is 0.

103. $[7-2(8\div(-2))]x=0$

Since $7-2(8\div(-2))\neq 0$ and the product on the left side of the equation is 0, then x must be 0.

105.
$$\frac{5x+3}{4}+\frac{25}{12}=\frac{5+2x}{3}$$
$$12\left(\frac{5x+3}{4}+\frac{25}{12}\right)=12\left(\frac{5+2x}{3}\right)$$
$$12\left(\frac{5x+3}{4}\right)+12\cdot\frac{25}{12}=4(5+2x)$$
$$3(5x+3)+25=4(5+2x)$$
$$15x+9+25=20+8x$$
$$15x+34=20+8x$$
$$7x=-14$$
$$x=-2$$

The solution is -2.

Exercise Set 2.3

1. We substitute 10 for t and calculate M.

$$M=\frac{1}{5}t=\frac{1}{5}\cdot 10=2$$

The storm is 2 miles away.

3. We substitute 21,345 for n and calculate f.

$$f=\frac{n}{15}=\frac{21,345}{15}=1423$$

There are 1423 full-time equivalent students.

5. Substitute 1800 for a and calculate B.

$$B=30a=30\cdot 1800=54,000$$

The minimum furnace output is 54,000 Btu's.

7. Substitute 1 for t and calculate n.

$$n=0.5t^4+3.45t^3-96.65t^2+347.7t$$
$$=0.5(1)^4+3.45(1)^3-96.65(1)^2+347.7(1)$$
$$=0.5+3.45-96.65+347.7$$
$$=255$$

255 mg of ibuprofen remains in the bloodstream.

9.
$$A=bh$$
$$\frac{A}{h}=\frac{bh}{h}\quad\text{Dividing both sides by } h$$
$$\frac{A}{h}=b$$

11.
$$d=rt$$
$$\frac{d}{t}=\frac{rt}{t}\quad\text{Dividing both sides by } t$$
$$\frac{d}{t}=r$$

13.
$$I=Prt$$
$$\frac{I}{rt}=\frac{Prt}{rt}\quad\text{Dividing both sides by } rt$$
$$\frac{I}{rt}=P$$

15.
$$H=65-m$$
$$H+m=65\quad\text{Adding } m \text{ to both sides}$$
$$m=65-H\quad\text{Subtracting } H \text{ from both sides}$$

17.
$$P=2l+2w$$
$$P-2w=2l+2w-2w\quad\text{Subtracting } 2w \text{ from both sides}$$
$$P-2w=2l$$
$$\frac{P-2w}{2}=\frac{2l}{2}\quad\text{Dividing both sides by 2}$$
$$\frac{P-2w}{2}=l,\text{ or}$$
$$\frac{P}{2}-w=l$$

19.
$$A=\pi r^2$$
$$\frac{A}{r^2}=\frac{\pi r^2}{r^2}$$
$$\frac{A}{r^2}=\pi$$

21.
$$A=\frac{1}{2}bh$$
$$2A=2\cdot\frac{1}{2}bh\quad\text{Multiplying both sides by 2}$$
$$2A=bh$$
$$\frac{2A}{b}=\frac{bh}{b}\quad\text{Dividing both sides by } h$$
$$\frac{2A}{b}=h$$

23.
$$E=mc^2$$
$$\frac{E}{c^2}=\frac{mc^2}{c^2}\quad\text{Dividing both sides by } c^2$$
$$\frac{E}{c^2}=m$$

25.
$$Q=\frac{c+d}{2}$$
$$2Q=2\cdot\frac{c+d}{2}\quad\text{Multiplying both sides by 2}$$
$$2Q=c+d$$
$$2Q-c=c+d-c\quad\text{Subtracting } c \text{ from both sides}$$
$$2Q-c=d$$

27.
$$A=\frac{a+b+c}{3}$$
$$3A=3\cdot\frac{a+b+c}{3}\quad\text{Multiplying both sides by 3}$$
$$3A=a+b+c$$
$$3A-a-c=a+b+c-a-c\quad\text{Subtracting } a \text{ and } c \text{ from both sides}$$
$$3A-a-c=b$$

29. $M = \frac{A}{s}$

$s \cdot M = s \cdot \frac{A}{s}$ Multiplying both sides by s

$sM = A$

31. $F = \frac{9}{5}C + 32$

$F - 32 = \frac{9}{5}C$

$\frac{5}{9}(F - 32) = \frac{5}{9} \cdot \frac{9}{5}C$

$\frac{5}{9}(F - 32) = C$

33. $A = at + bt$

$A = t(a + b)$ Factoring

$\frac{A}{a + b} = t$ Dividing both sides by $a + b$

35. $A = \frac{1}{2}ah + \frac{1}{2}bh$

$2A = 2\left(\frac{1}{2}ah + \frac{1}{2}bh\right)$

$2A = ah + bh$

$2A = h(a + b)$

$\frac{2A}{a + b} = h$

37. $R = r + \frac{400(W - L)}{N}$

$N \cdot R = N\left(r + \frac{400(W - L)}{N}\right)$ Multiplying both sides by N

$NR = Nr + 400(W - L)$

$NR = Nr + 400W - 400L$

$NR + 400L = Nr + 400W$ Adding $400L$ to both sides

$400L = Nr + 400W - NR$ Adding $-NR$ to both sides

$L = \frac{Nr + 400W - NR}{400}$

39. *Writing Exercise*

41. $0.79(38.4)0$

One factor is 0, so the product is 0.

43. $20 \div (-4) \cdot 2 - 3$

$= -5 \cdot 2 - 3$ Dividing and

$= -10 - 3$ multiplying from left to right

$= -13$ Subtracting

45. *Writing Exercise*

47. $K = 19.18w + 7h - 9.52a + 92.4$

$2627 = 19.18(82) + 7(185) - 9.52a + 92.4$

$2627 = 1572.76 + 1295 - 9.52a + 92.4$

$2627 = 2960.16 - 9.52a$

$-333.16 = -9.52a$

$35 \approx a$

The man is about 35 years old.

49. First we substitute 54 for A and solve for s to find the length of a side of the cube.

$A = 6s^2$

$54 = 6s^2$

$9 = s^2$

$3 = s$ Taking the positive square root

Now we substitute 3 for s in the formula for the volume of a cube and compute the volume.

$V = s^3 = 3^3 = 27$

The volume of the cube is 27 in^3.

51. $c = \frac{w}{a} \cdot d$

$ac = a \cdot \frac{w}{a} \cdot d$

$ac = wd$

$a = \frac{wd}{c}$

53. $ac = bc + d$

$ac - bc = d$

$c(a - b) = d$

$c = \frac{d}{a - b}$

55. $3a = c - a(b + d)$

$3a = c - ab - ad$

$3a + ab + ad = c$

$a(3 + b + d) = c$

$a = \frac{c}{3 + b + d}$

57. $K = 917 + 6(2.2046w + 0.3937h - a)$

$K = 917 + 13.2276w + 2.3622h - 6a$

Exercise Set 2.4

1. "What percent of 57 is 23?" can be translated as $n \cdot 57 = 23$, so choice (d) is correct.

3. "23 is 57% of what number?" can be translated as $23 = 0.57y$, so choice (e) is correct.

5. "57 is what percent of 23?" can be translated as $n \cdot 23 = 57$, so choice (c) is correct.

7. "What is 23% of 57?" can be translated as $a = (0.23)57$, so choice (f) is correct.

9. "23% of what number is 57?" can be translated as $57 = 0.23y$, so choice (b) is correct.

11. $30\% = 30.0\%$

30%   0.30.0

Move the decimal point 2 places to the left.

$30\% = 0.30$, or 0.3

13. $2\% = 2.0\%$

2%   0.02.0

Move the decimal point 2 places to the left.

$2\% = 0.02$

15. $77\% = 77.0\%$

77%   0.77.0

Move the decimal point 2 places to the left.

$77\% = 0.77$

17. $9\% = 9.0\%$

9%   0.09.0

Move the decimal point 2 places to the left.

$9\% = 0.09$

19. 62.58%   0.62.58

Move the decimal point 2 places to the left.

$62.58\% = 0.6258$

21. 0.7%   0.00.7

Move the decimal point 2 places to the left.

$0.7\% = 0.007$

23. 125%=125.0%   1.25.0

Move the decimal point 2 places to the left.

$125\% = 1.25$

25. 0.64

First move the decimal point two places to the right; then write a % symbol: 0.64.   64%

27. 0.106

First move the decimal point two places to the right; then write a % symbol: 0.10.6   10.6%

29. 0.42

First move the decimal point two places to the right; then write a % symbol: 0.42.   42%

31. 0.9

First move the decimal point two places to the right; then write a % symbol: 0.90.   90%

33. 0.0049

First move the decimal point two places to the right; then write a % symbol: 0.00. 49   0.49%

35. 1.08

First move the decimal point two places to the right; then write a % symbol: 1.08.   108%

37. 2.3

First move the decimal point two places to the right; then write a % symbol: 2.30.   230%

39. $\frac{4}{5}$ $\left(\text{Note: } \frac{4}{5} = 0.8\right)$

Move the decimal point two places to the right; then write a % symbol: 0.80.   80%

41. $\frac{8}{25}$ $\left(\text{Note: } \frac{8}{25} = 0.32\right)$

First move the decimal point two places to the right; then write a % symbol: 0.32.   32%

43. ***Translate.***

What percent of 68 is 17?

$y \cdot 68 = 17$

We solve the equation and then convert to percent notation.

$$y \cdot 68 = 17$$
$$y = \frac{17}{68}$$
$$y = 0.25 = 25\%$$

The answer is 25%.

45. ***Translate.***

What percent of 125 is 30?

$y \cdot 125 = 30$

We solve the equation and then convert to percent notation.

$$y \cdot 125 = 30$$
$$y = \frac{30}{125}$$
$$y = 0.24 = 24\%$$

The answer is 24%.

47. ***Translate.***

$$\underbrace{14}_{} \text{ is } 30\% \text{ of } \underbrace{\text{what number}}?$$

$$14 = 30\% \cdot y$$

We solve the equation.

$$14 = 0.3y \quad (30\% = 0.3)$$

$$\frac{14}{0.3} = y$$

$$46.\overline{6} = y$$

The answer is $46.\overline{6}$, or $46\frac{2}{3}$, or $\frac{140}{3}$.

49. ***Translate.***

$$0.3 \text{ is } 12\% \text{ of } \underbrace{\text{what number}}?$$

$$0.3 = 12\% \cdot y$$

We solve the equation.

$$0.3 = 0.12y \quad (12\% = 0.12)$$

$$\frac{0.3}{0.12} = y$$

$$2.5 = y$$

The answer is 2.5.

51. ***Translate.***

$$\underbrace{\text{What number}} \text{ is } 35\% \text{ of } 240?$$

$$y = 35\% \cdot 240$$

We solve the equation.

$$y = 0.35 \cdot 240 \quad (35\% = 0.35)$$

$$y = 84 \quad \text{Multiplying}$$

The answer is 84.

53. ***Translate.***

$$\underbrace{\text{What percent}} \text{ of } 60 \text{ is } 75?$$

$$y \cdot 60 = 75$$

We solve the equation and then convert to percent notation.

$$y \cdot 60 = 75$$

$$y = \frac{75}{60}$$

$$y = 1.25 = 125\%$$

The answer is 125%.

55. ***Translate.***

$$\text{What is } 2\% \text{ of } 40?$$

$$x = 2\% \cdot 40$$

We solve the equation.

$$x = 0.02 \cdot 40 \quad (2\% = 0.02)$$

$$x = 0.8 \quad \text{Multiplying}$$

The answer is 0.8.

57. Observe that 25 is half of 50. Thus, the answer is 0.5, or 50%. We could also do this exercise by translating to an equation.

Translate.

$$25 \text{ is } \underbrace{\text{what percent}} \text{ of } 50?$$

$$25 = y \cdot 50$$

We solve the equation and convert to percent notation.

$$25 = y \cdot 50$$

$$\frac{25}{50} = y$$

$$0.5 = y, \text{ or } 50\% = y$$

The answer is 50%.

59. First we reword and translate, letting p represent the price of a dog.

$$\text{What is } 3\% \text{ of } \$6600?$$

$$p = 0.03 \cdot 6600$$

$$p = 0.03 \cdot 6600 = 198$$

The price of the dog is $198.

61. First we reword and translate, letting v represent the amount spent on veterinary care.

$$\text{What is } 24\% \text{ of } \$6600?$$

$$v = 0.24 \cdot 6600$$

$$v = 0.24 \cdot 6600 = 1584$$

Veterinarian expenses are $1584.

63. First we reword and translate, letting s represent the cost of supplies.

$$\text{What is } 8\% \text{ of } \$6600?$$

$$s = 0.08 \cdot 6600$$

$$s = 0.08 \cdot 6600 = 528$$

The cost of supplies is $528.

65. First we reword and translate, letting c represent the number of credits Frank has completed.

$$\text{What is } 60\% \text{ of } 125?$$

$$c = 0.6 \cdot 125$$

$$c = 0.6 \cdot 125 = 75$$

Frank has completed 75 credits.

67. First we reword and translate, letting b represent the number of at-bats.

$$194 \text{ is } 31\% \text{ of } \underbrace{\text{what number}}?$$

$$194 = 0.31 \cdot b$$

$$\frac{194}{0.31} = b$$

$$626 \approx b$$

Ichiro Suzuki had 626 at-bats.

69. a) First we reword and translate, letting p represent the unknown percent.

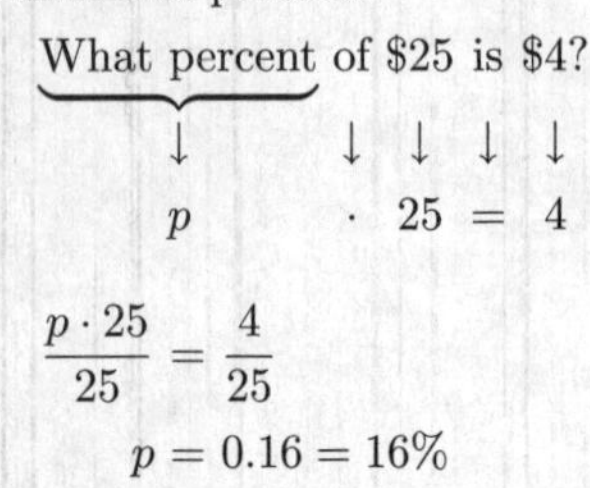

$$\frac{p \cdot 25}{25} = \frac{4}{25}$$
$$p = 0.16 = 16\%$$

The tip was 16% of the cost of the meal.

b) We add to find the total cost of the meal, including tip:

$$\$25 + \$4 = \$29$$

71. To find the percent of cars manufactured in the U.S., we first reword and translate, letting p represent the unknown percent.

$\underbrace{\text{6.0 million}}$ is $\underbrace{\text{what percent}}$ of $\underbrace{\text{8.3 million}}$?

$$\downarrow \quad \downarrow \quad \downarrow \quad \downarrow \quad \downarrow$$
$$6.0 \quad = \quad p \quad \cdot \quad 8.3$$

$$\frac{6.0}{8.3} = p$$
$$0.72 \approx p$$
$$72\% \approx p$$

About 72% of the cars were manufactured in the U.S.

To find the percent of cars manufactured outside the U.S., we subtract:

$$100\% - 72\% = 28\%.$$

About 28% of the cars were manufactured outside the U.S.

73. Let I = the amount of interest Sarah will pay. Then we have:

I is 8% of \$3500.
$$\downarrow \; \downarrow \;\; \downarrow \;\; \downarrow \;\;\; \downarrow$$
$$I = 0.08 \cdot \$3500$$
$$I = \$280$$

Sarah will pay \$280 interest.

75. If n = the number of women who had babies in good or excellent health, we have:

n is 95% of 300.
$$\downarrow \; \downarrow \;\; \downarrow \;\; \downarrow \;\; \downarrow$$
$$n = 0.95 \cdot 300$$
$$n = 285$$

285 women had babies in good or excellent health.

77. A self-employed person must earn 120% as much as a non-self-employed person. Let a = the amount Joy would need to earn, in dollars per hour, on her own for a comparable income. Then we have:

a is 120% of \$15.
$$\downarrow \; \downarrow \;\; \downarrow \;\; \downarrow \;\; \downarrow$$
$$a = 1.2 \cdot 15$$
$$a = 18$$

Joy would need to earn \$18 per hour on her own.

79. First we subtract to find the amount of the increase.

$$40,000 - 16,000 = 24,000$$

Then we reword and translate.

$\underbrace{\text{What percent}}$ of 16,000 is 24,000?
$$\downarrow \quad \downarrow \quad \downarrow \quad \downarrow \quad \downarrow$$
$$p \quad \cdot \quad 16,000 = 24,000$$

$$\frac{p \cdot 16,000}{16,000} = \frac{24,000}{16,000}$$
$$p = 1.5 = 150\%$$

The number of USA Triathlon members increased by 150% from 1993 to 2002.

81. When the sales tax is 5%, the total amount paid is 105% of the cost of the merchandise. Let c = the cost of the merchandise. Then we have:

\$37.80 is 105% of c.
$$\downarrow \quad \downarrow \quad \downarrow \quad \downarrow \; \downarrow$$
$$37.80 = 1.05 \cdot c$$
$$\frac{37.80}{1.05} = c$$
$$36 = c$$

The price of the merchandise was \$36.

83. When the sales tax is 5%, the total amount paid is 105% of the cost of the merchandise. Let c = the amount the school group owes, or the cost of the software without tax. Then we have:

\$157.41 is 106% of c.
$$\downarrow \quad \downarrow \quad \downarrow \quad \downarrow \; \downarrow$$
$$157.41 = 1.06 \cdot c$$
$$\frac{157.41}{1.06} = c$$
$$148.5 = c$$

The school group owes \$148.50.

85. First we reword and translate.

What is 16.5% of 191?
$$\downarrow \quad \downarrow \quad \downarrow \quad \downarrow \quad \downarrow$$
$$a = 0.165 \cdot 191$$

Solve. We convert 16.5% to decimal notation and multiply.

$$a = 0.165 \cdot 191$$
$$a = 31.515 \approx 31.5$$

About 31.5 lb of the author's body weight is fat.

87. Let b = the number of brochures the business can expect to be opened and read. Then we have:

b is 78% of 9500.
$$\downarrow \; \downarrow \;\; \downarrow \;\; \downarrow \;\; \downarrow$$
$$b = 0.78 \cdot 9500$$
$$b = 7410$$

The business can expect 7410 brochures to be opened and read.

89. The number of calories in a serving of Light Style Bread is 85% of the number of calories in a serving of regular bread.

Let c = the number of calories in a serving of regular bread. Then we have:

$$\underbrace{\text{140 calories}}_{140} \text{ is } 85\% \text{ of } c.$$

$$140 = 0.85 \cdot c$$

$$\frac{140}{0.85} = c$$

$$165 \approx c$$

There are about 165 calories in a serving of regular bread.

91. *Writing Exercise*

93. Let n represent "some number." Then we have $n+5$, or $5+n$.

95. $8 \cdot 2a$, or $2a \cdot 8$.

97. *Writing Exercise*

99. Let p = the population of Bardville. Then we have:

1332 is 15% of 48% of $\underbrace{\text{the population.}}_{p}$

$$1332 = 0.15 \cdot 0.48 \cdot p$$

$$\frac{1332}{0.15(0.48)} = p$$

$$18,500 = p$$

The population of Bardville is 18,500.

101. Since 4 ft $= 4 \times 1$ ft $= 4 \times 12$ in. $= 48$ in., we can express 4 ft 8 in. as 48 in. + 8 in., or 56 in. We reword and translate. Let a = Dana's final adult height.

56 in. is 84.4% of $\underbrace{\text{adult height}}_{a}$.

$$56 = 0.844 \cdot a$$

$$\frac{56}{0.844} = \frac{0.844 \cdot a}{0.844}$$

$$66 \approx a$$

Note that 66 in. = 60 in. + 6 in. = 5 ft 6 in. Dana's final adult height will be about 5 ft 6 in.

103. Using the formula for the area A of a rectangle with length l and width w, $A = l \cdot w$, we first find the area of the photo.

$$A = 8 \text{ in.} \times 6 \text{ in.} = 48 \text{ in}^2$$

Next we find the area of the photo that will be visible using a mat intended for a 5-in. by 7-in. photo.

$$A = 7 \text{ in.} \times 5 \text{ in.} = 35 \text{ in}^2$$

Then the area of the photo that will be hidden by the mat is $48 \text{ in}^2 - 35 \text{ in}^2$, or 13 in^2.

We find what percentage of the area of the photo this represents.

$\underbrace{\text{What percent}}_{p}$ of $\underbrace{48 \text{ in}^2}_{48}$ is $\underbrace{13 \text{ in}^2}_{13}$?

$$p \cdot 48 = 13$$

$$\frac{p \cdot 48}{48} = \frac{13}{48}$$

$$p \approx 0.27$$

$$p \approx 27\%$$

The mat will hide about 27% of the photo.

105. *Writing Exercise*

Exercise Set 2.5

1. ***Familiarize.*** Let n = the number. Then two fewer than ten times the number is $10n - 2$.

Translate.

$\underbrace{\text{Two fewer than ten times a number}}_{10n-2}$ is 78.

$$10n - 2 = 78$$

Carry out. We solve the equation.

$$10n - 2 = 78$$
$$10n = 80 \quad \text{Adding 2}$$
$$n = 8 \quad \text{Dividing by 10}$$

Check. Ten times 8 is 80 and two fewer than 80 is 78. The answer checks.

State. The number is 8.

3. ***Familiarize.*** Let a = the number. Then "five times the sum of 3 and some number" translates to $5(a+3)$.

Translate.

$\underbrace{\text{Five times the sum of 3 and some number}}_{5(a+3)}$ is 70.

$$5(a+3) = 70$$

Carry out. We solve the equation.

$$5(a+3) = 70$$
$$5a + 15 = 70 \quad \text{Using the distributive law}$$
$$5a = 55 \quad \text{Subtracting 15}$$
$$a = 11 \quad \text{Dividing by 5}$$

Check. The sum of 3 and 11 is 14, and $5 \cdot 14 = 70$. The answer checks.

State. The number is 11.

5. ***Familiarize.*** Let p = the regular price of the shoes. At 15% off, Amy paid (100−15)%, or 85% of the regular price.

Translate.

\$72.25 is 85% of $\underbrace{\text{the regular price.}}_{p}$

$$72.25 = 0.85 \cdot p$$

Carry out. We solve the equation.

$$72.25 = 0.85p$$
$$\frac{72.25}{0.85} = p \quad \text{Dividing both sides by 0.85}$$
$$85 = p$$

Check. 85% of \$85, or 0.85(\$85), is \$72.25. The answer checks.

State. The regular price was \$85.

7. ***Familiarize.*** Let c = the price of the graphing calculator itself. When the sales tax rate is 5%, the tax paid on the calculator is 5% of c, or $0.05c$.

Translate.

Price of calculator plus sales tax is \$89.25.

$$c \quad + \quad 0.05b \quad = \quad 89.25$$

Carry out. We solve the equation.

$$\begin{aligned} c + 0.05c &= 89.25 \\ 1.05c &= 89.25 \\ c &= \frac{89.25}{1.05} \\ c &= 85 \end{aligned}$$

Check. 5% of \$85, or 0.05(\$85), is \$4.25 and \$85 + \$4.25 is \$89.25, the total cost. The answer checks.

State. The graphing calculator itself cost \$85.

9. ***Familiarize.*** Let $d =$ Kouros' distance, in miles, from the start after 8 hr. Then the distance from the finish line is $2d$.

Translate.

Distance from start plus distance from finish is 188 mi.

$$d \quad + \quad 2d \quad = \quad 188$$

Carry out. We solve the equation.

$$\begin{aligned} d + 2d &= 188 \\ 3d &= 188 \\ d &= \frac{188}{3}, \text{ or } 62\frac{2}{3} \end{aligned}$$

Check. If Kouros is $\frac{188}{3}$ mi from the start, then he is $2 \cdot \frac{188}{3}$, or $\frac{376}{3}$ mi from the finish. Since $\frac{188}{3} + \frac{376}{3} = \frac{564}{3} = 188$, the total distance run, the answer checks.

State. Kouros had run approximately $62\frac{2}{3}$ mi.

11. ***Familiarize.*** Let $d =$ the distance Wheldon had traveled, in miles, at the given point. This is the distance from the start. The corresponding distance from the finish was $300 - d$ miles.

Translate. We reword and translate.

Distance to finish plus 80 mi more was distance to start.

$$300 - d \quad + \quad 80 \quad = \quad d$$

Carry out. We solve the equation.

$$\begin{aligned} 300 - d + 80 &= d \\ 380 - d &= d \\ 380 &= 2d \\ 190 &= d \end{aligned}$$

Check. If Wheldon was 190 mi from the start, he was 300 − 190, or 110 mi, from the finish. Since 190 is 80 more than 110, the answer checks.

State. Wheldon had traveled 190 mi at the given point.

13. ***Familiarize.*** Let $n =$ the number of Joan's apartment. Then $n+1 =$ the number of her next-door neighbor's apartment.

Translate.

Joan's number plus neighbor's number is 2409.

$$n \quad + \quad (n+1) \quad = 2409$$

Carry out. We solve the equation.

$$\begin{aligned} n + (n+1) &= 2409 \\ 2n + 1 &= 2409 \\ 2n &= 2408 \\ n &= 1204 \end{aligned}$$

If Joan's apartment number is 1204, then her next-door neighbor's number is 1204 + 1, or 1205.

Check. 1204 and 1205 are consecutive numbers whose sum is 2409. The answer checks.

State. The apartment numbers are 1204 and 1205.

15. ***Familiarize.*** Let $n =$ the smaller house number. Then $n + 2 =$ the larger number.

Translate.

Smaller number plus larger number is 794.

$$n \quad + \quad (n+2) \quad = 794$$

Carry out. We solve the equation.

$$\begin{aligned} n + (n+2) &= 794 \\ 2n + 2 &= 794 \\ 2n &= 792 \\ n &= 396 \end{aligned}$$

If the smaller number is 396, then the larger number is 396 + 2, or 398.

Check. 396 and 398 are consecutive even numbers and 396 + 398 = 794. The answer checks.

State. The house numbers are 396 and 398.

17. ***Familiarize.*** Let $x =$ the first page number. Then $x + 1 =$ the second page number, and $x + 2 =$ the third page number.

Translate.

The sum of three consecutive page numbers is 60.

$$x + (x+1) + (x+2) \quad = 60$$

Carry out. We solve the equation.

$$\begin{aligned} x + (x+1) + (x+2) &= 60 \\ 3x + 3 &= 60 \quad \text{Combining like terms} \\ 3x &= 57 \quad \text{Subtracting 3 from both sides} \\ x &= 19 \quad \text{Dividing both sides by 3} \end{aligned}$$

If x is 19, then $x + 1$ is 20 and $x + 2 = 21$.

Check. 19, 20, and 21 are consecutive integers, and $19 + 20 + 21 = 60$. The result checks.

State. The page numbers are 19, 20, and 21.

19. ***Familiarize***. Let $g =$ the groom's age. Then $g + 19 =$ the bride's age.

Translate.

Groom's age plus bride's age is 185.

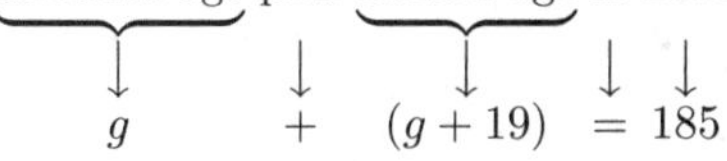

$g \quad + \quad (g+19) \quad = \quad 185$

Carry out. We solve the equation.

$$g + (g + 19) = 185$$
$$2g + 19 = 185$$
$$2g = 166$$
$$g = 83$$

If g is 83, then $g + 19$ is 102.

Check. 102 is 19 more than 83, and $83 + 102 = 185$. The answer checks.

State. The groom was 83 yr old, and the bride was 102 yr old.

21. ***Familiarize***. Let $a =$ the amount spent to remodel bathrooms, in billions of dollars. Then $2a =$ the amount spent to remodel kitchens. The sum of these two amounts is \$35 billion.

Translate.

Amount spent on bathrooms plus amount spent on kitchens is \$35 billion.

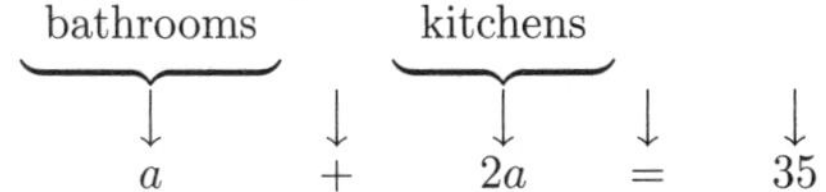

$a \quad + \quad 2a \quad = \quad 35$

Carry out. We solve the equation.

$$a + 2a = 35$$
$$3a = 35 \quad \text{Combining like terms}$$
$$a = \frac{35}{3}, \text{ or } 11\frac{2}{3}$$

If $a = \frac{35}{3}$, then $2a = 2 \cdot \frac{35}{3} = \frac{70}{3} = 23\frac{1}{3}$.

Check. $\frac{70}{3}$ is twice $\frac{35}{3}$, and $\frac{35}{3} + \frac{70}{3} = \frac{105}{3} = 35$. The answer checks.

State. \$$11\frac{2}{3}$ billion was spent to remodel bathrooms, and \$$23\frac{1}{3}$ billion was spent to remodel kitchens.

23. ***Familiarize.*** The page numbers are consecutive integers. If we let $x =$ the smaller number, then $x + 1 =$ the larger number.

Translate. We reword the problem.

First integer + Second integer = 281

$x \quad + \quad (x+1) \quad = 281$

Carry out. We solve the equation.

$$x + (x + 1) = 281$$
$$2x + 1 = 281 \quad \text{Combining like terms}$$
$$2x = 280 \quad \text{Adding } -1 \text{ on both sides}$$
$$x = 140 \quad \text{Dividing on both sides by 2}$$

Check. If $x = 140$, then $x + 1 = 141$. These are consecutive integers, and $140 + 141 = 281$. The answer checks.

State. The page numbers are 140 and 141.

25. ***Familiarize***. We draw a picture. Let $w =$ the width of the rectangle, in feet. Then $w + 60 =$ the length.

w

$w + 60$

The perimeter is twice the length plus twice the width, and the area is the product of the length and the width.

Translate.

Twice the length plus twice the width is 520 ft.

$2(w+60) \quad + \quad 2w \quad = \quad 520$

Carry out. We solve the equation.

$$2(w + 60) + 2w = 520$$
$$2w + 120 + 2w = 520$$
$$4w + 120 = 520$$
$$4w = 400$$
$$w = 100$$

Then $w + 60 = 100 + 60 = 160$, and the area is $160 \text{ ft} \cdot 100 \text{ ft} = 16,000 \text{ ft}^2$.

Check. The length, 160 ft, is 60 ft more than the width, 100 ft. The perimeter is $2 \cdot 160 \text{ ft} + 2 \cdot 100 \text{ ft}$, or 320 ft + 200 ft, or 520 ft. We can check the area by doing the calculation again. The answer checks.

State. The length is 160 ft, the width is 100 ft, and the area is 16,000 ft^2.

27. ***Familiarize***. We draw a picture. Let $w =$ the width of the court, in feet. Then $w + 34 =$ the length.

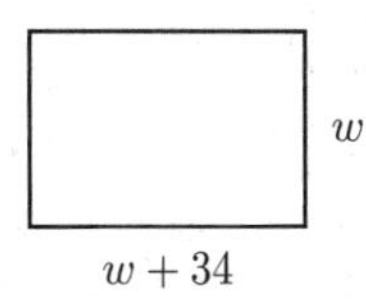

w

$w + 34$

The perimeter is twice the length plus twice the width.

Translate.

Twice the length plus twice the width is 268 ft.

$2(w+34) \quad + \quad 2w \quad = \quad 268$

Carry out. We solve the equation.

$$\begin{aligned}2(w+34)+2w&=268\\2w+68+2w&=268\\4w+68&=268\\4w&=200\\w&=50\end{aligned}$$

Then $w+34=50+34=84$.

Check. The length, 84 ft, is 34 ft more than the width, 50 ft. The perimeter is $2\cdot 84\text{ ft}+2\cdot 50\text{ ft}=168\text{ ft}+100\text{ ft}=268\text{ ft}$. The answer checks.

State. The length of the court is 84 ft, and the width is 50 ft.

29. ***Familiarize.*** Let $w=$ the width, in inches. Then $2w=$ the length. The perimeter is twice the length plus twice the width. We express $10\frac{1}{2}$ as 10.5.

Translate.

Twice the length plus twice the width is 10.5 in.

$$2\cdot 2w \quad + \quad 2w \quad = \quad 10.5$$

Carry out. We solve the equation.

$$\begin{aligned}2\cdot 2w+2w&=10.5\\4w+2w&=10.5\\6w&=10.5\\w&=1.75,\text{ or }1\frac{3}{4}\end{aligned}$$

Then $2w=2(1.75)=3.5$, or $3\frac{1}{2}$.

Check. The length, $3\frac{1}{2}$ in., is twice the width, $1\frac{3}{4}$ in. The perimeter is $2\left(3\frac{1}{2}\text{ in.}\right)+2\left(1\frac{3}{4}\text{ in.}\right)=$ $7\text{ in.}+3\frac{1}{2}\text{ in.}=10\frac{1}{2}\text{ in.}$ The answer checks.

State. The actual dimensions are $3\frac{1}{2}$ in. by $1\frac{3}{4}$ in.

31. ***Familiarize.*** We draw a picture. We let $x=$ the measure of the first angle. Then $3x=$ the measure of the second angle, and $x+30=$ the measure of the third angle.

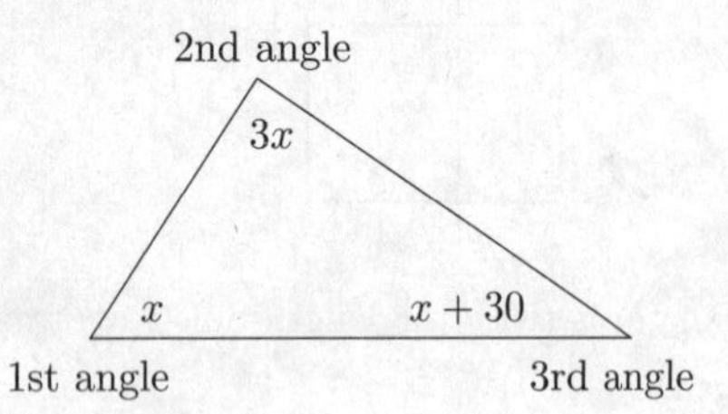

Recall that the measures of the angles of any triangle add up to 180°.

Translate.

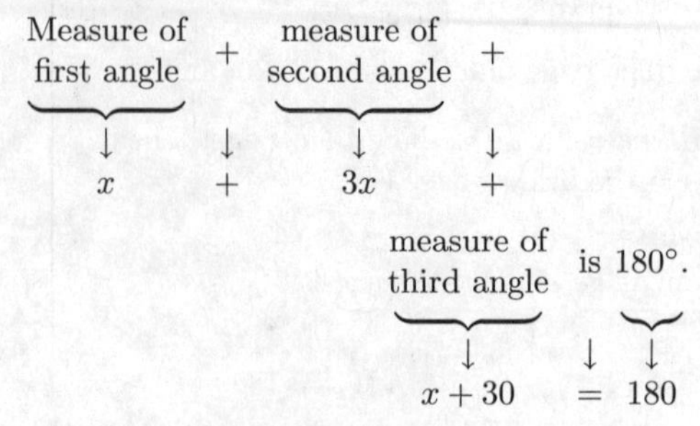

Carry out. We solve the equation.

$$\begin{aligned}x+3x+(x+30)&=180\\5x+30&=180\\5x&=150\\x&=30\end{aligned}$$

Possible answers for the angle measures are as follows:

First angle: $x=30^\circ$

Second angle: $3x=3(30)^\circ=90^\circ$

Third angle: $x+30^\circ=30^\circ+30^\circ=60^\circ$

Check. Consider 30°, 90°, and 60°. The second angle is three times the first, and the third is 30° more than the first. The sum of the measures of the angles is 180°. These numbers check.

State. The measure of the first angle is 30°, the measure of the second angle is 90°, and the measure of the third angle is 60°.

33. ***Familiarize.*** Let $x=$ the measure of the first angle. Then $3x=$ the measure of the second angle, and $x+3x+10=4x+10=$ the measure of the third angle. Recall that the sum of the measures of the angles of a triangle is 180°.

Translate.

Measure of first angle + measure of second angle + measure of third angle is 180°.

$$x \quad + \quad 3x \quad + \quad (4x+10) \quad = \quad 180$$

Carry out. We solve the equation.

$$\begin{aligned}x+3x+(4x+10)&=180\\8x+10&=180\\8x&=170\\x&=21.25\end{aligned}$$

If x is 21.25, then $3x$ is 63.75, and $4x+10$ is 95.

Check. Consider 21.25°, 63.75°, and 95°. The second is three times the first, and the third is 10° more than the sum of the other two. The sum of the measures of the angles is 180°. These numbers check.

State. The measure of the third angle is 95°.

35. ***Familiarize.*** Let b = the length of the bottom section of the rocket, in feet. Then $\frac{1}{6}b$ = the length of the top section, and $\frac{1}{2}b$ = the length of the middle section.

Translate.

Length of top section + length of middle section + length of bottom section is 240 ft.

$$\frac{1}{6}b + \frac{1}{2}b + b = 240$$

Carry out. We solve the equation. First we multiply by 6 on both sides to clear the fractions.

$$\frac{1}{6}b + \frac{1}{2}b + b = 240$$
$$6\left(\frac{1}{6}b + \frac{1}{2}b + b\right) = 6 \cdot 240$$
$$6 \cdot \frac{1}{6}b + 6 \cdot \frac{1}{2}b + 6 \cdot b = 1440$$
$$b + 3b + 6b = 1440$$
$$10b = 1440$$
$$b = 144$$

Then $\frac{1}{6}b = \frac{1}{6} \cdot 144 = 24$ and $\frac{1}{2}b = \frac{1}{2} \cdot 144 = 72$.

Check. 24 ft is $\frac{1}{6}$ of 144 ft, and 72 ft is $\frac{1}{2}$ of 144 ft. The sum of the lengths of the sections is
24 ft + 72 ft + 144 ft = 240 ft. The answer checks.

State. The length of the top section is 24 ft, the length of the middle section is 72 ft, and the length of the bottom section is 144 ft.

37. ***Familiarize.*** Let m = the number of miles that can be traveled on a \$18 budget. Then the total cost of the taxi ride, in dollars, is $1.90 + 1.60m$, or $1.9 + 1.6m$.

Translate.

Cost of taxi ride is \$18.

$$1.9 + 1.6m = 18$$

Carry out. We solve the equation.

$$1.9 + 1.6m = 18$$
$$1.6m = 16.1$$
$$m = \frac{16.1}{1.6} = \frac{161}{16} = 10\frac{1}{16}$$

Check. The mileage change is $\$1.60\left(10\frac{1}{16}\right)$, or \$16.10, and the total cost of the ride is
\$1.90 + \$16.10 = \$18. The answer checks.

State. Debbie and Alex can travel $10\frac{1}{16}$ mi on their budget.

39. ***Familiarize.*** The total cost is the daily charge plus the mileage charge. Let d = the distance that can be traveled, in miles, in one day for \$100. The mileage charge is the cost per mile times the number of miles traveled, or $0.39d$.

Translate.

Daily rate plus mileage charge is \$100.

$$49.95 + 0.39d = 100$$

Carry out. We solve the equation.

$$49.95 + 0.39d = 100$$
$$0.39d = 50.05$$
$$d = 128.\overline{3}, or 128\frac{1}{3}$$

Check. For a trip of $128\frac{1}{3}$ mi, the mileage charge is $\$0.39\left(128\frac{1}{3}\right)$, or \$50.05, and \$49.95 + \$50.05 = \$100. The answer checks.

State. They can travel $128\frac{1}{3}$ mi in one day and stay within their budget.

41. ***Familiarize.*** Let x = the measure of one angle. Then $90 - x$ = the measure of its complement.

Translate.

Measure of one angle is 15° more than twice the measure of its complement.

$$x = 15 + 2(90 - x)$$

Carry out. We solve the equation.

$$x = 15 + 2(90 - x)$$
$$x = 15 + 180 - 2x$$
$$x = 195 - 2x$$
$$3x = 195$$
$$x = 65$$

If x is 65, then $90 - x$ is 25.

Check. The sum of the angle measures is 90°. Also, 65° is 15° more than twice its complement, 25°. The answer checks.

State. The angle measures are 65° and 25°.

43. ***Familiarize.*** Let l = the length of the paper, in cm. Then $l - 6.3$ = the width. The perimeter is twice the length plus twice the width.

Translate.

Twice the length plus twice the width is 99 cm.

$$2l + 2(l - 6.3) = 99$$

Carry out. We solve the equation.

$$2l + 2(l - 6.3) = 99$$
$$2l + 2l - 12.6 = 99$$
$$4l - 12.6 = 99$$
$$4l = 111.6$$
$$l = 27.9$$

Then $l - 6.3 = 27.9 - 6.3 = 21.6$.

Check. The width, 21.6 cm, is 6.3 cm less than the length, 27.9 cm. The perimeter is 2(27.9 cm) + 2(21.6 cm) = 55.8 cm + 43.2 cm = 99 cm. The answer checks.

State. The length of the paper is 27.9 cm, and the width is 21.6 cm.

45. ***Familiarize.*** Let a = the amount Sharon invested. Then the simple interest for one year is 6% $\cdot a$, or $0.06a$.

Translate.

$$\underbrace{\text{Amount invested}}_{a}\ \underbrace{\text{plus}}_{+}\ \underbrace{\text{interest}}_{0.06a}\ \underbrace{\text{is}}_{=}\ \underbrace{\$6996.}_{6996}$$

Carry out. We solve the equation.

$$a + 0.06a = 6996$$
$$1.06a = 6996$$
$$a = 6600$$

Check. An investment of \$6600 at 6% simple interest earns 0.06(\$6600), or \$396, in one year. Since \$6600 + \$396 = \$6996, the answer checks.

State. Sharon invested \$6600.

47. ***Familiarize.*** Let w = the winning score. Then $w - 796$ = the losing score.

Translate.

$$\underbrace{\text{Winning score}}_{w}\ \underbrace{\text{plus}}_{+}\ \underbrace{\text{losing score}}_{w-796}\ \underbrace{\text{was}}_{=}\ \underbrace{\text{1302 points}}_{1302}.$$

Carry out. We solve the equation.

$$w + w - 796 = 1302$$
$$2w - 796 = 1302$$
$$2w = 2098$$
$$w = 1049$$

Then $w - 796 = 1049 - 796 = 253$.

Check. The winning score, 1049, is 796 points more than the losing score, 253. The total of the two scores is 1049 + 253, or 1302 points. The answer checks.

State. The winning score was 1049 points.

49. ***Familiarize.*** We will use the equation

$$T = \frac{1}{4}N + 40.$$

Translate. We substitute 80 for T.

$$80 = \frac{1}{4}N + 40$$

Carry out. We solve the equation.

$$80 = \frac{1}{4}N + 40$$
$$40 = \frac{1}{4}N$$
$$160 = N \quad \text{Multiplying by 4 on both sides}$$

Check. When $N = 160$, we have $T = \frac{1}{4} \cdot 160 + 40 = 40 + 40 = 80$. The answer checks.

State. A cricket chirps 160 times per minute when the temperature is 80°F.

51. *Writing Exercise*

53. Since -9 is to the left of 5 on the number line, we have $-9 < 5$.

55. Since -4 is to the left of 7 on the number line, we have $-4 < 7$.

57. *Writing Exercise*

59. ***Familiarize.*** Let c = the amount the meal originally cost. The 15% tip is calculated on the original cost of the meal, so the tip is $0.15c$.

Translate.

$$\underbrace{\text{Original cost}}_{c}\ \underbrace{\text{plus}}_{+}\ \underbrace{\text{tip}}_{0.15c}\ \underbrace{\text{less}}_{-}\ \underbrace{\$10}_{10}\ \underbrace{\text{is}}_{=}\ \underbrace{\$32.55.}_{32.55}$$

Carry out. We solve the equation.

$$c + 0.15c - 10 = 32.55$$
$$1.15c - 10 = 32.55$$
$$1.15c = 42.55$$
$$c = 37$$

Check. If the meal originally cost \$37, the tip was 15% of \$37, or 0.15(\$37), or \$5.55. Since \$37 + \$5.55 − \$10 = \$32.55, the answer checks.

State. The meal originally cost \$37.

61. ***Familiarize.*** Let s = one score. Then four score = $4s$ and four score and seven = $4s + 7$.

Translate. We reword .

$$\underbrace{1776}_{1776}\ \underbrace{\text{plus}}_{+}\ \underbrace{\text{four score and seven}}_{(4s+7)}\ \underbrace{\text{is}}_{=}\ \underbrace{1863}_{1863}$$

Carry out. We solve the equation.

$$1776 + (4s + 7) = 1863$$
$$4s + 1783 = 1863$$
$$4s = 80$$
$$s = 20$$

Check. If a score is 20 years, then four score and seven represents 87 years. Adding 87 to 1776 we get 1863. This checks.

State. A score is 20.

63. ***Familiarize.*** Let n = the number of half dollars. Then the number of quarters is $2n$; the number of dimes is $2 \cdot 2n$, or $4n$; and the number of nickels is $3 \cdot 4n$, or $12n$. The total value of each type of coin, in dollars, is as follows.

Half dollars: $0.5n$

Quarters: $0.25(2n)$, or $0.5n$

Dimes: $0.1(4n)$, or $0.4n$

Nickels: $0.05(12n)$, or $0.6n$

Then the sum of these amounts is $0.5n+0.5n+0.4n+0.6n$, or $2n$.

Translate.

$$\underbrace{\text{Total amount of change}}_{\downarrow\; 2n} \;\underset{\downarrow\; =}{\text{is}}\; \underset{\downarrow\; 10}{\$10.}$$

Carry out. We solve the equation.

$$2n = 10$$
$$n = 5$$

Then $2n = 2\cdot5 = 10$, $4n = 4\cdot5 = 20$, and $12n = 12\cdot5 = 60$.

Check. If there are 5 half dollars, 10 quarters, 20 dimes, and 60 nickels, then there are twice as many quarters as half dollars, twice as many dimes as quarters, and 3 times as many nickels as dimes. The total value of the coins is $\$0.5(5)+\$0.25(10)+\$0.1(20)+\$0.05(60) = \$2.50+\$2.50+\$2+\$3 = \$10$. The answer checks.

State. The shopkeeper got 5 half dollars, 10 quarters, 20 dimes, and 60 nickels.

65. ***Familiarize.*** Let $a =$ the original number of apples in the basket.

Translate.

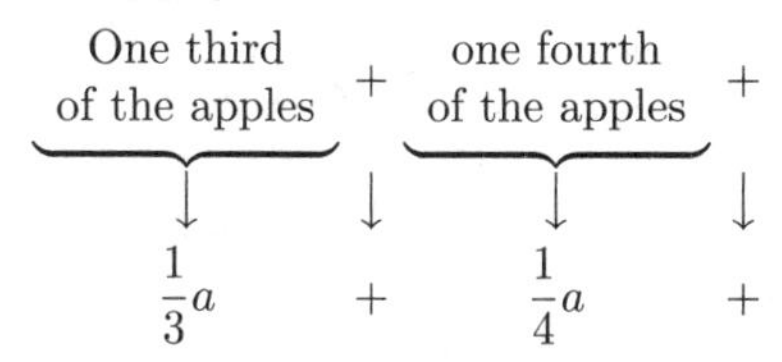

$$\frac{1}{3}a \quad + \quad \frac{1}{4}a \quad +$$

$$\underbrace{\text{one eighth of the apples}}_{\downarrow\;\frac{1}{8}a} \;\underset{\downarrow\;+}{+}\; \underbrace{\text{one fifth of the apples}}_{\downarrow\;\frac{1}{5}a} \;\underset{\downarrow\;+}{+}\; \underbrace{\text{10 apples}}_{10} \;+$$

$$\underbrace{\text{1 apple}}_{\downarrow\;1} \;\underset{\downarrow\;=}{\text{is}}\; \underbrace{\text{the original number of apples.}}_{\downarrow\;a}$$

Carry out. We solve the equation. Note that the LCD is 120.

$$\frac{1}{3}a+\frac{1}{4}a+\frac{1}{8}a+\frac{1}{5}a+10+1 = a$$
$$\frac{1}{3}a+\frac{1}{4}a+\frac{1}{8}a+\frac{1}{5}a+11 = a$$
$$120\left(\frac{1}{3}a+\frac{1}{4}a+\frac{1}{8}a+\frac{1}{5}a+11\right) = 120\cdot a$$
$$40a+30a+15a+24a+1320 = 120a$$
$$109a+1320 = 120a$$
$$1320 = 11a$$
$$120 = a$$

Check. $\frac{1}{3}\cdot120 = 40$, $\frac{1}{4}\cdot120 = 30$, $\frac{1}{8}\cdot120 = 15$, and $\frac{1}{5}\cdot120 = 24$. Then $40+30+15+24+10+1 = 120$. The result checks.

State. There were originally 120 apples in the basket.

67. ***Familiarize.*** Let $x =$ the number of additional games the Falcons will have to play. Then $\frac{x}{2} =$ the number of those games they will win, $15+\frac{x}{2} =$ the total number of games won, and $20+x =$ the total number of games played.

Translate.

$$\underbrace{\text{Number of games won}}_{\downarrow\;15+\frac{x}{2}} \;\underset{\downarrow\;=\;\downarrow\;0.6\;\downarrow\;\cdot}{\text{is 60\% of}}\; \underbrace{\text{total number of games.}}_{\downarrow\;20+x}$$

Carry out. We solve the equation.

$$15+\frac{x}{2} = 0.6(20+x)$$
$$15+0.5x = 12+0.6x \quad \left(\frac{x}{2} = \frac{1}{2}x = 0.5x\right)$$
$$15 = 12+0.1x$$
$$3 = 0.1x$$
$$30 = x$$

Check. If the Falcons play an additional 30 games, then they play a total of $20+30$, or 50, games. If they win half of the 30 additional games, or 15 games, then their wins total $15+15$, or 30. Since 60% of 50 is 30, the answer checks.

State. The Falcons will have to play 30 more games in order to win 60% of the total number of games.

69. ***Familiarize.*** Let $s =$ Ella's score on the third test. Her average score on the first two tests is 85, so she had a total of $2\cdot85$ points on those tests.

Translate. The average score on the three tests is the sum of the three scores divided by 3.

$$\frac{2\cdot85+s}{3} = 82$$

Carry out. We solve the equation.

$$\frac{2\cdot85+s}{3} = 82$$
$$\frac{170+s}{3} = 82$$
$$170+s = 246 \quad \text{Multiplying by 3}$$
$$s = 76$$

Check. If the score on the third test is 76, Ella's average score is $\frac{2\cdot85+76}{3} = \frac{246}{3} = 82$. The answer checks.

State. Ella's score on the third test was 76.

71. *Writing Exercise*

73. ***Familiarize.*** Let $w =$ the width of the rectangle, in cm. Then $w+4.25 =$ the length.

Translate.

$$\underbrace{\text{The perimeter}}_{\downarrow\;2(w+4.25)+2w} \;\underset{\downarrow\;=}{\text{is}}\; \underbrace{\text{101.74 cm.}}_{\downarrow\;101.74}$$

Carry out. We solve the equation.

$$\begin{aligned} 2(w+4.25)+2w &= 101.74 \\ 2w+8.5+2w &= 101.74 \\ 4w+8.5 &= 101.74 \\ 4w &= 93.24 \\ w &= 23.31 \end{aligned}$$

Then $w+4.25 = 23.31+4.25 = 27.56$.

Check. The length, 27.56 cm, is 4.25 cm more than the width, 23.31 cm. The perimeter is 2(27.56) cm + 2(23.31 cm) = 55.12 cm + 46.62 cm = 101.74 cm. The answer checks.

State. The length of the rectangle is 27.56 cm, and the width is 23.31 cm.

Exercise Set 2.6

1. $-5x \le 30$

$x \ge -6$ Dividing by -5 and reversing the inequality symbol

3. $-2t > -14$

$t < 7$ Dividing by -2 and reversing the inequality symbol

5. $x < -2$ and $-2 > x$ are equivalent.

7. If we add 1 to both sides of $-4x-1 \le 15$, we get $-4x \le 16$. The two given inequalities are equivalent.

9. $x > -2$

a) Since $5 > -2$ is true, 5 is a solution.

b) Since $0 > -2$ is true, 0 is a solution.

c) Since $-1.9 > -2$ is true, -1.9 is a solution.

d) Since $-7.3 > -2$ is false, -7.3 is not a solution.

e) Since $1.6 > -2$ is true, 1.6 is a solution.

11. $x \ge 6$

a) Since $-6 \ge 6$ is false, -6 is not a solution.

b) Since $0 \ge 6$ is false, 0 is not a solution.

c) Since $6 \ge 6$ is true, 6 is a solution.

d) Since $6.01 \ge 6$ is true, 6.01 is a solution.

e) Since $-3\frac{1}{2} \ge 6$ is false, $-3\frac{1}{2}$ is not a solution.

13. The solutions of $y < 2$ are those numbers less than 2. They are shown on the graph by shading all points to the left of 2. The open circle at 2 indicates that 2 is not part of the graph.

$y < 2$

−4 −2 0 2 4

15. The solutions of $y > 4$ are those numbers greater than 4. They are shown on the graph by shading all points to the right of 4. The open circle at 4 indicates that 4 is not part of the graph.

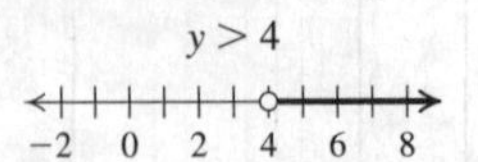

17. The solutions of $0 \le t$, or $t \ge 0$, are those numbers greater than or equal to zero. They are shown on the graph by shading the point 0 and all points to the right of 0. The closed circle at 0 indicates that 0 is part of the graph.

$0 \le t$

0

19. In order to be solution of the inequality $-5 \le x < 2$, a number must be a solution of both $-5 \le x$ and $x < 2$. The solution set is graphed as follows:

$-5 \le x < 2$

−6 −5 −4 −3 −2 −1 0 1 2 3 4

The closed circle at -5 means that -5 is part of the graph. The open circle at 2 means that 2 is not part of the graph.

21. In order to be a solution of the inequality $-5 \le x \le 0$, a number must be a solution of both $-5 \le x$ and $x \le 0$. The solution set is graphed as follows:

$-5 \le x \le 0$

−7 −6 −5 −4 −3 −2 −1 0 1 2 3

The closed circles at -5 and 0 mean that -5 and 0 are both part of the graph.

23. All points to the right of -4 are shaded. The open circle at -4 indicates that -4 is not part of the graph. Using set-builder notation we have $\{x|x > -4\}$.

25. The point 2 and all points to the left of 2 are shaded. Using set-builder notation we have $\{x|x \le 2\}$.

27. All points to the left of -1 are shaded. The open circle at -1 indicates that -1 is not part of the graph. Using set-builder notation we have $\{x|x < -1\}$.

29. The point 0 and all points to the right of 0 are shaded. Using set-builder notation we have $\{x|x \ge 0\}$.

31.

$$\begin{aligned} y+6 &> 9 \\ y+6-6 &> 9-6 \quad \text{Adding } -6 \text{ to both sides} \\ y &> 3 \quad \text{Simplifying} \end{aligned}$$

The solution set is $\{y|y > 3\}$. The graph is as follows:

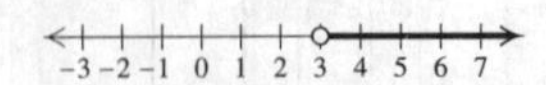

33.

$$\begin{aligned} x+9 &\le -12 \\ x+9-9 &\le -12-9 \quad \text{Adding } -9 \text{ to both sides} \\ x &\le -21 \quad \text{Simplifying} \end{aligned}$$

The solution set is $\{x|x \le -21\}$. The graph is as follows:

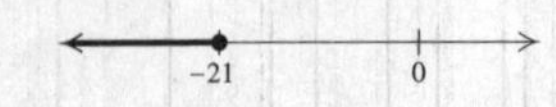

35.
$$\begin{aligned} x-3&<14 \\ x-3+3&<14+3 \quad \text{Adding 3 to both sides} \\ x&<17 \quad \text{Simplifying} \end{aligned}$$

The solution set is $\{x|x<17\}$. The graph is as follows:

0 17

37.
$$\begin{aligned} y-10&>-16 \\ y-10+10&>-16+10 \\ y&>-6 \end{aligned}$$

The solution set is $\{y|y>-6\}$. The graph is as follows:

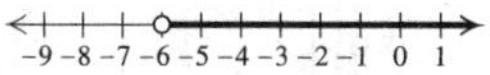

−9 −8 −7 −6 −5 −4 −3 −2 −1 0 1

39.
$$\begin{aligned} 2x&\le x+9 \\ 2x-x&\le x+9-x \\ x&\le 9 \end{aligned}$$

The solution set is $\{x|x\le 9\}$. The graph is as follows:

0 9

41.
$$\begin{aligned} y+\frac{1}{3}&\le\frac{5}{6} \\ y+\frac{1}{3}-\frac{1}{3}&\le\frac{5}{6}-\frac{1}{3} \\ y&\le\frac{5}{6}-\frac{2}{6} \\ y&\le\frac{3}{6} \\ y&\le\frac{1}{2} \end{aligned}$$

The solution set is $\left\{y\middle|y\le\frac{1}{2}\right\}$. The graph is as follows:

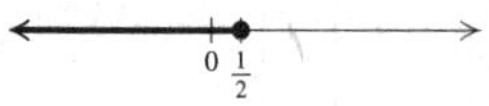

0 $\frac{1}{2}$

43.
$$\begin{aligned} t-\frac{1}{8}&>\frac{1}{2} \\ t-\frac{1}{8}+\frac{1}{8}&>\frac{1}{2}+\frac{1}{8} \\ t&>\frac{4}{8}+\frac{1}{8} \\ t&>\frac{5}{8} \end{aligned}$$

The solution set is $\left\{t\middle|t>\frac{5}{8}\right\}$. The graph is as follows:

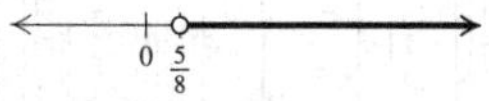

0 $\frac{5}{8}$

45.
$$\begin{aligned} -9x+17&>17-8x \\ -9x+17-17&>17-8x-17 \quad \text{Adding } -17 \\ -9x&>-8x \\ -9x+9x&>-8x+9x \quad \text{Adding } 9x \\ 0&>x \end{aligned}$$

The solution set is $\{x|x<0\}$. The graph is as follows:

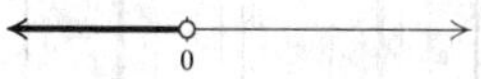

0

47. $-23<-t$

The inequality states that the opposite of 23 is less than the opposite of t. Thus, t must be less than 23, so the solution set is $\{t|t<23\}$. To solve this inequality using the addition principle, we would proceed as follows:

$$\begin{aligned} -23&<-t \\ t-23&<0 \quad \text{Adding } t \text{ to both sides} \\ t&<23 \quad \text{Adding 23 to both sides} \end{aligned}$$

The solution set is $\{t|t<23\}$. The graph is as follows:

0 23

49.
$$\begin{aligned} 5x&<35 \\ \frac{1}{5}\cdot 5x&<\frac{1}{5}\cdot 35 \quad \text{Multiplying by } \frac{1}{5} \\ x&<7 \end{aligned}$$

The solution set is $\{x|x<7\}$. The graph is as follows:

0 7

51.
$$\begin{aligned} -7x&<13 \\ -\frac{1}{7}\cdot(-7x)&>-\frac{1}{7}\cdot 13 \quad \text{Multiplying by } -\frac{1}{7} \end{aligned}$$

↑ The symbol has to be reversed.

$$x>-\frac{13}{7} \quad \text{Simplifying}$$

The solution set is $\left\{x\middle|x>-\frac{13}{7}\right\}$.

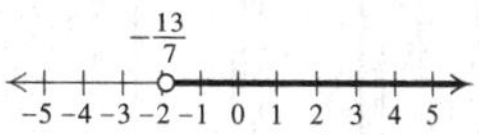

$-\frac{13}{7}$

−5 −4 −3 −2 −1 0 1 2 3 4 5

53. $-24>8t$

$-3>t$

The solution set is $\{t|t<-3\}$.

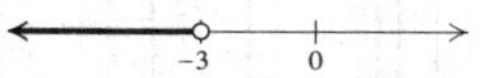

−3 0

55.
$$\begin{aligned} 7y&\ge -2 \\ \frac{1}{7}\cdot 7y&\ge\frac{1}{7}(-2) \quad \text{Multiplying by } \frac{1}{7} \\ y&\ge -\frac{2}{7} \end{aligned}$$

The solution set is $\left\{y\middle|y\ge -\frac{2}{7}\right\}$.

$-\frac{2}{7}$ 0

57. $-2y \le \frac{1}{5}$

$-\frac{1}{2}\cdot(-2y) \ge -\frac{1}{2}\cdot\frac{1}{5}$ ← The symbol has to be reversed.

$y \ge -\frac{1}{10}$

The solution set is $\left\{y\middle|y \ge -\frac{1}{10}\right\}$.

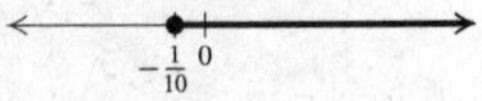

[Number line: closed dot at $-\frac{1}{10}$, shaded to the right; 0 marked]

59. $-\frac{8}{5} > -2x$

$-\frac{1}{2}\cdot\left(-\frac{8}{5}\right) < -\frac{1}{2}\cdot(-2x)$

$\frac{8}{10} < x$

$\frac{4}{5} < x$, or $x > \frac{4}{5}$

The solution set is $\left\{x\middle|\frac{4}{5} < x\right\}$, or $\left\{x\middle|x > \frac{4}{5}\right\}$.

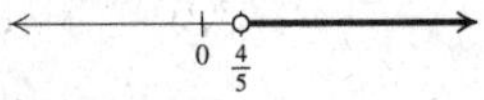

[Number line: open dot at $\frac{4}{5}$, shaded to the right; 0 marked]

61. $7 + 3x < 34$

$7 + 3x - 7 < 34 - 7$ Adding -7 to both sides

$3x < 27$ Simplifying

$x < 9$ Multiplying both sides by $\frac{1}{3}$

The solution set is $\{x|x < 9\}$.

63. $6 + 5y \ge 26$

$6 + 5y - 6 \ge 26 - 6$ Adding -6

$5y \ge 20$

$y \ge 4$ Multiplying by $\frac{1}{5}$

The solution set is $\{y|y \ge 4\}$.

65. $4t - 5 \le 23$

$4t - 5 + 5 \le 23 + 5$ Adding 5 to both sides

$4t \le 28$

$\frac{1}{4}\cdot 4t \le \frac{1}{4}\cdot 28$ Multiplying both sides by $\frac{1}{4}$

$t \le 7$

The solution set is $\{t|t \le 7\}$.

67. $16 < 4 - 3y$

$16 - 4 < 4 - 3y - 4$ Adding -4 to both sides

$12 < -3y$

$-\frac{1}{3}\cdot 12 > -\frac{1}{3}\cdot(-3y)$ Multiplying by $-\frac{1}{3}$

← The symbol has to be reversed.

$-4 > y$

The solution set is $\{y|-4 > y\}$, or $\{y|y < -4\}$.

69. $39 > 3 - 9x$

$39 - 3 > 3 - 9x - 3$ Adding -3

$36 > -9x$

$-\frac{1}{9}\cdot 36 < -\frac{1}{9}\cdot(-9x)$ Multiplying by $-\frac{1}{9}$

← The symbol has to be reversed.

$-4 < x$

The solution set is $\{x|-4 < x\}$, or $\{x|x > -4\}$.

71. $5 - 6y > 25$

$-5 + 5 - 6y > -5 + 25$

$-6y > 20$

$-\frac{1}{6}\cdot(-6y) < -\frac{1}{6}\cdot 20$ ← The symbol has to be reversed.

$y < -\frac{20}{6}$

$y < -\frac{10}{3}$

The solution set is $\left\{y\middle|y < -\frac{10}{3}\right\}$.

73. $-3 < 8x + 7 - 7x$

$-3 < x + 7$ Collecting like terms

$-3 - 7 < x + 7 - 7$

$-10 < x$

The solution set is $\{x|-10 < x\}$, or $\{x|x > -10\}$.

75. $6 - 4y > 4 - 3y$

$6 - 4y + 4y > 4 - 3y + 4y$ Adding $4y$

$6 > 4 + y$

$-4 + 6 > -4 + 4 + y$ Adding -4

$2 > y$, or $y < 2$

The solution set is $\{y|2 > y\}$, or $\{y|y < 2\}$.

77. $7 - 9y \le 4 - 8y$

$7 - 9y + 9y \le 4 - 8y + 9y$

$7 \le 4 + y$

$-4 + 7 \le -4 + 4 + y$

$3 \le y$, or $y \ge 3$

The solution set is $\{y|3 \le y\}$, or $\{y|y \ge 3\}$.

79. $33 - 12x < 4x + 97$

$33 - 12x - 97 < 4x + 97 - 97$

$-64 - 12x < 4x$

$-64 - 12x + 12x < 4x + 12x$

$-64 < 16x$

$-4 < x$

The solution set is $\{x|-4 < x\}$, or $\{x|x > -4\}$.

81. $2.1x + 43.2 > 1.2 - 8.4x$

$10(2.1x + 43.2) > 10(1.2 - 8.4x)$ Multiplying by 10 to clear decimals

$21x + 432 > 12 - 84x$

$21x + 84x > 12 - 432$ Adding $84x$ and -432

$105x > -420$

$x > -4$ Multiplying by $\frac{1}{105}$

The solution set is $\{x|x > -4\}$.

83. $0.7n - 15 + n \geq 2n - 8 - 0.4n$

$$\begin{aligned} 1.7n - 15 &\geq 1.6n - 8 && \text{Collecting like terms} \\ 10(1.7n - 15) &\geq 10(1.6n - 8) && \text{Multiplying by 10} \\ 17n - 150 &\geq 16n - 80 \\ 17n - 16n &\geq -80 + 150 && \text{Adding } -16n \text{ and } 150 \\ n &\geq 70 \end{aligned}$$

The solution set is $\{n|n \geq 70\}$

85.

$$\begin{aligned} \frac{x}{3} - 4 &\leq 1 \\ 3\left(\frac{x}{3} - 4\right) &\leq 3 \cdot 1 && \text{Multiplying by 3 to to clear the fraction} \\ x - 12 &\leq 3 && \text{Simplifying} \\ x &\leq 15 && \text{Adding 12} \end{aligned}$$

The solution set is $\{x|x \leq 15\}$.

87.

$$\begin{aligned} 3 &< 5 - \frac{t}{7} \\ -2 &< -\frac{t}{7} \\ -7(-2) &> -7\left(-\frac{t}{7}\right) \\ 14 &> t \end{aligned}$$

The solution set is $\{t|t < 14\}$.

89. $4(2y - 3) < 36$

$$\begin{aligned} 8y - 12 &< 36 && \text{Removing parentheses} \\ 8y &< 48 && \text{Adding 12} \\ y &< 6 && \text{Multiplying by } \frac{1}{8} \end{aligned}$$

The solution set is $\{y|y < 6\}$.

91. $3(t - 2) \geq 9(t + 2)$

$$\begin{aligned} 3t - 6 &\geq 9t + 18 \\ 3t - 9t &> 18 + 6 \\ -6t &\geq 24 \\ t &\leq -4 && \text{Multiplying by } -\frac{1}{6} \text{ and reversing the symbol} \end{aligned}$$

The solution set is $\{t|t \leq -4\}$.

93. $3(r - 6) + 2 < 4(r + 2) - 21$

$$\begin{aligned} 3r - 18 + 2 &< 4r + 8 - 21 \\ 3r - 16 &< 4r - 13 \\ -16 + 13 &< 4r - 3r \\ -3 &< r, \text{ or } r > -3 \end{aligned}$$

The solution set is $\{r|r > -3\}$.

95.

$$\begin{aligned} \frac{2}{3}(2x - 1) &\geq 10 \\ \frac{3}{2} \cdot \frac{2}{3}(2x - 1) &\geq \frac{3}{2} \cdot 10 && \text{Multiplying by } \frac{3}{2} \\ 2x - 1 &\geq 15 \\ 2x &\geq 16 \\ x &\geq 8 \end{aligned}$$

The solution set is $\{x|x \geq 8\}$.

97.

$$\begin{aligned} \frac{3}{4}\left(3x - \frac{1}{2}\right) - \frac{2}{3} &< \frac{1}{3} \\ \frac{3}{4}\left(3x - \frac{1}{2}\right) &< 1 && \text{Adding } \frac{2}{3} \\ \frac{9}{4}x - \frac{3}{8} &< 1 && \text{Removing parentheses} \\ 8 \cdot \left(\frac{9}{4}x - \frac{3}{8}\right) &< 8 \cdot 1 && \text{Clearing fractions} \\ 18x - 3 &< 8 \\ 18x &< 11 \\ x &< \frac{11}{18} \end{aligned}$$

The solution set is $\left\{x \middle| x < \frac{11}{18}\right\}$.

99. *Writing Exercise*

101. Let n represent "some number." Then we have $n + 3$, or $3 + n$.

103. Let x represent "a number." Then we have $2x - 3$.

105. *Writing Exercise*

107. $x < x + 1$

When any real number is increased by 1, the result is greater than the original number. Thus the solution set is $\{x|x \text{ is a real number}\}$.

109.

$$\begin{aligned} 27 - 4[2(4x - 3) + 7] &\geq 2[4 - 2(3 - x)] - 3 \\ 27 - 4[8x - 6 + 7] &\geq 2[4 - 6 + 2x] - 3 \\ 27 - 4[8x + 1] &\geq 2[-2 + 2x] - 3 \\ 27 - 32x - 4 &\geq -4 + 4x - 3 \\ 23 - 32x &\geq -7 + 4x \\ 23 + 7 &= 4x + 32x \\ 30 &\geq 36x \\ \frac{5}{6} &\geq x \end{aligned}$$

The solution set is $\left\{x \middle| x \leq \frac{5}{6}\right\}$.

111.

$$\begin{aligned} \frac{1}{2}(2x + 2b) &> \frac{1}{3}(21 + 3b) \\ x + b &> 7 + b \\ x + b - b &> 7 + b - b \\ x &> 7 \end{aligned}$$

The solution set is $\{x|x > 7\}$.

113.

$$\begin{aligned} y &< ax + b && \text{Assume } a < 0. \\ y - b &< ax \\ \frac{y - b}{a} &> x && \text{Since } a < 0 \text{, the inequality symbol must be reversed.} \end{aligned}$$

The solution set is $\left\{x \middle| x < \frac{y - b}{a}\right\}$.

115. $|x| > -3$

Since absolute value is always nonnegative, the absolute value of any real number will be greater than -3. Thus, the solution set is $\{x|x \text{ is a real number}\}$.

Exercise Set 2.7

1. a is at least b can be translated as $b \leq a$.

3. a is at most b can be translated as $a \leq b$.

5. b is no more than a can be translated as $b \leq a$.

7. b is less than a can be translated as $b < a$.

9. Let n represent the number. Then we have

$$n \geq 8.$$

11. Let t represent the temperature. Then we have

$$t \leq -3.$$

13. Let p represent the price of Pat's PT Cruiser. Then we have

$$p > 21,900.$$

15. Let d represent the distance to Normandale Community College. Then we have

$$d \leq 15.$$

17. Let n represent the number. Then we have

$$n > -2.$$

19. Let p represent the number of people attending the Million Man March. Then we have

$$400,000 < p < 1,200,000.$$

21. ***Familiarize.*** Let s = the length of the service call, in hours. The total charge is \$25 plus \$30 times the number of hours RJ's was there.

Translate.

$$\underbrace{\text{\$25 charge}}_{25} \quad \underbrace{\text{plus}}_{+} \quad \underbrace{\text{hourly rate}}_{30} \quad \underbrace{\text{times}}_{\cdot} \quad \underbrace{\text{number of hours}}_{s} \quad \underbrace{\text{is greater than}}_{>} \quad \underbrace{\text{\$100}}_{100}.$$

Carry out. We solve the inequality.

$$25 + 30x > 100$$
$$30x > 75$$
$$s > 2.5$$

Check. As a partial check, we show that the cost of a 2.5 hour service call is \$100.

$$\$25 + \$30(2.5) = \$25 + \$75 = \$100$$

State. The length of the service call was more than 2.5 hr.

23. ***Familiarize.*** Let t = the number of one-way trips made per month.

Translate.

$$\underbrace{\text{Cost per trip}}_{1 \cdot 15} \quad \underbrace{\text{times}}_{\cdot} \quad \underbrace{\text{number of trips}}_{t} \quad \underbrace{\text{is greater than}}_{>} \quad \underbrace{\text{\$21}}_{21}.$$

Carry out. We solve the inequality.

$$1.15t > 21$$
$$t > 18.3 \quad \text{Rounding}$$

Check. As a partial check we show that the total cost of 18 trips is less than \$21 and the total cost of 19 trips is more than \$21. For 18 trips the cost is \$1.15(18) = \$20.70. For 19 trips the cost is \$1.15(19) = \$21.85.

State. Gail should make more than 18 one-way trips per month in order for the pass to save her money.

25. ***Familiarize.*** The average of the five scores is their sum divided by the number of tests, 5. We let s represent Rod's score on the last test.

Translate. The average of the five scores is given by

$$\frac{73 + 75 + 89 + 91 + s}{5}.$$

Since this average must be at least 85, this means that it must be greater than or equal to 85. Thus, we can translate the problem to the inequality

$$\frac{73 + 75 + 89 + 91 + s}{5} \geq 85.$$

Carry out. We first multiply by 5 to clear the fraction.

$$5\left(\frac{73 + 75 + 89 + 91 + s}{5}\right) \geq 5 \cdot 85$$
$$73 + 75 + 89 + 91 + s \geq 425$$
$$328 + s \geq 425$$
$$s \geq 97$$

Check. As a partial check, we show that Rod can get a score of 97 on the fifth test and have an average of at least 85:

$$\frac{73 + 75 + 89 + 91 + 97}{5} = \frac{425}{5} = 85.$$

State. Scores of 97 and higher will earn Rod an average quiz grade of at least 85.

27. ***Familiarize.*** Let c = the number of credits Millie must complete in the fourth quarter.

Translate.

$$\underbrace{\text{Average number of credits}}_{\frac{5+7+8+c}{4}} \quad \underbrace{\text{is at least}}_{\geq} \quad \underbrace{7}_{7}.$$

Carry out. We solve the inequality.

$$\frac{5 + 7 + 8 + c}{4} \geq 7$$
$$4\left(\frac{5 + 7 + 8 + c}{4}\right) \geq 4 \cdot 7$$
$$5 + 7 + 8 + c \geq 28$$
$$20 + c \geq 28$$
$$c \geq 8$$

Check. As a partial check, we show that Millie can complete 8 credits in the fourth quarter and average 7 credits per quarter.

$$\frac{5 + 7 + 8 + 8}{4} = \frac{28}{4} = 7$$

State. Millie must complete 8 credits or more in the fourth quarter.

29. ***Familiarize.*** The average number of calls per week is the sum of the calls for the three weeks divided by the number of weeks, 3. We let c represent the number of calls made during the third week.

Translate. The average of the three weeks is given by

$$\frac{17+22+c}{3}.$$

Since the average must be at least 20, this means that it must be greater than or equal to 20. Thus, we can translate the problem to the inequality

$$\frac{17+22+c}{3} \geq 20.$$

Carry out. We first multiply by 3 to clear the fraction.

$$3\left(\frac{17+22+c}{3}\right) \geq 3\cdot 20$$
$$17+22+c \geq 60$$
$$39+c \geq 60$$
$$c \geq 21$$

Check. Suppose c is a number greater than or equal to 21. Then by adding 17 and 22 on both sides of the inequality we get

$$17+22+c \geq 17+22+21$$
$$17+22+c \geq 60$$

so

$$\frac{17+22+c}{3} \geq \frac{60}{3}, \text{ or } 20.$$

State. 21 calls or more will maintain an average of at least 20 for the three-week period.

31. ***Familiarize.*** We first make a drawing. We let b represent the length of the base. Then the lengths of the other sides are $b-2$ and $b+3$.

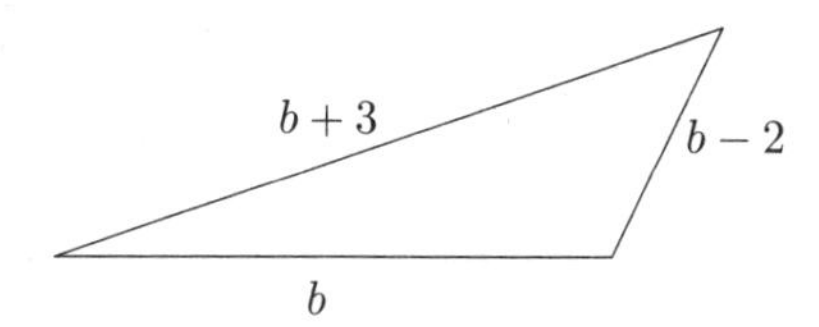

The perimeter is the sum of the lengths of the sides or $b+b-2+b+3$, or $3b+1$.

Translate.

The perimeter	is greater than	19 cm.
↓	↓	↓
$3b+1$	$>$	19

Carry out.

$$3b+1 > 19$$
$$3b > 18$$
$$b > 6$$

Check. We check to see if the solution seems reasonable.

When $b=5$, the perimeter is $3\cdot 5+1$, or 16 cm.

When $b=6$, the perimeter is $3\cdot 6+1$, or 19 cm.

When $b=7$, the perimeter is $3\cdot 7+1$, or 22 cm.

From these calculations, it would appear that the solution is correct.

State. For lengths of the base greater than 6 cm the perimeter will be greater than 19 cm.

33. ***Familiarize.*** Let d = the depth of the well, in feet. Then the cost on the pay-as-you-go plan is $\$500+\$8d$. The cost of the guaranteed-water plan is \$4000. We want to find the values of d for which the pay-as-you-go plan costs less than the guaranteed-water plan.

Translate.

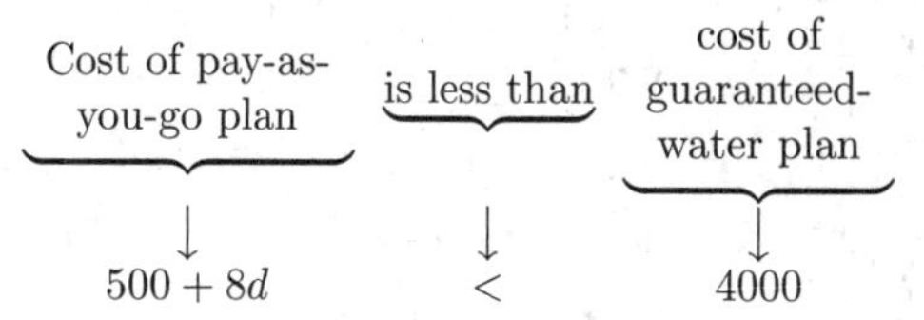

Carry out.

$$500+8d < 4000$$
$$8d < 3500$$
$$d < 437.5$$

Check. We check to see that the solution is reasonable.

When $d=437$, $\$500+\$8\cdot 437 = \$3996 < \$4000$

When $d=437.5$, $\$500+\$8(437.5) = \$4000$

When $d=438$, $\$500+\$8(438) = \$4004 > \$4000$

From these calculations, it appears that the solution is correct.

State. It would save a customer money to use the pay-as-you-go plan for a well of less than 437.5 ft.

35. ***Familiarize.*** Let v = the blue book value of the car. Since the car was repaired, we know that \$8500 does not exceed $0.8v$ or, in other words, $0.8v$ is at least \$8500.

Translate.

80% of the blue book value	is at least	\$8500.
↓	↓	↓
$0.8v$	$\geq$	8500

Carry out.

$$0.8v \geq 8500$$
$$v \geq \frac{8500}{0.8}$$
$$v \geq 10,625$$

Check. As a partial check, we show that 80% of \$10,625 is at least \$8500:

$$0.8(\$10,625) = \$8500$$

State. The blue book value of the car was at least \$10,625.

37. ***Familiarize.*** As in the drawing in the text, we let L = the length of the envelope. Recall that the area of a rectangle is the product of the length and the width.

Translate.

Length	times	width	is at least	$17\frac{1}{2}$ in^2
↓	↓	↓	↓	↓
L	$\cdot$	$3\frac{1}{2}$	$\geq$	$17\frac{1}{2}$

Carry out.

$$L \cdot 3\frac{1}{2} \geq 17\frac{1}{2}$$
$$L \cdot \frac{7}{2} \geq \frac{35}{2}$$
$$L \cdot \frac{7}{2} \cdot \frac{2}{7} \geq \frac{35}{2} \cdot \frac{2}{7}$$
$$L \geq 5$$

The solution set is $\{L|L \geq 5\}$.

Check. We can obtain a partial check by substituting a number greater than or equal to 5 in the inequality. For example, when $L = 6$:

$$L \cdot 3\frac{1}{2} = 6 \cdot 3\frac{1}{2} = 6 \cdot \frac{7}{2} = 21 \geq 17\frac{1}{2}$$

The result appears to be correct.

State. Lengths of 5 in. or more will satisfy the constraints. The solution set is $\{L|L \geq 5 \text{ in.}\}$.

39. ***Familiarize.*** We will use the formula $F = \frac{9}{5}C + 32$.

Translate.

$\underbrace{\text{Fahrenheit temperature}}$ $\underbrace{\text{is above}}$ 98.6°.

$$F \quad > \quad 98.6$$

Substituting $\frac{9}{5}C + 32$ for F, we have

$$\frac{9}{5}C + 32 > 98.6.$$

Carry out. We solve the inequality.

$$\frac{9}{5}C + 32 > 98.6$$
$$\frac{9}{5}C > 66.6$$
$$C > \frac{333}{9}$$
$$C > 37$$

Check. We check to see if the solution seems reasonable.

When $C = 36$, $\frac{9}{5} \cdot 36 + 32 = 96.8$.

When $C = 37$, $\frac{9}{5} \cdot 37 + 32 = 98.6$.

When $C = 38$, $\frac{9}{5} \cdot 38 + 32 = 100.4$.

It would appear that the solution is correct, considering that rounding occurred.

State. The human body is feverish for Celsius temperatures greater than 37°.

41. ***Familiarize.*** Let h = the height of the triangle, in ft. Recall that the formula for the area of a triangle with base b and height h is $A = \frac{1}{2}bh$.

Translate.

Area $\underbrace{\text{is at least}}$ $\underbrace{3 \text{ ft}^2}$.

$$\frac{1}{2}\left(1\frac{1}{2}\right)h \quad \geq \quad 3$$

Carry out. We solve the inequality.

$$\frac{1}{2}\left(1\frac{1}{2}\right)h \geq 3$$
$$\frac{1}{2} \cdot \frac{3}{2} \cdot h \geq 3$$
$$\frac{3}{4}h \geq 3$$
$$h \geq \frac{4}{3} \cdot 3$$
$$h \geq 3$$

Check. As a partial check, we show that the area of the triangle is 3 ft^2 when the height is 4 ft.

$$\frac{1}{2}\left(1\frac{1}{2}\right)(4) = \frac{1}{2} \cdot \frac{3}{2} \cdot \frac{4}{1} = 3$$

State. The height should be at least 4 ft.

43. ***Familiarize.*** Let r = the amount of fat in a serving of the regular peanut butter, in grams. If reduced fat peanut butter has at least 25% less fat than regular peanut butter, then it has at most 75% as much fat as the regular peanut butter.

Translate.

$\underbrace{\text{12 g of fat}}$ $\underbrace{\text{is at most}}$ 75% of $\underbrace{\text{the amount of fat in regular peanut butter.}}$

$$12 \quad \leq \quad 0.75 \quad \cdot \quad r$$

Carry out.

$$12 \leq 0.75r$$
$$16 \leq r$$

Check. As a partial check, we show that 12 g of fat does not exceed 75% of 16 g of fat:

$$0.75(16) = 12$$

State. Regular peanut butter contains at least 16 g of fat per serving.

45. ***Familiarize.*** Let d = the number of days after September 5.

Translate.

$\underbrace{\text{Weight on September 5}}$ plus $\underbrace{\text{26 lb per day}}$ times

$$532 \quad + \quad 26 \quad \cdot$$

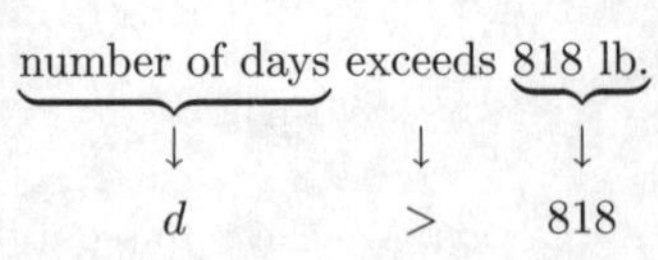

$\underbrace{\text{number of days}}$ exceeds $\underbrace{\text{818 lb.}}$

$$d \quad > \quad 818$$

Carry out. We solve the inequality.

$$532 + 26d > 818$$
$$26d > 286$$
$$d > 11$$

Check. As a partial check, we can show that the weight of the pumpkin is 818 lb 11 days after September 5.

$$532 + 26 \cdot 11 = 532 + 286 = 818 \text{ lb}$$

State. The pumpkin's weight will exceed 818 lb more than 11 days after September 5, or on dates after September 16.

47. ***Familiarize.*** Let c = the number of copies Myra has made. The total cost of the copies is the setup fee of \$6 plus \$4 times the number of copies, or $\$4 \cdot c$.

Translate.

$$\underbrace{\text{Setup fee}}_{\downarrow\atop 6}\ \underbrace{\text{plus}}_{\downarrow\atop +}\ \underbrace{\text{copying cost}}_{\downarrow\atop 4c}\ \underbrace{\text{cannot exceed}}_{\downarrow\atop \le}\ \underbrace{\$65.}_{\downarrow\atop 65}$$

Carry out. We solve the inequality.

$$6 + 4c \le 65$$
$$4c \le 59$$
$$c \le 14.75$$

Check. As a partial check, we show that Myra can have 14 copies made and not exceed her \$65 budget.

$$\$6 + \$4 \cdot 14 = \$6 + \$56 = \$62$$

State. Myra can have 14 or fewer copies made and stay within her budget.

49. ***Familiarize.*** We will use the formula $R = -0.0065t + 4.3222$.

Translate.

$$\underbrace{\text{The world record}}_{\downarrow\atop -0.0065t+4.3222}\ \underbrace{\text{is less than}}_{\downarrow\atop <}\ \underbrace{\text{3.7 minutes}}_{\downarrow\atop 3.7}.$$

Carry out. We solve the inequality.

$$-0.0065t + 4.3222 < 3.7$$
$$-0.0065t < -0.6222$$
$$t > 95.7$$

Check. As a partial check, we can show that the record is more than 3.7 min 95 yr after 1900 and is less than 3.7 min 96 yr after 1900.

For $t = 95$, $R = -0.0065(95) + 4.3222 = 3.7047$.

For $t = 96$, $R = -0.0065(96) + 4.3222 = 3.6982$.

State. The world record in the mile run is less than 3.7 min more than 95 yr after 1900, or in years after 1995.

51. ***Familiarize.*** We will use the equation $y = 0.03x + 0.21$.

Translate.

$$\underbrace{\text{The cost}}_{\downarrow\atop 0.03x+0.21}\ \underbrace{\text{is at most}}_{\downarrow\atop \le}\ \underbrace{\$6.}_{\downarrow\atop 6}$$

Carry out. We solve the inequality.

$$0.03x + 0.21 \le 6$$
$$0.03x \le 5.79$$
$$x \le 193$$

Check. As a partial check, we show that the cost for driving 193 mi is \$6.

$$0.03(193) + 0.21 = 6$$

State. The cost will be at most \$6 for mileages less than or equal to 193 mi.

53. *Writing Exercise*

55. $\dfrac{9-5}{6-4} = \dfrac{4}{2} = 2$

57. $\dfrac{8-(-2)}{1-4} = \dfrac{10}{-3}$, or $-\dfrac{10}{3}$

59. *Writing Exercise*

61. ***Familiarize.*** We use the formula $F = \frac{9}{5}C + 32$.

Translate. We are interested in temperatures such that $5° < F < 15°$. Substituting for F, we have:

$$5 < \frac{9}{5}C + 32 < 15$$

Solve.

$$5 < \frac{9}{5}C + 32 < 15$$
$$5 \cdot 5 < 5\left(\frac{9}{5}C + 32\right) < 5 \cdot 15$$
$$25 < 9C + 160 < 75$$
$$-135 < 9C < -85$$
$$-15 < C < -9\frac{4}{9}$$

Check. The check is left to the student.

State. Green ski wax works best for temperatures between $-15°$C and $-9\frac{4}{9}°$C.

63. Since $8^2 = 64$, the length of a side must be less than or equal to 8 cm (and greater than 0 cm, of course). We can also use the five-step problem-solving procedure.

Familiarize. Let s represent the length of a side of the square. The area s is the square of the length of a side, or s^2.

Translate.

$$\underbrace{\text{The area}}_{\downarrow\atop s^2}\ \underbrace{\text{is no more than}}_{\downarrow\atop \le}\ \underbrace{64\text{ cm}^2.}_{\downarrow\atop 64}$$

Carry out.

$$s^2 \le 64$$
$$s^2 - 64 \le 0$$
$$(s+8)(s-8) \le 0$$

We know that $(s+8)(s-8) = 0$ for $s = -8$ or $s = 8$. Now $(s+8)(s-8) < 0$ when the two factors have opposite signs. That is:

$s+8>0$ *and* $s-8<0$ *or* $s+8<0$ *and* $s-8>0$

$s>-8$ *and* $s<8$ *or* $s<-8$ *and* $s>8$

This can be expressed as $-8 < s < 8$. This is not possible.

Then $(s+8)(s-8) \le 0$ for $-8 \le s \le 8$.

Check. Since the length of a side cannot be negative we only consider positive values of s, or $0 < s \le 8$. We check to see if this solution seems reasonable.

When $s = 7$, the area is 7^2, or 49 cm^2.

When $s = 8$, the area is 8^2, or 64 cm^2.

When $s = 9$, the area is 9^2, or 81 cm^2.

From these calculations, it appears that the solution is correct.

State. Sides of length 8 cm or less will allow an area of no more than 64 cm^2. (Of course, the length of a side must be greater than 0 also.)

65. ***Familiarize.*** Let f = the fat content of a serving of regular tortilla chips, in grams. A product that contains 60% less fat than another product has 40% of the fat content of that product. If Reduced Fat Tortilla Pops cannot be labeled lowfat, then they contain at least 3 g of fat.

Translate.

40%	of	the fat content of regular tortilla chips	is at least	3 grams of fat
↓	↓	↓	↓	↓
0.4	·	f	$\geq$	3

Carry out.

$$0.4f \geq 3$$
$$f \geq 7.5$$

Check. As a partial check, we show that 40% of 7.5 g is not less than 3 g.

$$0.4(7.5) = 3$$

State. A serving of regular tortilla chips contains at least 7.5 g of fat.

67. ***Familiarize.*** Let p = the price of Neoma's tenth book. If the average price of each of the first 9 books is \$12, then the total price of the 9 books is $9 \cdot \$12$, or \$108. The average price of the first 10 books will be $\dfrac{\$108 + p}{10}$.

Translate.

The average price of 10 books	is at least	\$15.
↓	↓	↓
$\dfrac{108 + p}{10}$	$\geq$	15

Carry out. We solve the inequality.

$$\frac{108 + p}{10} \geq 15$$
$$108 + p \geq 150$$
$$p \geq 42$$

Check. As a partial check, we show that the average price of the 10 books is \$15 when the price of the tenth book is \$42.

$$\frac{\$108 + \$42}{10} = \frac{\$150}{10} = \$15$$

State. Neoma's tenth book should cost at least \$42 if she wants to select a \$15 book for her free book.

69. *Writing Exercise*

Chapter 3

Introduction to Graphing

Exercise Set 3.1

1. The x-values extend from -9 to 1 and the y-values range from -1 to 5, so (a) is the best choice.

3. The x-values extend from -2 to 4 and the y-values range from -9 to 1, so (b) is the best choice.

5. We go to the top of the bar that is above the body weight 100 lb. Then we move horizontally from the top of the bar to the vertical scale listing numbers of drinks. It appears that consuming approximately 2 drinks in one hour will give a 100 lb person a blood-alcohol level of 0.08%.

7. From 4 on the vertical scale we move horizontally until we reach a bar whose top is above the horizontal line on which we are moving. The first such bar corresponds to a body weight of 220 lb. This means that for body weights represented by bars to the left of this one, consuming 4 drinks will yield a blood-alcohol level of 0.08%. The bar immediately to the left of the 220-pound bar represents 200 pounds. Thus, we can conclude an individual weighs more than 200 lb if 4 drinks are consumed in one hour without reaching a blood-alcohol level of 0.08%.

9. ***Familiarize.*** Since there are 292 million Americans and about one-third of them live in the South, there are about $\frac{1}{3}\cdot 292$, or $\frac{292}{3}$ million Southerners. The pie chart indicates that 3% of Americans choose brown as their favorite color. Let $b =$ the number of Southerners, in millions, who choose brown as their favorite color.

Translate. We reword and translate the problem.

What is 3% of $\frac{292}{3}$ million?

$$b = 3\% \cdot \frac{292}{3}$$

Carry out. We solve the equation.

$$b = 0.03 \cdot \frac{292}{3} = 2.92$$

Check. We repeat the calculations. The answer checks.

State. About 2.92 million, or 2,920,000 Southerners choose brown as their favorite color.

11. ***Familiarize.*** Since there are 292 million Americans and about one-eighth are senior citizens, there are about $\frac{1}{8}\cdot 292$ million, or 36.5 million senior citizens. The pie chart indicates that 4% of Americans choose black as their favorite color. Let $b =$ the number of senior citizens who choose black as their favorite color.

Translate. We reword and translate the problem.

What is 4% of 36.5 million?

$$b = 4\% \cdot 36,500,000$$

Carry out. We solve the equation.

$$b = 0.04 \cdot 36,500,000 = 1,460,000$$

Check. We repeat the calculations. The answer checks.

State. About 1,460,000 senior citizens choose black as their favorite color.

13. ***Familiarize.*** From the pie chart we see that 10.7% of solid waste is plastic. We let $x =$ the amount of plastic, in millions of tons, in the waste generated in 2000.

Translate. We reword the problem.

What is 10.7% of 231.9?

$$x = 10.7\% \cdot 231.9$$

Carry out.

$$x = 0.107 \cdot 231.9 \approx 24.8$$

Check. We can repeat the calculation. The result checks.

State. In 2000, about 24.8 million tons of waste was plastic.

15. ***Familiarize.*** From the pie chart we see that 5.5% of solid waste is glass. From Exercise 13 we know that Americans generated 231.9 million tons of waste in 2000. Then the amount of this that is glass is

0.055(231.9), or about 12.8 million tons

We let $x =$ the amount of glass, in millions of tons, that Americans recycled in 2000.

Translate. We reword the problem.

What is 22.7% of 12.8 million tons?

$$x = 22.7\% \cdot 12.8$$

Carry out.

$$x = 0.227(12.8) \approx 2.9$$

Check. We go over the calculations again. The result checks.

State. Americans recycled about 2.9 million tons of glass in 2000.

17. We locate 2003 on the horizontal axis and then move up to the line. From there we move left to the vertical axis and read the number of cell phones, in millions. We estimate that about 120,000,000 cell phones had Internet access in 2003.

19. We locate 150 on the vertical axis and move right to the line. From there we move down to the horizontal scale and read the year. We see that approximately 150 million cell phones had Internet access in 2004.

21. Starting at the origin:

(1,2) is 1 unit right and 2 units up;

$(-2, 3)$ is 2 units left and 3 units up;

$(4, -1)$ is 4 units right and 1 unit down;

$(-5, -3)$ is 5 units left and 3 units down;

(4,0) is 4 units right and 0 units up or down;

$(0, -2)$ is 0 units right or left and 2 units down.

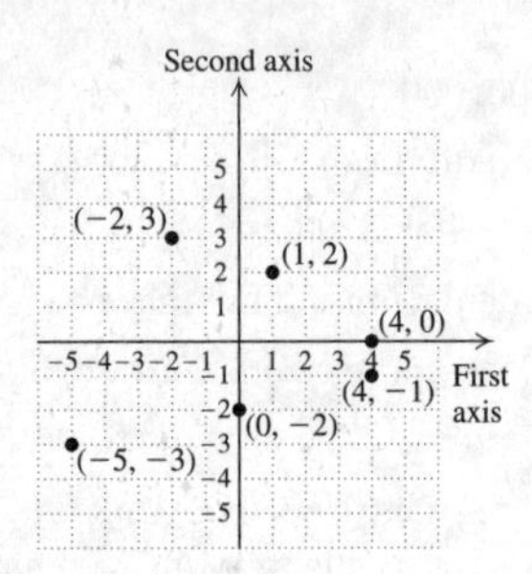

23. Starting at the origin:

(4,4) is 4 units right and 4 units up;

$(-2, 4)$ is 2 units left and 4 units up;

$(5, -3)$ is 5 units right and 3 units down;

$(-5, -5)$ is 5 units left and 5 units down;

(0,4) is 0 units right or left and 4 units up;

$(0, -4)$ is 0 units right or left and 4 units down;

(3,0) is 3 units right and 0 units up or down;

$(-4, 0)$ is 4 units left and 0 units up or down.

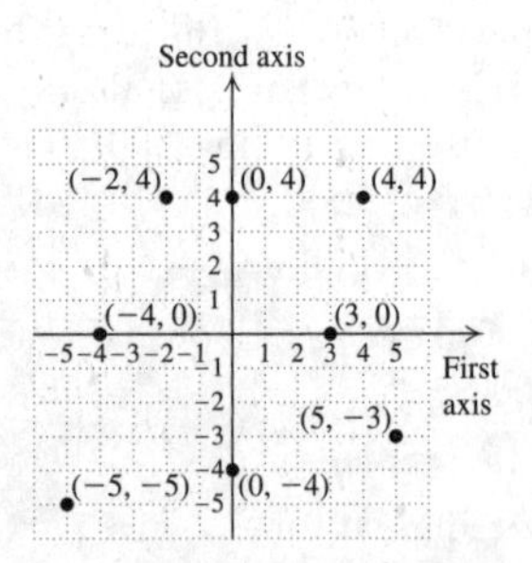

25. We plot the points (2001, 119.8), (2002, 135.3), (2003, 150.2), (2004, 163.8), and (2005, 176.9) and connect adjacent points with line segments.

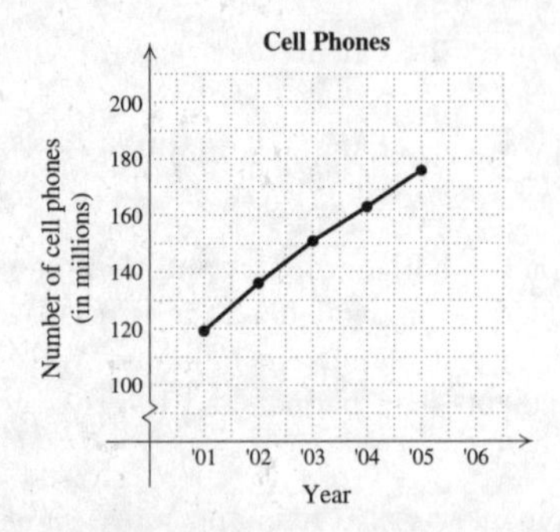

27.

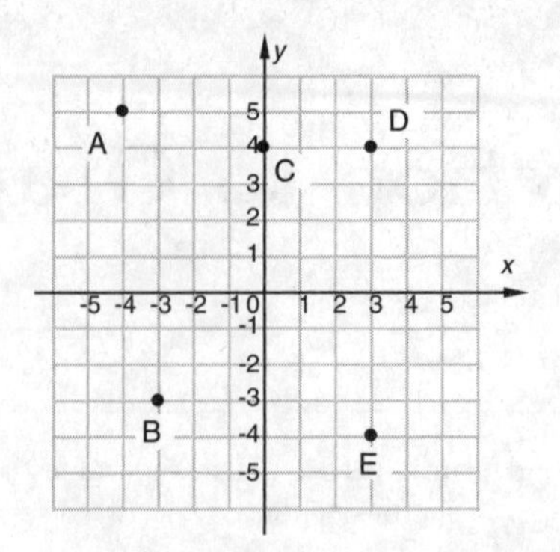

Point A is 4 units left and 5 units up. The coordinates of A are $(-4, 5)$.

Point B is 3 units left and 3 units down. The coordinates of B are $(-3, -3)$.

Point C is 0 units right or left and 4 units up. The coordinates of C are (0,4).

Point D is 3 units right and 4 units up. The coordinates of D are (3,4).

Point E is 3 units right and 4 units down. The coordinates of E are $(3, -4)$.

29.

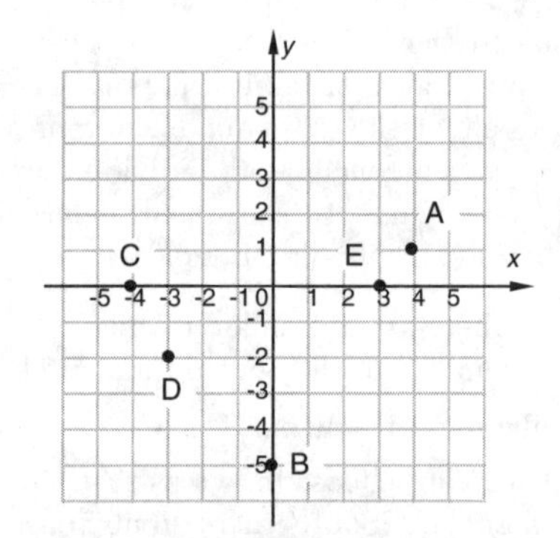

Point A is 4 units right and 1 unit up. The coordinates of A are (4,1).

Point B is 0 units right or left and 5 units down. The coordinates of B are $(0, -5)$.

Point C is 4 units left and 0 units up or down. The coordinates of C are $(-4, 0)$.

Point D is 3 units left and 2 units down. The coordinates of D are $(-3, -2)$.

Point E is 3 units right and 0 units up or down. The coordinates of E are (3,0).

31. Since the x-values range from -75 to 9, the 10 horizontal squares must span $9 - (-75)$, or 84 units. Since 84 is close to 100 and it is convenient to count by 10's, we can count backward from 0 eight squares to -80 and forward from 0 two squares to 20 for a total of $8 + 2$, or 10 squares.

Since the y-values range from -4 to 5, the 10 vertical squares must span $5 - (-4)$, or 9 units. It will be convenient to count by 2's in this case. We count down from 0 five squares to -10 and up from 0 five squares to 10 for a total of $5 + 5$, or 10 squares. (Instead, we might have chosen to count by 1's from -5 to 5.)

Then we plot the points $(-75, 5)$, $(-18, -2)$, and $(9, -4)$.

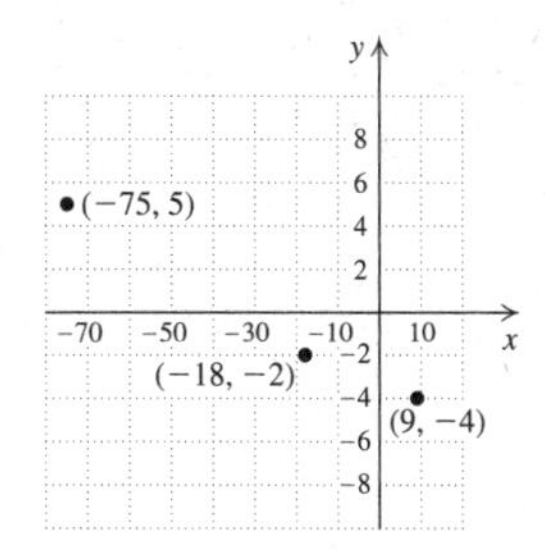

33. Since the x-values range from -5 to 5, the 10 horizontal squares must span $5 - (-5)$, or 10 units. It will be convenient to count by 2's in this case. We count backward from 0 five squares to -10 and forward from 0 five squares to 10 for a total of $5 + 5$, or 10 squares.

Since the y-values range from -14 to 83, the 10 vertical squares must span $83 - (-14)$, or 97 units. To include both -14 and 83, the squares should extend from about -20 to 90, or $90 - (-20)$, or 110 units. We cannot do this counting by 10's, so we use 20's instead. We count down from 0 four units to -80 and up from 0 six units to 120 for a total of $4 + 6$, or 10 units. There are other ways to cover the values from -14 to 83 as well.

Then we plot the points $(-1, 83)$, $(-5, -14)$, and $(5, 37)$.

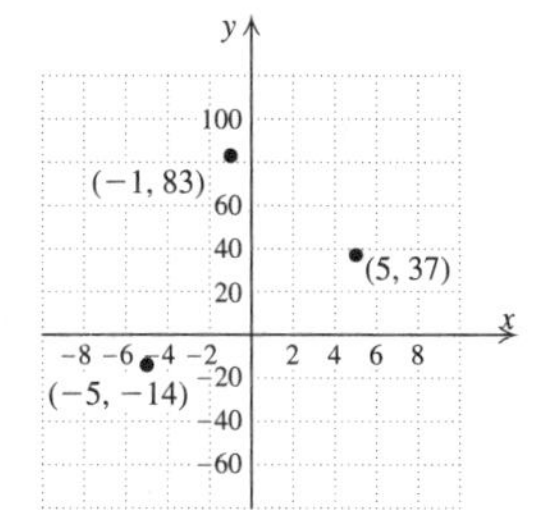

35. Since the x-values range from -16 to 3, the 10 horizontal squares must span $3-(-16)$, or 19 units. We could number by 2's or 3's. We number by 3's, going backward from 0 eight squares to -24 and forward from 0 two squares to 6 for a total of $8 + 2$, or 10 squares.

Since the y-values range from -4 to 15, the 10 vertical squares must span $15 - (-4)$, or 19 units. We will number the vertical axis by 3's as we did the horizontal axis. We go down from 0 four squares to -12 and up from 0 six squares to 18 for a total of $4 + 6$, or 10 squares.

Then we plot the points $(-10, -4)$, $(-16, 7)$, and $(3, 15)$.

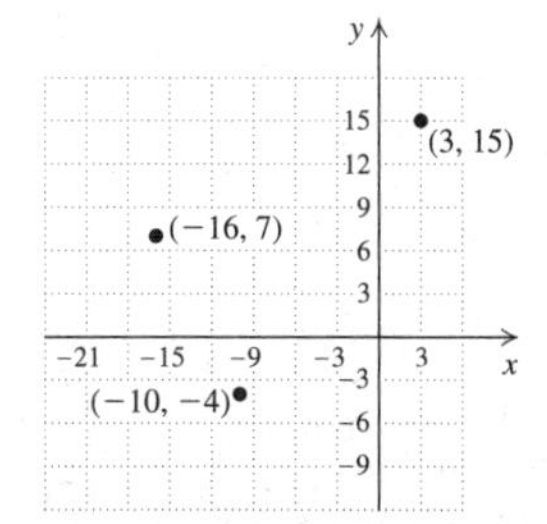

37. Since the x-values range from -100 and 800, the 10 horizontal squares must span $800-(-100)$, or 900 units. Since 900 is close to 1000 we can number by 100's. We go backward from 0 two squares to -200 and forward from 0 eight squares to 800 for a total of 2+8, or 10 squares. (We could have numbered from -100 to 900 instead.)

Since the y-values range from -5 to 37, the 10 vertical squares must span $37 - (-5)$, or 42 units. Since 42 is close to 50, we can count by 5's. We go down from 0 two squares to -10 and up from 0 eight squares to 40 for a total of 2+8, or 10 squares.

Then we plot the points $(-100, -5)$, $(350, 20)$, and $(800, 37)$.

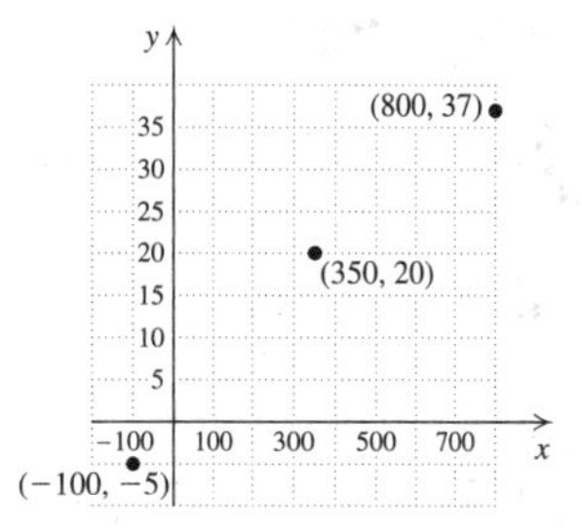

39. Since the x-values range from -124 to 54, the 10 horizontal squares must span $54 - (-124)$, or 178 units. We can number by 25's. We go backward from 0 six squares to -150 and forward from 0 four squares to 100 for a total of $6 + 4$, or 10 squares.

Since the y-values range from -238 to 491, the 10 vertical squares must span $491 - (-238)$, or 729 units. We can number by 100's. We go down from 0 four squares to -400 and up from 0 six squares to 600 for a total of $4+6$, or 10 squares.

Then we plot the points $(-83, 491)$, $(-124, -95)$, and $(54, -238)$.

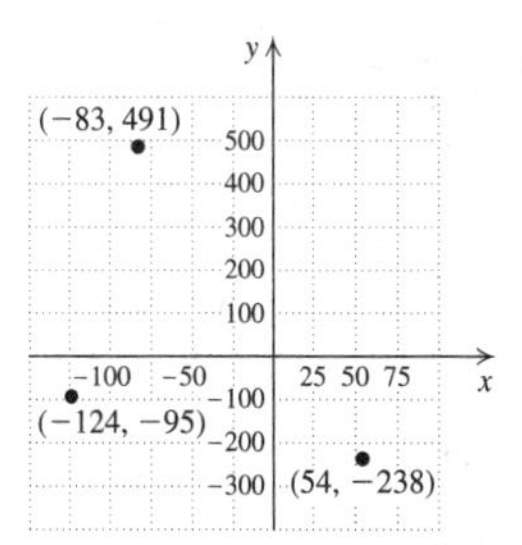

41. Since the first coordinate is positive and the second coordinate negative, the point $(7, -2)$ is located in quadrant IV.

43. Since both coordinates are negative, the point $(-4, -3)$ is in quadrant III.

45. Since both coordinates are positive, the point $(2, 1)$ is in quadrant I.

47. Since the first coordinate is negative and the second coordinate is positive, the point $(-4.9, 8.3)$ is in quadrant II.

49. First coordinates are positive in the quadrants that lie to the right of the origin, or in quadrants I and IV.

51. Points for which both coordinates are positive lie in quadrant I, and points for which both coordinates are negative life in quadrant III. Thus, both coordinates have the same sign in quadrants I and III.

53. *Writing Exercise*

55. $4 \cdot 3 - 6 \cdot 5 = 12 - 30 = -18$

57. $-\frac{1}{2}(-6) + 3 = 3 + 3 = 6$

59.
$$\begin{aligned} 3x - 2y &= 6 \\ -2y &= -3x + 6 \quad \text{Adding } -3x \text{ to both sides} \\ -\frac{1}{2}(-2y) &= -\frac{1}{2}(-3x + 6) \\ y &= -\frac{1}{2}(-3x) - \frac{1}{2}(6) \\ y &= \frac{3}{2}x - 3 \end{aligned}$$

61. *Writing Exercise*

63. The coordinates have opposite signs, so the point could be in quadrant II or quadrant IV.

65. 

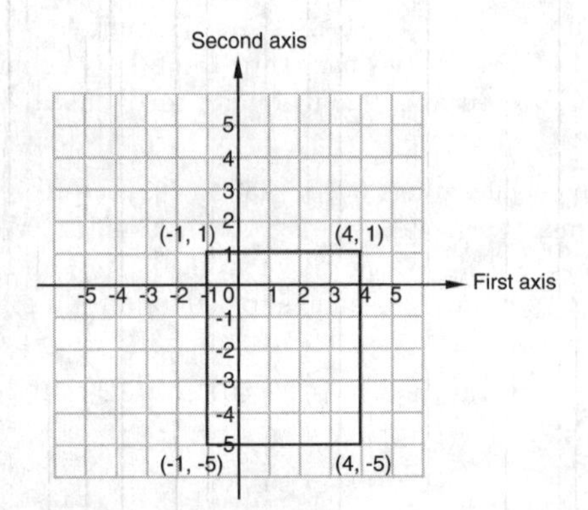

The coordinates of the fourth vertex are $(-1, -5)$.

67. Answers may vary.

We select eight points such that the sum of the coordinates for each point is 7.

$$\begin{array}{ll} (0,7) & 0+7=7 \\ (1,6) & 1+6=7 \\ (2,5) & 2+5=7 \\ (3,4) & 3+4=7 \\ (4,3) & 4+3=7 \\ (5,2) & 5+2=7 \\ (6,1) & 6+1=7 \\ (7,0) & 7+0=7 \end{array}$$

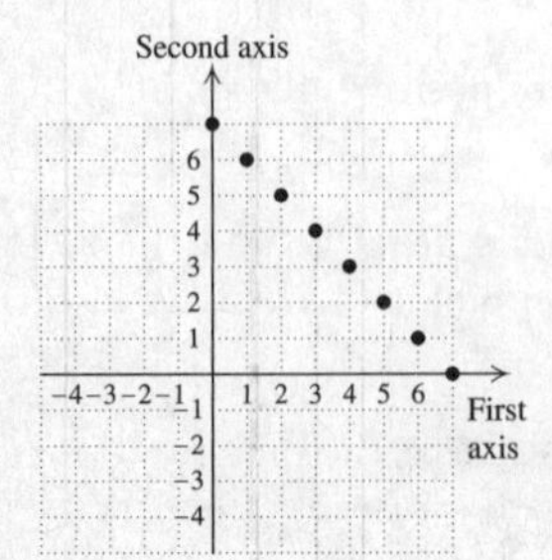

69.

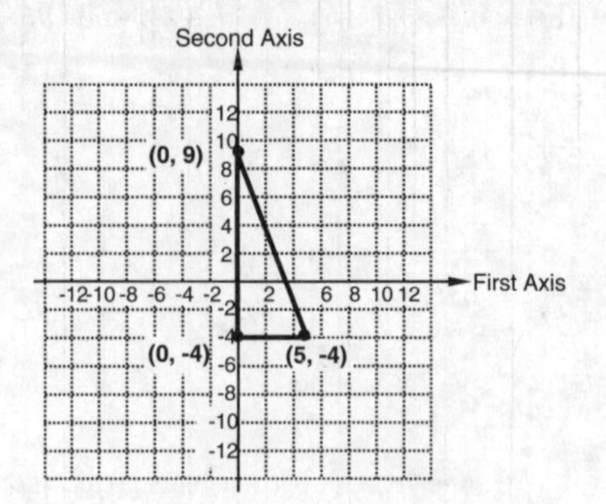

The base is 5 units and the height is 13 units.

$A = \frac{1}{2}bh = \frac{1}{2} \cdot 5 \cdot 13 = \frac{65}{2}$ sq units, or $32\frac{1}{2}$ sq units

71. Latitude 27° North,
Longitude 81° West

73. *Writing Exercise*

Exercise Set 3.2

1. We substitute 0 for x and 2 for y (alphabetical order of variables).

$$\begin{array}{c|l} \multicolumn{2}{c}{y = 5x + 1} \\ \hline 2 & 5 \cdot 0 + 1 \\ & 0 + 1 \end{array}$$

$2 \stackrel{?}{=} 1$ FALSE

Since $2 = 1$ is false, the pair $(0, 2)$ is not a solution.

3. We substitute 4 for x and 2 for y.

$$\begin{array}{c|c} \multicolumn{2}{c}{3y + 2x = 12} \\ \hline 3 \cdot 2 + 2 \cdot 4 & 12 \\ 6 + 8 & \end{array}$$

$14 \stackrel{?}{=} 12$ FALSE

Since $14 = 12$ is false, the pair (4,2) is not a solution.

5. We substitute 2 for a and -1 for b.

$$\begin{array}{c|c} \multicolumn{2}{c}{4a - 3b = 11} \\ \hline 4 \cdot 2 - 3(-1) & 11 \\ 8 + 3 & \end{array}$$

$11 \stackrel{?}{=} 11$ TRUE

Since $11 = 11$ is true, the pair $(2, -1)$ is a solution.

7. To show that a pair is a solution, we substitute, replacing x with the first coordinate and y with the second coordinate in each pair.

$$\begin{array}{c|l} \multicolumn{2}{c}{y = x + 3} \\ \hline 2 & -1 + 3 \end{array} \qquad \begin{array}{c|l} \multicolumn{2}{c}{y = x + 3} \\ \hline 7 & 4 + 3 \end{array}$$

$2 \stackrel{?}{=} 2$ TRUE $\qquad 7 \stackrel{?}{=} 7$ TRUE

In each case the substitution results in a true equation. Thus, $(-1, 2)$ and $(4, 7)$ are both solutions of $y = x + 3$.

We graph these points and sketch the line passing through them.

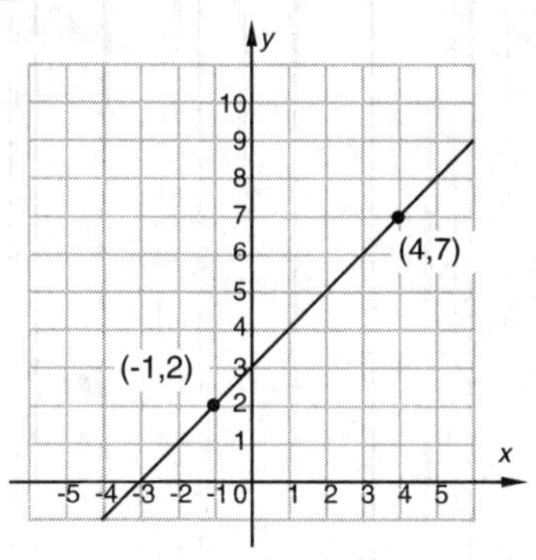

The line appears to pass through $(0, 3)$ also. We check to determine if $(0, 3)$ is a solution of $y = x + 3$.

$$\begin{array}{c|c} \multicolumn{2}{c}{y = x + 3} \\ \hline 3 & 0 + 3 \end{array}$$

$3 \stackrel{?}{=} 3$ TRUE

Thus, $(0, 3)$ is another solution. There are other correct answers, including $(-5, -2)$, $(-4, -1)$, $(-3, 0)$, $(-2, 1)$, $(1, 4)$, $(2, 5)$, and $(3, 6)$.

9. To show that a pair is a solution, we substitute, replacing x with the first coordinate and y with the second coordinate in each pair.

$$\begin{array}{c|c} \multicolumn{2}{c}{y = \frac{1}{2}x + 3} \\ \hline 5 & \frac{1}{2} \cdot 4 + 3 \\ & 2 + 3 \end{array} \qquad \begin{array}{c|c} \multicolumn{2}{c}{y = \frac{1}{2}x + 3} \\ \hline 2 & \frac{1}{2}(-2) + 3 \\ & -1 + 3 \end{array}$$

$5 \stackrel{?}{=} 5$ TRUE $\quad 2 \stackrel{?}{=} 2$ TRUE

In each case the substitution results in a true equation. Thus, $(4, 5)$ and $(-2, 2)$ are both solutions of $y = \frac{1}{2}x + 3$. We graph these points and sketch the line passing through them.

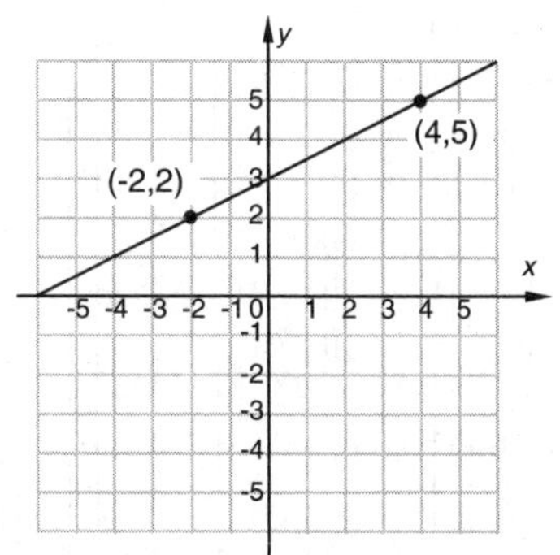

The line appears to pass through $(0, 3)$ also. We check to determine if $(0, 3)$ is a solution of $y = \frac{1}{2}x + 3$.

$$\begin{array}{c|c} \multicolumn{2}{c}{y = \frac{1}{2}x + 3} \\ \hline 3 & \frac{1}{2} \cdot 0 + 3 \end{array}$$

$3 \stackrel{?}{=} 3$ TRUE

Thus, $(0, 3)$ is another solution. There are other correct answers, including $(-6, 0)$, $(-4, 1)$, $(2, 4)$, and $(6, 6)$.

11. To show that a pair is a solution, we substitute, replacing x with the first coordinate and y with the second coordinate in each pair.

$$\begin{array}{c|c} \multicolumn{2}{c}{y + 3x = 7} \\ \hline 1 + 3 \cdot 2 & 7 \\ 1 + 6 & \end{array} \qquad \begin{array}{c|c} \multicolumn{2}{c}{y + 3x = 7} \\ \hline -5 + 3 \cdot 4 & 7 \\ -5 + 12 & \end{array}$$

$7 \stackrel{?}{=} 7$ TRUE $\quad 7 \stackrel{?}{=} 7$ TRUE

In each case the substitution results in a true equation. Thus, $(2, 1)$ and $(4, -5)$ are both solutions of $y + 3x = 7$. We graph these points and sketch the line passing through them.

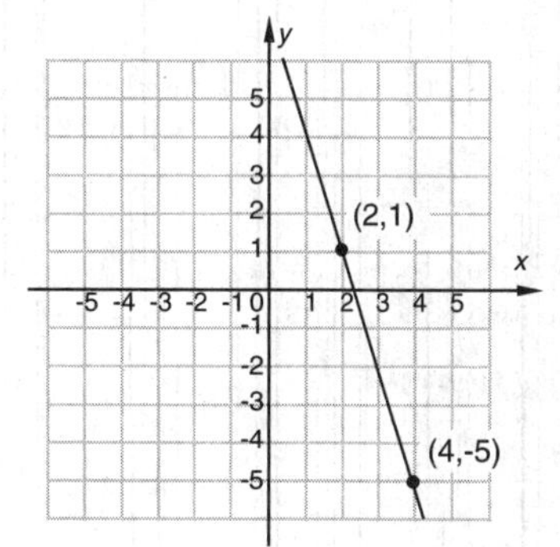

The line appears to pass through $(1, 4)$ also. We check to determine if $(1, 4)$ is a solution of $y + 3x = 7$.

$$\begin{array}{c|c} \multicolumn{2}{c}{y + 3x = 7} \\ \hline 4 + 3 \cdot 1 & 7 \\ 4 + 3 & \end{array}$$

$7 \stackrel{?}{=} 7$ TRUE

Thus, $(1, 4)$ is another solution. There are other correct answers, including $(3, -2)$.

13. To show that a pair is a solution, we substitute, replacing x with the first coordinate and y with the second coordinate in each pair.

$$\begin{array}{c|c} \multicolumn{2}{c}{4x - 2y = 10} \\ \hline 4 \cdot 0 - 2(-5) & 10 \end{array}$$

$10 \stackrel{?}{=} 10$ TRUE

$$\begin{array}{c|c} \multicolumn{2}{c}{4x - 2y = 10} \\ \hline 4 \cdot 4 - 2 \cdot 3 & 10 \\ 16 - 6 & \end{array}$$

$10 \stackrel{?}{=} 10$ TRUE

In each case the substitution results in a true equation. Thus, $(0, -5)$ and $(4, 3)$ are both solutions of $4x - 2y = 10$.

We graph these points and sketch the line passing through them.

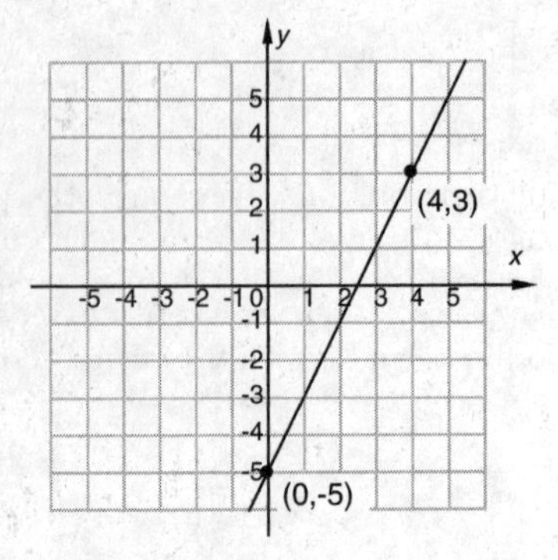

The line appears to pass through $(2,-1)$ also. We check to determine if $(2,-1)$ is a solution of $4x-2y=10$.

$4x - 2y = 10$	
$4\cdot 2-2(-1)$	10
$8+2$	

$10 \stackrel{?}{=} 10$ TRUE

Thus, $(2,-1)$ is another solution. There are other correct answers, including $(1,-3)$, $(2,-1)$, $(3,1)$, and $(5,5)$.

15. $y = x+1$

The equation is in the form $y = mx+b$. The y-intercept is $(0,1)$. We find two other pairs.

When $x=3$, $\quad y = 3+1 = 4$.
When $x=-5$, $\quad y=-5+1=-4$.

x	y
0	1
3	4
-5	-4

Plot these points, draw the line they determine, and label the graph $y=x+1$.

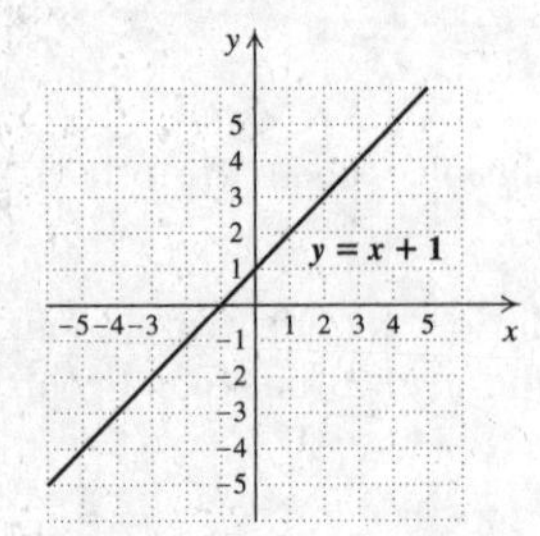

17. $y=-x$

The equation is equivalent to $y=-x+0$. The y-intercept is $(0,0)$. We find two other points.

When $x=-2$, $\quad y=-(-2)=2$.
When $x=3$, $\quad y=-3$.

x	y
0	0
-2	2
3	-3

Plot these points, draw the line they determine, and label the graph $y=-x$.

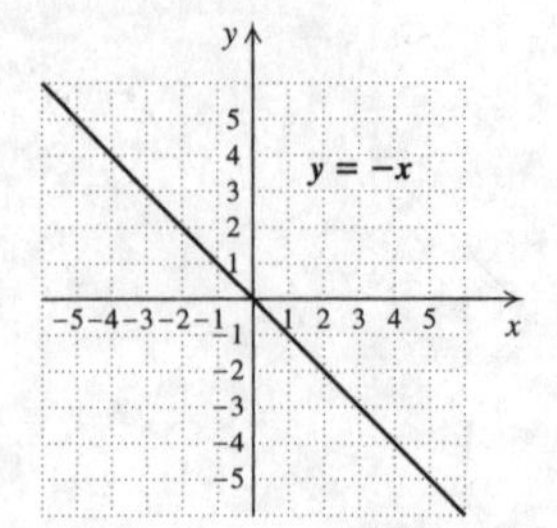

19. $y=\frac{1}{3}x$

The equation is equivalent to $y=\frac{1}{3}x+0$. The y-intercept is $(0,0)$. We find two other points, using multiples of 3 for x to avoid fractions.

When $x=-3$, $y=\frac{1}{3}(-3)=-1$.

When $x=3$, $y=\frac{1}{3}\cdot 3 = 1$.

x	y
0	0
-3	-1
3	1

Plot these points, draw the line they determine, and label the graph $y=\frac{1}{3}x$.

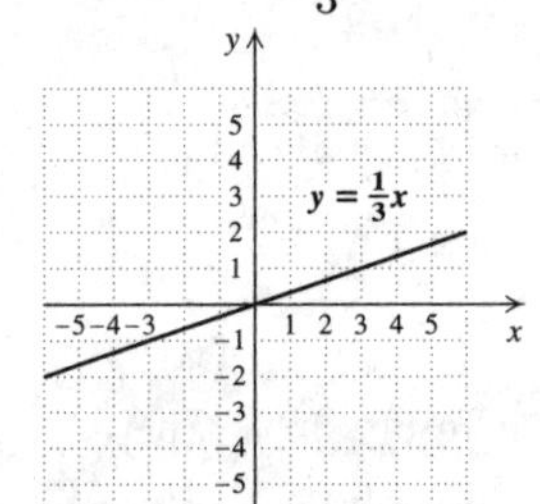

21. $y=x+3$

The equation is in the form $y=mx+b$. The y-intercept is $(0,3)$. We find two other points.

When $x=-4$, $y=-4+3=-1$.

When $x=2$, $y=2+3=5$.

x	y
0	3
-4	-1
2	5

Plot these points, draw the line they determine, and label the graph $y=x+3$.

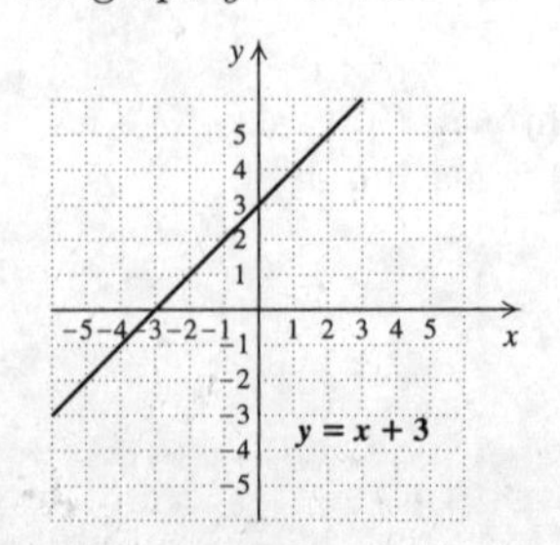

23. $y = 2x + 2$

The y-intercept is $(0, 2)$. We find two other points.

When $x = -3$, $y = 2(-3) + 2 = -6 + 2 = -4$.

When $x = 1$, $y = 2 \cdot 1 + 2 = 2 + 2 = 4$.

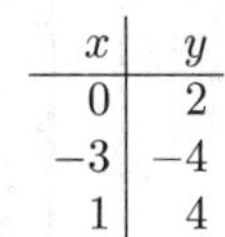

x	y
0	2
-3	-4
1	4

Plot these points, draw the line they determine, and label the graph $y = 2x + 2$.

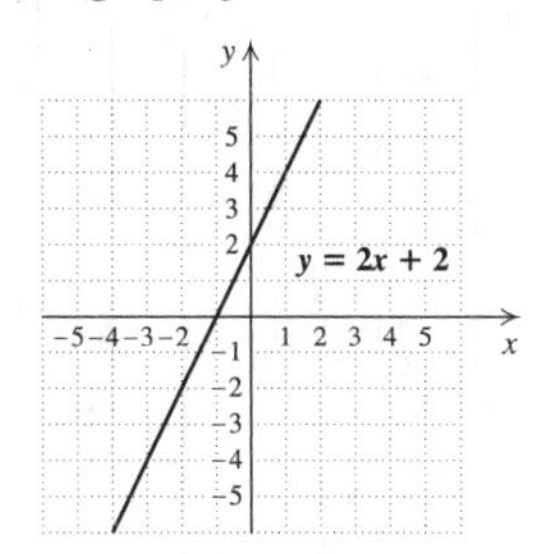

25. $y = \frac{1}{3}x - 4 = \frac{1}{3}x + (-4)$

The y-intercept is $(0, -4)$. We find two other points, using multiples of 3 for x to avoid fractions.

When $x = -3$, $y = \frac{1}{3}(-3) - 4 = -1 - 4 = -5$.

When $x = 3$, $y = \frac{1}{3} \cdot 3 - 4 = 1 - 4 = -3$.

x	y
0	-4
-3	-5
3	-3

Plot these points, draw the line they determine, and label the graph $y = \frac{1}{3}x - 4$.

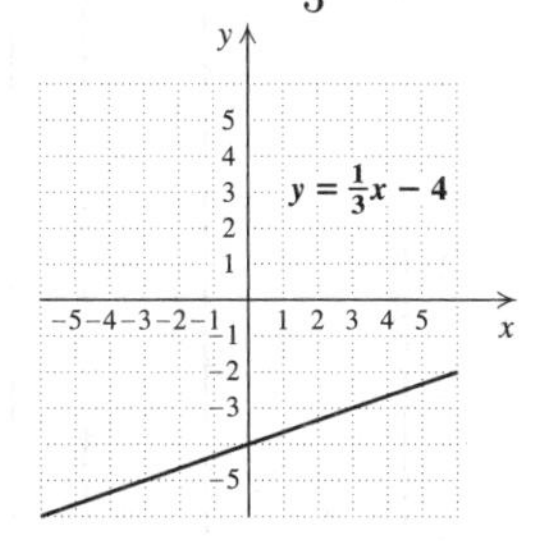

27. $x + y = 4$

$y = -x + 4$

The y-intercept is $(0, 4)$. We find two other points.

When $x = -1$, $y = -(-1) + 4 = 1 + 4 = 5$.

When $x = 2$, $y = -2 + 4 = 2$.

x	y
0	4
-1	5
2	2

Plot these points, draw the line they determine, and label the graph $x + y = 4$.

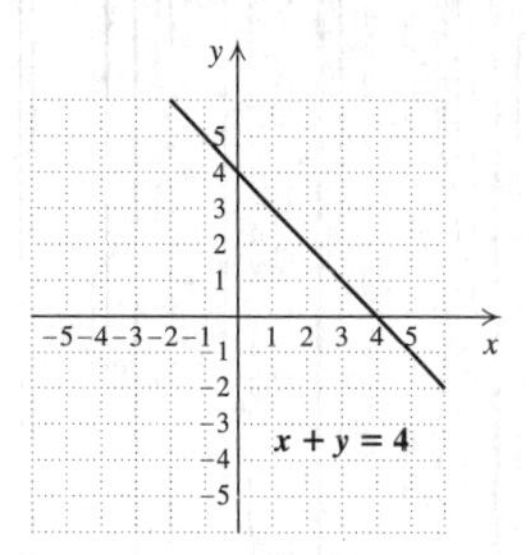

29. $y = \frac{5}{2}x + 3$

The y-intercept is $(0, 3)$. We find two other points, using multiples of 2 for x to avoid fractions.

When $x = -4$, $y = \frac{5}{2}(-4) + 3 = -10 + 3 = -7$.

When $x = -2$, $y = \frac{5}{2}(-2) + 3 = -5 + 3 = -2$.

x	y
0	3
-4	-7
-2	-2

Plot these points, draw the line they determine, and label the graph $y = \frac{5}{2}x + 3$.

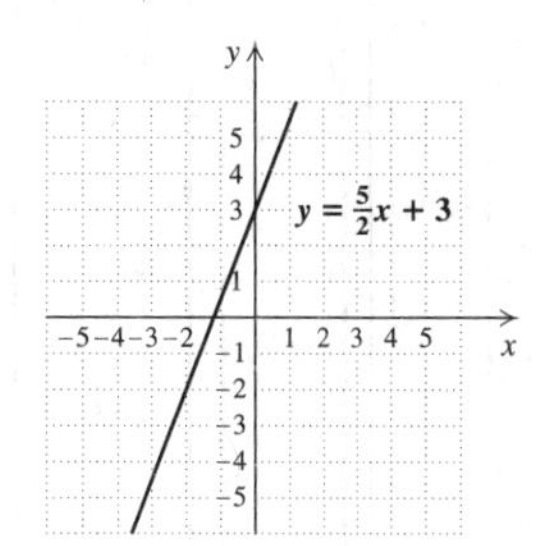

31.
$$
\begin{aligned}
x + 2y &= -6 \\
2y &= -x - 6 \\
y &= -\frac{1}{2}x - 3 \\
y &= -\frac{1}{2}x + (-3)
\end{aligned}
$$

The y-intercept is $(0, -3)$. We find two other points, using multiples of 2 for x to avoid fractions.

When $x = -4$, $y = -\frac{1}{2}(-4) - 3 = 2 - 3 = -1$.

When $x = 2$, $y = -\frac{1}{2} \cdot 2 - 3 = -1 - 3 = -4$.

x	y
0	-3
-4	-1
2	-4

Plot these points, draw the line they determine, and label the graph $x + 2y = -6$.

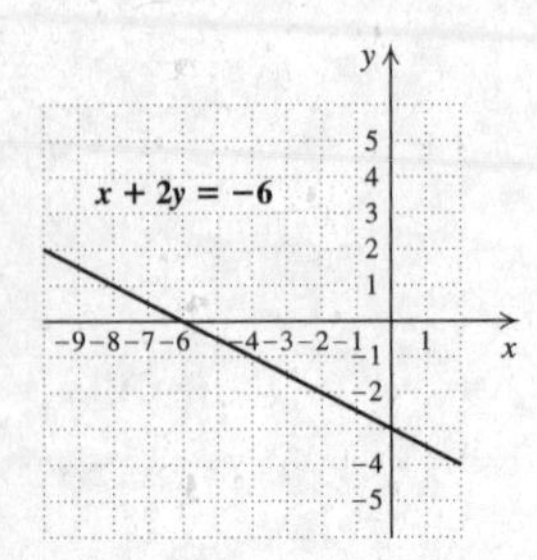

33. $y = -\dfrac{2}{3} + 4$

The y-intercept is $(0, 4)$. We find two other points, using multiples of 3 for x to avoid fractions.

When $x = 3$, $y = -\dfrac{2}{3} \cdot 3 + 4 = -2 + 4 = 2$.

When $x = 6$, $y = -\dfrac{2}{3} \cdot 6 + 4 = -4 + 4 = 0$.

x	y
0	4
3	2
6	0

Plot these points, draw the line they determine, and label the graph $y = -\dfrac{2}{3}x + 4$.

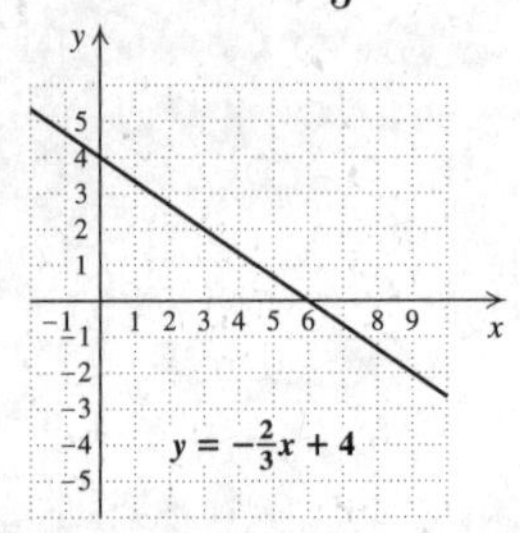

35. $8x - 4y = 12$

$-4y = -8x + 12$

$y = 2x - 3$

$y = 2x + (-3)$

The y-intercept is $(0, -3)$. We find two other points.

When $x = -1$, $y = 2(-1) - 3 = -2 - 3 = -5$.

When $x = 3$, $y = 2 \cdot 3 - 3 = 6 - 3 = 3$.

x	y
0	-3
-1	-5
3	3

Plot these points, draw the line they determine, and label the graph $8x - 4y = 12$.

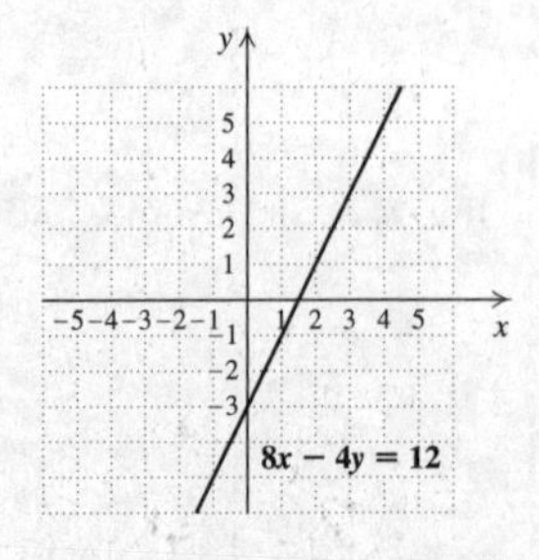

37. $6y + 2x = 8$

$6y = -2x + 8$

$y = -\dfrac{1}{3}x + \dfrac{4}{3}$

The y-intercept is $\left(0, \dfrac{4}{3}\right)$. We find two other points.

When $x = -2$, $y = -\dfrac{1}{3}(-2) + \dfrac{4}{3} = \dfrac{2}{3} + \dfrac{4}{3} = 2$.

When $x = 1$, $y = -\dfrac{1}{3} \cdot 1 + \dfrac{4}{3} = -\dfrac{1}{3} + \dfrac{4}{3} = 1$.

x	y
0	$\dfrac{4}{3}$
-2	2
1	1

Plot these points, draw the line they determine, and label the graph $6y + 2x = 8$.

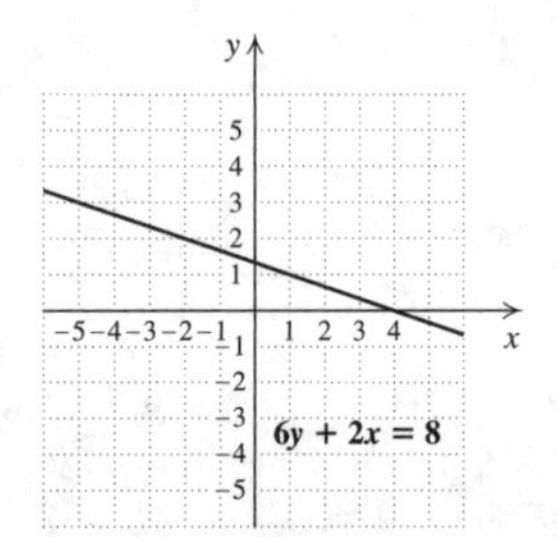

39. We graph $n = 0.9t + 19$ by selecting values for t and then calculating the associated values for n.

If $t = 0$, $n = 0.9(0) + 19 = 19$.

If $t = 3$, $n = 0.9(3) + 19 = 2.7 + 19 = 21.7$.

If $t = 7$, $n = 0.9(7) + 19 = 6.3 + 19 = 25.3$.

t	n
0	19
3	21.7
7	25.3

We plot the points and draw the graph.

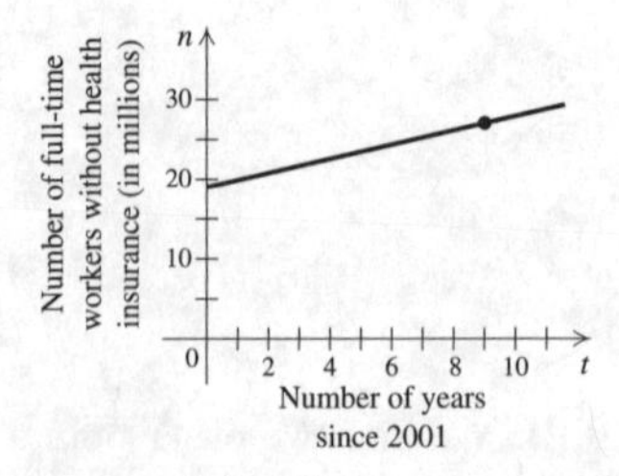

Since $2010 - 2001 = 9$, the year 2010 is 9 years after 2001. Thus, to estimate the number of uninsured full-time workers in 2010, we find the second coordinate associated with 9. Locate the point on the line that is above 9 and then find the value on the vertical axis that corresponds to that point. That value is about 27, so we estimate that there will be about 27 million uninsured full-time workers in 2010.

41. We graph $t + w = 15$, or $w = -t + 15$. Since time cannot be negative in this application, we select only nonnegative values for t.

If $t = 0$, $w = -0 + 15 = 15$.

If $t = 2$, $w = -2 + 15 = 13$.

If $t = 5$, $w = -5 + 15 = 10$.

t	w
0	15
2	13
5	10

We plot the points and draw the graph. Since the likelihood of death cannot be negative, the graph stops at the horizontal axis.

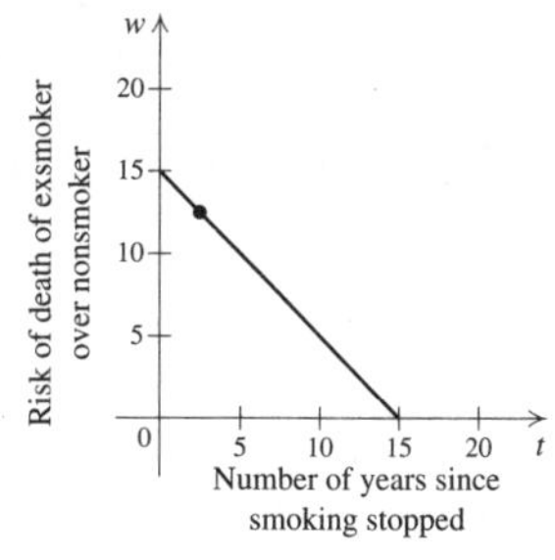

To estimate how much more likely it is for Sandy to die from lung cancer than Polly, we find the second coordinate associated with $2\frac{1}{2}$. Locate the point on the line that is above $2\frac{1}{2}$ and then find the value on the vertical axis that corresponds to that point. That value is about $12\frac{1}{2}$, so it is $12\frac{1}{2}$ times more likely for Sandy to die from lung cancer than Polly.

43. We graph $v = -\frac{3}{4}t + 6$. Since time cannot be negative in this application, we select only nonnegative values for t.

If $t = 0$, $v = -\frac{3}{4} \cdot 0 + 6 = 6$.

If $t = 2$, $v = -\frac{3}{4} \cdot 2 + 6 = -\frac{3}{2} + 6 = \frac{9}{2}$, or $4\frac{1}{2}$.

If $t = 8$, $v = -\frac{3}{4} \cdot 8 + 6 = -6 + 6 = 0$.

t	v
0	6
2	$4\frac{1}{2}$
8	0

We plot the points and draw the graph.

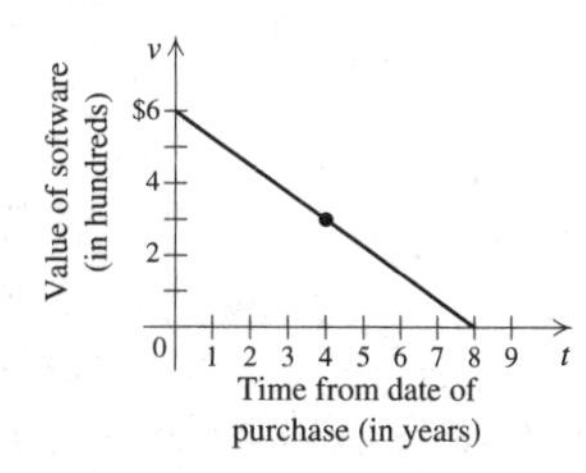

To estimate what the program is worth 4 yr after it was purchased, we find the second coordinate associated with 4. Locate the point on the line that is above 4 and then find the value on the vertical axis that corresponds to that point. That value is 3, so the program is worth \$300 after 4 years.

45. We graph $T = \frac{3}{4}c + 1$. Since the number of credits cannot be negative, we select only nonnegative values for c.

If $c = 4$, $T = \frac{3}{4} \cdot 4 + 1 = 3 + 1 = 4$.

If $c = 8$, $T = \frac{3}{4} \cdot 8 + 1 = 6 + 1 = 7$.

If $c = 16$, $T = \frac{3}{4} \cdot 16 = 12 + 1 = 13$.

c	T
4	4
8	7
16	13

We plot the points and draw the graph.

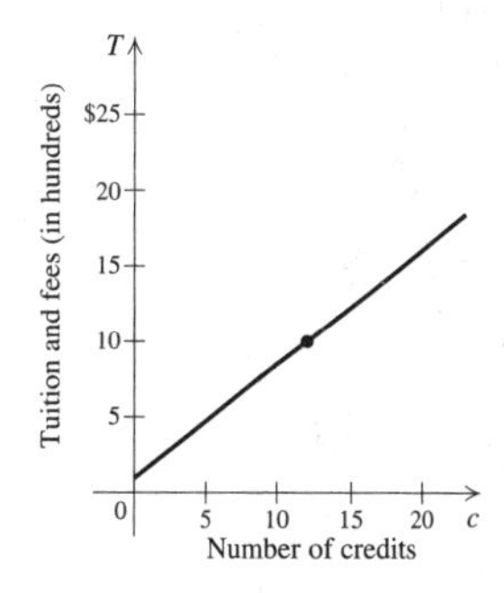

Four three-credit courses total $4 \cdot 3$, or 12, credits. To estimate the cost of tuition and fees for a student who is registered for 12 credits, we find the second coordinate associated with 12. Locate the point on the line that is above 12 and then find the value on the vertical axis that corresponds to that point. That value is about 10, so tuition and fees will cost \$10 hundred, or \$1000.

47. We graph $T = -2m + 54$.

If $m = 0$, $T = -2 \cdot 0 + 54 = 54$.

If $m = 10$, $T = -2 \cdot 10 + 54 = -20 + 54 = 34$.

If $m = 20$, $T = -2 \cdot 20 + 54 = -40 + 54 = 14$.

m	T
0	54
10	34
20	14

We plot the points and draw the graph.

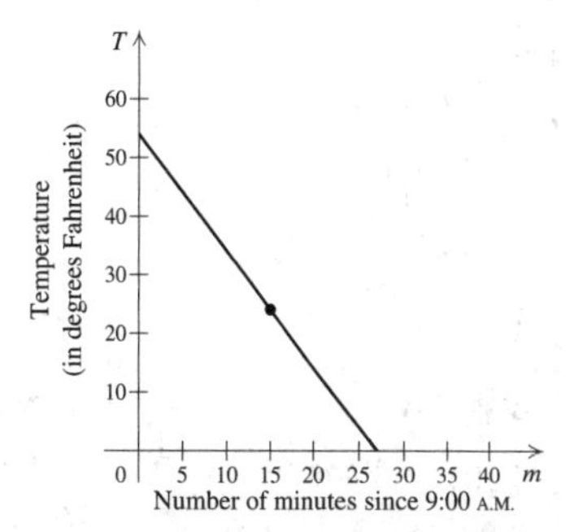

To estimate the temperature at 9:15 A.M., first note that 9:15 is 15 minutes after 9:00 A.M. Then find the second coordinate associated with 15. Locate the point on the line that is above 15 and find the corresponding value on the vertical axis. It appears that the temperature was about 24°F at 9:15 A.M.

49. *Writing Exercise*

51. $5x + 3 \cdot 0 = 12$

$5x + 0 = 12$

$5x = 12$

$x = \frac{12}{5}$

Check: $5x + 3 \cdot 0 = 12$

$5 \cdot \frac{12}{5} + 3 \cdot 0 \;|\; 12$

$12 + 0$

$\overset{?}{}$

$12 = 12$ TRUE

The solution is $\frac{12}{5}$.

53. $7 \cdot 0 - 4y = 10$

$0 - 4y = 10$

$y = -\frac{5}{2}$

Check: $7 \cdot 0 - 4y = 10$

$7 \cdot 0 - 4\left(-\frac{5}{2}\right) \;|\; 10$

$0 + 10$

$\overset{?}{}$

$10 = 10$ TRUE

The solution is $-\frac{5}{2}$.

55. $Ax + By = C$

$By = C - Ax$ Subtracting Ax

$y = \frac{C - Ax}{B}$ Dividing by B

57. *Writing Exercise*

59. Let s represent the gear that Lauren uses on the southbound portion of her ride and n represent the gear she uses on the northbound portion. Then we have $s + n = 18$. We graph this equation, using only positive integer values for s and n.

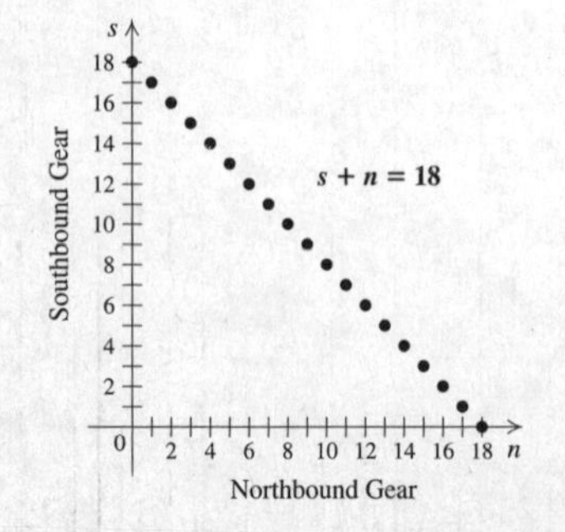

61. Note that the sum of the coordinates of each point on the graph is 2. Thus, we have $x + y = 2$, or $y = -x + 2$.

63. Note that when $x = 0$, $y = -5$ and when $y = 0$, $x = 3$. An equation that fits this situation is $5x - 3y = 15$, or $y = \frac{5}{3}x - 5$.

65. The equation is $25d + 5l = 225$.

Since the number of dinners cannot be negative, we choose only nonnegative values of d when graphing the equation. The graph stops at the horizontal axis since the number of lunches cannot be negative.

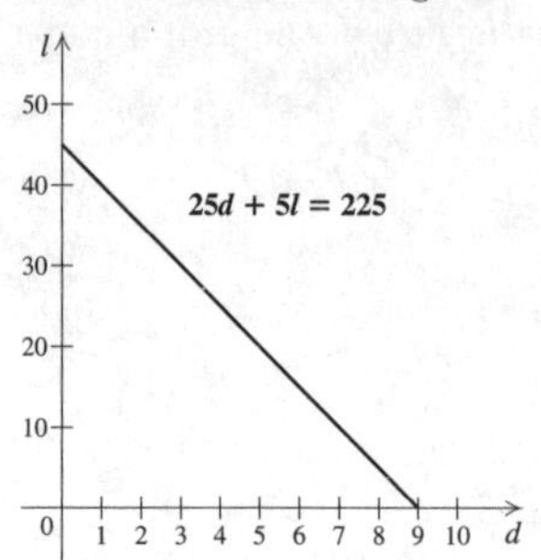

We see that three points on the graph are $(1, 40)$, $(5, 20)$, and $(8, 5)$. Thus, three combinations of dinners and lunches that total $225 are

1 dinner, 40 lunches,

5 dinners, 20 lunches,

8 dinners, 5 lunches.

67. $y = -|x|$

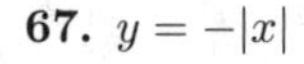

x	y
−3	−3
−2	−2
−1	−1
0	0
1	−1
2	−2
3	−3

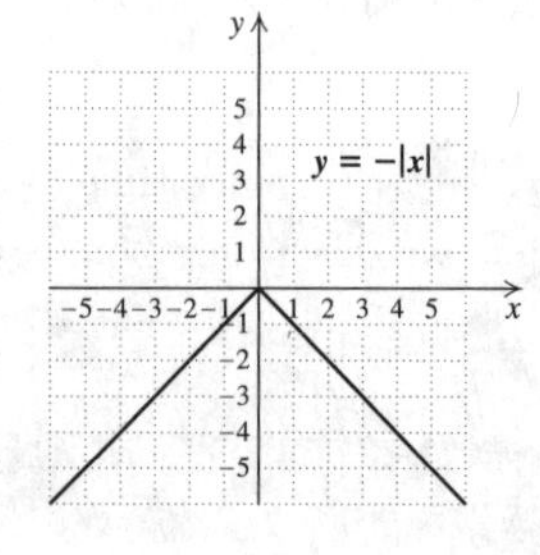

69. $y = -|x| + 2$

x	y
−3	−1
−2	0
−1	1
0	2
1	1
2	0
3	−1

71.

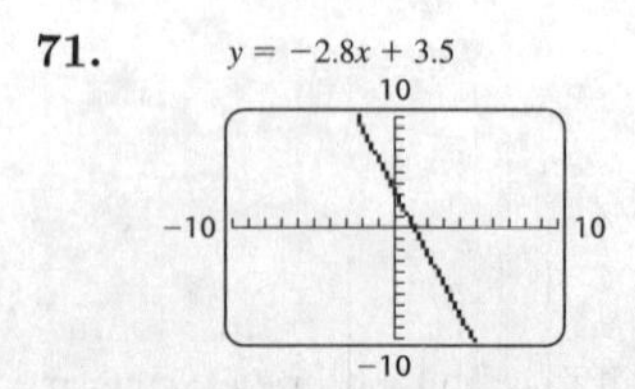

73.

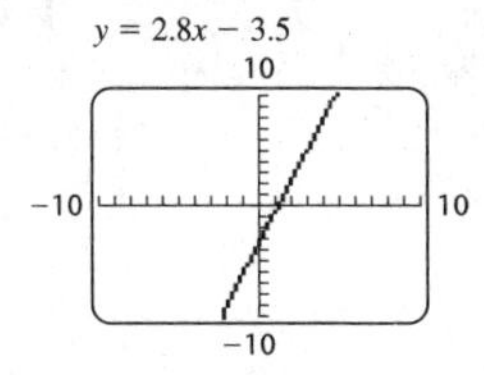

75.

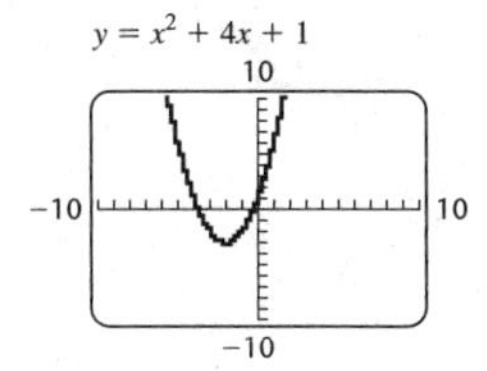

77. *Writing Exercise*

Exercise Set 3.3

1. The graph of $x = -4$ is a vertical line, so (f) is the most appropriate choice.

3. The point $(0, 2)$ lies on the y-axis, so (d) is the most appropriate choice.

5. The point $(3, -2)$ does not lie on an axis, so it could be used as a check when we graph using intercepts. Thus (b) is the most appropriate choice.

7. (a) The graph crosses the y-axis at $(0, 5)$, so the y-intercept is $(0, 5)$.

(b) The graph crosses the x-axis at $(2, 0)$, so the x-intercept is $(2, 0)$.

9. (a) The graph crosses the y-axis at $(0, -4)$, so the y-intercept is $(0, -4)$.

(b) The graph crosses the x-axis at $(3, 0)$, so the x-intercept is $(3, 0)$.

11. (a) The graph crosses the y-axis at $(0, -2)$, so the y-intercept is $(0, -2)$.

(b) The graph crosses the x-axis at $(-3, 0)$, so the x-intercept is $(-3, 0)$.

13. $5x + 3y = 15$

(a) To find the y-intercept, let $x = 0$. This is the same as temporarily ignoring the x-term and then solving.

$$3y = 15$$
$$y = 5$$

The y-intercept is $(0, 5)$.

(b) To find the x-intercept, let $y = 0$. This is the same as temporarily ignoring the y-term and then solving.

$$5x = 15$$
$$x = 3$$

The x-intercept is $(3, 0)$.

15. $7x - 2y = 28$

(a) To find the y-intercept, let $x = 0$. This is the same as temporarily ignoring the x-term and then solving.

$$-2y = 28$$
$$y = -14$$

The y−intercept is $(0, -14)$.

(b) To find the x-intercept, let $y = 0$. This is the same as temporarily ignoring the y-term and then solving.

$$7x = 28$$
$$x = 4$$

The x-intercept is $(4, 0)$.

17. $-4x + 3y = 150$

(a) To find the y-intercept, let $x = 0$. This is the same as temporarily ignoring the x-term and then solving.

$$3y = 150$$
$$y = 50$$

The y-intercept is $(0, 50)$.

(b) To find the x-intercept, let $y = 0$. This is the same as temporarily ignoring the y-term and then solving.

$$-4x = 150$$
$$x = -\frac{75}{2}$$

The x-intercept is $\left(-\frac{75}{2}, 0\right)$.

19. $y = 9$

Observe that this is the equation of a horizontal line 9 units above the x-axis. Thus, (a) the y-intercept is $(0, 9)$ and (b) there is no x-intercept.

21. $x = -7$

Observe that this is the equation of a vertical line 7 units to the left of the y-axis. Thus, (a) there is no y-intercept and (b) the x-intercept is $(-7, 0)$.

23. $3x + 2y = 12$

Find the y-intercept:

$$2y = 12 \quad \text{Ignoring the } x\text{-term}$$
$$y = 6$$

The y-intercept is $(0, 6)$.

Find the x-intercept:

$$3x = 12 \quad \text{Ignoring the } y\text{-term}$$
$$x = 4$$

The x-intercept is $(4, 0)$.

To find a third point we replace x with 2 and solve for y.

$$3 \cdot 2 + 2y = 12$$
$$6 + 2y = 12$$
$$2y = 6$$
$$y = 3$$

The point $(2, 3)$ appears to line up with the intercepts, so we draw the graph.

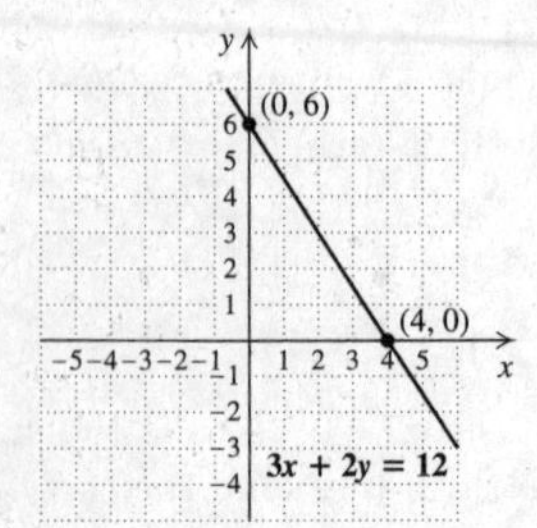

25. $x + 3y = 6$

Find the y-intercept:

$3y = 6$ Ignoring the x-term
$y = 2$

The y-intercept is $(0, 2)$.

Find the x-intercept:

$x = 6$ Ignoring the y-term

The x-intercept is $(6, 0)$.

To find a third point we replace x with 3 and solve for y.

$$3 + 3y = 6$$
$$3y = 3$$
$$y = 1$$

The point $(3, 1)$ appears to line up with the intercepts, so we draw the graph.

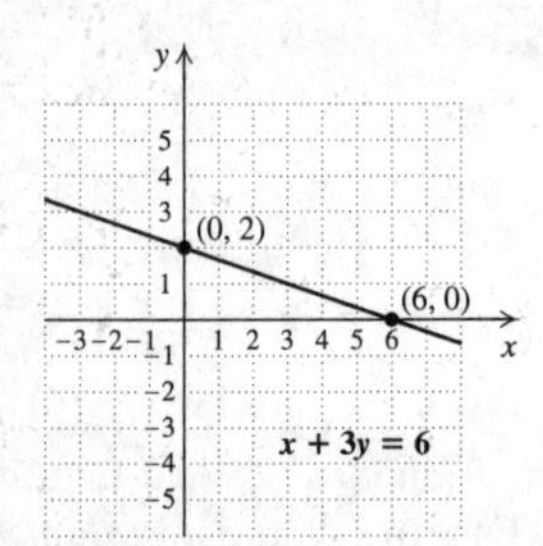

27. $-x + 2y = 8$

Find the y-intercept:

$2y = 8$ Ignoring the x-term
$y = 4$

The y-intercept is $(0, 4)$.

Find the x-intercept:

$-x = 8$ Ignoring the y-term
$x = -8$

The x-intercept is $(-8, 0)$.

To find a third point we replace x with 4 and solve for y.

$$-4 + 2y = 8$$
$$2y = 12$$
$$y = 6$$

The point $(4, 6)$ appears to line up with the intercepts, so we draw the graph.

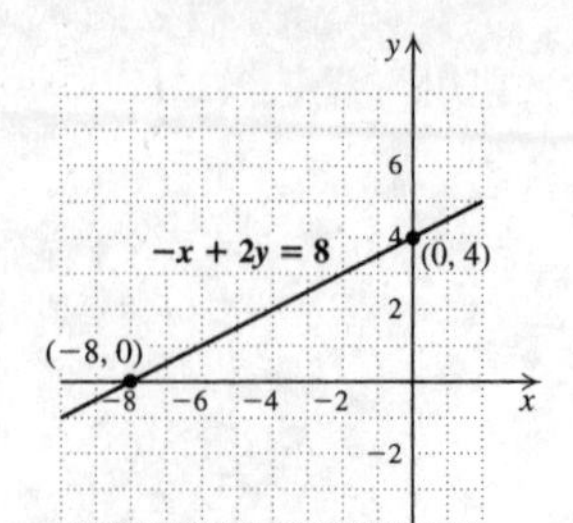

29. $3x + y = 9$

Find the y-intercept:

$y = 9$ Ignoring the x-term

The y-intercept is $(0, 9)$.

Find the $x-$intercept:

$3x = 9$ Ignoring the y-term
$x = 3$

The x-intercept is $(3, 0)$.

To find a third point we replace x with 2 and solve for y.

$$3 \cdot 2 + y = 9$$
$$6 + y = 9$$
$$y = 3$$

The point $(2, 3)$ appears to line up with the intercepts, so we draw the graph.

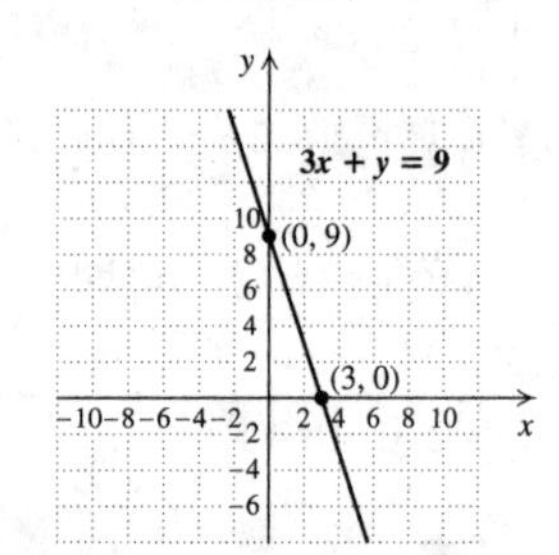

31. $y = 2x - 6$

Find the y-intercept:

$y = -6$ Ignoring the x-term

The y-intercept is $(0, -6)$.

Find the $x-$intercept:

$0 = 2x - 6$ Replacing y with 0
$6 = 2x$
$3 = x$

The x-intercept is $(3, 0)$.

To find a third point we replace x with 2 and find y.

$$y = 2 \cdot 2 - 6 = 4 - 6 = -2$$

The point $(2, -2)$ appears to line up with the intercepts, so we draw the graph.

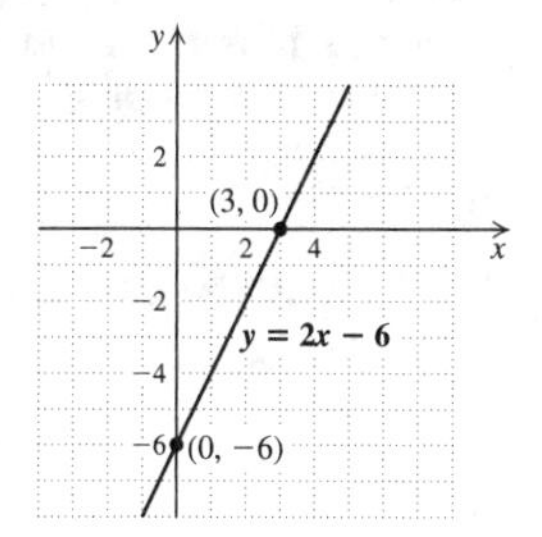

33. $3x - 9 = 3y$

Find the y-intercept:

$$\begin{aligned} -9 &= 3y \quad \text{Ignoring the } x\text{-term} \\ -3 &= y \end{aligned}$$

The y-intercept is $(0, -3)$.

To find the x-intercept, let $y = 0$.

$$\begin{aligned} 3x - 9 &= 3 \cdot 0 \\ 3x - 9 &= 0 \\ 3x &= 9 \\ x &= 3 \end{aligned}$$

The x-intercept is $(3, 0)$.

To find a third point we replace x with 1 and solve for y.

$$\begin{aligned} 3 \cdot 1 - 9 &= 3y \\ 3 - 9 &= 3y \\ -6 &= 3y \\ -2 &= y \end{aligned}$$

The point $(1, -2)$ appears to line up with the intercepts, so we draw the graph.

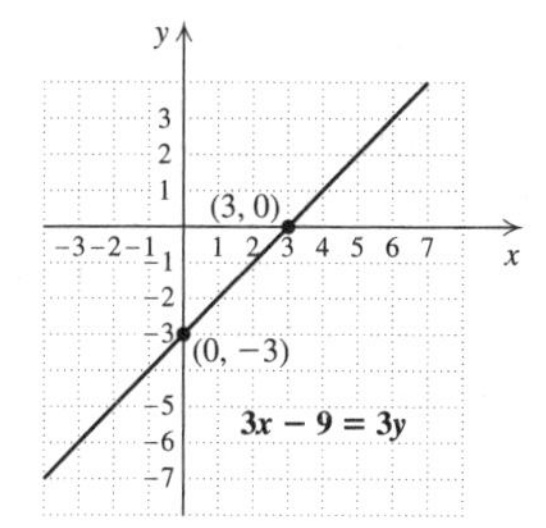

35. $2x - 3y = 6$

Find the y-intercept:

$$\begin{aligned} -3y &= 6 \quad \text{Ignoring the } x\text{-term} \\ y &= -2 \end{aligned}$$

The y-intercept is $(0, -2)$.

Find the x-intercept:

$$\begin{aligned} 2x &= 6 \quad \text{Ignoring the } y\text{-term} \\ x &= 3 \end{aligned}$$

The x-intercept is $(3, 0)$.

To find a third point we replace x with -3 and solve for y.

$$\begin{aligned} 2(-3) - 3y &= 6 \\ -6 - 3y &= 6 \\ -3y &= 12 \\ y &= -4 \end{aligned}$$

The point $(-3, -4)$ appears to line up with the intercepts, so we draw the graph.

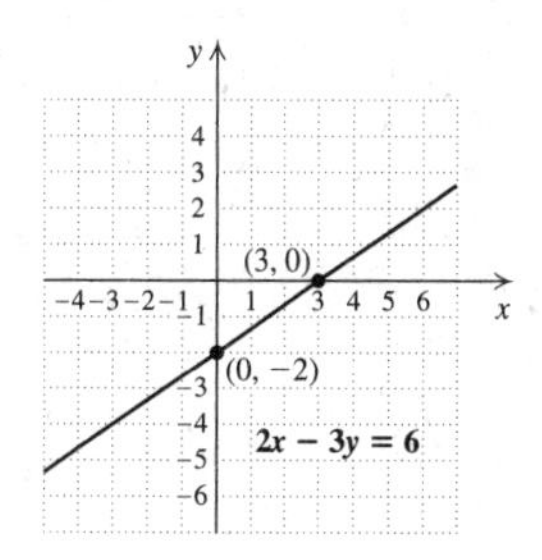

37. $4x + 5y = 20$

Find the y-intercept:

$$\begin{aligned} 5y &= 20 \quad \text{Ignoring the } x\text{-term} \\ y &= 4 \end{aligned}$$

The y-intercept is $(0, 4)$.

Find the x-intercept:

$$\begin{aligned} 4x &= 20 \quad \text{Ignoring the } y\text{-term} \\ x &= 5 \end{aligned}$$

The x-intercept is $(5, 0)$.

To find a third point we replace x with 4 and solve for y.

$$\begin{aligned} 4 \cdot 4 + 5y &= 20 \\ 16 + 5y &= 20 \\ 5y &= 4 \\ y &= \frac{4}{5} \end{aligned}$$

The point $\left(4, \frac{4}{5}\right)$ appears to line up with the intercepts, so we draw the graph.

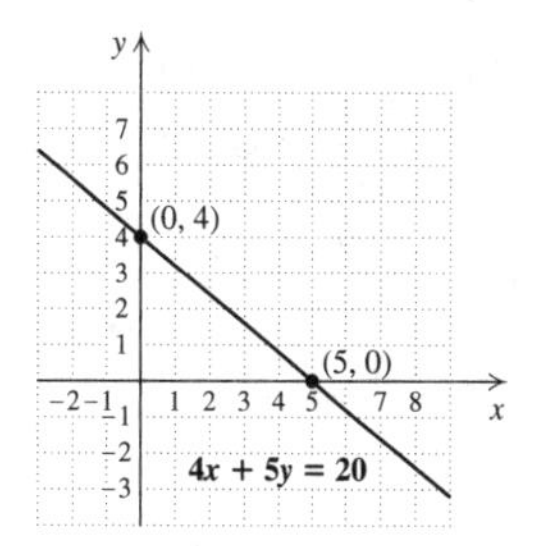

39. $3x + 2y = 8$

Find the y-intercept:

$$\begin{aligned} 2y &= 8 \quad \text{Ignoring the } x\text{-term} \\ y &= 4 \end{aligned}$$

The y-intercept is $(0, 4)$.

Find the x-intercept:

$$\begin{aligned} 3x &= 8 \quad \text{Ignoring the } y\text{-term} \\ x &= \frac{8}{3} \end{aligned}$$

The x-intercept is $\left(\frac{8}{3}, 0\right)$.

To find a third point we replace x with 2 and solve for y.

$$\begin{aligned} 3 \cdot 2 + 2y &= 8 \\ 6 + 2y &= 8 \\ 2y &= 2 \\ y &= 1 \end{aligned}$$

The point $(2, 1)$ appears to line up with the intercepts, so we draw the graph.

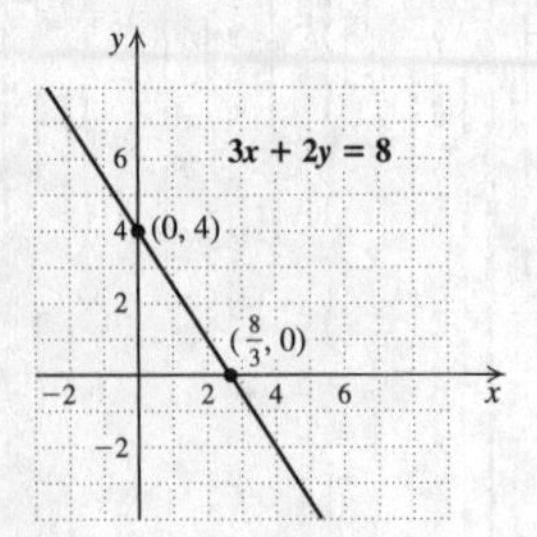

41. $2x + 4y = 6$

Find the y-intercept:

$$4y = 6 \quad \text{Ignoring the } x\text{-term}$$
$$y = \frac{3}{2}$$

The y-intercept is $\left(0, \frac{3}{2}\right)$.

Find the x-intercept:

$$2x = 6 \quad \text{Ignoring the } y\text{-term}$$
$$x = 3$$

The x-intercept is $(3, 0)$.

To find a third point we replace x with -3 and solve for y.

$$2(-3) + 4y = 6$$
$$-6 + 4y = 6$$
$$4y = 12$$
$$y = 3$$

The point $(-3, 3)$ appears to line up with the intercepts, so we draw the graph.

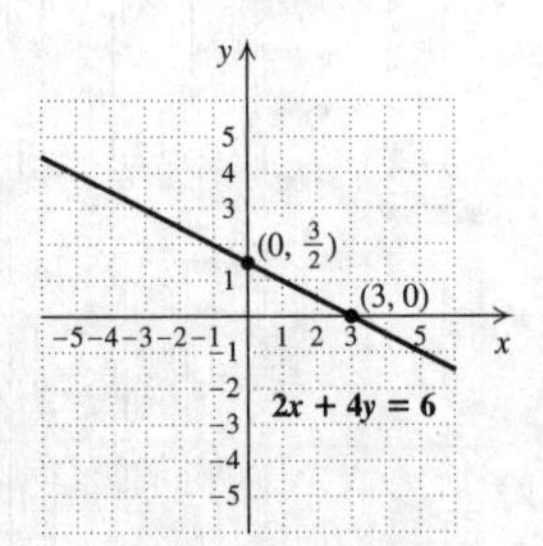

43. $5x + 3y = 180$

Find the y-intercept:

$$3y = 180 \quad \text{Ignoring the } x\text{-term}$$
$$y = 60$$

The y-intercept is $(0, 60)$.

Find the x-intercept:

$$5x = 180 \quad \text{Ignoring the } y\text{-term}$$
$$x = 36$$

The x-intercept is $(36, 0)$.

To find a third point we replace x with 6 and solve for y.

$$5 \cdot 6 + 3y = 180$$
$$30 + 3y = 180$$
$$3y = 150$$
$$y = 50$$

This means that $(6, 50)$ is on the graph.

To graph all three points, the y-axis must go to at least 60 and the x-axis must go to at least 36. Using a scale of 10 units per square allows us to display both intercepts and $(6, 50)$ as well as the origin.

The point $(6, 50)$ appears to line up with the intercepts, so we draw the graph.

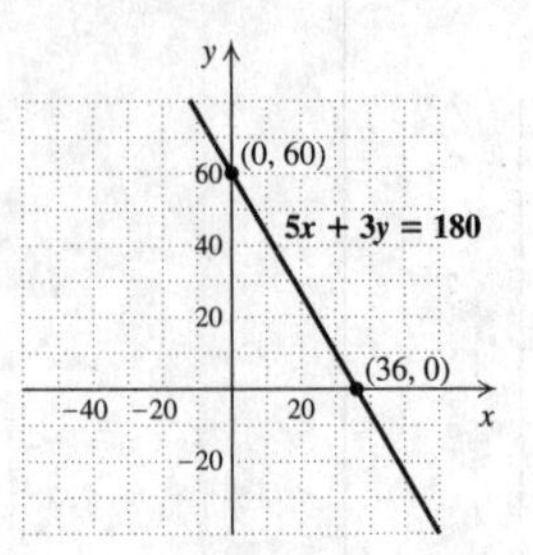

45. $y = -30 + 3x$

Find the y-intercept:

$$y = -30 \quad \text{Ignoring the } x\text{-term}$$

The y-intercept is $(0, -30)$.

To find the x-intercept, let $y = 0$.

$$0 = -30 + 3x$$
$$30 = 3x$$
$$10 = x$$

The x-intercept is $(10, 0)$.

To find a third point we replace x with 5 and solve for y.

$$y = -30 + 3 \cdot 5$$
$$y = -30 + 15$$
$$y = -15$$

This means that $(5, -15)$ is on the graph.

To graph all three points, the y-axis must go to at least -30 and the x-axis must go to at least 10. Using a scale of 5 units per square allows us to display both intercepts and $(5, -15)$ as well as the origin.

The point $(5, -15)$ appears to line up with the intercepts, so we draw the graph.

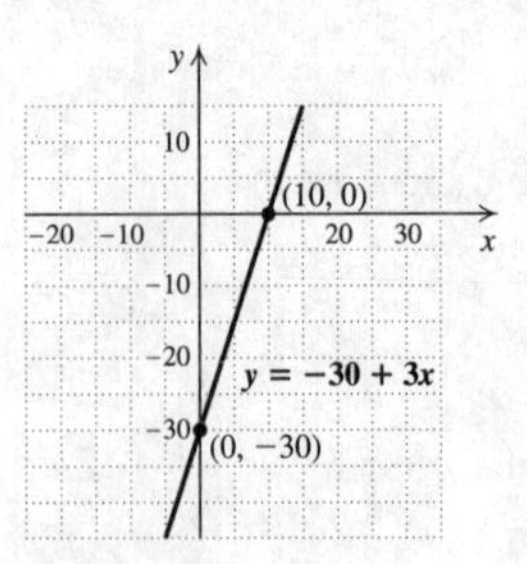

47. $-4x = 20y + 80$

To find the y-intercept, we let $x = 0$.

$$-4 \cdot 0 = 20y + 80$$
$$0 = 20y + 80$$
$$-80 = 20y$$
$$-4 = y$$

The y-intercept is $(0, -4)$.

Find the x-intercept:

$$-4x = 80 \quad \text{Ignoring the } y\text{-term}$$
$$x = -20$$

The x-intercept is $(-20, 0)$.

To find a third point we replace x with -40 and solve for y.

$$-4(-40) = 20y + 80$$
$$160 = 20y + 80$$
$$80 = 20y$$
$$4 = y$$

This means that $(-40, 4)$ is on the graph.

To graph all three points, the y-axis must go at least from -4 to 4 and the x-axis must go at least from -40 to -20. Since we also want to include the origin we can use a scale of 10 units per square on the x-axis and 1 unit per square on the y-axis.

The point $(-40, 4)$ appears to line up with the intercepts, so we draw the graph.

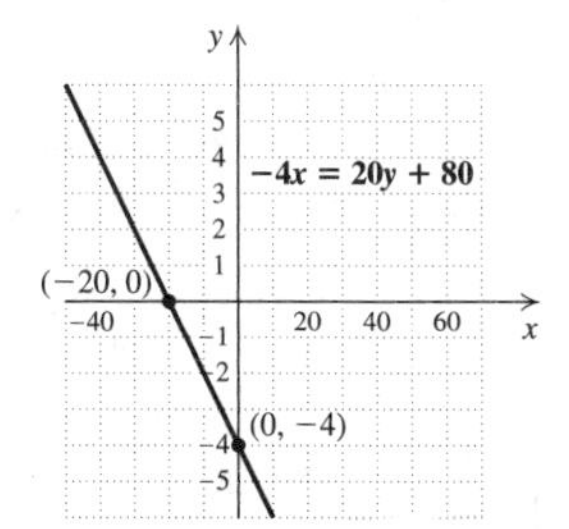

49. $y - 3x = 0$

Find the y-intercept:

$$y = 0 \quad \text{Ignoring the } x\text{-term}$$

The y-intercept is $(0, 0)$. Note that this is also the x-intercept.

In order to graph the line, we will find a second point.

When $x = 1$, $y - 3 \cdot 1 = 0$
$$y - 3 = 0$$
$$y = 3$$

To find a third point we replace $x = -1$ and solve for y.

$$y - 3(-1) = 0$$
$$y + 3 = 0$$
$$y = -3$$

The point $(-1, -3)$ appears to line up with the other two points, so we draw the graph.

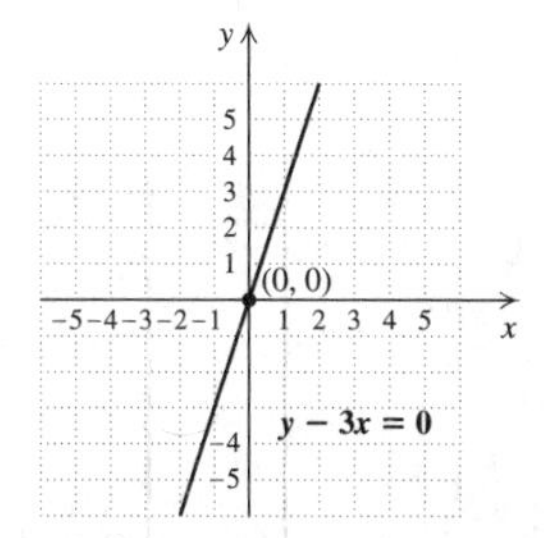

51. $y = 5$

Any ordered pair $(x, 5)$ is a solution. The variable y must be 5, but the x variable can be any number we choose. A few solutions are listed below. Plot these points and draw the line.

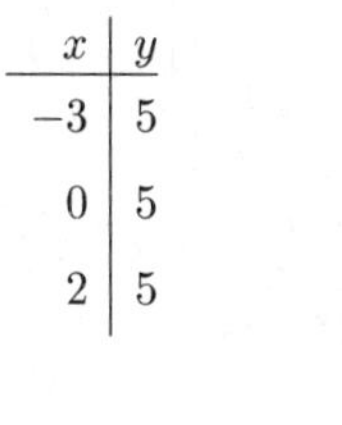

x	y
-3	5
0	5
2	5

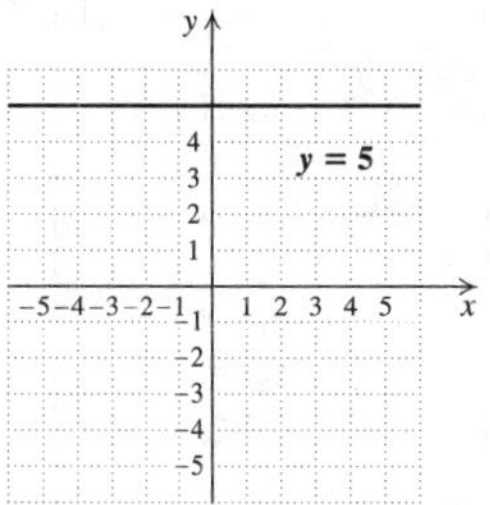

53. $x = 4$

Any ordered pair $(4, y)$ is a solution. The variable x must be 4, but the y variable can be any number we choose. A few solutions are listed below. Plot these points and draw the line.

x	y
4	-2
4	0
4	4

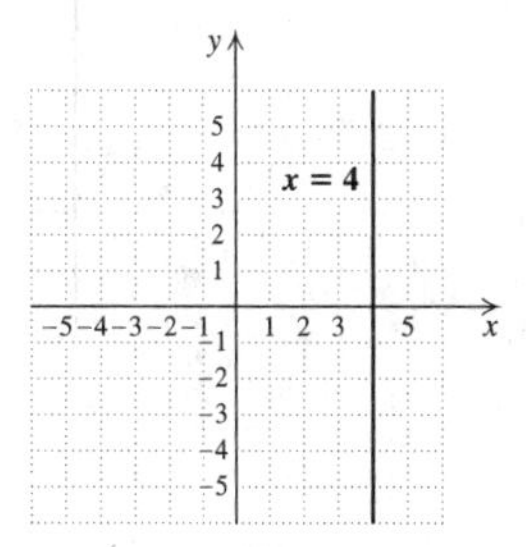

55. $y = -2$

Any ordered pair $(x, -2)$ is a solution. The variable y must be -2, but the x variable can be any number we choose. A few solutions are listed below. Plot these points and draw the line.

x	y
-3	-2
0	-2
4	-2

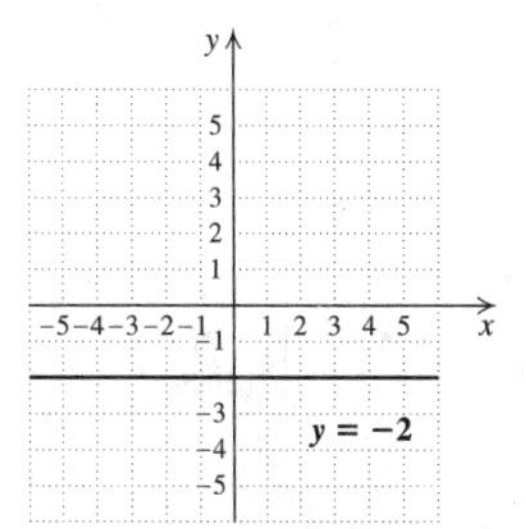

57. $x = -1$

Any ordered pair $(-1, y)$ is a solution. The variable x must be -1, but the y variable can be any number we choose. A few solutions are listed below. Plot these points and draw the line.

x	y
-1	-3
-1	0
-1	2

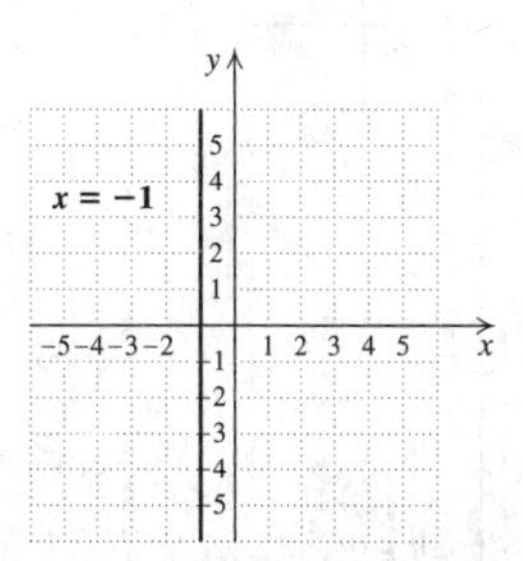

59. $x = 18$

Any ordered pair $(18, y)$ is a solution. A few solutions are listed below. Plot these points and draw the line choosing an appropriate scale.

x	y
18	-1
18	4
18	5

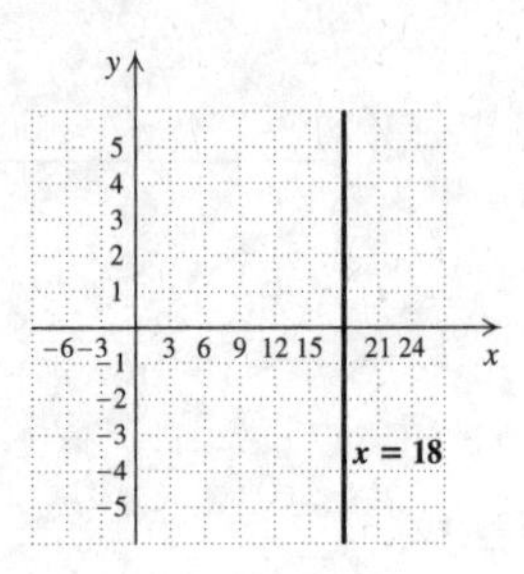

61. $y = 0$

Any ordered pair $(x, 0)$ is a solution. A few solutions are listed below. Plot these points and draw the line.

x	y
-4	0
0	0
2	0

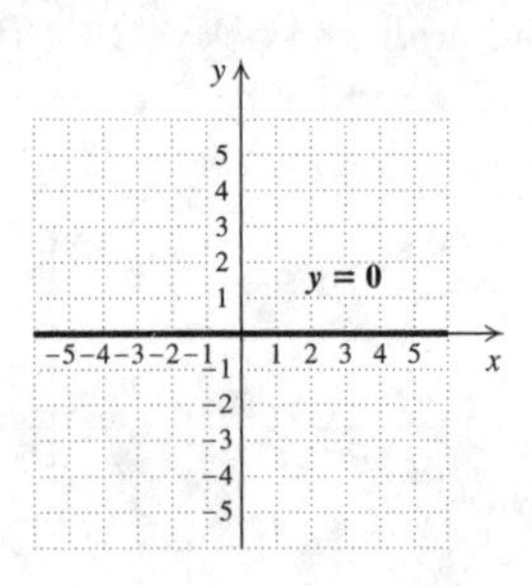

63. $x = -\frac{5}{2}$

Any ordered pair $\left(-\frac{5}{2}, y\right)$ is a solution. A few solutions are listed below. Plot these points and draw the line.

x	y
$-\frac{5}{2}$	-3
$-\frac{5}{2}$	0
$-\frac{5}{2}$	5

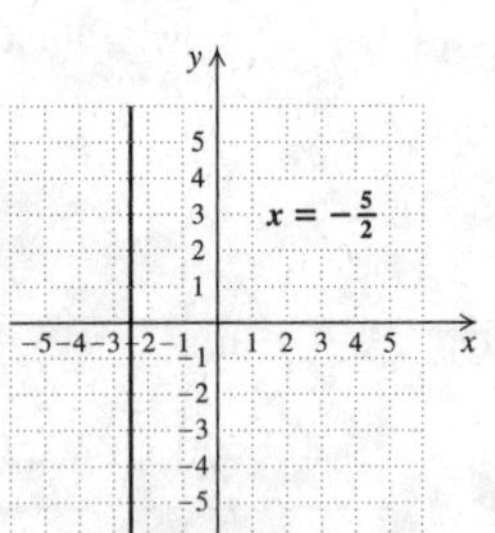

65. $-5y = -300$

$y = 60$ Dividing by -5

The graph is a horizontal line 60 units above the x-axis.

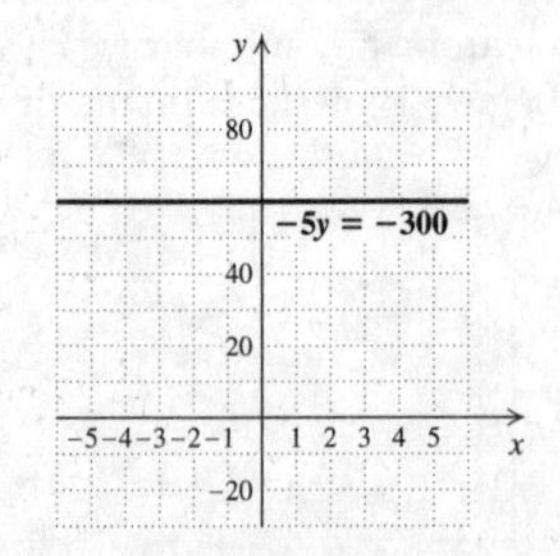

67.

$$35 + 7y = 0$$
$$7y = -35$$
$$y = -5$$

The graph is a horizontal line 5 units below the x-axis.

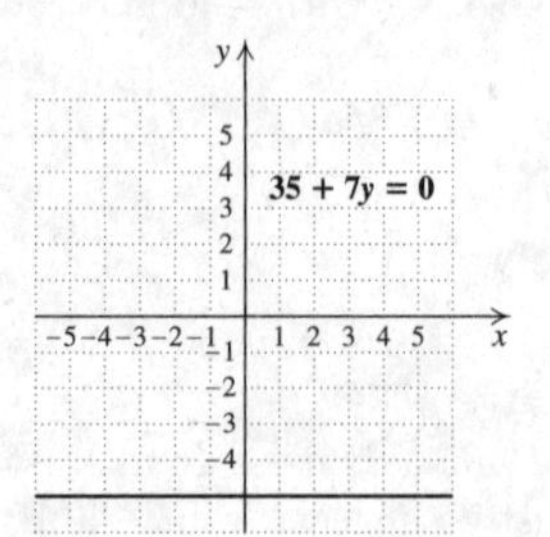

69. Note that every point on the horizontal line passing through $(0, -1)$ has -1 as the y-coordinate. Thus, the equation of the line is $y = -1$.

71. Note that every point on the vertical line passing through $(4, 0)$ has 4 as the x-coordinate. Thus, the equation of the line is $x = 4$.

73. Note that every point on the horizontal line passing through $(0, 0)$ has 0 as the y-coordinate. Thus, the equation of the line is $y = 0$.

75. *Writing Exercise*

77. $d - 7$

79. Let x represent the number. Then we have $2 + x$, or $x + 2$.

81. Let x and y represent the numbers. Then we have $2(x+y)$.

83. *Writing Exercise*

85. The x-axis is a horizontal line, so it is of the form $y = b$. All points on the x-axis are of the form $(x, 0)$, so b must be 0 and the equation is $y = 0$.

87. A line parallel to the y-axis has an equation of the form $x = a$. Since the x-coordinate of one point on the line is -2, then $a = -2$ and the equation is $x = -2$.

89. Since the x-coordinate of the point of intersection must be -3 and y must equal x, the point of intersection is $(-3, -3)$.

91. The y-intercept is $(0, 5)$, so we have $y = mx + 5$. Another point on the line is $(-3, 0)$ so we have

$$0 = m(-3) + 5$$
$$-5 = -3m$$
$$\frac{5}{3} = m$$

The equation is $y = \frac{5}{3}x + 5$, or $5x - 3y = -15$, or $-5x + 3y = 15$.

93. Substitute 0 for x and -8 for y.

$$4 \cdot 0 = C - 3(-8)$$
$$0 = C + 24$$
$$-24 = C$$

95. Find the y-intercept:

$2y = 50$ Covering the x-term

$y = 25$

The y-intercept is $(0, 25)$.

Find the x-intercept:

$3x = 50$ Covering the y-term

$x = \frac{50}{3} = 16.\overline{6}$

The x-intercept is $\left(\frac{50}{3}, 0\right)$, or $(16.\overline{6}, 0)$.

97. From the equation we see that the y-intercept is $(0, -9)$.

To find the x-intercept, let $y = 0$.

$0 = 0.2x - 9$

$9 = 0.2x$

$45 = x$

The x-intercept is $(45, 0)$.

99. Find the y-intercept.

$-20y = 1$ Covering the x-term

$y = -\frac{1}{20}$, or -0.05

The y-intercept is $\left(0, -\frac{1}{20}\right)$, or $(0, -0.05)$.

Find the x-intercept:

$25x = 1$ Covering the y-term

$x = \frac{1}{25}$, or 0.04

The x-intercept is $\left(\frac{1}{25}, 0\right)$, or $(0.04, 0)$.

Exercise Set 3.4

1. a) We divide the number of miles traveled by the number of gallons of gas used for that amount of driving.

Rate, in miles per gallon

$= \frac{14,014 \text{ mi} - 13,741 \text{ mi}}{13 \text{ gal}}$

$= \frac{273 \text{ mi}}{13 \text{ gal}}$

$= 21$ mi/gal

$= 21$ miles per gallon

b) We divide the cost of the rental by the number of days. From June 5 to June 8 is $8 - 5$, or 3 days.

Average cost, in dollars per day

$= \frac{118 \text{ dollars}}{3 \text{ days}}$

≈ 39.33 dollars/day

$\approx \$39.33$ per day

c) We divide the number of miles traveled by the number of days. In part (a) we found that the van was driven 273 mi, and in part (b) we found that it was rented for 3 days.

Rate, in miles per day

$= \frac{273 \text{ mi}}{3 \text{ days}}$

$= 91$ mi/day

$= 91$ mi per day

d) Note that $\$118 = 11,800¢$. From part (a) we know that the van was driven 273 mi.

Rate, in cents per mile $= \frac{11,800¢}{273 \text{ mi}}$

$\approx 43¢$ per mi

3. a) From 2:00 to 5:00 is $5 - 2$, or 3 hr.

Average speed, in miles per hour

$= \frac{18 \text{ mi}}{3 \text{ hr}}$

$= 6$ mph

b) From part (a) we know that the bike was rented for 3 hr.

Rate, in dollars per hour $= \frac{\$12}{3 \text{ hr}}$

$= \$4$ per hr

c) Rate, in dollars per mile $= \frac{\$12}{18 \text{ mi}}$

$\approx \$0.67$ per mile

5. a) It is 3 hr from 9:00 A.M. to noon and 5 more hours from noon to 5:00 P.M., so the typist worked $3 + 5$, or 8 hr.

Rate, in dollars per hour $= \frac{\$128}{8 \text{ hr}}$

$= \$16$ per hr

b) The number of pages typed is $48 - 12$, or 36.

In part (a) we found that the typist worked 8 hr.

Rate, in pages per hour $= \frac{36 \text{ pages}}{8 \text{ hr}}$

$= 4.5$ pages per hr

c) In part (b) we found that 36 pages were typed.

Rate, in dollars per page $= \frac{\$128}{36 \text{ pages}}$

$\approx \$3.56$ per page

7. The tuition increased $\$1359 - \$1327$, or $32, in $2001 - 1999$ or 2 yr.

Rate of increase $= \frac{\text{Change in tuition}}{\text{Change in time}}$

$= \frac{\$32}{2 \text{ yr}}$

$= \$16$ per yr

9. a) The elevator traveled $34 - 5$, or 29 floors in 2:40 – 2:38, or 2 min.

Average rate of travel $= \frac{29 \text{ floors}}{2 \text{ min}}$

$= 14.5$ floors per min

b) In part (a) we found that the elevator traveled 29 floors in 2 min. Note that 2 min = 2×1 min = 2×60 sec = 120 sec.

$$\text{Average rate of travel} = \frac{120 \text{ sec}}{29 \text{ floors}}$$
$$\approx 4.14 \text{ sec per floor}$$

11. a) Krakauer ascended 29,028 ft − 27,600 ft, or 1428 ft. From 7:00 A.M. to noon it is 5 hr = 5×1 hr = 5×60 min = 300 min. From noon to 1:25 P.M. is another 1 hr, 25 min, or 1 hr + 25 min = 60 min + 25 min = 85 min. The total time of the ascent is 300 min + 85 min, or 385 min.

$$\text{Rate, in feet per minute} = \frac{1428 \text{ ft}}{385 \text{ min}}$$
$$\approx 3.71 \text{ ft per min}$$

b) We use the information found in part (a).

$$\text{Rate, in minutes per foot} = \frac{385 \text{ min}}{1428 \text{ ft}}$$
$$\approx 0.27 \text{ min per ft}$$

13. The rate of increase of the average copayment is given in dollars per year, so we list Average copayment on the vertical axis and Year on the horizontal axis. If we count by increments of 10 on the vertical axis we can easily reach 19 and beyond. We label the units on the vertical axis in dollars. We list the years on the horizontal axis, beginning with 2003.

We plot the point (2003, \$19). Then, to display the rate of growth, we move from that point to a point that represents a copayment of \$2 more one year later. The coordinates of this point are (2003 + 1, \$19 + \$2), or (2004, \$21). Finally, we draw a line through the two points.

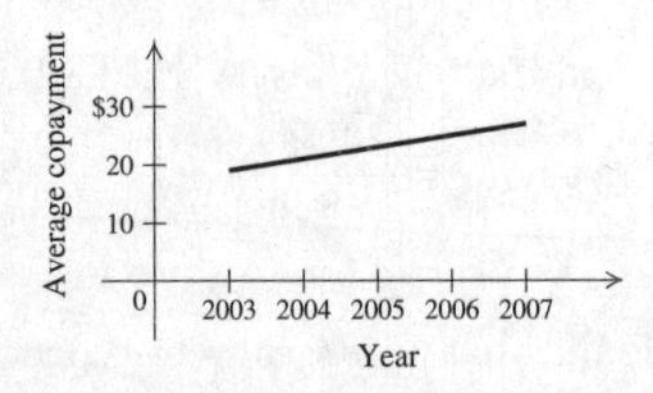

15. The rate is given in millions of crimes per year, so we list Number of crimes, in millions, on the vertical axis and Year on the horizontal axis. If we count by 10's of millions on the vertical axis we can easily reach 26 million and beyond. We plot the point (2000, 26 million). Then, to display the rate of growth, we move from that point to a point that represents 1.2 million fewer crimes 1 year later. The coordinates of this point are (2000 + 1, 26 − 1.2 million), or (2001, 24.8 million). Finally, we draw a line through the two points.

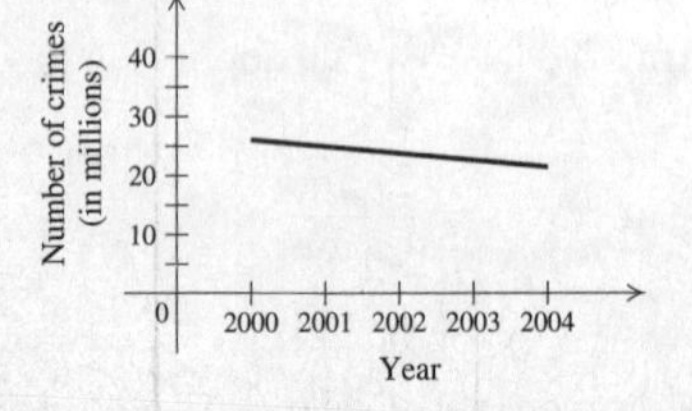

17. The rate is given in miles per hour, so we list the number of miles traveled on the vertical axis and the time of day on the horizontal axis. If we count by 100's of miles on the vertical axis we can easily reach 230 without needing a terribly large graph. We plot the point (3:00, 230). Then to display the rate of travel, we move from that point to a point that represents 90 more miles traveled 1 hour later. The coordinates of this point are (3:00 + 1 hr,230 + 90), or (4:00, 320). Finally, we draw a line through the two points.

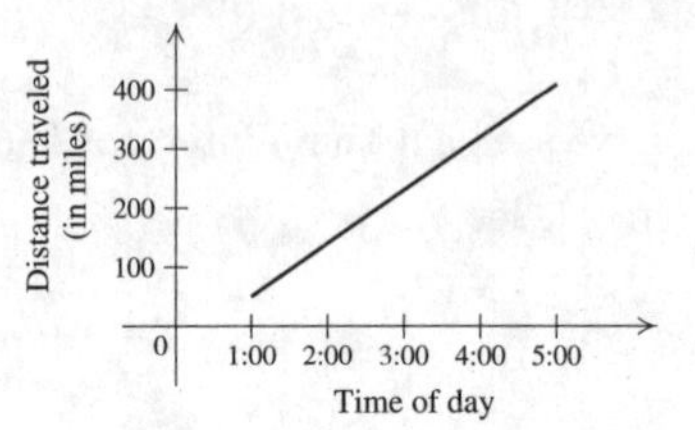

19. The rate is given in dollars per hour so we list money earned on the vertical axis and the time of day on the horizontal axis. We can count by \$20 on the vertical axis and reach \$50 without needing a terribly large graph. Next we plot the point (2:00 P.M., \$50). To display the rate we move from that point to a point that represents \$15 more 1 hour later. The coordinates of this point are (2+1, \$50+\$15), or (3:00 P.M., \$65). Finally, we draw a line through the two points.

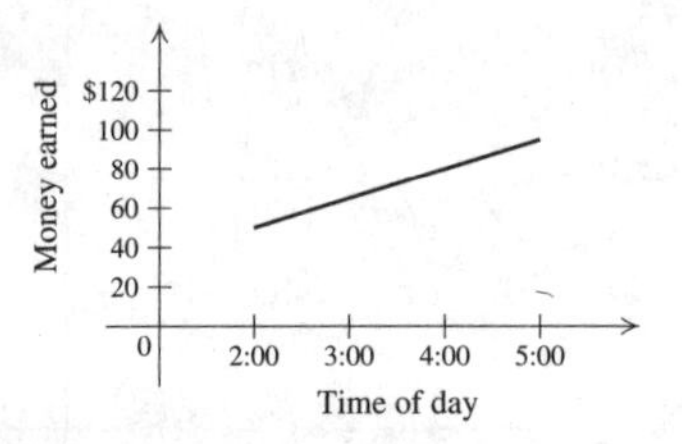

21. The rate is given in cost per minute so we list the amount of the telephone bill on the vertical axis and the number of additional minutes on the horizontal axis. We begin with \$7.50 on the vertical axis and count by \$0.50. A jagged line at the base of the axis indicates that we are not showing amounts smaller than \$7.50. We begin with 0 additional minutes on the horizontal axis and plot the point (0, \$7.50). We move from there to a point that represents \$0.10 more 1 minute later. The coordinates of this point are (0 + 1 min, \$7.50 + \$0.10), or (1 min, \$7.60). Then we draw a line through the two points.

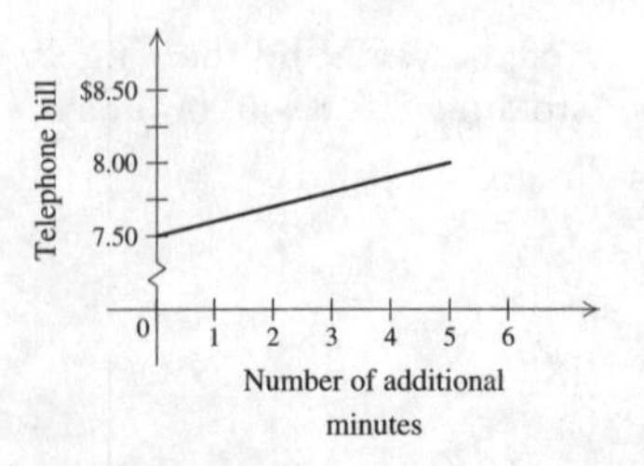

23. The points (2:00, 7 haircuts) and (4:30, 12 haircuts) are on the graph. This tells us that in the 2.5 hr between 2:00 and 4:30 there were 12 − 7 = 5 haircuts completed. The rate is

$$\frac{5 \text{ haircuts}}{2.5 \text{ hr}} = 2 \text{ haircuts per hour.}$$

25. The points (12:00, 100 mi) and (2:00, 250 mi) are on the graph. This tells us that in the 2 hr between 12:00 and 2:00 the train traveled $250 - 100 = 150$ mi. The rate is

$$\frac{150 \text{ mi}}{2\text{hr}} = 75 \text{ mi per hr.}$$

27. The points (5 min, 35¢) and (10 min, 70¢) are on the graph. This tells us that in $10 - 5 = 5$ min the cost of the call increased 70¢ − 35¢ = 35¢. The rate is

$$\frac{35¢}{5 \text{ min}} = 7¢ \text{ per min.}$$

29. The points (2 yr, \$2000) and (4 yr, \$1000) are on the graph. This tells us that in $4 - 2 = 2$ yr the value of the copier changes \$1000 − \$2000 = −\$1000. The rate is

$$\frac{-\$1000}{2 \text{ yr}} = -\$500 \text{ per yr.}$$

This means that the value of the copier is decreasing at a rate of \$500 per yr.

31. The points (50 mi, 2 gal) and (200 mi, 8 gal) are on the graph. This tells us that when driven $200 - 50 = 150$ mi the vehicle consumed $8 - 2 = 6$ gal of gas. The rate is

$$\frac{6 \text{ gal}}{150 \text{ mi}} = 0.04 \text{ gal per mi.}$$

33. Since swimming is the slowest of the three sports and biking is the fastest, the slope of the line representing swimming speed will be the least steep of the three and that representing biking speed will be the steepest. The second segment of graph (e) rises most steeply and the third segment is the least steep of the three segments. Thus this graph represents running followed by biking and then swimming.

35. Since swimming is the slowest of the three sports and biking is the fastest, the slope of the line representing swimming speed will be the least steep of the three and that representing biking speed will be the steepest. The first segment of graph (d) is the least steep and the second segment is the steepest of the three segments. Thus this graph represents swimming followed by biking and then running.

37. Since swimming is the slowest of the three sports and biking is the fastest, the slope of the line representing swimming speed will be the least steep of the three and that representing biking speed will be the steepest. The first segment of graph (b) is the steepest and the second segment is the least steep of the three segments. Thus this graph represents biking followed by swimming and then running.

39. *Writing Exercise*

41. $-2 - (-7) = -2 + 7 = 5$

43. $\dfrac{5 - (-4)}{-2 - 7} = \dfrac{9}{-9} = -1$

45. $\dfrac{-4 - 8}{7 - (-2)} = \dfrac{-12}{9} = -\dfrac{4}{3}$

47. *Writing Exercise*

49. Let t = flight time and a = altitude. While the plane is climbing at a rate of 6500 ft/min, the equation $a = 6500t$ describes the situation. Solving $34,000 = 6500t$, we find that the cruising altitude of 34,000 ft is reached after about 5.23 min. Thus we graph $a = 6500t$ for $0 \le t \le 5.23$.

The plane cruises at 34,000 ft for 3 min, so we graph $a = 34,000$ for $5.23 < t \le 8.23$. After 8.23 min the plane descends at a rate of 3500 ft/min and lands. The equation $a = 34,000 - 3500(t - 8.23)$, or $a = -3500t + 62,805$, describes this situation. Solving $0 = -3500t + 62,805$, we find that the plane lands after about 17.94 min. Thus we graph $a = -3500t + 62,805$ for $8.23 < t \le 17.94$. The entire graph is show below.

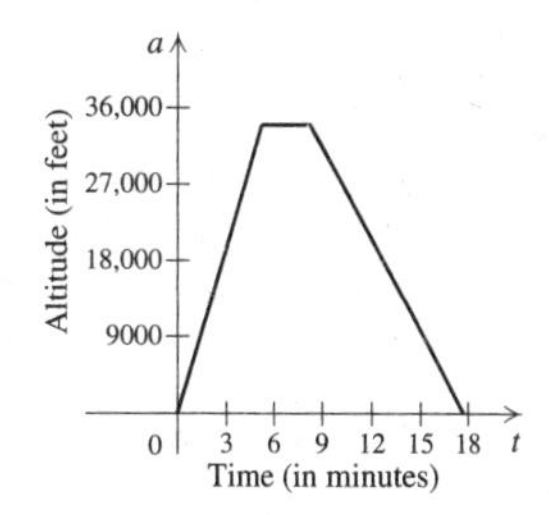

51. Let the horizontal axis represent the distance traveled, in miles, and let the vertical axis represent the fare, in dollars. Use increments of 1/5, or 0.2 mi, on the horizontal axis and of \$1 on the vertical axis. The fare for traveling 0.2 mi is \$2 + \$0.50 · 1, or \$2.50 and for 0.4 mi, or 0.2 mi × 2, we have \$2 + \$0.50(2), or \$3. Plot the points (0.2 mi, \$2.50) and (0.4 mi, \$3) and draw the line through them.

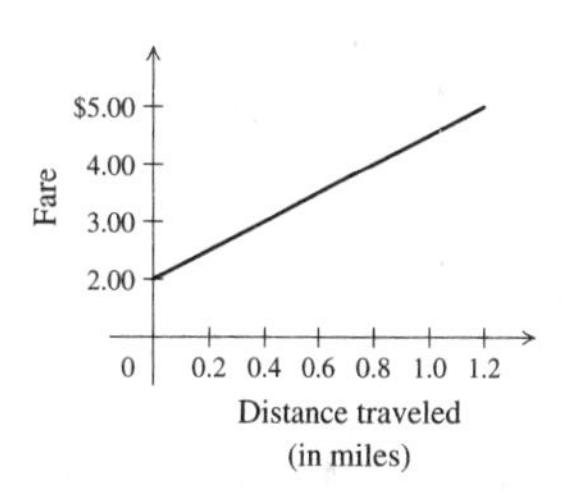

53. Penny walks forward at a rate of $\dfrac{24 \text{ ft}}{3 \text{ sec}}$, or 8 ft per sec. In addition, the boat is traveling at a rate of 5 ft per sec. Thus, with respect to land, Penny is traveling at a rate of $8 + 5$, or 13 ft per sec.

55. First we find Annette's speed in minutes per kilometer.

$$\text{Speed} = \frac{15.5 \text{ min}}{7 \text{ km} - 4 \text{ km}} = \frac{15.5 \text{ min}}{3 \text{ km}}$$

Now we convert min/km to min/mi.

$$\frac{15.5 \text{ min}}{3 \text{ km}} \approx \frac{15.5 \text{ min}}{3 \cancel{\text{km}}} \cdot \frac{1 \cancel{\text{km}}}{0.621 \text{ min}} \approx \frac{15.5 \text{ min}}{1.863 \text{ mi}}$$

At a rate of $\dfrac{15.5 \text{ min}}{1.863 \text{ mi}}$, to run a 5-mi race it would take $\dfrac{15.5 \text{ min}}{1.863 \cancel{\text{mi}}} \cdot 5 \cancel{\text{mi}} \approx 41.6$ min.

(Answers may vary slightly depending on the conversion factor used.)

57. In the 2 hours from 3 P.M. to 5 P.M., the number of candles made was 100 – 46, or 54. Then candles were being made at the rate of

$$\frac{54 \text{ candles}}{2 \text{ hr}}, \text{ or 27 candles per hour.}$$

Since 82 – 46 = 36, the length of time it took to make the 82nd candle after the 46th candle was made is

$$\frac{36 \text{ candles}}{27 \text{ candles per hour}} = \frac{4}{3} \text{ hr, or 1 hr 20 min.}$$

Thus the 82nd candle was made 1 hr 20 min after 3 P.M., or at 4:20 P.M.

Exercise Set 3.5

1. A teenager's height increases over time, so the rate is positive.

3. The water level decreases during a drought, so the rate is negative.

5. The number of people present increases as the opening tipoff approaches, so the rate is positive.

7. A person's IQ does not change during sleep, so the rate is zero.

9. The inventory decreases as the sale progresses, so the rate is negative.

11. The rate can be found using the coordinates of any two points on the line. We use (2001, 285 million) and (2003, 290 million).

$$\begin{aligned}\text{Rate} &= \frac{\text{change in population}}{\text{corresponding change in time}}\\ &= \frac{290 \text{ million} - 285 \text{ million}}{2003 - 2001}\\ &= \frac{5 \text{ million}}{2 \text{ yr}}\\ &= 2.5 \text{ million people per year}\end{aligned}$$

13. The rate can be found using the coordinates of any two points on the line. We use (1992, 61%) and (2002, 74%).

$$\begin{aligned}\text{Rate} &= \frac{\text{change in percent}}{\text{corresponding change in time}}\\ &= \frac{74\% - 61\%}{2002 - 1992}\\ &= \frac{13\%}{10 \text{ yr}}\\ &= 1.3\% \text{ per year}\end{aligned}$$

15. The rate can be found using the coordinates of any two points on the line. We use (35, 490) and (45, 500), where 35 and 45 are in $1000's.

$$\begin{aligned}\text{Rate} &= \frac{\text{change in score}}{\text{corresponding change in income}}\\ &= \frac{500 - 490 \text{ points}}{45 - 35}\\ &= \frac{10 \text{ points}}{10}\\ &= 1 \text{ point per \$1000 income}\end{aligned}$$

17. The rate can be found using the coordinates of any two points on the line. We use (0 min, 54°) and (27 min, –4°).

$$\begin{aligned}\text{Rate} &= \frac{\text{change in temperature}}{\text{corresponding change in time}}\\ &= \frac{-4^\circ - 54^\circ}{27 \text{ min} - 0 \text{ min}}\\ &= \frac{-58^\circ}{27 \text{ min}}\\ &\approx -2.1^\circ \text{per min}\end{aligned}$$

19. We can use any two points on the line, such as (0, 1) and (4, 4).

$$\begin{aligned}m &= \frac{\text{change in } y}{\text{change in } x}\\ &= \frac{4-1}{4-0} = \frac{3}{4}\end{aligned}$$

21. We can use any two points on the line, such as (1, 0) and (3, 3).

$$\begin{aligned}m &= \frac{\text{change in } y}{\text{change in } x}\\ &= \frac{3-0}{3-1} = \frac{3}{2}\end{aligned}$$

23. We can use any two points on the line, such as (–3, –4) and (0, –3).

$$\begin{aligned}m &= \frac{\text{change in } y}{\text{change in } x}\\ &= \frac{-3-(-4)}{0-(-3)} = \frac{1}{3}\end{aligned}$$

25. We can use any two points on the line, such as (0, 2) and (2, 0).

$$\begin{aligned}m &= \frac{\text{change in } y}{\text{change in } x}\\ &= \frac{2-0}{0-2} = \frac{2}{-2} = -1\end{aligned}$$

27. This is the graph of a horizontal line. Thus, the slope is 0.

29. We can use any two points on the line, such as (0, 2) and (3, 1).

$$\begin{aligned}m &= \frac{\text{change in } y}{\text{change in } x}\\ &= \frac{1-2}{3-0} = -\frac{1}{3}\end{aligned}$$

31. This is the graph of a vertical line. Thus, the slope is undefined.

33. We can use any two points on the line, such as (–2, 3) and (2, 2).

$$\begin{aligned}m &= \frac{\text{change in } y}{\text{change in } x}\\ &= \frac{2-3}{2-(-2)} = -\frac{1}{4}\end{aligned}$$

35. We can use any two points on the line, such as $(-2,-3)$ and $(2,3)$.

$$m = \frac{\text{change in } y}{\text{change in } x} = \frac{3-(-3)}{2-(-2)} = \frac{6}{4} = \frac{3}{2}$$

37. This is the graph of a horizontal line, so the slope is 0.

39. We can use any two points on the line, such as $(-3,5)$ and $(0,-4)$.

$$m = \frac{\text{change in } y}{\text{change in } x} = \frac{-4-5}{0-(-3)} = \frac{-9}{3} = -3$$

41. $(1,2)$ and $(5,8)$

$$m = \frac{8-2}{5-1} = \frac{6}{4} = \frac{3}{2}$$

43. $(-2,4)$ and $(3,0)$

$$m = \frac{4-0}{-2-3} = \frac{4}{-5} = -\frac{4}{5}$$

45. $(-4,0)$ and $(5,7)$

$$m = \frac{7-0}{5-(-4)} = \frac{7}{9}$$

47. $(0,8)$ and $(-3,10)$

$$m = \frac{8-10}{0-(-3)} = \frac{8-10}{0+3} = \frac{-2}{3} = -\frac{2}{3}$$

49. $(-2,3)$ and $(-6,5)$

$$m = \frac{5-3}{-6-(-2)} = \frac{2}{-6+2} = \frac{2}{-4} = -\frac{1}{2}$$

51. $\left(-2,\frac{1}{2}\right)$ and $\left(-5,\frac{1}{2}\right)$

Observe that the points have the same y-coordinate. Thus, they lie on a horizontal line and its slope is 0. We could also compute the slope.

$$m = \frac{\frac{1}{2}-\frac{1}{2}}{-2-(-5)} = \frac{\frac{1}{2}-\frac{1}{2}}{-2+5} = \frac{0}{3} = 0$$

53. $(3,4)$ and $(9,-7)$

$$m = \frac{-7-4}{9-3} = \frac{-11}{6} = -\frac{11}{6}$$

55. $(6,-4)$ and $(6,5)$

Observe that the points have the same x-coordinate. Thus, they lie on a vertical line and its slope is undefined. We could also compute the slope.

$$m = \frac{-4-5}{6-6} = \frac{-9}{0}, \text{ undefined}$$

57. The line $x = -3$ is a vertical line. The slope is undefined.

59. The line $y = 4$ is a horizontal line. A horizontal line has slope 0.

61. The line $x = 9$ is a vertical line. The slope is undefined.

63. The line $y = -9$ is a horizontal line. A horizontal line has slope 0.

65. The grade is expressed as a percent.

$$m = \frac{106}{1325} = 0.08 = 8\%$$

67. The grade is expressed as a percent.

$$m = \frac{1}{12} = 0.08\overline{3} = 8.\overline{3}\%$$

69. 2 ft 5 in. $= 2\cdot 12$ in. $+ 5$ in. $= 24$ in. $+ 5$ in. $= 29$ in.

8 ft 2 in. $= 8\cdot 12$ in. $+ 2$ in. $= 96$ in. $+ 2$ in. $= 98$ in.

$$m = \frac{29}{98}, \text{ or about } 30\%$$

71. Longs Peak rises $14,255 - 9600 = 4655$ ft.

$$m = \frac{4655}{15,840} \approx 0.29 \approx 29\%$$

73. *Writing Exercise*

75.

$$\begin{aligned} ax + by &= c \\ by &= c - ax && \text{Adding } -ax \text{ to both sides} \\ y &= \frac{c-ax}{b} && \text{Dividing both sides by } b \end{aligned}$$

77.

$$\begin{aligned} ax - by &= c \\ -by &= c - ax && \text{Adding } -ax \text{ to both sides} \\ y &= \frac{c-ax}{-b} && \text{Dividing both sides by } -b \end{aligned}$$

We could also express this result as $y = \dfrac{ax-c}{b}$.

79.

$$\begin{aligned} \frac{2}{3}x - 5 &= \frac{2}{3}\cdot 12 - 5 && \text{Substituting} \\ &= 8 - 5 \\ &= 3 \end{aligned}$$

81. *Writing Exercise*

83. From the dimensions on the drawing, we see that the ramps labeled A have a rise of 61 cm and a run of 167.6 cm.

$$m = \frac{61 \text{ cm}}{167.6 \text{ cm}} \approx 0.364, \text{ or } 36.4\%$$

85. If the line passes through $(2,5)$ and never enters the second quadrant, then it slants up from left to right or is vertical. This means that its slope is positive. The line slants least steeply if it passes through $(0,0)$. In this case, $m = \dfrac{5-0}{2-0} = \dfrac{5}{2}$. Thus, the numbers the line could have for it slope are $\left\{m \middle| m \ge \dfrac{5}{2}\right\}$.

87. Let t = the number of units each tick mark on the vertical axis represents. Note that the graph drops 4 units for every 3 units of horizontal change. Then we have:

$$\begin{aligned} \frac{-4t}{3} &= -\frac{2}{3} \\ -4t &= -2 && \text{Multiplying by 3} \\ t &= \frac{1}{2} && \text{Dividing by } -4 \end{aligned}$$

Each tick mark on the vertical axis represents $\frac{1}{2}$ unit.

Exercise Set 3.6

1. $y = 2x - 3 = 2x + (-3)$

   The y-intercept is $(0, -3)$, so choice (c) is correct.

3. We can read the slope, 3, directly from the equation. Choice (f) is correct.

5. We can read the slope, $\frac{2}{3}$, directly from the equation. Choice (d) is correct.

7. Slope $\frac{2}{5}$; y-intercept $(0, 1)$

   We plot $(0, 1)$ and from there move up 2 units and right 5 units. This locates the point $(5, 3)$. We plot $(5, 3)$ and draw a line passing through $(0, 1)$ and $(5, 3)$.

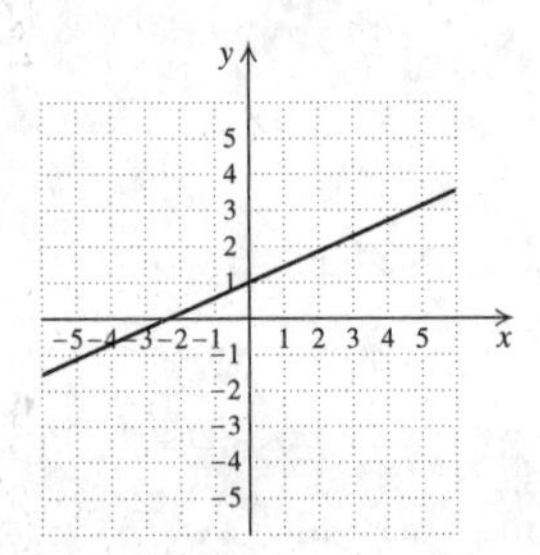

9. Slope $\frac{5}{3}$; y-intercept $(0, -2)$

   We plot $(0, -2)$ and from there move up 5 units and right 3 units. This locates the point $(3, 3)$. We plot $(3, 3)$ and draw a line passing through $(0, -2)$ and $(3, 3)$.

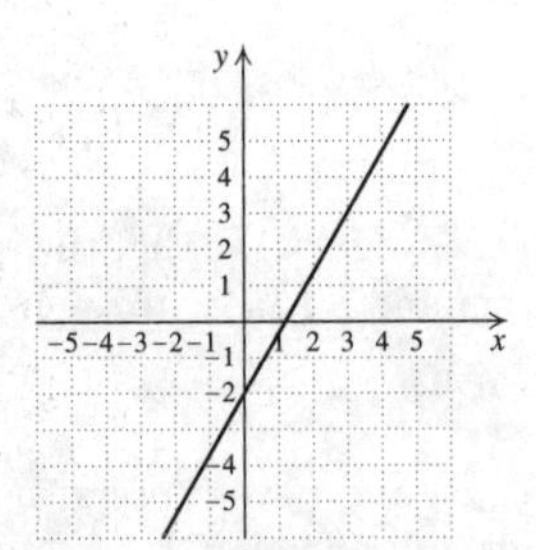

11. Slope $-\frac{3}{4}$; y-intercept $(0, 5)$

    We plot $(0, 5)$. We can think of the slope as $\frac{-3}{4}$, so from $(0, 5)$ we move down 3 units and right 4 units. This locates the point $(4, 2)$. We plot $(4, 2)$ and draw a line passing through $(0, 5)$ and $(4, 2)$.

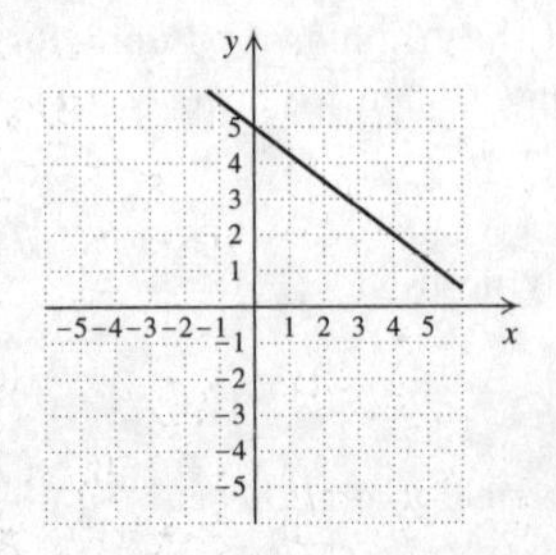

13. Slope 2; y-intercept $(0, -4)$

    We plot $(0, -4)$. We can think of the slope as $\frac{2}{1}$, so from $(0, -4)$ we move up 2 units and right 1 unit. This locates the point $(1, -2)$. We plot $(1, -2)$ and draw a line passing through $(0, -4)$ and $(1, -2)$.

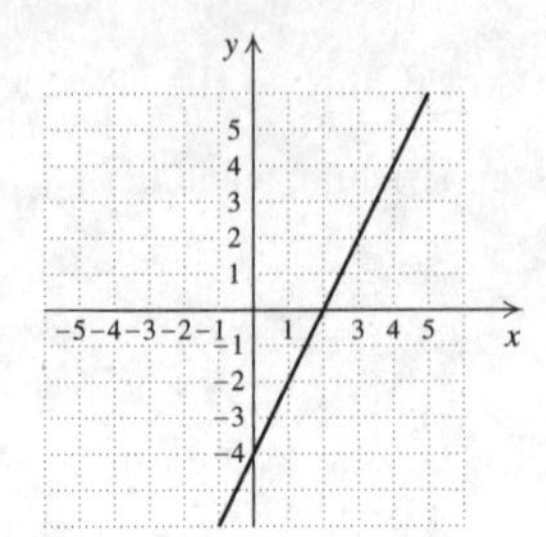

15. Slope -3; y-intercept $(0, 2)$

    We plot $(0, 2)$. We can think of the slope as $\frac{-3}{1}$, so from $(0, 2)$ we move down 3 units and right 1 unit. This locates the point $(1, -1)$. We plot $(1, -1)$ and draw a line passing through $(0, 2)$ and $(1, -1)$.

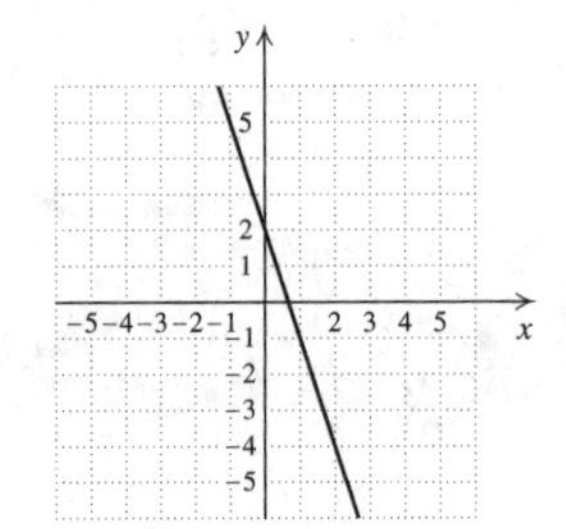

17. We read the slope and y-intercept from the equation.

    $$y = -\frac{2}{7}x + 5$$

    The slope is $-\frac{2}{7}$. The y-intercept is $(0, 5)$.

19. We read the slope and y-intercept from the equation.

    $$y = -\frac{5}{8}x + 3$$

    The slope is $-\frac{5}{8}$. The y-intercept is $(0, 3)$.

21. $y = \frac{9}{5}x - 4$

    $y = \frac{9}{5}x + (-4)$

    The slope is $\frac{9}{5}$, and the y-intercept is $(0, -4)$.

23. We solve for y to rewrite the equation in the form $y = mx + b$.

    $-3x + y = 7$

    $y = 3x + 7$

    The slope is 3, and the y-intercept is $(0, 7)$.

25. $5x + 2y = 8$

$$2y = -5x + 8$$
$$y = \frac{1}{2}(-5x + 8)$$
$$y = -\frac{5}{2}x + 4$$

The slope is $-\frac{5}{2}$, and the y-intercept is $(0, 4)$.

27. Observe that this is the equation of a horizontal line that lies 4 units above the x-axis. Thus, the slope is 0, and the y-intercept is $(0, 4)$. We could also write the equation in slope-intercept form.

$$y = 4$$
$$y = 0x + 4$$

The slope is 0, and the y-intercept is $(0, 4)$.

29. $2x - 5y = -8$

$$-5y = -2x - 8$$
$$y = -\frac{1}{5}(-2x - 8)$$
$$y = \frac{2}{5}x + \frac{8}{5}$$

The slope in $\frac{2}{5}$, and the y-intercept is $\left(0, \frac{8}{5}\right)$.

31. We use the slope-intercept equation, substituting 3 for m and 7 for b:

$$y = mx + b$$
$$y = 3x + 7$$

33. We use the slope-intercept equation, substituting $\frac{7}{8}$ for m and -1 for b:

$$y = mx + b$$
$$y = \frac{7}{8}x - 1$$

35. We use the slope-intercept equation, substituting $-\frac{5}{3}$ for m and -8 for b:

$$y = mx + b$$
$$y = -\frac{5}{3}x - 8$$

37. Since the slope is 0, we know that the line is horizontal. Its y-intercept is $(0, 3)$, so the equation of the line must be $y = 3$.

We could also use the slope-intercept equation, substituting 0 for m and 3 for b.

$$y = mx + b$$
$$y = 0 \cdot x + 3$$
$$y = 3$$

39. From the graph we see that the y-intercept is $(0, 9)$. We also see that the point $(9, 17)$ is on the graph. We find the slope:

$$m = \frac{17 - 9}{9 - 0} = \frac{8}{9}$$

Substituting $\frac{8}{9}$ for m and 9 for b in the slope-intercept equation $y = mx + b$, we have

$$y = \frac{8}{9}x + 9,$$

where y is the number of gallons of bottled water consumed per person and x is the number of years since 1990.

41. From the graph we see that the y-intercept is $(0, 250)$. We also see that the point $(10, 400)$ is on the graph. We find the slope:

$$m = \frac{400 - 250}{10 - 0} = \frac{150}{10} = 15$$

Substituting 15 for m and 250 for b in the slope-intercept equation $y = mx + b$, we have

$$y = 15x + 250,$$

where y is the number of jobs for medical assistants, in thousands, and x is the number of years since 1998.

43. $y = \frac{3}{5}x + 2$

First we plot the y-intercept $(0, 2)$. We can start at the y-intercept and use the slope, $\frac{3}{5}$, to find another point. We move up 3 units and right 5 units to get a new point $(5, 5)$. Thinking of the slope as $\frac{-3}{-5}$ we can start at $(0, 2)$ and move down 3 units and left 5 units to get another point $(-5, -1)$.

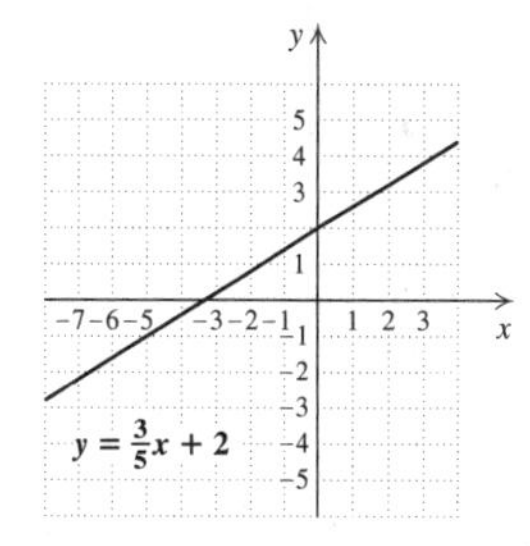

45. $y = -\frac{3}{5}x + 1$

First we plot the y-intercept $(0, 1)$. We can start at the y-intercept and, thinking of the slope as $\frac{-3}{5}$, find another point by moving down 3 units and right 5 units to the point $(5, -2)$. Thinking of the slope as $\frac{3}{-5}$ we can start at $(0, 1)$ and move up 3 units and left 5 units to get another point $(-5, 4)$.

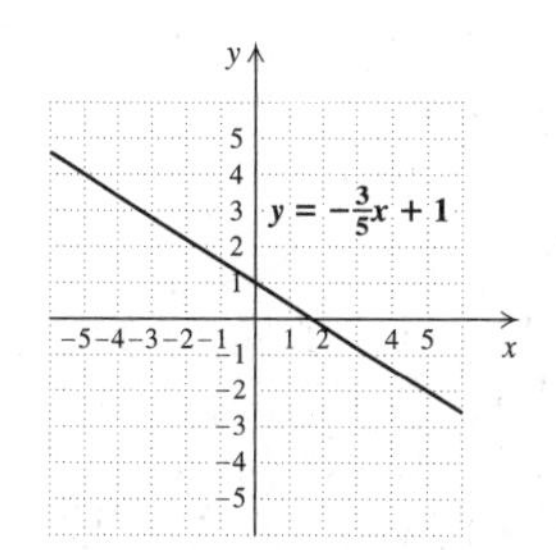

47. $y = \frac{5}{3}x + 3$

First we plot the y-intercept $(0, 3)$. We can start at the y-intercept and use the slope, $\frac{5}{3}$, to find another point. We move up 5 units and right 3 units to get a new point $(3, 8)$. Thinking of the slope as $\frac{-5}{-3}$ we can start at $(0, 3)$ and move down 5 units and left 3 units to get another point $(-3, -2)$.

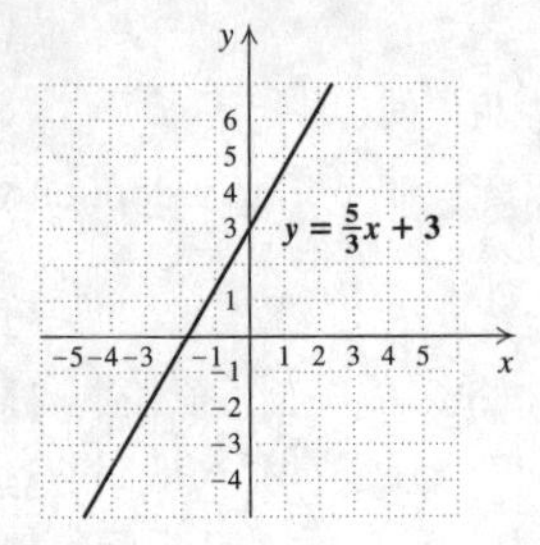

49. $y = -\frac{3}{2}x - 2$

First we plot the y-intercept $(0, -2)$. We can start at the y-intercept and, thinking of the slope as $\frac{-3}{2}$, find another point by moving down 3 units and right 2 units to the point $(2, -5)$. Thinking of the slope as $\frac{3}{-2}$ we can start at $(0, -2)$ and move up 3 units and left 2 units to get another point $(-2, 1)$.

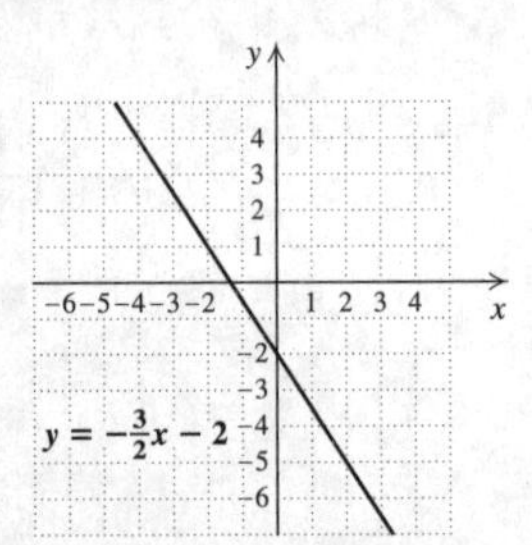

51. We first rewrite the equation in slope-intercept form.

$$2x + y = 1$$
$$y = -2x + 1$$

Now we plot the y-intercept $(0, 1)$. We can start at the y-intercept and, thinking of the slope as $\frac{-2}{1}$, find another point by moving down 2 units and right 1 unit to the point $(1, -1)$. In a similar manner, we can move from the point $(1, -1)$ to find a third point $(2, -3)$.

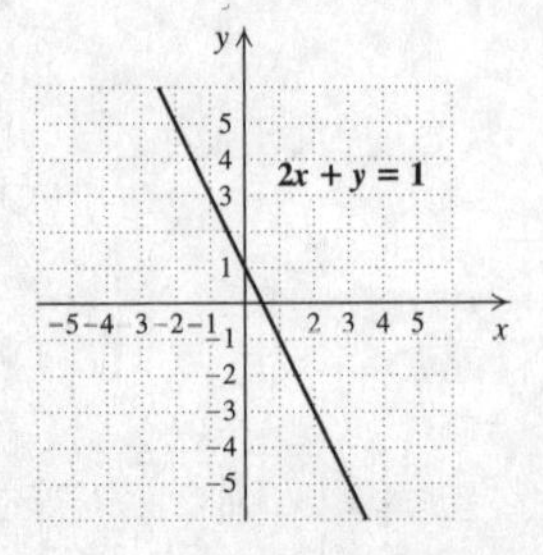

53. We first rewrite the equation in slope-intercept form.

$$3x + y = 0$$
$$y = -3x, \text{ or } y = -3x + 0$$

Now we plot the y-intercept $(0, 0)$. We can start at the y-intercept and, thinking of the slope as $\frac{-3}{1}$, find another point by moving down 3 units and right 1 unit to the point $(1, -3)$. Thinking of the slope as $\frac{3}{-1}$ we can start at $(0, 0)$ and move up 3 units and left 1 unit to get another point $(-1, 3)$.

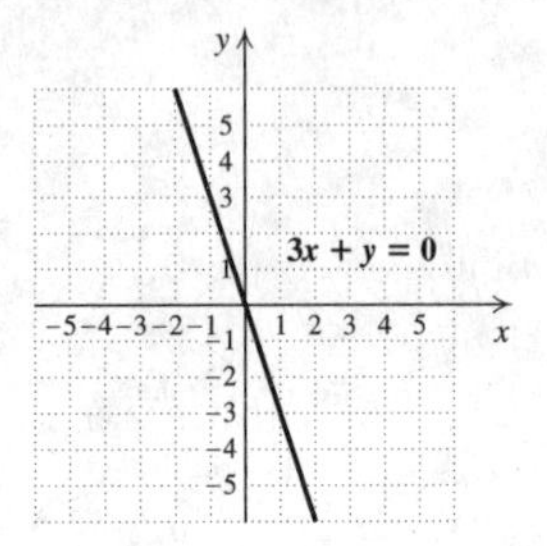

55. We first rewrite the equation in slope-intercept form.

$$2x + 3y = 9$$
$$3y = -2x + 9$$
$$y = \frac{1}{3}(-2x + 9)$$
$$y = -\frac{2}{3}x + 3$$

Now we plot the y-intercept $(0, 3)$. We can start at the y-intercept and, thinking of the slope as $\frac{-2}{3}$, find another point by moving down 2 units and right 3 units to the point $(3, 1)$. Thinking of the slope as $\frac{2}{-3}$ we can start at $(0, 3)$ and move up 2 units and left 3 units to get another point $(-3, 5)$.

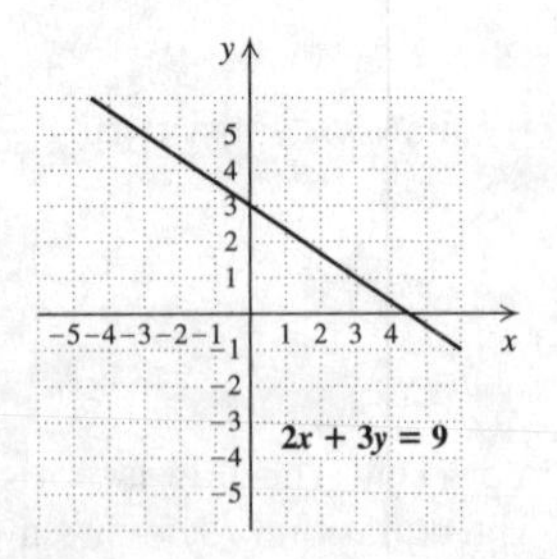

57. We first rewrite the equation in slope-intercept form.

$$x - 4y = 12$$
$$-4y = -x + 12$$
$$y = -\frac{1}{4}(-x + 12)$$
$$y = \frac{1}{4}x - 3$$

Now we plot the y-intercept $(0, -3)$. We can start at the y-intercept and use the slope, $\frac{1}{4}$, to find another point. We move up 1 unit and right 4 units to the point $(4, -2)$. Thinking of the slope as $\frac{-1}{-4}$ we can start at $(0, -3)$ and move down 1 unit and left 4 units to get another point $(-4, -4)$.

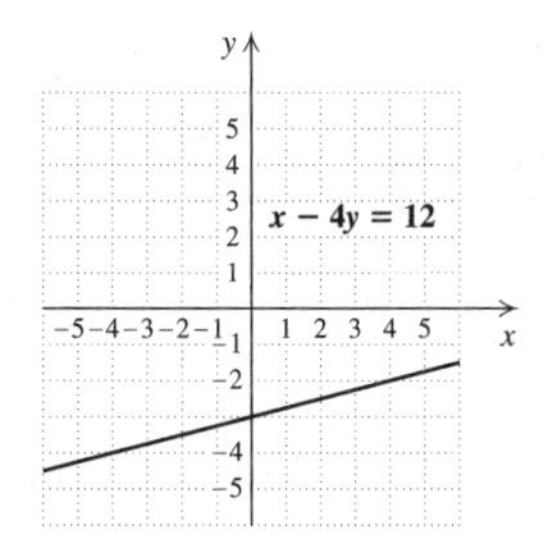

59. $y = \frac{2}{3}x + 7$: The slope is $\frac{2}{3}$, and the y-intercept is $(0, 7)$.

$y = \frac{2}{3}x - 5$: The slope is $\frac{2}{3}$, and the y-intercept is $(0, -5)$.

Since both lines have slope $\frac{2}{3}$ but different y-intercepts, their graphs are parallel.

61. The equation $y = 2x - 5$ represents a line with slope 2 and y-intercept $(0, -5)$. We rewrite the second equation in slope-intercept form.

$$\begin{aligned} 4x + 2y &= 9 \\ 2y &= -4x + 9 \\ y &= \frac{1}{2}(-4x + 9) \\ y &= -2x + \frac{9}{2} \end{aligned}$$

The slope is -2 and the y-intercept is $\left(0, \frac{9}{2}\right)$. Since the lines have different slopes, their graphs are not parallel.

63. Rewrite each equation in slope-intercept form.

$$\begin{aligned} 3x + 4y &= 8 \\ 4y &= -3x + 8 \\ y &= \frac{1}{4}(-3x + 8) \\ y &= -\frac{3}{4}x + 2 \end{aligned}$$

The slope is $-\frac{3}{4}$, and the y-intercept is $(0, 2)$.

$$\begin{aligned} 7 - 12y &= 9x \\ -12y &= 9x - 7 \\ y &= -\frac{1}{12}(9x - 7) \\ y &= -\frac{3}{4}x + \frac{7}{12} \end{aligned}$$

The slope is $-\frac{3}{4}$, and the y-intercept is $\left(0, \frac{7}{12}\right)$.

Since both lines have slope $-\frac{3}{4}$ but different y-intercepts, their graphs are parallel.

65. *Writing Exercise*

67. $y - k = m(x - h)$

$y = m(x - h) + k$ Adding k to both sides

69. $-5 - (-7) = -5 + 7 = 2$

71. $-3 - 6 = -3 + (-6) = -9$

73. *Writing Exercise*

75. See the answer section in the text.

77. Rewrite each equation in slope-intercept form.

$$\begin{aligned} 2x - 6y &= 10 \\ -6y &= -2x + 10 \\ y &= \frac{1}{3}x - \frac{5}{3} \end{aligned}$$

The slope of the line is $\frac{1}{3}$.

$$\begin{aligned} 9x + 6y &= 18 \\ 6y &= -9x + 18 \\ y &= -\frac{3}{2}x + 3 \end{aligned}$$

The y-intercept of the line is $(0, 3)$.

The equation of the line is $y = \frac{1}{3}x + 3$.

79. Rewrite each equation in slope-intercept form.

$$\begin{aligned} 4x + 5y &= 9 \\ 5y &= -4x + 9 \\ y &= \frac{1}{5}(-4x + 9) \\ y &= -\frac{4}{5}x + \frac{9}{5} \end{aligned}$$

The slope of the line is $-\frac{4}{5}$.

$$\begin{aligned} 2x + 3y &= 12 \\ 3y &= -2x + 12 \\ y &= \frac{1}{3}(-2x + 12) \\ y &= -\frac{2}{3}x + 4 \end{aligned}$$

The $y-$intercept of the line is $(0, 4)$.

The equation of the line is $y = -\frac{4}{5}x + 4$.

81. Rewrite each equation in slope-intercept form.

$$\begin{aligned} -4x + 8y &= 5 \\ 8y &= 4x + 5 \\ y &= \frac{1}{8}(4x + 5) \\ y &= \frac{1}{2}x + \frac{5}{8} \end{aligned}$$

The slope is $\frac{1}{2}$.

$$4x - 3y = 0$$
$$-3y = -4x$$
$$y = -\frac{1}{3}(-4x)$$
$$y = \frac{4}{3}x, \text{ or } y = \frac{4}{3}x + 0$$

The y–intercept is $(0, 0)$.

The equation of the line is $y = \frac{1}{2}x + 0$, or $y = \frac{1}{2}x$.

83. Rewrite each equation in slope-intercept form.

$$y + 3x = 10$$
$$y = -3x + 10$$

The slope is -3.

$$2x - 6y = 18$$
$$-6y = -2x + 18$$
$$y = -\frac{1}{6}(-2x + 18)$$
$$y = \frac{1}{3}x - 3$$

The slope is $\frac{1}{3}$.

Since $-3 \cdot \frac{1}{3} = -1$, the graphs of the equations are perpendicular.

85. Rewrite each equation in slope-intercept form.

$$10 - 4y = 7x$$
$$-4y = 7x - 10$$
$$y = -\frac{1}{4}(7x - 10)$$
$$y = -\frac{7}{4}x + \frac{5}{2}$$

The slope is $-\frac{7}{4}$.

$$7y + 21 = 4x$$
$$7y = 4x - 21$$
$$y = \frac{1}{7}(4x - 21)$$
$$y = \frac{4}{7}x - 3$$

The slope is $\frac{4}{7}$.

Since $-\frac{7}{4} \cdot \frac{4}{7} = -1$, the graphs of the equations are perpendicular.

87. The slope of $y = -2x$ is -2.

The slope of $x = \frac{1}{2}$ is undefined.

Since the product of the slopes is not -1, the graphs of the equations are not perpendicular.

89. Rewrite the first equation in slope-intercept form.

$$2x + 3y = 7$$
$$3y = -2x + 7$$
$$y = \frac{1}{3}(-2x + 7)$$
$$y = -\frac{2}{3}x + \frac{7}{3}$$

The slope of the line is $-\frac{2}{3}$.

The slope of a line perpendicular to this line is a number m such that

$$-\frac{2}{3}m = -1, \text{ or}$$
$$m = \frac{3}{2}.$$

Now rewrite the second equation in slope-intercept form.

$$5x + 2y = 10$$
$$2y = -5x + 10$$
$$y = \frac{1}{2}(-5x + 10)$$
$$y = -\frac{5}{2}x + 5$$

The y-intercept of the line is $(0, 5)$.

The equation of the line is $y = \frac{3}{2}x + 5$.

91. Rewrite $2x + 5y = 6$ in slope-intercept form.

$$2x + 5y = 6$$
$$5y = -2x + 6$$
$$y = \frac{1}{5}(-2x + 6)$$
$$y = -\frac{2}{5}x + \frac{6}{5}$$

The slope is $-\frac{2}{5}$.

The slope of a line perpendicular to this line is a number m such that

$$-\frac{2}{5}m = -1, \text{ or}$$
$$m = \frac{5}{2}.$$

We graph the line whose equation we want to find. First we plot the given point $(2, 6)$. Now think of the slope as $\frac{-5}{-2}$. From the point $(2, 6)$ go down 5 units and left 2 units to the point $(0, 1)$. Plot this point and draw the graph.

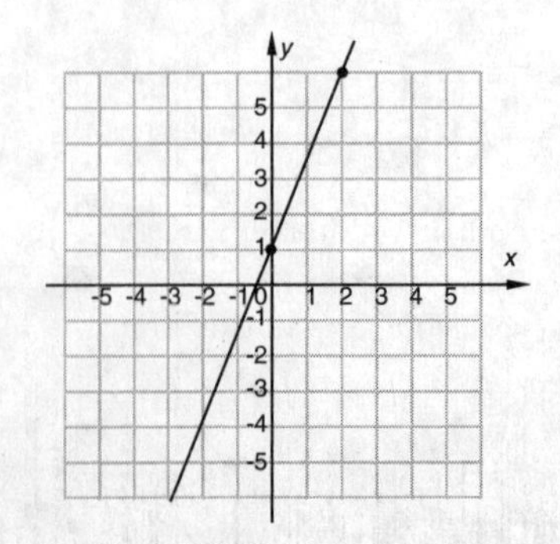

We see that the y-intercept is $(0, 1)$, so the desired equation is $y = \frac{5}{2}x + 1$.

Exercise Set 3.7

1. Substituting 5 for m, 2 for x_1, and 3 for y_1 in the point-slope equation $y - y_1 = m(x - x_1)$, we have $y - 3 = 5(x - 2)$. Choice (g) is correct.

3. Substituting -5 for m, -2 for x_1, and -3 for y_1 in the point-slope equation $y - y_1 = m(x - x_1)$, we have $y - (-3) = -5(x - (-2))$, or $y + 3 = -5(x + 2)$. Choice (e) is correct.

5. Substituting 5 for m, 3 for x_1, and 2 for y_1 in the point-slope equation $y - y_1 = m(x - x_1)$, we have $y - 2 = 5(x - 3)$. Choice (b) is correct.

7. Substituting -5 for m, -3 for x_1, and -2 for y_1 in the point-slope equation $y - y_1 = m(x - x_1)$, we have $y - (-2) = -5(x - (-3))$, or $y + 2 = -5(x + 3)$. Choice (f) is correct.

9. We see that the points $(1, -4)$ and $(-3, 2)$ are on the line. To go from $(1, -4)$ to $(-3, 2)$ we go up 6 units and left 4 units so the slope of the line is $\frac{6}{-4}$, or $-\frac{3}{2}$. Then, substituting $-\frac{3}{2}$ for m, 1 for x_1, and -4 for y_1 in the point-slope equation $y - y_1 = m(x - x_1)$, we have $y - (-4) = -\frac{3}{2}(x - 1)$, or $y + 4 = -\frac{3}{2}(x - 1)$. Choice (c) is correct.

11. We see that the points $(1, -4)$ and $(5, 2)$ are on the line. To go from $(1, -4)$ to $(5, 2)$ we go up 6 units and right 4 units so the slope of the line is $\frac{6}{4}$, or $\frac{3}{2}$. Then, substituting $\frac{3}{2}$ for m, 1 for x_1, and -4 for y_1 in the point-slope equation $y - y_1 = m(x - x_1)$, we have $y - (-4) = \frac{3}{2}(x - 1)$, or $y + 4 = \frac{3}{2}(x - 1)$. Choice (d) is correct.

13. $y - y_1 = m(x - x_1)$

We substitute 5 for m, 6 for x_1, and 2 for y_1.

$y - 2 = 5(x - 6)$

15. $y - y_1 = m(x - x_1)$

We substitute $\frac{3}{5}$ for m, 7 for x_1, and 3 for y_1.

$y - 3 = \frac{3}{5}(x - 7)$

17. $y - y_1 = m(x - x_1)$

We substitute -4 for m, 3 for x_1, and 1 for y_1.

$y - 1 = -4(x - 3)$

19. $y - y_1 = m(x - x_1)$

We substitute $\frac{3}{2}$ for m, 5 for x_1, and -4 for y_1.

$y - (-4) = \frac{3}{2}(x - 5)$

21. $y - y_1 = m(x - x_1)$

We substitute $\frac{5}{4}$ for m, -2 for x_1, and 6 for y_1.

$y - 6 = \frac{5}{4}(x - (-2))$

23. $y - y_1 = m(x - x_1)$

We substitute -2 for m, -4 for x_1, and -1 for y_1.

$y - (-1) = -2(x - (-4))$

25. $y - y_1 = m(x - x_1)$

We substitute 1 for m, -2 for x_1, and 8 for y_1.

$y - 8 = 1(x - (-2))$

27. First we write the equation in point-slope form.

$$y - y_1 = m(x - x_1)$$
$$y - 7 = 2(x - 5) \quad \text{Substituting}$$

Next we find an equivalent equation of the form $y = mx + b$.

$$y - 7 = 2(x - 5)$$
$$y - 7 = 2x - 10$$
$$y = 2x - 3$$

29. First we write the equation in point-slope form.

$$y - y_1 = m(x - x_1)$$
$$y - (-2) = \frac{7}{4}(x - 4) \quad \text{Substituting}$$

Next we find an equivalent equation of the form $y = mx + b$.

$$y - (-2) = \frac{7}{4}(x - 4)$$
$$y + 2 = \frac{7}{4}x - 7$$
$$y = \frac{7}{4}x - 9$$

31. First we write the equation in point-slope form.

$$y - y_1 = m(x - x_1)$$
$$y - 6 = -3(x - (-1))$$

Next we find an equivalent equation of the form $y = mx + b$.

$$y - 6 = -3(x - (-1))$$
$$y - 6 = -3(x + 1)$$
$$y - 6 = -3x - 3$$
$$y = -3x + 3$$

33. First we write the equation in point-slope form.

$$y - y_1 = m(x - x_1)$$
$$y - (-1) = -4(x - (-2))$$

Next we find an equivalent equation of the form $y = mx + b$.

$$y - (-1) = -4(x - (-2))$$
$$y + 1 = -4(x + 2)$$
$$y + 1 = -4x - 8$$
$$y = -4x - 9$$

35. First we write the equation in point-slope form.

$$y - y_1 = m(x - x_1)$$
$$y - 5 = \frac{2}{3}(x - 6)$$

Next we find an equivalent equation of the form $y = mx + b$.

$$y - 5 = \frac{2}{3}(x - 6)$$
$$y - 5 = \frac{2}{3}x - 4$$
$$y = \frac{2}{3}x + 1$$

37. The slope is $-\frac{5}{6}$ and the y-intercept is $(0, 4)$. Substituting $-\frac{5}{6}$ for m and 4 for b in the slope-intercept equation $y = mx + b$, we have $y = -\frac{5}{6}x + 4$.

39. We plot $(1, 2)$, move up 4 and to the right 3 to $(4, 6)$ and draw the line.

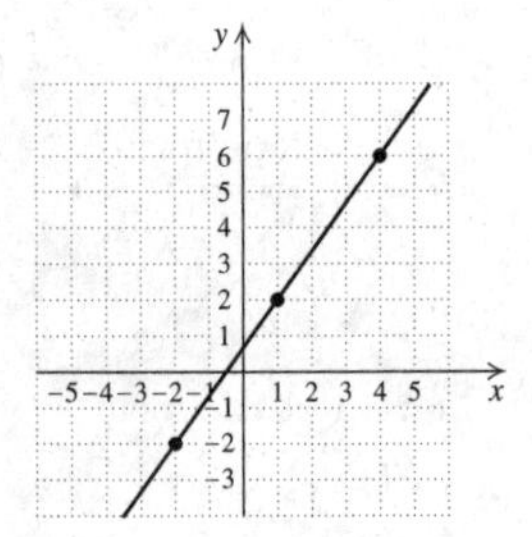

41. We plot $(2, 5)$, move down 3 and to the right 4 to $(6, 2)$ $\left(\text{since } -\frac{3}{4} = \frac{-3}{4}\right)$, and draw the line. We could also think of $-\frac{3}{4}$ and $\frac{3}{-4}$ and move up 3 and to the left 4 from the point $(2, 5)$ to $(-2, 8)$.

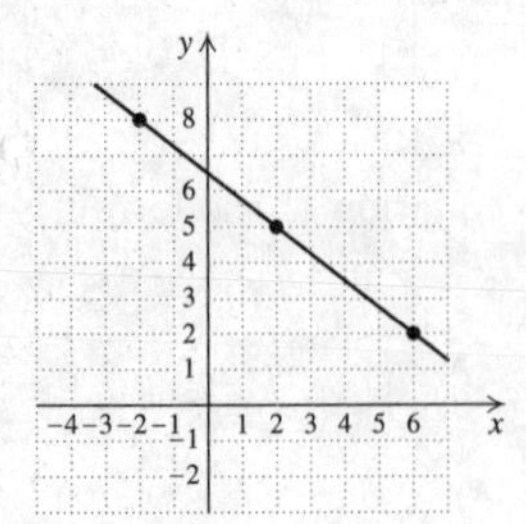

43. $y - 2 = \frac{1}{2}(x - 1)$ Point-slope form

The line has slope $\frac{1}{2}$ and passes through $(1, 2)$. We plot $(1, 2)$ and then find a second point by moving up 1 unit and right 2 units to $(3, 3)$. We draw the line through these points.

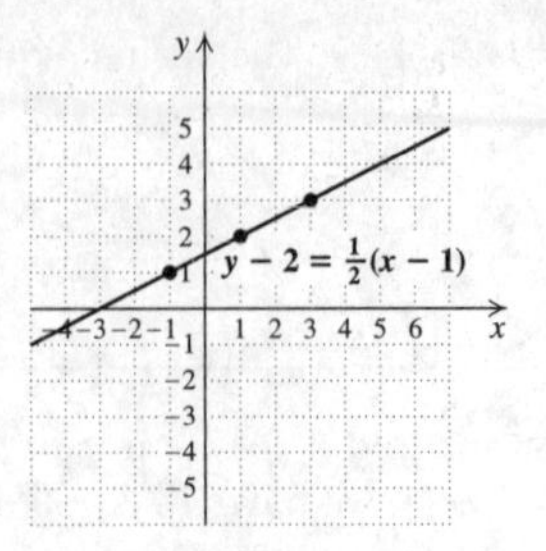

45. $y - 1 = -\frac{1}{2}(x - 3)$ Point-slope form

The line has slope $-\frac{1}{2}$, or $\frac{1}{-2}$ passes through $(3, 1)$. We plot $(3, 1)$ and then find a second point by moving up 1 unit and left 2 units to $(1, 2)$. We draw the line through these points.

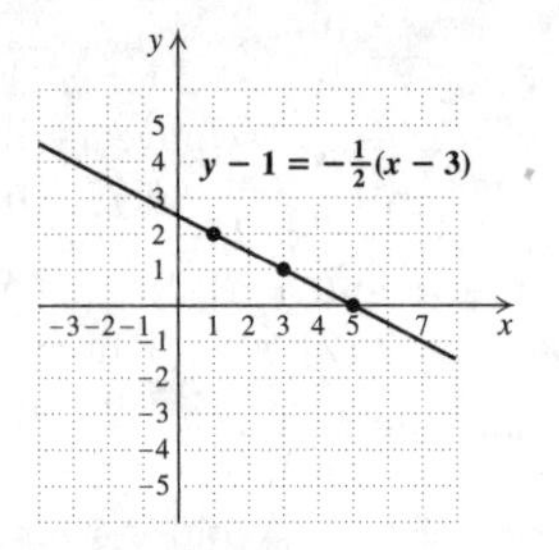

47. $y + 2 = \frac{1}{2}(x - 3)$, or $y - (-2) = \frac{1}{2}(x - 3)$

The line has slope $\frac{1}{2}$ and passes through $(3, -2)$. We plot $(3, -2)$ and then find a second point by moving up 1 unit and right 2 units to $(5, -1)$. We draw the line through these points.

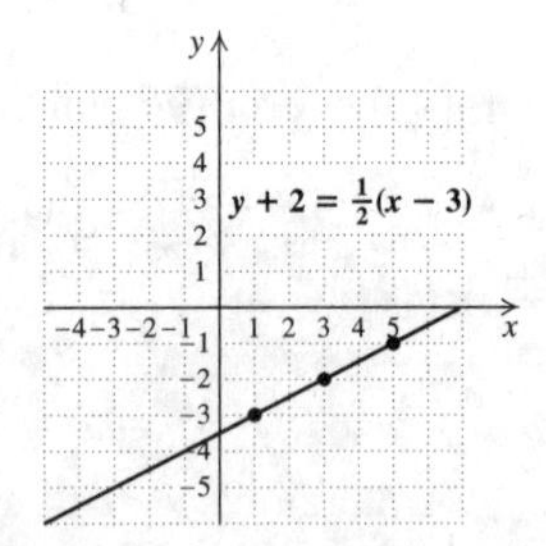

49. $y + 4 = 3(x + 1)$, or $y - (-4) = 3(x - (-1))$

The line has slope 3, or $\frac{3}{1}$, and passes through $(-1, -4)$. We plot $(-1, -4)$ and then find a second point by moving up 3 units and right 1 unit to $(0, -1)$. We draw the line through these points.

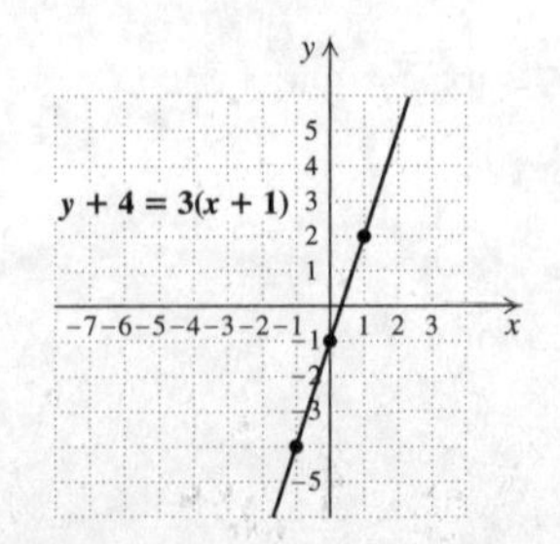

51. $y - 4 = -2(x + 1)$, or $y - 4 = -2(x - (-1))$

The line has slope -2, or $\frac{-2}{1}$, and passes through $(-1, 4)$. We plot $(-1, 4)$ and then find a second point by moving down 2 units and right 1 unit to $(0, 2)$. We draw the line through these points.

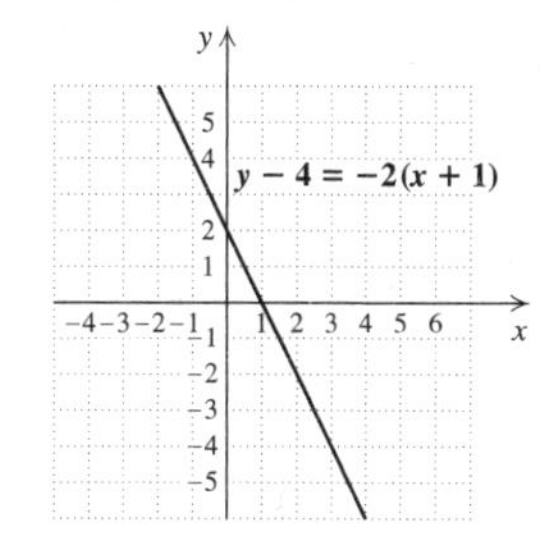

53. $y + 3 = -(x + 2)$, or $y - (-3) = -1(x - (-2))$

The line has slope -1, or $\frac{-1}{1}$, and passes through $(-2, -3)$. We plot $(-2, -3)$ and then find a second point by moving down 1 unit and right 1 unit to $(-1, -4)$. We draw the line through these points.

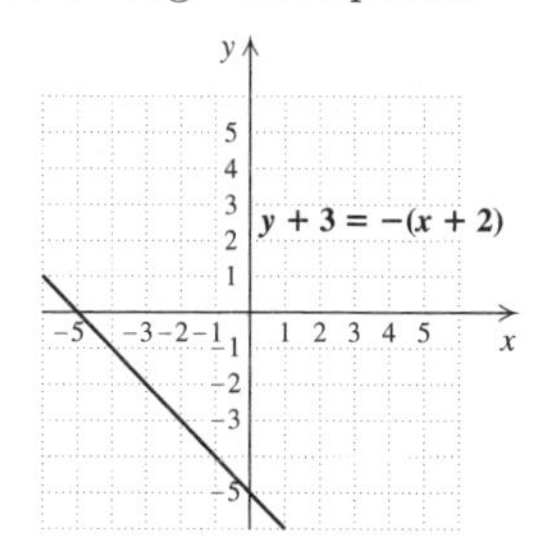

55. $y + 1 = -\frac{3}{5}(x + 2)$, or $y - (-1) = -\frac{3}{5}(x - (-2))$

The line has slope $-\frac{3}{5}$, or $\frac{-3}{5}$ and passes through $(-2, -1)$. We plot $(-2, -1)$ and then find a second point by moving down 3 units and right 5 units to $(3, -4)$, and draw the line.

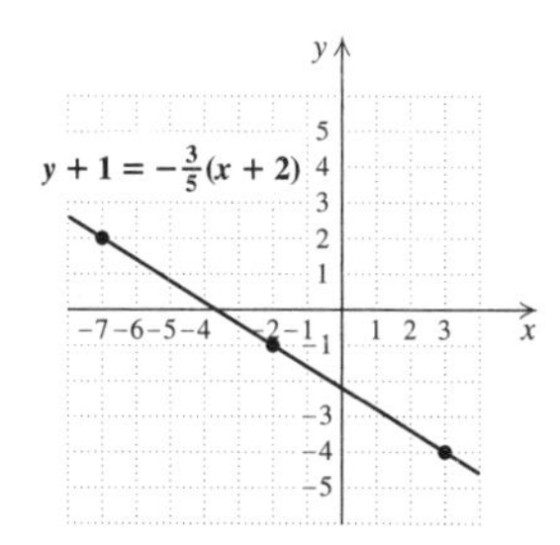

57. a) First find the slope of the line passing through the points $(1, 62.1)$ and $(11, 45.9)$.

$$m = \frac{45.9 - 62.1}{11 - 1} = \frac{-16.2}{10} = -1.62$$

Now write an equation of the line. We use $(1, 62.1)$ in the point-slope equation and then write an equivalent slope-intercept equation.

$$y - y_1 = m(x - x_1)$$
$$y - 62.1 = -1.62(x - 1)$$
$$y - 62.1 = -1.62x + 1.62$$
$$y = -1.62x + 63.72$$

Since 1999 is 9 yr after 1990, we substitute 9 for x to calculate the birth rate in 1999.

$$y = -1.62(9) + 63.72 = -14.58 + 63.72 = 49.14$$

In 1999, there were 49.14 births per 1000 females age 15 to 19.

b) 2008 is 18 yr after 1990 $(2008 - 1990 = 18)$, so we substitute 18 for x.

$$y = -1.62(18) + 63.72 = -29.16 + 63.72 = 34.56$$

We predict that the birth rate among teenagers will be 34.56 births per 1000 females in 2008.

59. a) First find the slope of the line passing through the points $(0, 45.7)$ and $(3, 32.5)$. In each case, we let the first coordinate represent the number of years after 1985.

$$m = \frac{32.5 - 45.7}{13 - 0} = \frac{-13.2}{13} \approx -1.02$$

The y-intercept of the line is $(0, 45.7)$. We write the slope-intercept equation: $y = -1.02x + 45.7$. Since 1990 is 5 yr after 1985, we substitute 5 for x to calculate the percentage in 1990.

$$y = -1.02(5) + 45.7 = -5.1 + 45.7 = 40.6\%$$

b) Since $2008 - 1985 = 23$, we substitute 23 for x to find the percentage in 2008.

$$y = -1.02(23) + 45.7 = -23.46 + 45.7 \approx 22.2\%$$

(Answers will vary depending on how the slope was rounded in part (a).)

61. a) First find the slope of the line passing through $(0, 60.3)$ and $(10, 68.3)$. In each case, we let the first coordinate represent the number of years after 1990. The second coordinate represents the enrollment in millions.

$$m = \frac{68.3 - 60.3}{10 - 0} = \frac{8}{10} = 0.8$$

The y-intercept is $(0, 60.3)$. We write the slope-intercept equation: $y = 0.8x + 60.3$.

Since 1996 is 6 yr after 1990, we substitute 6 for x to find the enrollment in 1996.

$y = 0.8(6) + 60.3 = 4.8 + 60.3 = 65.1$ million students

b) Since 2005 is 15 yr after 1990, we substitute 15 for x to find the enrollment in 2005.

$y = 0.8(15) + 60.3 = 12 + 60.3 = 72.3$ million students

63. a) First find the slope of the line through $(0, 31)$ and $(12, 35.6)$. In each case, we let the first coordinate represent the number of years after 1990 and the second millions of residents.

$$m = \frac{35.6 - 31}{12 - 0} = \frac{4.6}{12} \approx 0.38$$

The y-intercept is $(0, 31)$. We write the slope-intercept equation: $y = 0.38x + 31$.

Since 1996 is 6 yr after 1990, we substitute 6 for x to find the number of U.S. residents over the age of 65 in 1996.

$y = 0.38(6)+31 = 2.28+31 = 33.28 \approx 33.3$ million residents

b) Since 2010 is 20 yr after 1990, we substitute 20 for x to find the number of U.S. residents over the age of 65 in 2010.

$y = 0.38(20)+31 = 7.6+31 = 38.6$ million residents

(Answers will vary depending on how the slope is rounded.)

65. $(1,5)$ and $(4,2)$

First we find the slope.

$$m = \frac{5-2}{1-4} = \frac{3}{-3} = -1$$

Then we write an equation of the line in point-slope form using either of the points above.

$$y-5 = -1(x-1)$$

Finally, we find an equivalent equation in slope-intercept form.

$$y-5 = -1(x-1)$$
$$y-5 = -x+1$$
$$y = -x+6$$

67. $(-3,1)$ and $(3,5)$

First we find the slope.

$$m = \frac{1-5}{-3-3} = \frac{-4}{-6} = \frac{2}{3}$$

Then we write an equation of the line in point-slope form using either of the points above.

$$y-5 = \frac{2}{3}(x-3)$$

Finally, we find an equivalent equation in slope-intercept form.

$$y-5 = \frac{2}{3}(x-3)$$
$$y-5 = \frac{2}{3}x-2$$
$$y = \frac{2}{3}x+3$$

69. $(5,0)$ and $(0,-2)$

First we find the slope.

$$m = \frac{0-(-2)}{5-0} = \frac{2}{5}$$

Then we write an equation of the line in point-slope form using either of the points above.

$$y-0 = \frac{2}{5}(x-5)$$

Finally, we find an equivalent equation in slope-intercept form.

$$y-0 = \frac{2}{5}(x-5)$$
$$y = \frac{2}{5}x-2$$

71. $(-2,-4)$ and $(2,-1)$

First we find the slope.

$$m = \frac{-4-(-1)}{-2-2} = \frac{-4+1}{-2-2} = \frac{-3}{-4} = \frac{3}{4}$$

Then we write an equation of the line in point-slope form using either of the points above.

$$y-(-4) = \frac{3}{4}(x-(-2))$$

Finally, we find an equivalent equation in slope-intercept form.

$$y-(-4) = \frac{3}{4}(x-(-2))$$
$$y+4 = \frac{3}{4}(x+2)$$
$$y+4 = \frac{3}{4}x+\frac{3}{2}$$
$$y = \frac{3}{4}x-\frac{5}{2}$$

73. *Writing Exercise*

75. $(-5)^3 = (-5)(-5)(-5) = -125$

77. $3\cdot 2^4 - 5\cdot 2^3$

$= 3\cdot 16 - 5\cdot 8$ Evaluating the exponential expressions

$= 48-40$ Multiplying

$= 8$ Subtracting

79. $(-2)^3(-3)^2 = -8\cdot 9 = -72$

81. *Writing Exercise*

83. $y-3 = 0(x-52)$

Observe that the slope is 0. Then this is the equation of a horizontal line that passes through $(52,3)$. Thus, its graph is a horizontal line 3 units above the x-axis.

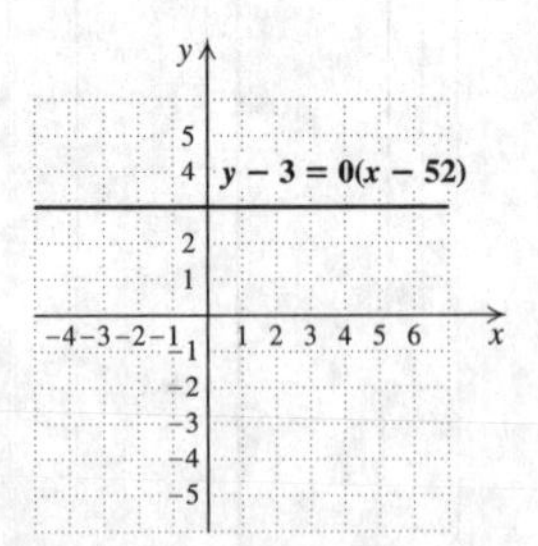

85. First we find the slope of the line using any two points on the line. We will use $(3,-3)$ and $(4,-1)$.

$$m = \frac{-3-(-1)}{3-4} = \frac{-2}{-1} = 2$$

Then we write an equation of the line in point-slope form using either of the points above.

$$y-(-3) = 2(x-3)$$

Finally, we find an equivalent equation in slope-intercept form.

$$y-(-3) = 2(x-3)$$
$$y+3 = 2x-6$$
$$y = 2x-9$$

87. First we find the slope of the line using any two points on the line. We will use $(2,5)$ and $(5,1)$.

$$m = \frac{5-1}{2-5} = \frac{4}{-3} = -\frac{4}{3}$$

Then we write an equation of the line in point-slope form using either of the points above.

$$y - 5 = -\frac{4}{3}(x-2)$$

Finally, we find an equivalent equation in slope-intercept form.

$$y - 5 = -\frac{4}{3}(x-2)$$
$$y - 5 = -\frac{4}{3}x + \frac{8}{3}$$
$$y = -\frac{4}{3}x + \frac{23}{3}$$

89. First find the slope of $2x + 3y = 11$.

$$2x + 3y = 11$$
$$3y = -2x + 11$$
$$y = -\frac{2}{3}x + \frac{11}{3}$$

The slope is $-\frac{2}{3}$.

Then write a point-slope equation of the line containing $(-4,7)$ and having slope $-\frac{2}{3}$.

$$y - 7 = -\frac{2}{3}(x-(-4))$$

91. The slope of $y = 3-4x$ is -4. We are given the y-intercept of the line, so we use slope-intercept form. The equation is $y = -4x + 7$.

93. First find the slope of the line passing through $(2,7)$ and $(-1,-3)$.

$$m = \frac{-3-7}{-1-2} = \frac{-10}{-3} = \frac{10}{3}$$

Now find an equation of the line containing the point $(-1,5)$ and having slope $\frac{10}{3}$.

$$y - 5 = \frac{10}{3}(x-(-1))$$
$$y - 5 = \frac{10}{3}(x+1)$$
$$y - 5 = \frac{10}{3}x + \frac{10}{3}$$
$$y = \frac{10}{3}x + \frac{25}{3}$$

95. *Writing Exercise*

Chapter 4

Polynomials

Exercise Set 4.1

1. By the power rule on page 230, choice (b) is correct.

3. By the rule for raising a product to a power on page 230, choice (e) is correct.

5. By the definition of 0 as an exponent on page 229, choice (g) is correct.

7. By the rule for raising a quotient to a power on page 231, choice (c) is correct.

9. $r^4 \cdot r^6 = r^{4+6} = r^{10}$

11. $9^5 \cdot 9^3 = 9^{5+3} = 9^8$

13. $a^6 \cdot a = a^6 \cdot a^1 = a^{6+1} = a^7$

15. $8^4 \cdot 8^7 = 8^{4+7} = 8^{11}$

17. $(3y)^4(3y)^8 = (3y)^{4+8} = (3y)^{12}$

19. $(5t)(5t)^6 = (5t)^1(5t)^6 = (5t)^{1+6} = (5t)^7$

21. $(a^2b^7)(a^3b^2) = a^2b^7a^3b^2$ Using an associative law
$= a^2a^3b^7b^2$ Using a commutative law
$= a^5b^9$ Adding exponents

23. $(x+1)^5(x+1)^7 = (x+1)^{5+7} = (x+1)^{12}$

25. $r^3 \cdot r^7 \cdot r^0 = r^{3+7+0} = r^{10}$

27. $(xy^4)(xy)^3 = (xy^4)(x^3y^3)$
$= x \cdot x^3 \cdot y^4 \cdot y^3$
$= x^{1+3}y^{4+3}$
$= x^4y^7$

29. $\dfrac{7^5}{7^2} = 7^{5-2} = 7^3$ Subtracting exponents

31. $\dfrac{x^{15}}{x^3} = x^{15-3} = x^{12}$ Subtracting exponents

33. $\dfrac{t^5}{t} = \dfrac{t^5}{t^1} = t^{5-1} = t^4$

35. $\dfrac{(5a)^7}{(5a)^6} = (5a)^{7-6} = (5a)^1 = 5a$

37. $\dfrac{(x+y)^8}{(x+y)^8}$

Observe that we have an expression divided by itself. Thus, the result is 1.

We could also do this exercise as follows:

$\dfrac{(x+y)^8}{(x+y)^8} = (x+y)^{8-8} = (x+y)^0 = 1$

39. $\dfrac{6m^5}{8m^2} = \dfrac{6}{8}m^{5-2} = \dfrac{3}{4}m^3$

41. $\dfrac{8a^9b^7}{2a^2b} = \dfrac{8}{2} \cdot \dfrac{a^9}{a^2} \cdot \dfrac{b^7}{b^1} = 4a^{9-2}b^{7-1} = 4a^7b^6$

43. $\dfrac{m^9n^8}{m^0n^4} = \dfrac{m^9}{m^0} \cdot \dfrac{n^8}{n^4} = m^{9-0}n^{8-4} = m^9n^4$

45. When $x = 13$, $x^0 = 13^0 = 1$. (Any nonzero number raised to the 0 power is 1.)

47. When $x = -4$, $5x^0 = 5(-4)^0 = 5 \cdot 1 = 5$.

49. $7^0 + 4^0 = 1 + 1 = 2$

51. $(-3)^1 - (-3)^0 = -3 - 1 = -4$

53. $(x^4)^7 = x^{4 \cdot 7} = x^{28}$ Multiplying exponents

55. $(5^8)^2 = 5^{8 \cdot 2} = 5^{16}$ Multiplying exponents

57. $(m^7)^5 = m^{7 \cdot 5} = m^{35}$

59. $(t^{20})^4 = t^{20 \cdot 4} = t^{80}$

61. $(7x)^2 = 7^2 \cdot x^2 = 49x^2$

63. $(-2a)^3 = (-2)^3a^3 = -8a^3$

65. $(4m^3)^2 = 4^2(m^3)^2 = 16m^6$

67. $(a^2b)^7 = (a^2)^7(b^7) = a^{14}b^7$

69. $(x^3y)^2(x^2y^5) = (x^3)^2y^2x^2y^5 = x^6y^2x^2y^5 = x^8y^7$

71. $(2x^5)^3(3x^4) = 2^3(x^5)^3(3x^4) = 8x^{15} \cdot 3x^4 = 24x^{19}$

73. $\left(\dfrac{a}{4}\right)^3 = \dfrac{a^3}{4^3} = \dfrac{a^3}{64}$ Raising the numerator and the denominator to the third power

75. $\left(\dfrac{7}{5a}\right)^2 = \dfrac{7^2}{(5a)^2} = \dfrac{49}{5^2a^2} = \dfrac{49}{25a^2}$

77. $\left(\dfrac{a^4}{b^3}\right)^5 = \dfrac{(a^4)^5}{(b^3)^5} = \dfrac{a^{20}}{b^{15}}$

79. $\left(\dfrac{y^3}{2}\right)^2 = \dfrac{(y^3)^2}{2^2} = \dfrac{y^6}{4}$

81. $\left(\dfrac{x^2y}{z^3}\right)^4 = \dfrac{(x^2y)^4}{(z^3)^4} = \dfrac{(x^2)^4(y^4)}{z^{12}} = \dfrac{x^8y^4}{z^{12}}$

83. $\left(\dfrac{a^3}{-2b^5}\right)^4 = \dfrac{(a^3)^4}{(-2b^5)^4} = \dfrac{a^{12}}{(-2)^4(b^5)^4} = \dfrac{a^{12}}{16b^{20}}$

85. $\left(\dfrac{5x^7y}{2z^4}\right)^3 = \dfrac{(5x^7y)^3}{(2z^4)^3} = \dfrac{5^3(x^7)^3y^3}{2^3(z^4)^3} =$
$\dfrac{125x^{21}y^3}{8z^{12}}$

87. $\left(\dfrac{4x^3y^5}{3z^7}\right)^0$

Observe that for $x \neq 0$, $y \neq 0$, and $z \neq 0$, we have a nonzero number raised to the 0 power. Thus, the result is 1.

89. *Writing Exercise*

91. $3s - 3r + 3t = 3 \cdot s - 3 \cdot r + 3 \cdot t = 3(s - r + t)$

93. $9x + 2y - x - 2y = 9x - x + 2y - 2y =$
$(9-1)x + (2-2)y = 8x + 0y = 8x$

95. $2y + 3x$

97. *Writing Exercise*

99. *Writing Exercise*

101. Choose any number except 0.

For example, let $a = 1$. Then $(a+5)^2 = (1+5)^2 = 6^2 = 36$, but $a^2 + 5^2 = 1^2 + 5^2 = 1 + 25 = 26$.

103. Choose any number except $\dfrac{7}{6}$. For example let $a = 0$. Then $\dfrac{0+7}{7} = \dfrac{7}{7} = 1$, but $a = 0$.

105. $a^{10k} \div a^{2k} = a^{10k-2k} = a^{8k}$

107. $$\frac{\left(\frac{1}{2}\right)^3\left(\frac{2}{3}\right)^4}{\left(\frac{5}{6}\right)^3} = \frac{\frac{1}{8}\cdot\frac{16}{81}}{\frac{125}{216}} = \frac{1}{8}\cdot\frac{16}{81}\cdot\frac{216}{125} =$$

$$\frac{1\cdot 2\cdot \cancel{8}\cdot \cancel{27}\cdot 8}{\cancel{8}\cdot 3\cdot \cancel{27}\cdot 125} = \frac{16}{375}$$

109.
$$\frac{t^{26}}{t^x} = t^x$$
$$t^{26-x} = t^x$$
$$26 - x = x \quad \text{Equating exponents}$$
$$26 = 2x$$
$$13 = x$$

The solution is 13.

111. Since the bases are the same, the expression with the larger exponent is larger. Thus, $4^2 < 4^3$.

113. $4^3 = 64$, $3^4 = 81$, so $4^3 < 3^4$.

115.
$$25^8 = (5^2)^8 = 5^{16}$$
$$125^5 = (5^3)^5 = 5^{15}$$
$$5^{16} > 5^{15}, \text{ or } 25^8 > 125^5.$$

117. $2^{22} = 2^{10} \cdot 2^{10} \cdot 2^2 \approx 10^3 \cdot 10^3 \cdot 4 \approx 1000 \cdot 1000 \cdot 4 \approx 4,000,000$

Using a calculator, we find that $2^{22} = 4,194,304$. The difference between the exact value and the approximation is $4,194,304 - 4,000,000$, or 194,304.

119. $2^{31} = 2^{10} \cdot 2^{10} \cdot 2^{10} \cdot 2 \approx 10^3 \cdot 10^3 \cdot 10^3 \cdot 2 \approx$
$1000 \cdot 1000 \cdot 1000 \cdot 2 = 2,000,000,000$

Using a calculator, we find that $2^{31} = 2,147,483,648$. The difference between the exact value and the approximation is $2,147,483,648 - 2,000,000,000 = 147,483,648$.

121.
$$\begin{aligned} 1.5 \text{ MB} &= 1.5 \times 1000 \text{ KB} \\ &= 1.5 \times 1000 \times 1 \times 2^{10} \text{ bytes} \\ &= 1,536,000 \text{ bytes} \\ &\approx 1,500,000 \text{ bytes} \end{aligned}$$

Exercise Set 4.2

1. The only expression with 4 terms is (b).

3. Expression (d) is the only expression for which the degree of the leading term is 5.

5. Expression (g) has two terms, and the degree of the leading term is 7.

7. Expression (c) has three terms, but it is not a trinomial because $\dfrac{3}{x}$ is not a monomial.

9. $7x^4 + x^3 - 5x + 8 = 7x^4 + x^3 + (-5x) + 8$

The terms are $7x^4$, x^3, $-5x$, and 8.

11. $-t^4 + 7t^3 - 3t^2 + 6 = -t^4 + 7t^3 + (-3t^2) + 6$

The terms are $-t^4$, $7t^3$, $-3t^2$, and 6.

13. $4x^5 + 7x$

Term	Coefficient	Degree
$4x^5$	4	5
$7x$	7	1

15. $9t^2 - 3t + 4$

Term	Coefficient	Degree
$9t^2$	9	2
$-3t$	-3	1
4	4	0

17. $7a^4 + 9a + a^3$

Term	Coefficient	Degree
$7a^4$	7	4
$9a$	9	1
a^3	1	3

19. $x^4 - x^3 + 4x - 3$

Term	Coefficient	Degree
x^4	1	4
$-x^3$	-1	3
$4x$	4	1
-3	-3	0

21. $5x - 9x^2 + 3x^6$

a)

Term	$5x$	$-9x^2$	$3x^6$
Degree	1	2	6

b) The term of highest degree is $3x^6$. This is the leading term. Then the leading coefficient is 3.

c) Since the term of highest degree is $3x^6$, the degree of the polynomial is 6.

23. $3a^2 - 7 + 2a^4$

a)

Term	$3a^2$	-7	$2a^4$
Degree	2	0	4

b) The term of highest degree is $2a^4$. This is the leading term. Then the leading coefficient is 2.

c) Since the term of highest degree is $2a^4$, the degree of the polynomial is 4.

25. $8 + 6x^2 - 3x - x^5$

a)

Term	8	$6x^2$	$-3x$	$-x^5$
Degree	0	2	1	5

b) The term of highest degree is $-x^5$. This is the leading term. Then the leading coefficient is -1 since $-x^5 = -1 \cdot x^5$.

c) Since the term of highest degree is $-x^5$, the degree of the polynomial is 5.

27. $7x^2 + 8x^5 - 4x^3 + 6 - \frac{1}{2}x^4$

Term	Coefficient	Degree of Term	Degree of Polynomial
$8x^5$	8	5	5
$-\frac{1}{2}x^4$	$-\frac{1}{2}$	4	
$-4x^3$	-4	3	
$7x^2$	7	2	
6	6	0	

29. Three monomials are added, so $x^2 - 23x + 17$ is a trinomial.

31. The polynomial $x^3 - 7x^2 + 2x - 4$ is none of these because it is composed of four monomials.

33. Two monomials are added, so $8t^2 + 5t$ is a binomial.

35. The polynomial 17 is a monomial because it is the product of a constant and a variable raised to a whole number power. (In this case the variable is raised to the power 0.)

37. $7x^2 + 3x + 4x^2 = (7+4)x^2 + 3x = 11x^2 + 3x$

39. $3a^4 - 2a + 2a + a^4 = (3+1)a^4 + (-2+2)a =$
$4a^4 + 0a = 4a^4$

41. $2x^2 - 6x + 3x + 4x^2 = (2+4)x^2 + (-6+3)x =$
$6x^2 - 3x$

43. $9x^3 + 2x - 4x^3 + 5 - 3x = (9-4)x^3 + (2-3)x + 5 = 5x^3 - x + 5$

45. $10x^2 + 2x^3 - 3x^3 - 4x^2 - 6x^2 - x^4 =$
$-x^4 + (2-3)x^3 + (10-4-6)x^2 = -x^4 - x^3$

47. $\frac{1}{5}x^4 + 7 - 2x^2 + 3 - \frac{2}{15}x^4 + 2x^2 =$
$\left(\frac{1}{5} - \frac{2}{15}\right)x^4 + (-2+2)x^2 + (7+3) =$
$\left(\frac{3}{15} - \frac{2}{15}\right)x^4 + 0x^2 + 10 = \frac{1}{15}x^4 + 10$

49. $5.9x^2 - 2.1x + 6 + 3.4x - 2.5x^2 - 0.5 =$
$(5.9 - 2.5)x^2 + (-2.1 + 3.4)x + (6 - 0.5) =$
$3.4x^2 + 1.3x + 5.5$

51. For $x = 3$: $-7x + 4 = -7 \cdot 3 + 4$
$= -21 + 4$
$= -17$

For $x = -3$: $-7x + 4 = -7(-3) + 4$
$= 21 + 4$
$= 25$

53. For $x = 3$: $2x^2 - 3x + 7 = 2 \cdot 3^2 - 3 \cdot 3 + 7$
$= 2 \cdot 9 - 3 \cdot 3 + 7$
$= 18 - 9 + 7$
$= 16$

For $x = -3$: $2x^2 - 3x + 7 = 2(-3)^2 - 3(-3) + 7$
$= 2 \cdot 9 - 3(-3) + 7$
$= 18 + 9 + 7$
$= 34$

55. For $x = 3$:
$-2x^3 - 3x^2 + 4x + 2 = -2 \cdot 3^3 - 3 \cdot 3^2 + 4 \cdot 3 + 2$
$= -2 \cdot 27 - 3 \cdot 9 + 4 \cdot 3 + 2$
$= -54 - 27 + 12 + 2$
$= -67$

For $x = -3$:
$-2x^3 - 3x^2 + 4x + 2 = -2(-3)^3 - 3(-3)^2 + 4(-3) + 2$
$= -2(-27) - 3 \cdot 9 + 4(-3) + 2$
$= 54 - 27 - 12 + 2$
$= 17$

57. For $x = 3$: $\frac{1}{3}x^4 - 2x^3 = \frac{1}{3} \cdot 3^4 - 2 \cdot 3^3$
$= \frac{1}{3} \cdot 81 - 2 \cdot 27$
$= 27 - 54$
$= -27$

For $x = -3$: $\frac{1}{3}x^4 - 2x^3 = \frac{1}{3}(-3)^4 - 2(-3)^3$
$= \frac{1}{3} \cdot 81 - 2(-27)$
$= 27 + 54$
$= 81$

59. For $x = 3$: $-x^4 - x^3 - x^2 = -3^4 - 3^3 - 3^2$
$= -81 - 27 - 9$
$= -117$

For $x = -3$: $-x^4 - x^3 - x^2 = -(-3)^4 - (-3)^3 - (-3)^2$
$= -81 - (-27) - 9$
$= -81 + 27 - 9$
$= -63$

61. Since 2005 is 5 years after 2000, we evaluate the polynomial for $t = 5$.

$0.15t + 1.42 = 0.15(5) + 1.42$
$= 0.75 + 1.42$
$= 2.17$

The amount of consumer debt in 2005 is about \$2.17 trillion.

63. $11.12t^2 = 11.12(10)^2 = 11.12(100) = 1112$

A skydiver has fallen approximately 1112 ft 10 seconds after jumping from a plane.

65. $2\pi r = 2(3.14)(10)$ Substituting 3.14 for π and 10 for r
$= 62.8$

The circumference is 62.8 cm.

67. $\pi r^2 = 3.14(7)^2$ Substituting 3.14 for π and 7 for r
$= 3.14(49)$
$= 153.86$

The area is 153.86 m^2.

69. Since 2006 is 6 years after 2000, we first locate 6 on the horizontal axis. From there we move vertically to the graph and then horizontally to the E-axis. This locates an E-value of about 5. Thus U.S. electricity consumption in 2006 is about 5 million gigawatt hours.

71. Since 2003 is 3 years after 2000, we first locate 3 on the horizontal axis. From there we move vertically to the graph and then horizontally to the E-axis. This locates an E-value of about 4.5. Thus U.S. electricity consumption in 2003 was about 4.5 million gigawatt hours.

73. Locate 10 on the horizontal axis. From there move vertically to the graph and then horizontally to the M-axis. This locates an M-value of about 9. Thus, about 9 words were memorized in 10 minutes.

75. Locate 8 on the horizontal axis. From there move vertically to the graph and then horizontally to the M-axis. This locates an M-value of about 6. Thus, the value of $-0.001t^3 + 0.1t^2$ for $t = 8$ is approximately 6.

77. We evaluate the polynomial for $x = 20$:

$N = -0.00006(20)^3 + 0.006(20)^2 - 0.1(20) + 1.9$
$= -0.00006(8000) + 0.006(400) - 0.1(20) + 1.9$
$= -0.48 + 2.4 - 2.0 + 1.9$
$= 1.82 \approx 1.8$

There are about 1.8 million hearing-impaired Americans of age 20.

We evaluate the polynomial for $x = 50$:

$N = -0.00006(50)^3 + 0.006(50)^2 - 0.1(50) + 1.9$
$= -0.00006(125,000) + 0.006(2500) - 0.1(50) + 1.9$
$= -7.5 + 15 - 5 + 1.9$
$= 4.4$

There are about 4.4 million hearing-impaired Americans of age 50.

79. *Writing Exercise*

81. $-19 + 24$ A negative and a positive number. We subtract the absolute values: $24 - 19 = 5$. The positive number has the larger absolute value so the answer is positive.

$-19 + 24 = 5$

83. $5x + 15 = 5 \cdot x + 5 \cdot 3 = 5(x + 3)$

85. ***Familiarize***. Let x = the cost per mile of gasoline in dollars. Then the total cost of the gasoline for the year was $14,800x$.

Translate.

Cost of insurance	+	cost of registration and oil	+	cost of gasoline	=	\$2011.
↓	↓	↓	↓	↓	↓	↓
972	+	114	+	$14,800x$	=	2011

Carry out. We solve the equation.

$972 + 114 + 14,800x = 2011$
$1086 + 14,800x = 2011$
$14,800x = 925$
$x = 0.0625$

Check. If gasoline cost \$0.0625 per mile, then the total cost of the gasoline was 14,800(\$0.0625), or \$925. Then the total auto expense was \$972 + \$114 + \$925, or \$2011. The answer checks.

State. Gasoline cost \$0.0625, or 6.25¢ per mile.

87. *Writing Exercise*

89. Answers may vary. Choose an ax^5-term where a is an even integer. Then choose three other terms with different degrees, each less than degree 5, and coefficients $a+2$, $a+4$, and $a+6$, respectively, when the polynomial is written in descending order. One such polynomial is $2x^5 + 4x^4 + 6x^3 + 8$.

91. Find the total revenue from the sale of 30 monitors:

$250x - 0.5x^2 = 250(30) - 0.5(30)^2$
$= 250(30) - 0.5(900)$
$= 7500 - 450$
$= \$7050$

Find the total cost of producing 30 monitors:

$$\begin{aligned}4000 + 0.6x^2 &= 4000 + 0.6(30)^2\\ &= 4000 + 0.6(900)\\ &= 4000 + 540\\ &= \$4540\end{aligned}$$

Subtract the cost from the revenue to find the profit: $7050 − $4540 = $2510

93. $(3x^2)^3 + 4x^2 \cdot 4x^4 - x^4(2x)^2 + [(2x)^2]^3 -$
$100x^2(x^2)^2$
$= 27x^6 + 4x^2 \cdot 4x^4 - x^4 \cdot 4x^2 + (2x)^6 - 100x^2 \cdot x^4$
$= 27x^6 + 16x^6 - 4x^6 + 64x^6 - 100x^6$
$= 3x^6$

95. First locate 1.5 on the vertical axis. Then move horizontally to the graph. We meet the curve at 3 places. At each place move down vertically to the horizontal axis and read the corresponding x-value. We see that the ages at which 1.5 million Americans are hearing impaired are 5, 13, and 80.

97. We first find q, the quiz average, and t, the test average.

$$q = \frac{60+85+72+91}{4} = \frac{308}{4} = 77$$

$$t = \frac{89+93+90}{3} = \frac{272}{3} \approx 90.7$$

Now we substitute in the polynomial.

$$\begin{aligned}A &= 0.3q + 0.4t + 0.2f + 0.1h\\ &= 0.3(77) + 0.4(90.7) + 0.2(84) + 0.1(88)\\ &= 23.1 + 36.28 + 16.8 + 8.8\\ &= 84.98\\ &\approx 85.0\end{aligned}$$

99. When $t = 3$, $-t^2 + 10t - 18 = -3^2 + 10 \cdot 3 - 18 =$
$-9 + 30 - 18 = 3.$
When $t = 4$, $-t^2 + 10t - 18 = -4^2 + 10 \cdot 4 - 18 =$
$-16 + 40 - 18 = 6.$
When $t = 5$, $-t^2 + 10t - 18 = -5^2 + 10 \cdot 5 - 18 =$
$-25 + 50 - 18 = 7.$
When $t = 6$, $-t^2 + 10t - 18 = -6^2 + 10 \cdot 6 - 18 =$
$-36 + 60 - 18 = 6.$
When $t = 7$, $-t^2 + 10t - 18 = -7^2 + 10 \cdot 7 - 18 =$
$-49 + 70 - 18 = 3.$

We complete the table. Then we plot the points and connect them with a smooth curve.

t	$-t^2 + 10t - 18$
3	3
4	6
5	7
6	6
7	3

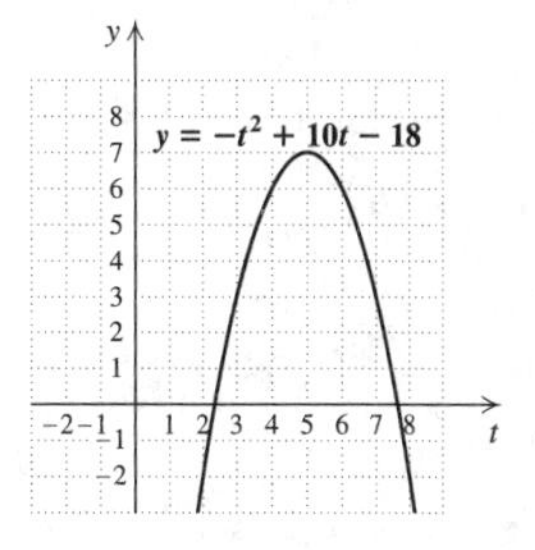

Exercise Set 4.3

1. $(3x + 2) + (-5x + 4) = (3 - 5)x + (2 + 4) = -2x + 6$

3. $(-6x + 2) + (x^2 + x - 3) =$
$x^2 + (-6 + 1)x + (2 - 3) = x^2 - 5x - 1$

5. $(7t^2 - 3t + 6) + (2t^2 + 8t - 9) =$
$(7 + 2)t^2 + (-3 + 8)t + (6 - 9) = 9t^2 + 5t - 3$

7. $(4m^3 - 4m^2 + m - 5) + (4m^3 + 7m^2 - 4m - 2) =$
$(7 + 4)m^3 + (-4 + 7)m^2 + (1 - 4)m + (-5 - 2) =$
$8m^3 + 3m^2 - 3m - 7$

9. $(3 + 6a + 7a^2 + 8a^3) + (4 + 7a - a^2 + 6a^3) =$
$(3 + 4) + (6 + 7)a + (7 - 1)a^2 + (8 + 6)a^3 =$
$7 + 13a + 6a^2 + 14a^3$

11. $(9x^8 - 7x^4 + 2x^2 + 5) + (8x^7 + 4x^4 - 2x) =$
$9x^8 + 8x^7 + (-7 + 4)x^4 + 2x^2 - 2x + 5 =$
$9x^8 + 8x^7 - 3x^4 + 2x^2 - 2x + 5$

13. $\left(\frac{1}{4}x^4 + \frac{2}{3}x^3 + \frac{5}{8}x^2 + 9\right) + \left(-\frac{3}{4}x^4 + \frac{3}{8}x^2 - 7\right) =$
$\left(\frac{1}{4} - \frac{3}{4}\right)x^4 + \frac{2}{3}x^3 + \left(\frac{5}{8} + \frac{3}{8}\right)x^2 + (9 - 7) =$
$-\frac{2}{4}x^4 + \frac{2}{3}x^3 + \frac{8}{8}x^2 + 2 =$
$-\frac{1}{2}x^4 + \frac{2}{3}x^3 + x^2 + 2$

15. $(5.3t^2 - 6.4t - 9.1) + (4.2t^3 - 1.8t^2 + 7.3) =$
$4.2t^3 + (5.3 - 1.8)t^2 - 6.4t + (-9.1 + 7.3) =$
$4.2t^3 + 3.5t^2 - 6.4t - 1.8$

17.

$$\begin{array}{r}-3x^4 + 6x^2 + 2x - 4\\ -3x^2 + 2x + 4\\ \hline -3x^4 + 3x^2 + 4x + 0\\ -3x^4 + 3x^2 + 4x\end{array}$$

19. Rewrite the problem so the coefficients of like terms have the same number of decimal places.

$$\begin{array}{rrrrr}0.15x^4 & +\ 0.10x^3 & -\ 0.90x^2 & & \\ & -\ 0.01x^3 & +\ 0.01x^2 & +\ x & \\ 1.25x^4 & & +\ 0.11x^2 & & +\ 0.01\\ & 0.27x^3 & & & +\ 0.99\\ -0.35x^4 & & +\ 15.00x^2 & & -\ 0.03\\ \hline 1.05x^4 & +\ 0.36x^3 & +\ 14.22x^2 & +\ x & +\ 0.97\end{array}$$

21. Two forms of the opposite of $-t^3 + 4t^2 - 9$ are

i) $-(-t^3 + 4t^2 - 9)$ and

ii) $t^3 - 4t^2 + 9$. (Changing the sign of every term)

23. Two forms for the opposite of $12x^4 - 3x^3 + 3$ are

i) $-(12x^4 - 3x^3 + 3)$ and

ii) $-12x^4 + 3x^3 - 3$. (Changing the sign of every term)

25. We change the sign of every term inside parentheses.

$-(8x - 9) = -8x + 9$

27. We change the sign of every term inside parentheses.

$-(3a^4 - 5a^2 + 9) = -3a^4 + 5a^2 - 9$

29. We change the sign of every term inside parentheses.

$-\left(-4x^4 + 6x^2 + \frac{3}{4}x - 8\right) = 4x^4 - 6x^2 - \frac{3}{4}x + 8$

31. $(7x + 4) - (-2x + 1)$

$= 7x + 4 + 2x - 1$ Changing the sign of every term inside parentheses

$= 9x + 3$

33. $(-5t+6) - (t^2+3t-1) = -5t+6-t^2-3t+1 = -t^2-8t+7$

35. $(6x^4 + 3x^3 - 1) - (4x^2 - 3x + 3)$

$= 6x^4 + 3x^3 - 1 - 4x^2 + 3x - 3$

$= 6x^4 + 3x^3 - 4x^2 + 3x - 4$

37. $(1.2x^3 + 4.5x^2 - 3.8x) - (-3.4x^3 - 4.7x^2 + 23)$

$= 1.2x^3 + 4.5x^2 - 3.8x + 3.4x^3 + 4.7x^2 - 23$

$= 4.6x^3 + 9.2x^2 - 3.8x - 23$

39. $(7x^3 - 2x^2 + 6) - (7x^3 - 2x^2 + 6)$

Observe that we are subtracting the polynomial $7x^3 - 2x^2 + 6$ from itself. The result is 0.

41. $(3 + 5a + 3a^2 - a^3) - (2 + 3a - 4a^2 + 2a^3) =$

$3 + 5a + 3a^2 - a^3 - 2 - 3a + 4a^2 - 2a^3 =$

$1 + 2a + 7a^2 - 3a^3$

43. $\frac{5}{8}x^3 - \frac{1}{4}x - \frac{1}{3} - \left(-\frac{1}{8}x^3 + \frac{1}{4}x - \frac{1}{3}\right)$

$= \frac{5}{8}x^3 - \frac{1}{4}x - \frac{1}{3} + \frac{1}{8}x^3 - \frac{1}{4}x + \frac{1}{3}$

$= \frac{6}{8}x^3 - \frac{2}{4}x$

$= \frac{3}{4}x^3 - \frac{1}{2}x$

45. $(0.07t^3 - 0.03t^2 + 0.01t) - (0.02t^3 + 0.04t^2 - 1) = 0.07t^3 - 0.03t^2 + 0.01t - 0.02t^3 - 0.04t^2 + 1 = 0.05t^3 - 0.07t^2 + 0.01t + 1$

47.

$$\begin{array}{r} x^2 + 5x + 6 \\ -(x^2 + 2x + 1) \\ \hline \end{array}$$

$$\begin{array}{rl} x^2 + 5x + 6 & \text{Changing signs and} \\ -x^2 - 2x - 1 & \text{removing parentheses} \\ \hline 3x + 5 & \text{Adding} \end{array}$$

49.

$$\begin{array}{r} 5x^4 + 6x^3 - 9x^2 \\ -(-6x^4 - 6x^3 + x^2) \\ \hline \end{array}$$

$$\begin{array}{rl} 5x^4 + 6x^3 - 9x^2 & \text{Changing signs and} \\ 6x^4 + 6x^3 - x^2 & \text{removing parentheses} \\ \hline 11x^4 + 12x^3 - 10x^2 & \text{Adding} \end{array}$$

51. a)

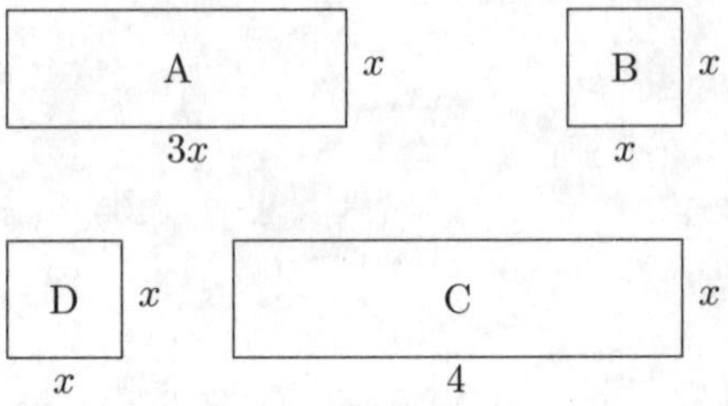

Familiarize. The area of a rectangle is the product of the length and the width.

Translate. The sum of the areas is found as follows:

Area of A	+	Area of B	+	Area of C	+	Area of D
$= 3x \cdot x$	+	$x \cdot x$	+	$4 \cdot x$	+	$x \cdot x$

Carry out. We collect like terms.

$3x^2 + x^2 + 4x + x^2 = 5x^2 + 4x$

Check. We can go over our calculations. We can also assign some value to x, say 2, and carry out the computation of the area in two ways.

Sum of areas: $3 \cdot 2 \cdot 2 + 2 \cdot 2 + 4 \cdot 2 + 2 \cdot 2 =$
$12 + 4 + 8 + 4 = 28$

Substituting in the polynomial:
$5(2)^2 + 4 \cdot 2 = 20 + 8 = 28$

Since the results are the same, our solution is probably correct.

State. A polynomial for the sum of the areas is $5x^2 + 4x$.

b) For $x = 5$: $5x^2 + 4x = 5 \cdot 5^2 + 4 \cdot 5 =$
$5 \cdot 25 + 4 \cdot 5 = 125 + 20 = 145$

When $x = 5$, the sum of the areas is 145 square units.

For $x = 7$: $5x^2 + 4x = 5 \cdot 7^2 + 4 \cdot 7 =$
$5 \cdot 49 + 4 \cdot 7 = 245 + 28 = 273$

When $x = 7$, the sum of the areas is 273 square units.

53. ***Familiarize.*** The perimeter is the sum of the lengths of the sides.

Translate. The sum of the lengths is found as follows:

$4a + 7 + a + \frac{1}{2}a + 5 + a + 2a + 3a$

Carry out. We combine like terms.

$4a + 7 + a + \frac{1}{2}a + 5 + a + 2a + 3a$

$= \left(4 + 1 + \frac{1}{2} + 1 + 2 + 3\right)a + (7 + 5)$

$= 11\frac{1}{2}a + 12$, or $\frac{23}{2}a + 12$

Check. We can go over our calculations. We can also perform a partial check by assigning some value to a, say 2, and carry out the computation of the perimeter in two ways.

Sum of lengths of sides:

$4 \cdot 2 + 7 + 2 + \frac{1}{2} \cdot 2 + 5 + 2 + 2 \cdot 2 + 3 \cdot 2 =$

$8 + 7 + 2 + 1 + 5 + 2 + 4 + 6 = 35$

Substituting in the polynomial we found:

$$\frac{23}{2} \cdot 2 + 12 = 23 + 12 = 35$$

Since the results are the same, our solution is probably correct.

State. A polynomial for the perimeter of the figure is $\frac{23}{2}a + 12$.

55.

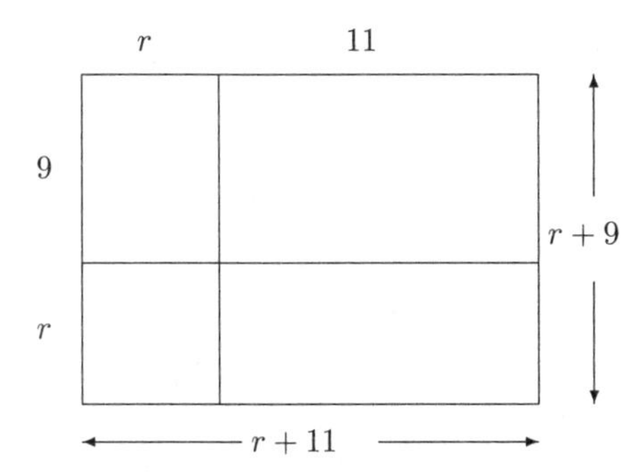

The length and width of the figure can be expressed as $r + 11$ and $r + 9$, respectively. The area of this figure (a rectangle) is the product of the length and width. An algebraic expression for the area is $(r + 11) \cdot (r + 9)$.

The algebraic expressions $9r + 99 + r^2 + 11r$ and $(r + 11) \cdot (r + 9)$ represent the same area.

	r	11	
9	A	B	9
r	C	D	r
	r	11	

The area of the figure can be found by adding the areas of the four rectangles A, B, C, and D. The area of a rectangle is the product of the length and the width.

	Area of A	+	Area of B	+	Area of C	+	Area of D
=	$9 \cdot r$	+	$11 \cdot 9$	+	$r \cdot r$	+	$11 \cdot r$
=	$9r$	+	99	+	r^2	+	$11r$

An algebraic expression for the area of the figure is $9r + 99 + r^2 + 11r$.

57.

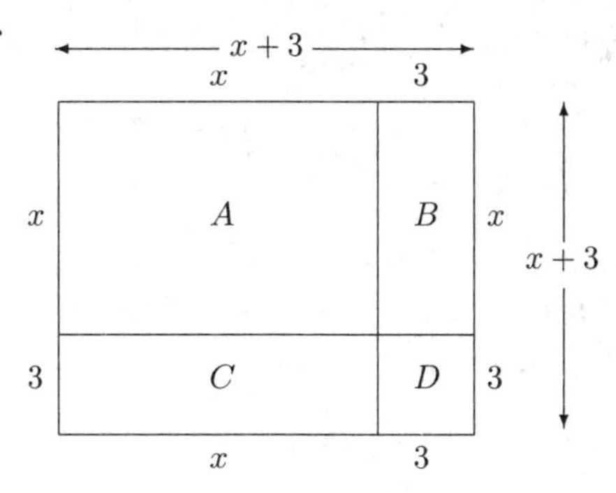

The length and width of the figure can each be expressed as $x + 3$. The area can be expressed as $(x + 3) \cdot (x + 3)$, or $(x + 3)^2$. Another way to express the area is to find an expression for the sum of the areas of the four rectangles A, B, C, and D. The area of each rectangle is the product of its length and width.

	Area of A	+	Area of B	+	Area of C	+	Area of D
=	$x \cdot x$	+	$3 \cdot x$	+	$3 \cdot x$	+	$3 \cdot 3$
=	x^2	+	$3x$	+	$3x$	+	9

The algebraic expressions $(x + 3)^2$ and $x^2 + 3x + 3x + 9$ represent the same area.

$$(x + 3)^2 = x^2 + 3x + 3x + 9$$

59.

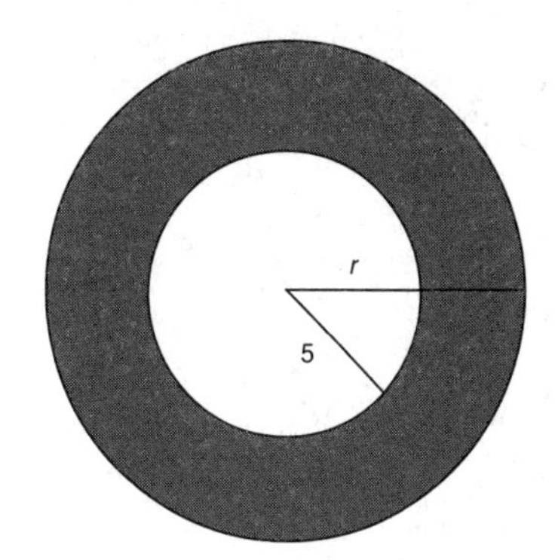

Familiarize. Recall that the area of a circle is the product of π and the square of the radius, r^2.

$$A = \pi r^2$$

Translate.

Area of circle with radius r	$-$	Area of circle with radius 5	=	Shaded area
$\pi \cdot r^2$	$-$	$\pi \cdot 5^2$	=	Shaded area

Carry out. We simplify the expression.

$$\pi \cdot r^2 - \pi \cdot 5^2 = \pi r^2 - 25\pi$$

Check. We can go over our calculations. We can also assign some value to r, say 7, and carry out the computation in two ways.

Difference of areas: $\pi \cdot 7^2 - \pi \cdot 5^2 = 49\pi - 25\pi = 24\pi$

Substituting in the polynomial: $\pi \cdot 7^2 - 25\pi = 49\pi - 25\pi = 24\pi$

Since the results are the same, our solution is probably correct.

State. A polynomial for the shaded area is $\pi r^2 - 25\pi$.

61. ***Familiarize.*** Recall that the area of a rectangle is the product of the length and the width. The shaded area is the area of the entire rectangle with length x and width y less the nonshaded area of a rectangle with length 7 and width 3.

Translate.

Area of entire rectangle	$-$	Area not shaded	is	Shaded area
$x \cdot y$	$-$	$7 \cdot 3$	$=$	Shaded area

Carry out. We simplify the expression.

$x \cdot y - 7 \cdot 3 = xy - 21$

Check. We go over the calculation. The answer checks.

State. A polynomial for the shaded area is $xy - 21$.

63. ***Familiarize.*** Recall that the area of a rectangle is the product of the length and the width and that, consequently, the area of a square with side s is s^2. The remaining floor area is the area of the entire floor less the area of the bath enclosure, in square feet.

Translate.

Area of entire floor	$-$	Area of bath enclosure	$=$	Remaining floor area
x^2	$-$	$2 \cdot 6$	$=$	Remaining floor area

Carry out. We simplify the expression.

$x^2 - 2 \cdot 6 = x^2 - 12$

Check. We go over the calculations. The answer checks.

State. A polynomial for the remaining floor area is $(x^2 - 12)$ ft^2.

65. ***Familiarize.*** Recall that the area of a circle with radius r is πr^2 and the area of a square with side s is s^2. The radius of the hot tub is half the width, $\frac{8 \text{ ft}}{2}$, or 4 ft. The remaining area of the patio is the entire area less the area of the hot tub, in square feet.

Translate.

Area of entire patio	$-$	Area of hot tub	is	Remaining patio area
z^2	$-$	$\pi \cdot 4^2$	$=$	Remaining patio area

Carry out. We simplify the expression.

$z^2 - \pi \cdot 4^2 = z^2 - \pi \cdot 16 = z^2 - 16\pi$

Check. We go over the calculations. The answer checks.

State. A polynomial for the remaining area of the patio is $(z^2 - 16\pi)$ ft^2.

67. ***Familiarize.*** Recall that the area of a square with side s is s^2 and the area of a circle with radius r is πr^2. The radius of the circle is half the diameter, or $\frac{d}{2}$ m. The area of the mat outside the circle is the area of the entire mat less the area of the circle, in square meters.

Translate.

Area of mat	$-$	Area of circle	is	Area outside the circle
12^2	$-$	$\pi \cdot \left(\frac{d}{2}\right)^2$	$=$	Area outside the circle

Carry out. We simplify the expression.

$$12^2 - \pi \cdot \left(\frac{d}{2}\right)^2 = 144 - \pi \cdot \frac{d^2}{4} = 144 - \frac{d^2}{4}\pi$$

Check. We go over the calculations. The answer checks.

State. A polynomial for the area of the mat outside the wrestling circle is $\left(144 - \frac{d^2}{4}\pi\right)$ m^2.

69. *Writing Exercise*

71.

$5(4+3) - 5 \cdot 4 - 5 \cdot 3$	
$= 5 \cdot 7 - 5 \cdot 4 - 5 \cdot 3$	Adding inside the parentheses
$= 35 - 20 - 15$	Multiplying
$= 0$	Subtracting

73. $2(5t+7) + 3t = 10t + 14 + 3t = 13t + 14$

75.

$2(x+3) > 5(x-3) + 7$	
$2x + 6 > 5x - 15 + 7$	Removing parentheses
$2x + 6 > 5x - 8$	Collecting like terms
$2x + 14 > 5x$	Adding 8 to both sides
$14 > 3x$	Adding $-2x$ to both sides
$\frac{14}{3} > x$	Dividing both sides by 3

The solution set is $\left\{x \middle| \frac{14}{3} > x\right\}$, or $\left\{x \middle| x < \frac{14}{3}\right\}$.

77. *Writing Exercise*

79.
$$(6t^2 - 7t) + (3t^2 - 4t + 5) - (9t - 6)$$
$$= 6t^2 - 7t + 3t^2 - 4t + 5 - 9t + 6$$
$$= 9t^2 - 20t + 11$$

81.
$$(-8y^2 - 4) - (3y + 6) - (2y^2 - y)$$
$$= -8y^2 - 4 - 3y - 6 - 2y^2 + y$$
$$= -10y^2 - 2y - 10$$

83.
$$(345.099x^3 - 6.178x) - (94.508x^3 - 8.99x)$$
$$= 345.099x^3 - 6.178x - 94.508x^3 + 8.99x$$
$$= 250.591x^3 + 2.812x$$

85. ***Familiarize.*** The surface area is $2lw + 2lh + 2wh$, where l = length, w = width, and h = height of the rectangular solid. Here we have $l = 3$, $w = w$, and $h = 7$.

Translate. We substitute in the formula above.

$2 \cdot 3 \cdot w + 2 \cdot 3 \cdot 7 + 2 \cdot w \cdot 7$

Carry out. We simplify the expression.

$$2 \cdot 3 \cdot w + 2 \cdot 3 \cdot 7 + 2 \cdot w \cdot 7$$
$$= 6w + 42 + 14w$$
$$= 20w + 42$$

Check. We can go over the calculations. We can also assign some value to w, say 6, and carry out the computation in two ways.

Using the formula: $2 \cdot 3 \cdot 6 + 2 \cdot 3 \cdot 7 + 2 \cdot 6 \cdot 7 =$
$36 + 42 + 84 = 162$

Substituting in the polynomial: $20 \cdot 6 + 42 =$
$120 + 42 = 162$

Since the results are the same, our solution is probably correct.

State. A polynomial for the surface area is $20w + 42$.

87. ***Familiarize***. The surface area is $2lw + 2lh + 2wh$, where $l =$ length, $w =$ width, and $h =$ height of the rectangular solid. Here we have $l = x$, $w = x$, and $h = 5$.

Translate. We substitute in the formula above.

$$2 \cdot x \cdot x + 2 \cdot x \cdot 5 + 2 \cdot x \cdot 5$$

Carry out. We simplify the expression.

$$\begin{aligned} &2 \cdot x \cdot x + 2 \cdot x \cdot 5 + 2 \cdot x \cdot 5 \\ &= 2x^2 + 10x + 10x \\ &= 2x^2 + 20x \end{aligned}$$

Check. We can go over the calculations. We can also assign some value to x, say 3, and carry out the computation in two ways.

Using the formula: $2 \cdot 3 \cdot 3 + 2 \cdot 3 \cdot 5 + 2 \cdot 3 \cdot 5 =$
$18 + 30 + 30 = 78$

Substituting in the polynomial: $2 \cdot 3^2 + 20 \cdot 3 =$
$2 \cdot 9 + 60 = 18 + 60 = 78$

Since the results are the same, our solution is probably correct.

State. A polynomial for the surface area is $2x^2 + 20x$.

89. Length of top edges: $x + 6 + x + 6$, or $2x + 12$

Length of bottom edges: $x + 6 + x + 6$, or $2x + 12$

Length of vertical edges: $4 \cdot x$, or $4x$

Total length of edges: $2x + 12 + 2x + 12 + 4x = 8x + 24$

91. *Writing Exercise*

Exercise Set 4.4

1. $4x^4 \cdot 2x^2 = (4 \cdot 2)(x^4 \cdot x^2) = 8x^6$

Choice (b) is correct.

3. $4x^3 \cdot 2x^5 = (4 \cdot 2)(x^3 \cdot x^5) = 8x^8$

Choice (d) is correct.

5. $(4x^3)9 = (4 \cdot 9)x^3 = 36x^3$

7. $(-x^3)(x^4) = (-1 \cdot x^3)(x^4) = -1(x^3 \cdot x^4) = -1 \cdot x^7 = -x^7$

9. $(-x^6)(-x^2) = (-1 \cdot x^6)(-1 \cdot x^2) = (-1)(-1)(x^6 \cdot x^2) = x^8$

11. $(8a^2)(3a^2) = (8 \cdot 3)(a^2 \cdot a^2) = 24a^4$

13. $(0.3x^3)(-0.4x^6) = 0.3(-0.4)(x^3 \cdot x^6) = -0.12x^9$

15. $\left(-\frac{1}{4}x^4\right)\left(\frac{1}{5}x^8\right) = \left(-\frac{1}{4} \cdot \frac{1}{5}\right)(x^4 \cdot x^8) = -\frac{1}{20}x^{12}$

17. $(-5n^3)(-1) = (-5)(-1)n^3 = 5n^3$

19. $(-4y^5)(6y^2)(-3y^3) = -4(6)(-3)(y^5 \cdot y^2 \cdot y^3) = 72y^{10}$

21. $2x(4x - 6) = 2x \cdot 4x + 2x(-6) = 8x^2 - 12x$

23. $3x(x + 2) = 3x \cdot x + 3x \cdot 2 = 3x^2 + 6x$

25. $(a + 9)3a = a \cdot 3a + 9 \cdot 3a = 3a^2 + 27a$

27. $\begin{aligned}[t] x^2(x^3 + 1) &= x^2(x^3) + x^2(1) \\ &= x^5 + x^2 \end{aligned}$

29. $\begin{aligned}[t] 3x(2x^2 - 6x + 1) &= 3x(2x^2) + 3x(-6x) + 3x(1) \\ &= 6x^3 - 18x^2 + 3x \end{aligned}$

31. $5t^2(3t + 6) = 5t^2(3t) + 5t^2(6) = 15t^3 + 30t^2$

33. $\begin{aligned}[t] -6x^2(x^2 + x) &= -6x^2(x^2) - 6x^2(x) \\ &= -6x^4 - 6x^3 \end{aligned}$

35. $\begin{aligned}[t] &\frac{2}{3}a^4\left(6a^5 - 12a^3 - \frac{5}{8}\right) \\ &= \frac{2}{3}a^4(6a^5) - \frac{2}{3}a^4(12a^3) - \frac{2}{3}a^4\left(\frac{5}{8}\right) \\ &= \frac{12}{3}a^9 - \frac{24}{3}a^7 - \frac{10}{24}a^4 \\ &= 4a^9 - 8a^7 - \frac{5}{12}a^4 \end{aligned}$

37. $\begin{aligned}[t] (x + 2)(x + 6) &= (x + 2)x + (x + 2)6 \\ &= x \cdot x + 2 \cdot x + x \cdot 6 + 2 \cdot 6 \\ &= x^2 + 2x + 6x + 12 \\ &= x^2 + 8x + 12 \end{aligned}$

39. $\begin{aligned}[t] (x + 5)(x - 2) &= (x + 5)x + (x + 5)(-2) \\ &= x \cdot x + 5 \cdot x + x(-2) + 5(-2) \\ &= x^2 + 5x - 2x - 10 \\ &= x^2 + 3x - 10 \end{aligned}$

41. $\begin{aligned}[t] (a - 6)(a - 7) &= (a - 6)a + (a - 6)(-7) \\ &= a \cdot a - 6 \cdot a + a(-7) + (-6)(-7) \\ &= a^2 - 6a - 7a + 42 \\ &= a^2 - 13a + 42 \end{aligned}$

43. $\begin{aligned}[t] (x + 3)(x - 3) &= (x + 3)x + (x + 3)(-3) \\ &= x \cdot x + 3 \cdot x + x(-3) + 3(-3) \\ &= x^2 + 3x - 3x - 9 \\ &= x^2 - 9 \end{aligned}$

45. $\begin{aligned}[t] (5 - x)(5 - 2x) &= (5 - x)5 + (5 - x)(-2x) \\ &= 5 \cdot 5 - x \cdot 5 + 5(-2x) - x(-2x) \\ &= 25 - 5x - 10x + 2x^2 \\ &= 25 - 15x + 2x^2 \end{aligned}$

47. $\begin{aligned}[t] \left(t + \frac{3}{2}\right)\left(t + \frac{4}{3}\right) &= \left(t + \frac{3}{2}\right)t + \left(t + \frac{3}{2}\right)\left(\frac{4}{3}\right) \\ &= t \cdot t + \frac{3}{2} \cdot t + t \cdot \frac{4}{3} + \frac{3}{2} \cdot \frac{4}{3} \\ &= t^2 + \frac{3}{2}t + \frac{4}{3}t + 2 \\ &= t^2 + \frac{9}{6}t + \frac{8}{6}t + 2 \\ &= t^2 + \frac{17}{6}t + 2 \end{aligned}$

49. $\left(\frac{1}{4}a+2\right)\left(\frac{3}{4}a-1\right)$

$= \left(\frac{1}{4}a+2\right)\left(\frac{3}{4}a\right)+\left(\frac{1}{4}a+2\right)(-1)$

$= \frac{1}{4}a\left(\frac{3}{4}a\right)+2\cdot\frac{3}{4}a+\frac{1}{4}a(-1)+2(-1)$

$= \frac{3}{16}a^2+\frac{3}{2}a-\frac{1}{4}a-2$

$= \frac{3}{16}a^2+\frac{6}{4}a-\frac{1}{4}a-2$

$= \frac{3}{16}a^2+\frac{5}{4}a-2$

51. Illustrate $x(x+5)$ as the area of a rectangle with width x and length $x+5$.

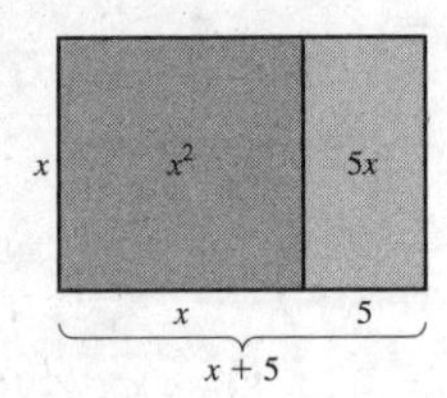

53. Illustrate $(x+1)(x+2)$ as the area of a rectangle with width $x+1$ and length $x+2$.

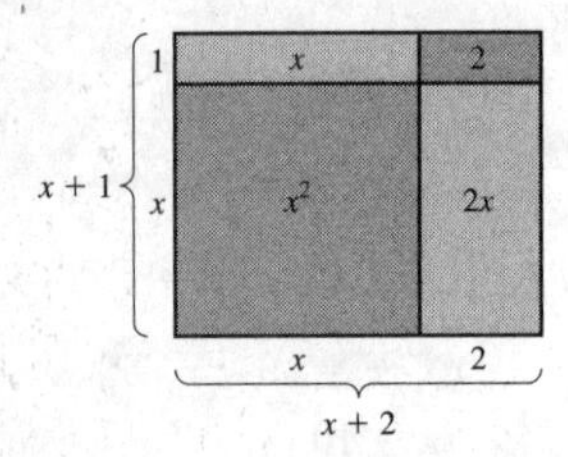

55. Illustrate $(x+5)(x+3)$ as the area of a rectangle with length $x+5$ and width $x+3$.

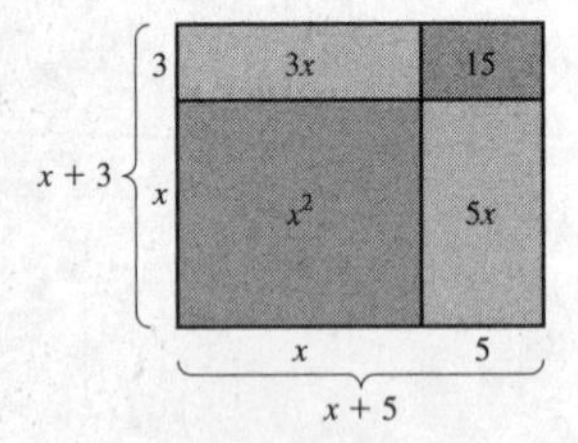

57. $(x^2-x+5)(x+1)$

$= (x^2-x+5)x+(x^2-x+5)1$

$= x^3-x^2+5x+x^2-x+5$

$= x^3+4x+5$

A partial check can be made by selecting a convenient replacement for x, say 1, and comparing the values of the original expression and the result.

$$\begin{array}{ll} (1^2-1+5)(1+1) & 1^3+4\cdot 1+5 \\ =(1-1+5)(1+1) & =1+4+5 \\ =5\cdot 2 & =10 \\ =10 & \end{array}$$

Since the value of both expressions is 10, the multiplication is very likely correct.

59. $(2a+5)(a^2-3a+2)$

$= (2a+5)a^2-(2a+5)(3a)+(2a+5)2$

$= 2a\cdot a^2+5\cdot a^2-2a\cdot 3a-5\cdot 3a+2a\cdot 2+5\cdot 2$

$= 2a^3+5a^2-6a^2-15a+4a+10$

$= 2a^3-a^2-11a+10$

A partial check can be made as in Exercise 57.

61. $(y^2-7)(2y^3+y+1)$

$= (y^2-7)(2y^3)+(y^2-7)y+(y^2-7)(1)$

$= y^2\cdot 2y^3-7\cdot 2y^3+y^2\cdot y-7\cdot y+y^2\cdot 1-7\cdot 1$

$= 2y^5-14y^3+y^3-7y+y^2-7$

$= 2y^5-13y^3+y^2-7y-7$

A partial check can be made as in Exercise 57.

63. $(3x+2)(5x+4x+7)$

$= (3x+2)(9x+7)$

$= (3x+2)(9x)+(3x+2)(7)$

$= 3x\cdot 9x+2\cdot 9x+3x\cdot 7+2\cdot 7$

$= 27x^2+18x+21x+14$

$= 27x^2+39x+14$

A partial check can be made as in Exercise 57.

65.

$$\begin{array}{rl} x^2-3x+2 & \text{Line up like terms} \\ x^2+x+1 & \text{in columns} \\ \hline x^2-3x+2 & \text{Multiplying by 1} \\ x^3-3x^2+2x & \text{Multiplying by } x \\ x^4-3x^3+2x^2 & \text{Multiplying by } x^2 \\ \hline x^4-2x^3-x+2 & \end{array}$$

A partial check can be made as in Exercise 57.

67.

$$\begin{array}{rl} 2t^2-5t-4 & \\ 3t^2-t+\frac{1}{2} & \\ \hline t^2-\frac{5}{2}t-2 & \text{Multiplying by } \frac{1}{2} \\ -2t^3+5t^2+4t & \text{Multiplying by } -t \\ 6t^4-15t^3-12t^2 & \text{Multiplying by } 3t^2 \\ \hline 6t^4-17t^3-6t^2+\frac{3}{2}t-2 & \end{array}$$

A partial check can be made as in Exercise 57.

69. We will multiply horizontally while still aligning like terms.

$(x+1)(x^3+7x^2+5x+4)$

$$\begin{array}{rl} = x^4+7x^3+5x^2+4x & \text{Multiplying by } x \\ +x^3+7x^2+5x+4 & \text{Multiplying by 1} \\ \hline = x^4+8x^3+12x^2+9x+4 & \end{array}$$

A partial check can be made as in Exercise 57.

71. *Writing Exercise*

73. $5-3\cdot 2+7=5-6+7=-1+7=6$

75.
$$\begin{aligned}&(8-2)(8+2)+2^2-8^2\\&=6\cdot 10+2^2-8^2\\&=6\cdot 10+4-64\\&=60+4-64\\&=64-64\\&=0\end{aligned}$$

77. *Writing Exercise*

79. The shaded area is the area of the large rectangle, $6y(14y-5)$ less the area of the unshaded rectangle, $3y(3y+5)$. We have:

$$\begin{aligned}&6y(14y-5)-3y(3y+5)\\&=84y^2-30y-9y^2-15y\\&=75y^2-45y\end{aligned}$$

81. Let n = the missing number. Label the figure with the known areas.

2	$2x$	$2n$
x	x^2	nx
	x	n

Then the area of the figure is $x^2+2x+nx+2n$. This is equivalent to $x^2+7x+10$, so we have $2x+nx=7x$ and $2n=10$. Solving either equation for n, we find that the missing number is 5.

83.

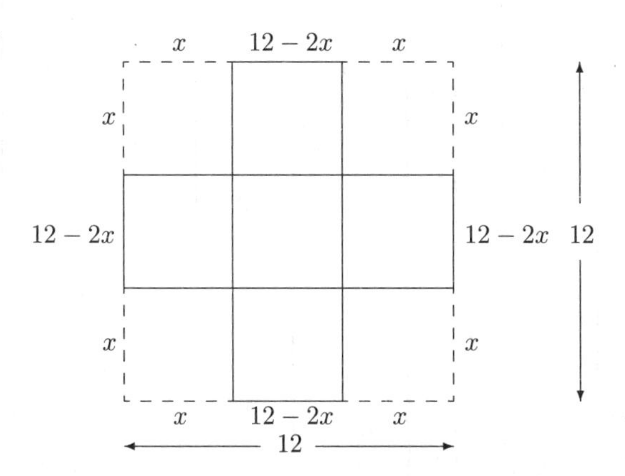

The dimensions, in inches, of the box are $12-2x$ by $12-2x$ by x. The volume is the product of the dimensions (volume = length × width × height):

$$\begin{aligned}\text{Volume}&=(12-2x)(12-2x)x\\&=(144-48x+4x^2)x\\&=144x-48x^2+4x^3\text{ in}^3,\text{ or}\\&\quad 4x^3-48x^2+144x\text{ in}^3\end{aligned}$$

The outside surface area is the sum of the area of the bottom and the areas of the four sides. The dimensions, in inches, of the bottom are $12-2x$ by $12-2x$, and the dimensions, in inches, of each side are x by $12-2x$.

$$\begin{aligned}\text{Surface area}&=\text{Area of bottom}+4\cdot\text{Area of each side}\\&=(12-2x)(12-2x)+4\cdot x(12-2x)\\&=144-24x-24x+4x^2+48x-8x^2\\&=144-48x+4x^2+48x-8x^2\\&=144-4x^2\text{ in}^2,\text{ or }-4x^2+144\text{ in}^2\end{aligned}$$

85. We have a rectangular solid with dimensions x m by x m by $x+2$ m with a rectangular solid piece with dimensions 6 m by 5 m by 7 m cut out of it.

$$\begin{aligned}\text{Volume}&=\text{Volume of large solid}-\text{Volume of small solid}\\&=(x\text{ m})(x\text{ m})(x+2\text{ m})-(6\text{ m})(5\text{ m})(7\text{ m})\\&=x^2(x+2)\text{ m}^3-210\text{ m}^3\\&=x^3+2x^2-210\text{ m}^3\end{aligned}$$

87. Let x = the width of the garden. Then $2x$ = the length of the garden.

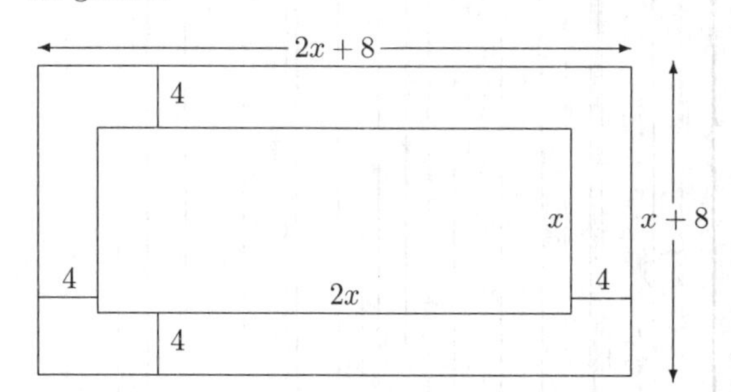

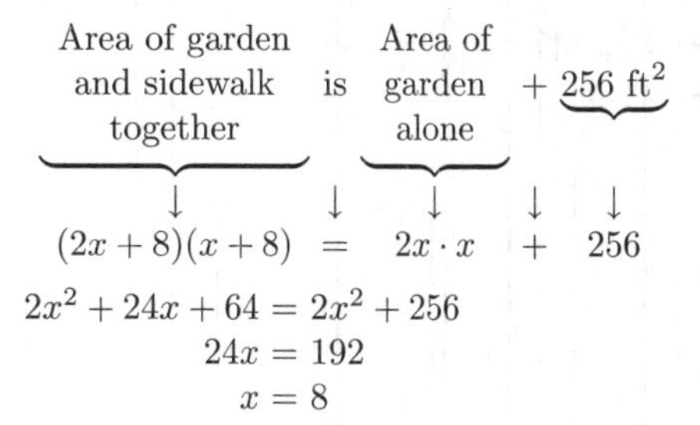

$$\begin{aligned}2x^2+24x+64&=2x^2+256\\24x&=192\\x&=8\end{aligned}$$

The dimensions are 8 ft by 16 ft.

89. $(x-2)(x-7)-(x-7)(x-2)$

First observe that, by the commutative law of multiplication, $(x-2)(x-7)$ and $(x-7)(x-2)$ are equivalent expressions. Then when we subtract $(x-7)(x-2)$ from $(x-2)(x-7)$, the result is 0.

91.
$$\begin{aligned}&(x-a)(x-b)\cdots(x-x)(x-y)(x-z)\\&=(x-a)(x-b)\cdots 0\cdot(x-y)(x-z)\\&=0\end{aligned}$$

Exercise Set 4.5

1. It is true that FOIL is simply a memory device for finding the product of two binomials.

3. This statement is false. See the material on squaring binomials at the bottom of page 264 in the text.

5. $(x+3)(x^2+5)$

F O I L

$= x\cdot x^2 + x\cdot 5 + 3\cdot x^2 + 3\cdot 5$

$= x^3+5x+3x^2+15$, or $x^3+3x^2+5x+15$

7. $(x^3+6)(x+2)$

F O I L

$= x^3\cdot x + x^3\cdot 2 + 6\cdot x + 6\cdot 2$

$= x^4+2x^3+6x+12$

9. $(y+2)(y-3)$

F O I L

$= y\cdot y + y\cdot(-3) + 2\cdot y + 2\cdot(-3)$

$= y^2-3y+2y-6$

$= y^2-y-6$

11. $(3x+2)(3x+5)$

F O I L

$= 3x\cdot 3x + 3x\cdot 5 + 2\cdot 3x + 2\cdot 5$

$= 9x^2+15x+6x+10$

$= 9x^2+21x+10$

13. $(5x-4)(x+2)$

F O I L

$= 5x\cdot x + 5x\cdot 2 + (-4)\cdot x + (-4)\cdot 2$

$= 5x^2+10x-4x-8$

$= 5x^2+6x-8$

15. $(1+3t)(2-3t)$

F O I L

$= 1\cdot 2 + 1(-3t) + 3t\cdot 2 + 3t(-3t)$

$= 2-3t+6t-9t^2$

$= 2+3t-9t^2$

17. $(2x-5)(x-4)$

F O I L

$= 2x\cdot x + 2x\cdot(-4) + (-5)\cdot x + (-5)\cdot(-4)$

$= 2x^2-8x-5x+20$

$= 2x^2-13x+20$

19. $\left(p-\frac{1}{4}\right)\left(p+\frac{1}{4}\right)$

F O I L

$= p\cdot p + p\cdot\frac{1}{4} + \left(-\frac{1}{4}\right)\cdot p + \left(-\frac{1}{4}\right)\cdot\frac{1}{4}$

$= p^2+\frac{1}{4}p-\frac{1}{4}p-\frac{1}{16}$

$= p^2-\frac{1}{16}$

21. $(x-0.1)(x+0.1)$

F O I L

$= x\cdot x + x\cdot(0.1) + (-0.1)\cdot x + (-0.1)(0.1)$

$= x^2+0.1x-0.1x-0.01$

$= x^2-0.01$

23. $(-2x+1)(x+6)$

F O I L

$= -2x^2 - 12x + x + 6$

$= -2x^2-11x+6$

25. $(a+9)(a+9)$

F O I L

$= a^2 + 9a + 9a + 81$

$= a^2+18a+81$

27. $(1+3t)(1-5t)$

F O I L

$= 1 - 5t + 3t - 15t^2$

$= 1-2t-15t^2$

29. $(x^2+3)(x^3-1)$

F O I L

$= x^5 - x^2 + 3x^3 - 3$, or $x^5+3x^3-x^2-3$

31. $(3x^2-2)(x^4-2)$

F O I L

$= 3x^6 - 6x^2 - 2x^4 + 4$, or $3x^6-2x^4-6x^2+4$

33. $(2t^3+5)(2t^3+3)$

F O I L

$= 4t^6 + 6t^3 + 10t^3 + 15$

$= 4t^6+16t^3+15$

35. $(8x^3+5)(x^2+2)$

F O I L

$= 8x^5 + 16x^3 + 5x^2 + 10$

37. $(4x^2+3)(x-3)$

F O I L

$= 4x^3 - 12x^2 + 3x - 9$

39. $(x+7)(x-7)$ Product of sum and difference of the same two terms

$= x^2-7^2$

$= x^2-49$

41. $(2x+1)(2x-1)$ Product of sum and difference of the same two terms

$= (2x)^2-1^2$

$= 4x^2-1$

43. $(5m-2)(5m+2)$ Product of sum and difference of the same two terms

$= (5m)^2-2^2$

$= 25m^2-4$

45. $(3x^4-1)(3x^4+1)$

$= (3x^4)^2-1^2$

$= 9x^8-1$

47. $(x^4+7)(x^4-7)$

$= (x^4)^2-7^2$

$= x^8-49$

49. $\left(t-\frac{3}{4}\right)\left(t+\frac{3}{4}\right)$

$= t^2-\left(\frac{3}{4}\right)^2$

$= t^2-\frac{9}{16}$

51. $(x+2)^2$
$= x^2 + 2 \cdot x \cdot 2 + 2^2$ Square of a binomial
$= x^2 + 4x + 4$

53. $(7x^3+1)^2$ Square of a binomial
$= (7x^3)^2 + 2 \cdot 7x^3 \cdot 1 + 1^2$
$= 49x^6 + 14x^3 + 1$

55. $\left(a - \frac{2}{5}\right)^2$ Square of a binomial
$= a^2 - 2 \cdot a \cdot \frac{2}{5} + \left(\frac{2}{5}\right)^2$
$= a^2 - \frac{4}{5}a + \frac{4}{25}$

57. $= (t^3+5)^2$ Square of a binomial
$= (t^3)^2 + 2 \cdot t^3 \cdot 5 + 5^2$
$= t^6 + 10t^3 + 25$

59. $(2-3x^4)^2 = 2^2 - 2 \cdot 2 \cdot 3x^4 + (3x^4)^2$
$= 4 - 12x^4 + 9x^8$

61. $(5+6t^2)^2 = 5^2 + 2 \cdot 5 \cdot 6t^2 + (6t^2)^2$
$= 25 + 60t^2 + 36t^4$

63. $(7x-0.3)^2 = (7x)^2 - 2(7x)(0.3) + (0.3)^2$
$= 49x^2 - 4.2x + 0.09$

65. $5a^3(2a^2-1)$
$= 5a^3 \cdot 2a^2 - 5a^3 \cdot 1$ Multiplying each term of the binomial by the monomial
$= 10a^5 - 5a^3$

67. $(a-3)(a^2+2a-4)$

$$\begin{array}{rrrr} = a^3 & +\ 2a^2 & -\ 4a & \\ & -\ 3a^2 & -\ 6a & +\ 12 \\ \hline = a^3 & -\ a^2 & -\ 10a & +\ 12 \end{array}$$

Multiplying horizontally and aligning like terms

69. $(3-2x^3)^2$
$= 3^2 - 2 \cdot 3 \cdot 2x^3 + (2x^3)^2$ Squaring a binomial
$= 9 - 12x^3 + 4x^6$

71. $5x(x^2+6x-2)$
$= 5x \cdot x^2 + 5x \cdot 6x + 5x(-2)$ Multiplying each term of the trinomial by the monomial
$= 5x^3 + 30x^2 - 10x$

73. $(-t^3+1)^2$
$= (-t^3)^2 + 2(-t)^3(1) + 1^2$ Squaring a binomial
$= t^6 - 2t^3 + 1$

75. $3t^2(5t^3 - t^2 + t)$
$= 3t^2 \cdot 5t^3 + 3t^2(-t^2) + 3t^2 \cdot t$ Multiplying each term of the trinomial by the monomial
$= 15t^5 - 3t^4 + 3t^3$

77. $(6x^4-3)^2$ Squaring a binomial
$= (6x^4)^2 - 2 \cdot 6x^4 \cdot 3 + 3^2$
$= 36x^8 - 36x^4 + 9$

79. $(3x+2)(4x^2+5)$ Product of two binomials; use FOIL
$= 3x \cdot 4x^2 + 3x \cdot 5 + 2 \cdot 4x^2 + 2 \cdot 5$
$= 12x^3 + 15x + 8x^2 + 10$, or
$12x^3 + 8x^2 + 15x + 10$

81. $(5-6x^4)^2$ Squaring a binomial
$= 5^2 - 2 \cdot 5 \cdot 6x^4 + (6x^4)^2$
$= 25 - 60x^4 + 36x^8$

83. $(a+1)(a^2-a+1)$

$$\begin{array}{rrrr} = a^3 & -\ a^2 & +\ a & \\ & a^2 & -\ a & +\ 1 \\ \hline a^3 & & & +\ 1 \end{array}$$

Multiplying horizontally and aligning like terms

85.

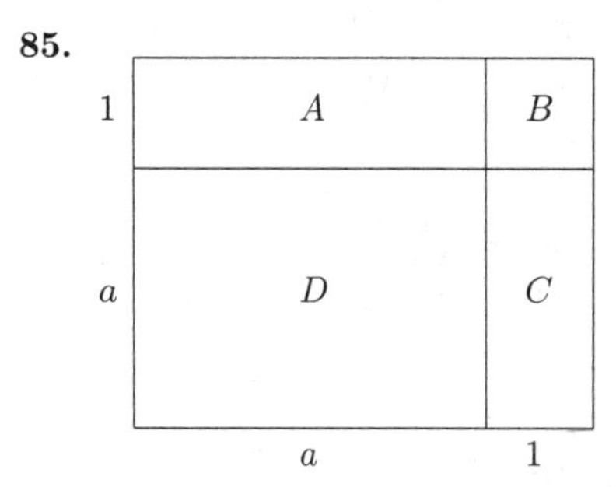

We can find the shaded area in two ways.

Method 1: The figure is a square with side $a+1$, so the area is $(a+1)^2 = a^2 + 2a + 1$.

Method 2: We add the areas of A, B, C, and D.

$1 \cdot a + 1 \cdot 1 + 1 \cdot a + a \cdot a = a + 1 + a + a^2 = a^2 + 2a + 1$.

Either way we find that the total shaded area is $a^2 + 2a + 1$.

87.

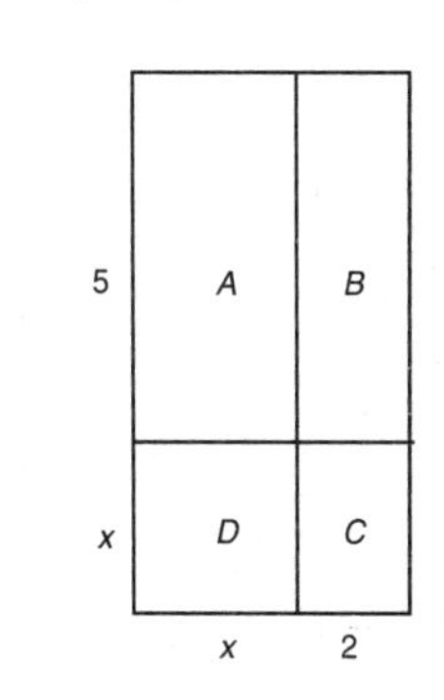

We can find the shaded area in two ways.

Method 1: The figure is a rectangle with dimensions $x+5$ by $x+2$, so the area is

$(x+5)(x+2) = x^2 + 2x + 5x + 10 = x^2 + 7x + 10$.

Method 2: We add the areas of A, B, C, and D.

$5 \cdot x + 2 \cdot 5 + 2 \cdot x + x \cdot x = 5x + 10 + 2x + x^2 = x^2 + 7x + 10$.

Either way, we find that the area is $x^2 + 7x + 10$.

89.

We can find the shaded area in two ways.

Method 1: The figure is a square with side $x + 7$, so the area is $(x + 7)^2 = x^2 + 14x + 49$.

Method 2: We add the areas of A, B, C, and D.

$x \cdot x + x \cdot 7 + 7 \cdot 7 + 7 \cdot x = x^2 + 7x + 49 + 7x = x^2 + 14x + 49$.

Either way, we find that the total shaded area is $x^2 + 14x + 49$.

91.

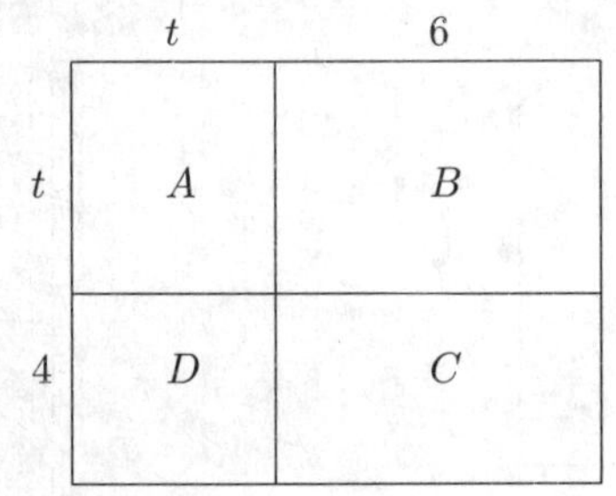

We can find the shaded area in two ways.

Method 1: The figure is a rectangle with dimensions $t + 6$ by $t + 4$, so the area is $(t + 6)(t + 4) =$
$t^2 + 4t + 6t + 24 = t^2 + 10t + 24$.

Method 2: We add the areas of A, B, C, and D.

$t \cdot t + t \cdot 6 + 6 \cdot 4 + 4 \cdot t = t^2 + 6t + 24 + 4t = t^2 + 10t + 24$.

Either way, we find that the total shaded area is $t^2 + 10t + 24$.

93.

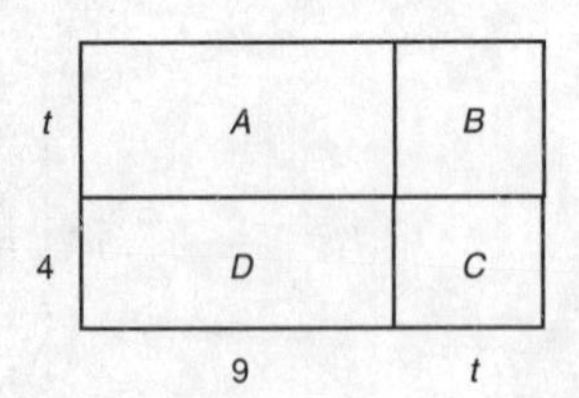

We can find the shaded area in two ways.

Method 1: The figure is a rectangle with dimensions $t + 9$ by $t + 4$, so the area is

$(t + 9)(t + 4) = t^2 + 4t + 9t + 36 = t^2 + 13t + 36$

Method 2: We add the areas of A, B, C, and D.

$9 \cdot t + t \cdot t + 4 \cdot t + 4 \cdot 9 = 9t + t^2 + 4t + 36 = t^2 + 13t + 36$.

Either way, we find that the total shaded area is $t^2 + 13t + 36$.

95.

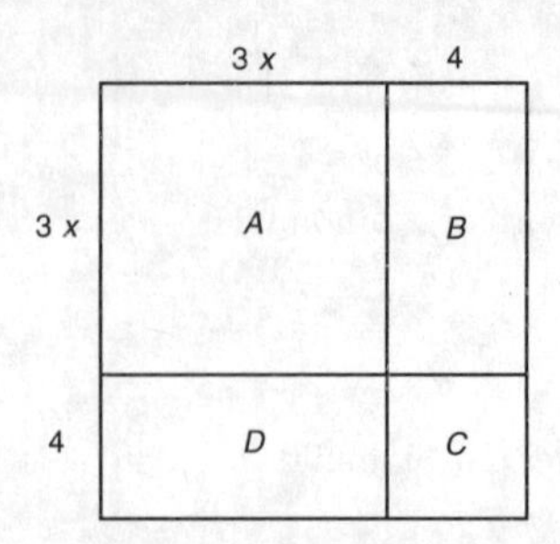

We can find the shaded area in two ways.

Method 1: The figure is a square with side $3x + 4$, so the area is $(3x + 4)^2 = 9x^2 + 24x + 16$.

Method 2: We add the areas of A, B, C, and D.

$3x \cdot 3x + 3x \cdot 4 + 4 \cdot 4 + 3x \cdot 4 = 9x^2 + 12x + 16 + 12x =$
$9x^2 + 24x + 16$.

Either way, we find that the total shaded area is $9x^2 + 24x + 16$.

97. We draw a square with side $x + 5$.

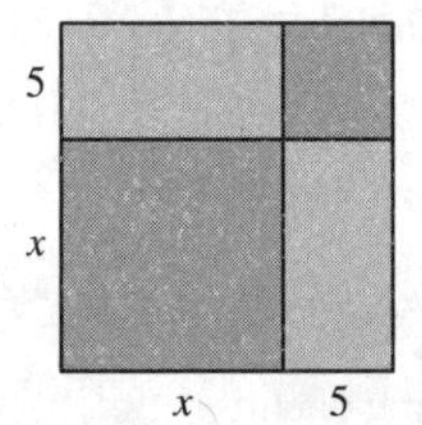

99. We draw a square with side $t + 9$.

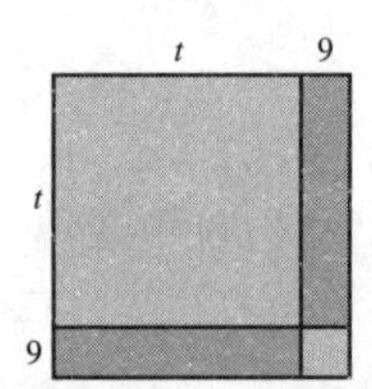

101. We draw a square with side $3 + x$.

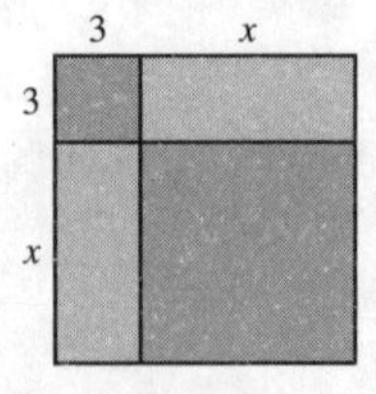

103. *Writing Exercise*

105. ***Familiarize.*** Let $t =$ the number of watts used by the television set. Then $10t =$ the number of watts used by the lamps, and $40t =$ the number of watts used by the air conditioner.

Translate.

$$\underbrace{\text{Lamp watts}}_{10t} + \underbrace{\text{Air conditioner watts}}_{40t} + \underbrace{\text{Television watts}}_{t} = \underbrace{\text{Total watts}}_{2550}$$

$$10t + 40t + t = 2550$$

Solve. We solve the equation.

$$10t + 40t + t = 2550$$
$$51t = 2550$$
$$t = 50$$

The possible solution is:

Television, t: 50 watts

Lamps, $10t$: $10 \cdot 50$, or 500 watts

Air conditioner, $40t$: $40 \cdot 50$, or 2000 watts

Check. The number of watts used by the lamps, 500, is 10 times 50, the number used by the television. The number of watts used by the air conditioner, 2000, is 40 times 50, the number used by the television. Also, $50+500+2000 = 2550$, the total wattage used.

State. The television uses 50 watts, the lamps use 500 watts, and the air conditioner uses 2000 watts.

107. $5xy = 8$

$y = \dfrac{8}{5x}$ Dividing both sides by $5x$

109. $ax - b = c$

$ax = b + c$ Adding b to both sides

$x = \dfrac{b+c}{a}$ Dividing both sides by a

111. *Writing Exercise*

113. $(4x^2 + 9)(2x + 3)(2x - 3)$

$= (4x^2 + 9)(4x^2 - 9)$

$= 16x^4 - 81$

115. $(3t - 2)^2(3t + 2)^2$

$= [(3t - 2)(3t + 2)]^2$

$= (9t^2 - 4)^2$

$= 81t^4 - 72t^2 + 16$

117. $(t^3 - 1)^4(t^3 + 1)^4$

$= [(t^3 - 1)(t^3 + 1)]^4$

$= (t^6 - 1)^4$

$= [(t^6 - 1)^2]^2$

$= (t^{12} - 2t^6 + 1)^2$

$= (t^{12} - 2t^6 + 1)(t^{12} - 2t^6 + 1)$

$= t^{24} - 2t^{18} + t^{12} - 2t^{18} + 4t^{12} - 2t^6 +$
$t^{12} - 2t^6 + 1$

$= t^{24} - 4t^{18} + 6t^{12} - 4t^6 + 1$

119. $18 \times 22 = (20 - 2)(20 + 2) = 20^2 - 2^2 =$
$400 - 4 = 396$

121. $(x + 2)(x - 5) = (x + 1)(x - 3)$

$x^2 - 5x + 2x - 10 = x^2 - 3x + x - 3$

$x^2 - 3x - 10 = x^2 - 2x - 3$

$-3x - 10 = -2x - 3$ Adding $-x^2$

$-3x + 2x = 10 - 3$ Adding $2x$ and 10

$-x = 7$

$x = -7$

The solution is -7.

123.

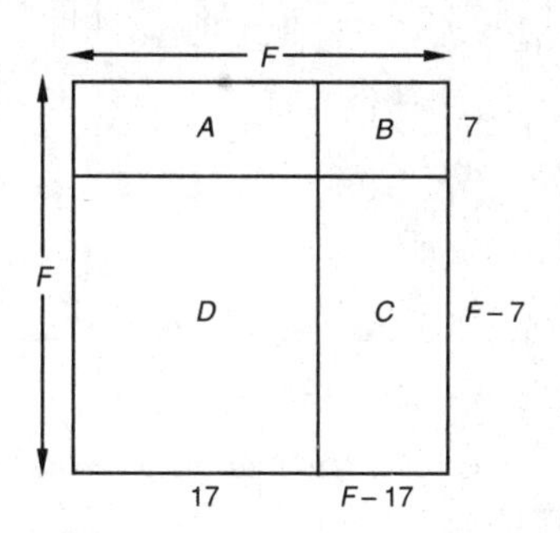

The area of the entire figure is F^2. The area of the unshaded region, C, is $(F-7)(F-17)$. Then one expression for the area of the shaded region is $F^2 - (F-7)(F-17)$.

To find a second expression we add the areas of regions A, B, and D. We have:

$17 \cdot 7 + 7(F - 17) + 17(F - 7)$

$= 119 + 7F - 119 + 17F - 119$

$= 24F - 119$

It is possible to find other equivalent expressions also.

125.

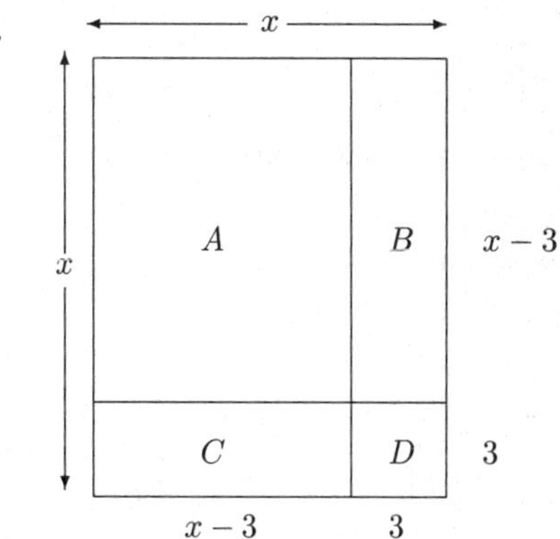

The area of the entire region is x^2. The unshaded area is $3 \cdot 3$, or 9, so an expression for the shaded area is $x^2 - 9$.

To find another expression we add the areas of regions A, B, and C. The dimensions of region A are $x - 3$ by $x - 3$ and regions B and C each have dimensions 3 by $x - 3$, so the sum of the areas is $(x - 3)^2 + 3(x - 3) + 3(x - 3)$. It is possible to find other equivalent expressions also.

127.

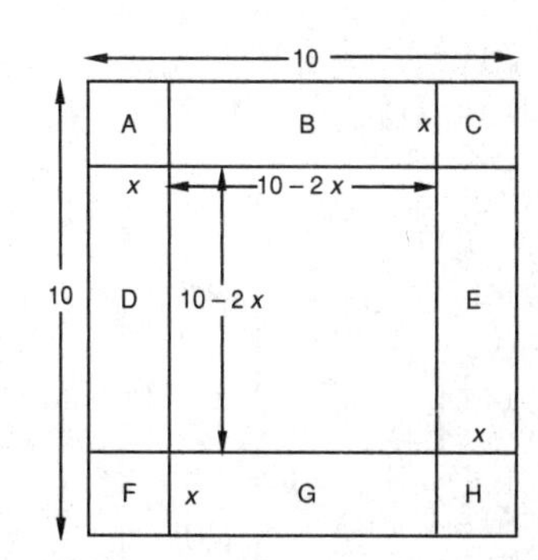

$(10 - 2x)^2$ is the area of the entire square less the areas of A, B, C, D, E, F, G, and H. The areas of A, C, F, and H are each $x \cdot x$, or x^2. The areas of B, D, E, and G are each $x(10 - 2x)$, or $10x - 2x^2$. We have:

$(10-2x)^2 = 10^2 - 4\cdot x^2 - 4(10x-2x^2)$
$= 100 - 4x^2 - 40x + 8x^2$
$= 100 - 40x + 4x^2$

129. 

Exercise Set 4.6

1. $(2x-7y)^2$ is the square of a binomial, choice (a).

3. $(5a+6b)(-6b+5a)$, or $(5a+6b)(5a-6b)$ is the product of the sum and difference of the same two terms, choice (b).

5. $(r-3s)(5r+3s)$ is neither the square of a binomial nor the product of the sum and difference of the same two terms, so choice (c) is appropriate.

7. $(4x-9y)(4x-9y)$, or $(4x-9y)^2$ is the square of a binomial, choice (a).

9. We replace x by 5 and y by -2.
$x^2 - 3y^2 + 2xy = 5^2 - 3(-2)^2 + 2\cdot 5(-2) =$
$25 - 12 - 20 = -7.$

11. We replace x by 2, y by -3, and z by -4.
$xyz^2 - z = 2(-3)(-4)^2 - (-4) = -96 + 4 = -92$

13. Evaluate the polynomial for $h = 160$ and $A = 50$.
$0.041h - 0.018A - 2.69$
$= 0.041(160) - 0.018(50) - 2.69$
$= 6.56 - 0.9 - 2.69$
$= 2.97$
The woman's lung capacity is 2.97 liters.

15. Evaluate the polynomial for $w = 87$, $h = 185$, and $a = 59$.
$19.18w + 7h - 9.52a + 92.4$
$= 19.18(87) + 7(185) - 9.52(59) + 92.4$
$= 1668.66 + 1295 - 561.68 + 92.4$
≈ 2494
The daily caloric needs are 2494 calories.

17. Evaluate the polynomial for $h = 4$, $r = \dfrac{3}{4}$, and $\pi \approx 3.14$.
$2\pi rh + \pi r^2$
$\approx 2(3.14)\left(\dfrac{3}{4}\right)(4) + 3.14\left(\dfrac{3}{4}\right)^2$
$\approx 18.84 + 1.76625$
≈ 20.60625
The surface area is about 20.60625 in^2.

19. Evaluate the polynomial for $h = 50$, $v = 18$, and $t = 2$.
$h + vt - 4.9t^2$
$= 50 + 18\cdot 2 - 4.9(2)^2$
$= 50 + 36 - 19.6$
$= 66.4$
The ball will be 66.4 m above the ground 2 seconds after it is thrown.

21. $x^3y - 2xy + 3x^2 - 5$

Term	Coefficient	Degree	
x^3y	1	4	(Think: $x^3y = x^3y^1$)
$-2xy$	-2	2	(Think: $-2xy = -2x^1y^1$)
$3x^2$	3	2	
-5	-5	0	(Think: $-5 = -5x^0$)

The degree of the polynomial is the degree of the term of highest degree. The term of highest degree is x^3y. Its degree is 4, so the degree of the polynomial is 4.

23. $17x^2y^3 - 3x^3yz - 7$

Term	Coefficient	Degree	
$17x^2y^3$	17	5	
$-3x^3yz$	-3	5	(Think: $-3x^3yz = -3x^3y^1z^1$)
-7	-7	0	(Think: $-7 = -7x^0$)

The terms of highest degree are $17x^2y^3$ and $-3x^3yz$. Each has degree 5. The degree of the polynomial is 5.

25. $7a + b - 4a - 3b = (7-4)a + (1-3)b = 3a - 2b$

27. $3x^2y - 2xy^2 + x^2 + 5x$
There are no like terms, so none of the terms can be collected.

29. $2u^2v - 3uv^2 + 6u^2v - 2uv^2 + 7u^2$
$= (2+6)u^2v + (-3-2)uv^2 + 7u^2$
$= 8u^2v - 5uv^2 + 7u^2$

31. $5a^2c - 2ab^2 + a^2b - 3ab^2 + a^2c - 2ab^2$
$= (5+1)a^2c + (-2-3-2)ab^2 + a^2b$
$= 6a^2c - 7ab^2 + a^2b$

33. $(4x^2 - xy + y^2) + (-x^2 - 3xy + 2y^2)$
$= (4-1)x^2 + (-1-3)xy + (1+2)y^2$
$= 3x^2 - 4xy + 3y^2$

35. $(3a^4 - 5ab + 6ab^2) - (9a^4 + 3ab - ab^2)$
$= 3a^4 - 5ab + 6ab^2 - 9a^4 - 3ab + ab^2$
Adding the opposite
$= (3-9)a^4 + (-5-3)ab + (6+1)ab^2$
$= -6a^4 - 8ab + 7ab^2$

37. $(5r^2 - 4rt + t^2) + (-6r^2 - 5rt - t^2) + (-5r^2 + 4rt - t^2)$
Observe that the polynomials $5r^2 - 4rt + t^2$ and $-5r^2 + 4rt - t^2$ are opposites. Thus, their sum is 0 and the sum in the exercise is the remaining polynomial, $-6r^2 - 5rt - t^2$.

39. $(x^3 - y^3) - (-2x^3 + x^2y - xy^2 + 2y^3)$
$= x^3 - y^3 + 2x^3 - x^2y + xy^2 - 2y^3$
$= 3x^3 - 3y^3 - x^2y + xy^2$, or
$3x^3 - x^2y + xy^2 - 3y^3$

41. $(2y^4x^2 - 5y^3x) + (5y^4x^2 - y^3x) + (3y^4x^2 - 2y^3x)$
$= (2+5+3)y^4x^2 + (-5-1-2)y^3x$
$= 10y^4x^2 - 8y^3x$

43. $(4x + 5y) + (-5x + 6y) - (7x + 3y)$
$= 4x + 5y - 5x + 6y - 7x - 3y$
$= (4 - 5 - 7)x + (5 + 6 - 3)y$
$= -8x + 8y$

45.
$$\begin{aligned} & \qquad\qquad\quad \text{F} \qquad \text{O} \qquad \text{I} \qquad \text{L} \\ (3z - u)(2z + 3u) &= 6z^2 + 9zu - 2uz - 3u^2 \\ &= 6z^2 + 7zu - 3u^2 \end{aligned}$$

47.
$$\begin{aligned} & \qquad\qquad \text{F} \qquad \text{O} \qquad \text{I} \qquad \text{L} \\ (xy + 7)(xy - 4) &= x^2y^2 - 4xy + 7xy - 28 - 28 \\ &= x^2y^2 + 3xy - 28 \end{aligned}$$

49. $(2a - b)(2a + b)$ $[(A + B)(A - B) = A^2 - B^2]$
$= 4a^2 - b^2$

51.
$$\begin{aligned} & \qquad\qquad \text{F} \qquad \text{O} \qquad \text{I} \qquad \text{L} \\ (5rt - 2)(3rt + 1) &= 15r^2t^2 + 5rt - 6rt - 2 \\ &= 15r^2t^2 - rt - 2 \end{aligned}$$

53. $(m^3n + 8)(m^3n - 6)$
$\quad$ F $\quad$ O $\quad$ I $\quad$ L
$= m^6n^2 - 6m^3n + 8m^3n - 48$
$= m^6n^2 + 2m^3n - 48$

55. $(6x - 2y)(5x - 3y)$
$\quad$ F $\quad$ O $\quad$ I $\quad$ L
$= 30x^2 - 18xy - 10xy + 6y^2$
$= 30x^2 - 28xy + 6y^2$

57. $(pq + 0.1)(-pq + 0.1)$
$= (0.1 + pq)(0.1 - pq)$ $[(A + B)(A - B) = A^2 - B^2]$
$= 0.01 - p^2q^2$

59. $(x + h)^2$
$= x^2 + 2xh + h^2$ $[(A + B)^2 = A^2 + 2AB + B^2]$

61. $(4a + 5b)^2$
$= 16a^2 + 40ab + 25b^2$ $[(A+B)^2 = A^2+2AB+B^2]$

63. $(c^2 - d)(c^2 + d) = (c^2)^2 - d^2$
$= c^4 - d^2$

65. $(ab + cd^2)(ab - cd^2) = (ab)^2 - (cd^2)^2$
$= a^2b^2 - c^2d^4$

67. $(a + b - c)(a + b + c)$
$= [(a + b) - c][(a + b) + c]$
$= (a + b)^2 - c^2$
$= a^2 + 2ab + b^2 - c^2$

69. $[a + b + c][a - (b + c)]$
$= [a + (b + c)][a - (b + c)]$
$= a^2 - (b + c)^2$
$= a^2 - (b^2 + 2bc + c^2)$
$= a^2 - b^2 - 2bc - c^2$

71. The figure is a rectangle with dimensions $a + b$ by $a + c$. Its area is $(a + b)(a + c) = a^2 + ac + ab + bc$.

73. The figure is a parallelogram with base $x + z$ and height $x - z$. Thus the area is $(x + z)(x - z) = x^2 - z^2$.

75. The figure is a square with side $x + y + z$. Thus the area is
$(x + y + z)^2$
$= [(x + y) + z]^2$
$= (x + y)^2 + 2(x + y)(z) + z^2$
$= x^2 + 2xy + y^2 + 2xz + 2yz + z^2$.

77. The figure is a triangle with base $x + 2y$ and height $x - y$. Thus the area is $\frac{1}{2}(x + 2y)(x - y) = \frac{1}{2}(x^2 + xy - 2y^2) = \frac{1}{2}x^2 + \frac{1}{2}xy - y^2$.

79. We draw a rectangle with dimensions $r + s$ by $u + v$.

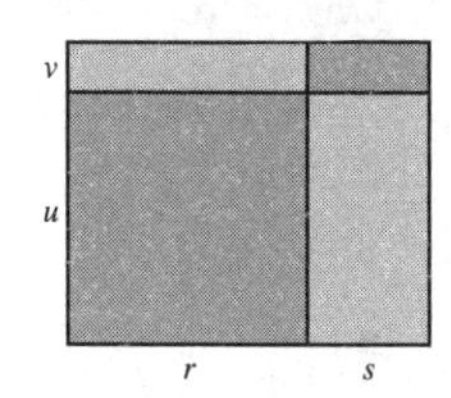

81. We draw a rectangle with dimensions $a+b+c$ by $a+d+f$.

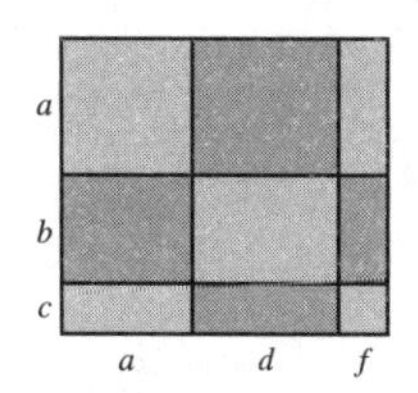

83. *Writing Exercise*

85. $5 + \dfrac{7 + 4 + 2 \cdot 5}{3}$

$= 5 + \dfrac{7 + 4 + 10}{3}$	Multiplying
$= 5 + \dfrac{21}{3}$	Adding in the numerator
$= 5 + 7$	Dividing
$= 12$	Adding

87. $(4 + 3 \cdot 5 + 8) \div 3 \cdot 3$

$= (4 + 15 + 8) \div 3 \cdot 3$	Multiplying inside the parentheses
$= 27 \div 3 \cdot 3$	Adding
$= 9 \cdot 3$	Dividing
$= 27$	Multiplying

89. $[3 \cdot 5 - 4 \cdot 2 + 7(-3)] \div (-2)$

$= (15 - 8 - 21) \div (-2)$	Multiplying
$= -14 \div (-2)$	Subtracting
$= 7$	Dividing

91. *Writing Exercise*

93. It is helpful to add additional labels to the figure.

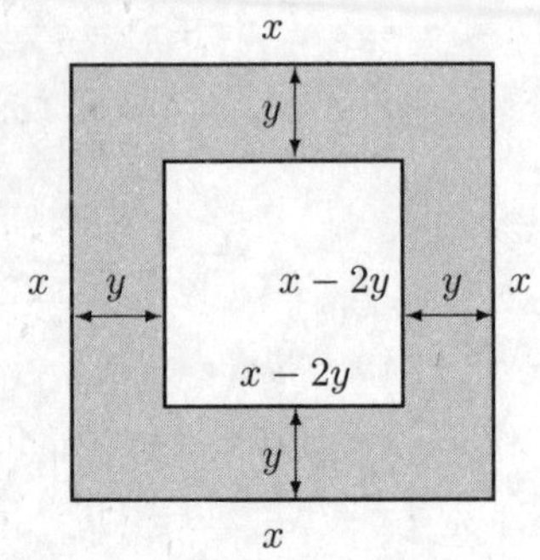

The area of the large square is $x \cdot x$, or x^2. The area of the small square is $(x - 2y)(x - 2y)$, or $(x - 2y)^2$.

$$\text{Area of shaded region} = \text{Area of large square} - \text{Area of small square}$$

$$\text{Area of shaded region} = x^2 - (x-2y)^2$$

$$= x^2 - (x^2 - 4xy + 4y^2)$$
$$= x^2 - x^2 + 4xy - 4y^2$$
$$= 4xy - 4y^2$$

95. It is helpful to add additional labels to the figure.

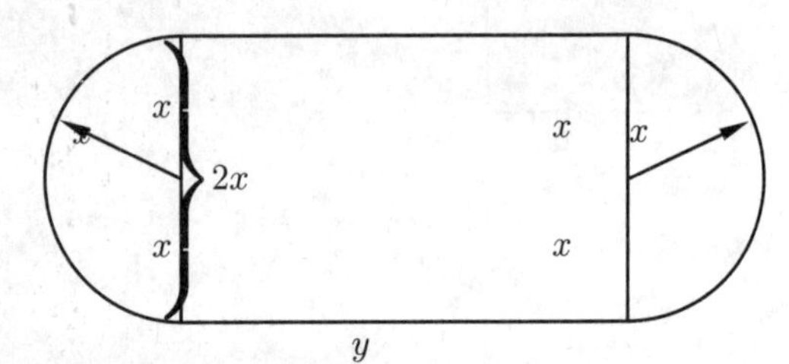

The two semicircles make a circle with radius x. The area of that circle is πx^2. The area of the rectangle is $2x \cdot y$. The sum of the two regions, $\pi x^2 + 2xy$, is the area of the shaded region.

97. The figure can be thought of as a cube with side x, a rectangular solid with dimensions x by x by y, a rectangular solid with dimensions x by y by y, and a rectangular solid with dimensions y by y by $2y$. Thus the volume is

$x^3 + x \cdot x \cdot y + x \cdot y \cdot y + y \cdot y \cdot 2y$, or

$x^3 + x^2y + xy^2 + 2y^3$.

99. The lateral surface area of the outer portion of the solid is the lateral surface area of a right circular cylinder with radius n and height h. The lateral surface area of the inner portion is the lateral surface area of a right circular cylinder with radius m and height h. Recall that the formula for the lateral surface area of a right circular cylinder with radius r and height h is $2\pi rh$.

The surface area of the top is the area of a circle with radius n less the area of a circle with radius m. The surface area of the bottom is the same as the surface area of the top.

Thus, the surface area of the solid is

$2\pi nh + 2\pi mh + 2\pi n^2 - 2\pi m^2$.

101. *Writing Exercise*

103. Replace t with 2 and multiply.

$$P(1+r)^2$$
$$= P(1 + 2r + r^2)$$
$$= P + 2Pr + Pr^2$$

105. Substitute \$10,400 for P, 8.5%, or 0.085 for r, and 5 for t.

$$P(1+r)^t$$
$$= \$10,400(1 + 0.085)^5$$
$$\approx \$15,638.03$$

Exercise Set 4.7

1. $\dfrac{32x^5 - 24x}{8} = \dfrac{32x^5}{8} - \dfrac{24x}{8}$

$= \dfrac{32}{8}x^5 - \dfrac{24}{8}x$ Dividing coefficients

$= 4x^5 - 3x$

To check, we multiply the quotient by 8:

$(4x^5 - 3x)8 = 32x^5 - 24x$

The answer checks.

3. $\dfrac{u - 2u^2 + u^7}{u}$

$= \dfrac{u}{u} - \dfrac{2u^2}{u} + \dfrac{u^7}{u}$

$= 1 - 2u + u^6$

Check: We multiply.

$u(1 - 2u + u^6) = u - 2u^2 + u^7$

5. $(18t^3 - 24t^2 + 6t) \div (3t)$

$= \dfrac{18t^3 - 24t^2 + 6t}{3t}$

$= \dfrac{18t^3}{3t} - \dfrac{24t^2}{3t} + \dfrac{6t}{3t}$

$= 6t^2 - 8t + 2$

Check: We multiply.

$3t(6t^2 - 8t + 2) = 18t^3 - 24t^2 + 6t$

7. $(25x^6 - 20x^4 - 5x^2) \div (-5x^2)$

$= \dfrac{25x^6 - 20x^4 - 5x^2}{-5x^2}$

$= \dfrac{25x^6}{-5x^2} - \dfrac{20x^4}{-5x^2} - \dfrac{5x^2}{-5x^2}$

$= -5x^4 - (-4x^2) - (-1)$

$= -5x^4 + 4x^2 + 1$

Check: We multiply.

$-5x^2(-5x^4 + 4x^2 + 1) = 25x^6 - 20x^4 - 5x^2$

9. $(24t^5 - 40t^4 + 6t^3) \div (4t^3)$

$= \dfrac{24t^5 - 40t^4 + 6t^3}{4t^3}$

$= \dfrac{24t^5}{4t^3} - \dfrac{40t^4}{4t^3} + \dfrac{6t^3}{4t^3}$

$= 6t^2 - 10t + \dfrac{3}{2}$

Check: We multiply.

$$4t^3\left(6t^2 - 10t + \frac{3}{2}\right) = 24t^5 - 40t^4 + 6t^3$$

11. $\dfrac{8x^2 - 10x + 1}{2}$

$= \dfrac{8x^2}{2} - \dfrac{10x}{2} + \dfrac{1}{2}$

$= 4x^2 - 5x + \dfrac{1}{2}$

Check: We multiply.

$$2\left(4x^2 - 5x + \frac{1}{2}\right) = 8x^2 - 10x + 1$$

13. $\dfrac{4x^7 + 6x^5 + 4x^2}{4x^3}$

$= \dfrac{4x^7}{4x^3} + \dfrac{6x^5}{4x^3} + \dfrac{4x^2}{4x^3}$

$= x^4 + \dfrac{3}{2}x^2 + \dfrac{1}{x}$

Check: We multiply.

$$4x^3\left(x^4 + \frac{3}{2}x^2 + \frac{1}{x}\right) = 4x^7 + 6x^5 + 4x^2$$

15. $\dfrac{9r^2s^2 + 3r^2s - 6rs^2}{-3rs}$

$= \dfrac{9r^2s^2}{-3rs} + \dfrac{3r^2s}{-3rs} - \dfrac{6rs^2}{-3rs}$

$= -3rs - r + 2s$

Check: We multiply.

$$-3rs(-3rs - r + 2s) = 9r^2s^2 + 3r^2s - 6rs^2$$

17.

$$\begin{array}{r l}
 & x + 6 \\
x - 2 \,\big)\!\! & x^2 + 4x - 12 \\
 & x^2 - 2x \\
 & \quad 6x - 12 \leftarrow (x^2 + 4x) - (x^2 - 2x) = 6x \\
 & \quad 6x - 12 \\
 & \qquad 0 \leftarrow (6x - 12) - (6x - 12) = 0
\end{array}$$

The answer is $x + 6$.

19.

$$\begin{array}{r l}
 & t - 5 \\
t - 5 \,\big)\!\! & t^2 - 10t - 20 \\
 & t^2 - 5t \\
 & \quad -5t - 20 \leftarrow (t^2 - 10t) - (t^2 - 5t) = -5t \\
 & \quad -5t + 25 \\
 & \qquad -45 \leftarrow (-5t - 20) - (-5t + 25) = -45
\end{array}$$

The answer is $t - 5 + \dfrac{-45}{t - 5}$.

21.

$$\begin{array}{r l}
 & 2x - 1 \\
x + 6 \,\big)\!\! & 2x^2 + 11x - 5 \\
 & 2x^2 + 12x \\
 & \quad -x - 5 \leftarrow (2x^2 + 11x) - (2x^2 + 12x) = -x \\
 & \quad -x - 6 \\
 & \qquad 1 \leftarrow (-x - 5) - (-x - 6) = 1
\end{array}$$

The answer is $2x - 1 + \dfrac{1}{x + 6}$.

23.

$$\begin{array}{r l}
 & a^2 - 2a + 4 \\
a + 2 \,\big)\!\! & a^3 + 0a^2 + 0a + 8 \leftarrow \text{Writing in the missing terms} \\
 & a^3 + 2a^2 \\
 & \quad -2a^2 + 0a \leftarrow a^3 - (a^3 + 2a^2) = -2a^2 \\
 & \quad -2a^2 - 4a \\
 & \qquad 4a + 8 \leftarrow -2a^2 - (-2a^2 - 4a) = 4a \\
 & \qquad 4a + 8 \\
 & \qquad\quad 0 \leftarrow (4a + 8) - (4a + 8) = 0
\end{array}$$

The answer is $a^2 - 2a + 4$.

25.

$$\begin{array}{r l}
 & t + 4 \\
t - 4 \,\big)\!\! & t^2 + 0t - 13 \leftarrow \text{Writing in the missing term} \\
 & t^2 - 4t \\
 & \quad 4t - 13 \leftarrow t^2 - (t^2 - 4t) = 4t \\
 & \quad 4t - 16 \\
 & \qquad 3 \leftarrow (4t - 13) - (4t - 16) = 3
\end{array}$$

The answer is $t + 4 + \dfrac{3}{t - 4}$.

27.

$$\begin{array}{r l}
 & x + 4 \\
3x - 1 \,\big)\!\! & 3x^2 + 11x - 4 \\
 & 3x^2 - x \\
 & \quad 12x - 4 \leftarrow (3x^2 + 11x) - (3x^2 - x) = 12x \\
 & \quad 12x - 4 \\
 & \qquad 0 \leftarrow (12x - 4) - (12x - 4) = 0
\end{array}$$

The answer is $x + 4$.

29.

$$\begin{array}{r l}
 & 3a + 1 \\
2a + 5 \,\big)\!\! & 6a^2 + 17a + 8 \\
 & 6a^2 + 15a \\
 & \quad 2a + 8 \leftarrow (6a^2 + 17a) - (6a^2 + 15a) = 2a + 8 \\
 & \quad 2a + 5 \\
 & \qquad 3 \leftarrow (2a + 8) - (2a + 5) = 3
\end{array}$$

The answer is $3a + 1 + \dfrac{3}{2a + 5}$.

31.

$$\begin{array}{r l}
 & t^2 - 3t + 1 \\
2t - 3 \,\big)\!\! & 2t^3 - 9t^2 + 11t - 3 \\
 & 2t^3 - 3t^2 \\
 & \quad -6t^2 + 11t \leftarrow (2t^3 - 9t^2) - (2t^3 - 3t^2) = -6t^2 \\
 & \quad -6t^2 + 9t \\
 & \qquad 2t - 3 \leftarrow (-6t^2 + 11t) - (-6t^2 + 9t) = 2t \\
 & \qquad 2t - 3 \\
 & \qquad\quad 0 \leftarrow (2t - 3) - (2t - 3) = 0
\end{array}$$

The answer is $t^2 - 3t + 1$.

33.

$$\begin{array}{r l}
 & t^2-2t+3 \\
t+1 \,\overline{)\, t^3-t^2+t-1} & \\
t^3+t^2 & \\
\hline
-2t^2+t & \leftarrow (t^3-t^2)-(t^3+t^2)=-2t^2 \\
-2t^2-2t & \\
\hline
3t-1 & \leftarrow (-2t^2+t)- \\
 & \quad (-2t^2-2t)=3t \\
3t+3 & \\
\hline
-4 &
\end{array}$$

The answer is $t^2-2t+3+\dfrac{-4}{t+1}$.

35.

$$\begin{array}{r l}
 & t^2 \quad +1 \\
t^2-3 \,\overline{)\, t^4+0t^3-2t^2+4t-5} & \leftarrow \text{Writing in the missing term} \\
t^4 \quad -3t^2 & \\
\hline
t^2+4t-5 & \leftarrow (t^4-2t^2)- \\
t^2 \quad -3 & \quad (t^4-3t^2)=t^2 \\
\hline
4t-2 & \leftarrow (t^2+4t-5)- \\
 & \quad (t^2-3)=4t-2
\end{array}$$

The answer is $t^2+1+\dfrac{4t-2}{t^2-3}$.

37.

$$\begin{array}{r l}
 & 2x^2 \quad +1 \\
2x^2-3 \,\overline{)\, 4x^4+0x^3-4x^2-x-3} & \leftarrow \text{Writing in the missing term} \\
4x^4 \quad -6x^2 & \\
\hline
2x^2-x-3 & \leftarrow (4x^4-4x^2)- \\
 & \quad (4x^4-6x^2)=2x^2 \\
2x^2 \quad -3 & \\
\hline
-x & \leftarrow (2x^2-x-3)- \\
 & \quad (2x^2-3)=-x
\end{array}$$

The answer is $2x^2+1+\dfrac{-x}{2x^2-3}$.

39. *Writing Exercise*

41. $-4+(-13)$ Two negative numbers. Add the absolute values, 4 and 13, to get 17. Make the answer negative.

$-4+(-13)=-17$

43. $-9-(-7)=-9+7=-2$

45. ***Familiarize.*** Let w = the width. Then $w+15$ = the length. We draw a picture.

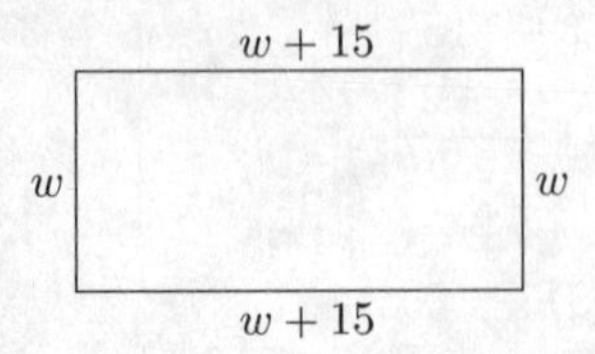

We will use the fact that the perimeter is 640 ft to find w (the width). Then we can find $w+15$ (the length) and multiply the length and the width to find the area.

Translate.

Width+Width+ Length + Length =Perimeter

$w \; + \; w \; +(w+15)+(w+15)= \; 640$

Carry out.

$$\begin{aligned}
w+w+(w+15)+(w+15) &= 640 \\
4w+30 &= 640 \\
4w &= 610 \\
w &= 152.5
\end{aligned}$$

If the width is 152.5, then the length is 152.5+15, or 167.5.

Check. The length, 167.5 ft, is 15 ft greater than the width, 152.5 ft. The perimeter is $152.5+152.5+167.5+167.5$, or 640 ft. The answer checks.

State. The length of the rectangle is 167.5 ft.

47. Graph: $3x-2y=12$.

We will graph the equation using intercepts. To find the y-intercept, we let $x=0$.

$-2y=12$ Ignoring the x-term

$y=-6$

The y-intercept is $(0,-6)$.

To find the x-intercept, we let $y=0$.

$3x=12$ Ignoring the y-term

$x=4$

The x-intercept is $(4,0)$.

To find a third point, replace x with -2 and solve for y:

$$\begin{aligned}
3(-2)-2y &= 12 \\
-6-2y &= 12 \\
-2y &= 18 \\
y &= -9
\end{aligned}$$

The point $(-2,-9)$ appears to line up with the intercepts, so we draw the graph.

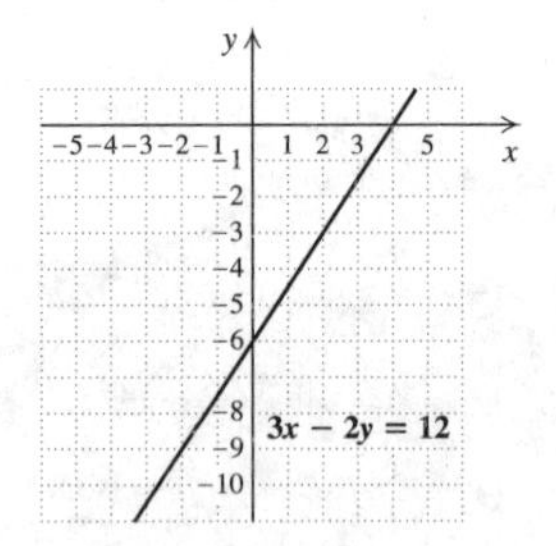

49. *Writing Exercise*

51.

$$\begin{aligned}
&(10x^{9k}-32x^{6k}+28x^{3k}) \div (2x^{3k}) \\
&= \frac{10x^{9k}-32x^{6k}+28x^{3k}}{2x^{3k}} \\
&= \frac{10x^{9k}}{2x^{3k}}-\frac{32x^{6k}}{2x^{3k}}+\frac{28x^{3k}}{2x^{3k}} \\
&= 5x^{9k-3k}-16x^{6k-3k}+14x^{3k-3k} \\
&= 5x^{6k}-16x^{3k}+14
\end{aligned}$$

53.

$$
\begin{array}{r}
3t^{2h}+2t^h-5 \qquad\qquad \\
2t^h+3\,\overline{\big)\,6t^{3h}+13t^{2h}-4t^h-15} \\
\underline{6t^{3h}+9t^{2h}} \qquad\qquad\qquad \\
4t^{2h}-4t^h \qquad\quad \\
\underline{4t^{2h}+6t^h} \qquad\quad \\
-10t^h-15 \\
\underline{-10t^h-15} \\
0
\end{array}
$$

The answer is $3t^{2h}+2t^h-5$.

55.

$$
\begin{array}{r}
a+3 \qquad\qquad \\
5a^2-7a-2\,\overline{\big)\,5a^3+8a^2-23a-1} \\
\underline{5a^3-7a^2-2a} \qquad \\
15a^2-21a-1 \\
\underline{15a^2-21a-6} \\
5
\end{array}
$$

The answer is $a+3+\dfrac{5}{5a^2-7a-2}$.

57.

$$
\begin{aligned}
&(4x^5-14x^3-x^2+3)+ \\
&\qquad (2x^5+3x^4+x^3-3x^2+5x) \\
&= 6x^5+3x^4-13x^3-4x^2+5x+3
\end{aligned}
$$

$$
\begin{array}{r}
2x^2+x-3 \qquad\qquad\qquad \\
3x^3-2x-1\,\overline{\big)\,6x^5+3x^4-13x^3-4x^2+5x+3} \\
\underline{6x^5 \qquad -4x^3-2x^2} \qquad\qquad \\
3x^4-9x^3-2x^2+5x \qquad \\
\underline{3x^4 \qquad\quad -2x^2-x} \qquad \\
-9x^3 \qquad +6x+3 \\
\underline{-9x^3 \qquad +6x+3} \\
0
\end{array}
$$

The answer is $2x^2+x-3$.

59.

$$
\begin{array}{r}
x-3 \qquad \\
x-1\,\overline{\big)\,x^2-4x+c} \\
\underline{x^2-x} \qquad \\
-3x+c \\
\underline{-3x+3} \\
c-3
\end{array}
$$

We set the remainder equal to 0.

$c-3=0$

$c=3$

Thus, c must be 3.

61.

$$
\begin{array}{r}
c^2x+(2c+c^2) \qquad\qquad \\
x-1\,\overline{\big)\,c^2x^2+2cx+1} \qquad\qquad \\
\underline{c^2x^2-c^2x} \qquad\qquad\qquad \\
(2c+c^2)x+1 \qquad\quad \\
\underline{(2c+c^2)x-(2c+c^2)} \\
1+(2c+c^2)
\end{array}
$$

We set the remainder equal to 0.

$c^2+2c+1=0$

$(c+1)^2=0$

$c+1=0$ *or* $c+1=0$

$c=-1$ *or* $c=-1$

Thus, c must be -1.

Exercise Set 4.8

1. $7^{-2}=\dfrac{1}{7^2}=\dfrac{1}{49}$

3. $10^{-4}=\dfrac{1}{10^4}=\dfrac{1}{10,000}$

5. $(-2)^{-6}=\dfrac{1}{(-2)^6}=\dfrac{1}{64}$

7. $x^{-8}=\dfrac{1}{x^8}$

9. $xy^{-2}=x\cdot\dfrac{1}{y^2}=\dfrac{x}{y^2}$

11. $r^{-5}t=\dfrac{1}{r^5}\cdot t=\dfrac{t}{r^5}$

13. $\dfrac{1}{t^{-8}}=t^8$

15. $\dfrac{1}{h^{-8}}=h^8$

17. $7^{-1}=\dfrac{1}{7^1}=\dfrac{1}{7}$

19. $\left(\dfrac{3}{5}\right)^{-2}=\left(\dfrac{5}{3}\right)^2=\dfrac{5^2}{3^2}=\dfrac{25}{9}$

21. $\left(\dfrac{a}{2}\right)^{-3}=\left(\dfrac{2}{a}\right)^3=\dfrac{2^3}{a^3}=\dfrac{8}{a^3}$

23. $\left(\dfrac{s}{t}\right)^{-7}=\left(\dfrac{t}{s}\right)^7=\dfrac{t^7}{s^7}$

25. $\dfrac{1}{6^2}=6^{-2}$

27. $\dfrac{1}{t^6}=t^{-6}$

29. $\dfrac{1}{a^4}=a^{-4}$

31. $\dfrac{1}{p^7}=p^{-7}$

33. $\dfrac{1}{5}=\dfrac{1}{5^1}=5^{-1}$

35. $\dfrac{1}{t}=\dfrac{1}{t^1}=t^{-1}$

37. $2^{-5}\cdot 2^8=2^{-5+8}=2^3$, or 8

39. $x^{-2}\cdot x^{-7}=x^{-2+(-7)}=x^{-9}=\dfrac{1}{x^9}$

41. $t^{-3}\cdot t=t^{-3}\cdot t^1=t^{-3+1}=t^{-2}=\dfrac{1}{t^2}$

43. $(a^{-2})^9=a^{-2\cdot 9}=a^{-18}=\dfrac{1}{a^{18}}$

45. $(t^{-3})^{-6}=t^{-3(-6)}=t^{18}$

47. $(t^4)^{-3}=t^{4(-3)}=t^{-12}=\dfrac{1}{t^{12}}$

49. $(x^{-2})^{-4} = x^{-2(-4)} = x^8$

51. $(ab)^{-3} = \frac{1}{(ab)^3} = \frac{1}{a^3b^3}$

53. $(mn)^{-7} = \frac{1}{(mn)^7} = \frac{1}{m^7n^7}$

55. $(3x^{-4})^2 = 3^2(x^{-4})^2 = 9x^{-8} = \frac{9}{x^8}$

57. $(5r^{-4}t^3)^2 = 5^2(r^{-4})^2(t^3)^2 = 25r^{-8}t^6 = \frac{25t^6}{r^8}$

59. $\frac{t^7}{t^{-3}} = t^{7-(-3)} = t^{10}$

61. $\frac{y^{-7}}{y^{-3}} = y^{-7-(-3)} = y^{-4} = \frac{1}{y^4}$

63. $\frac{12y^{-4}}{4y^{-9}} = 3y^{-4-(-9)} = 3y^5$

65. $\frac{2x^6}{x} = 2\frac{x^6}{x^1} = 2x^{6-1} = 2x^5$

67. $\frac{15a^{-7}}{10b^{-9}} = \frac{3b^9}{2a^7}$

69. $\frac{t^{-7}}{t^{-7}}$

Note that we have an expression divided by itself. Thus, the result is 1. We could also find this result as follows:

$$\frac{t^{-7}}{t^{-7}} = t^{-7-(-7)} = t^0 = 1.$$

71. $\frac{3x^{-5}}{y^{-6}z^{-2}} = \frac{3y^6z^2}{x^5}$

73. $\frac{3t^4}{s^{-2}u^{-4}} = 3s^2t^4u^4$

75. $(x^4y^5)^{-3} = (x^4)^{-3}(y^5)^{-3} = x^{-12}y^{-15} = \frac{1}{x^{12}y^{15}}$

77. $(x^{-6}y^{-2})^{-4} = (x^{-6})^{-4}(y^{-2})^{-4} = x^{24}y^8$

79. $(a^{-5}b^7c^{-2})(a^{-3}b^{-2}c^6) = a^{-5+(-3)}b^{7+(-2)}c^{-2+6} =$
$a^{-8}b^5c^4 = \frac{b^5c^4}{a^8}$

81. $\left(\frac{a^4}{3}\right)^{-2} = \left(\frac{3}{a^4}\right)^2 = \frac{3^2}{(a^4)^2} = \frac{9}{a^8}$

83. $\left(\frac{m^{-1}}{n^{-4}}\right)^3 = \frac{(m^{-1})^3}{(n^{-4})^3} = \frac{m^{-3}}{n^{-12}} = \frac{n^{12}}{m^3}$

85. $\left(\frac{2a^2}{3b^4}\right)^{-3} = \left(\frac{3b^4}{2a^2}\right)^3 = \frac{(3b^4)^3}{(2a^2)^3} = \frac{3^3(b^4)^3}{2^3(a^2)^3} = \frac{27b^{12}}{8a^6}$

87. $\left(\frac{5x^{-2}}{3y^{-2}z}\right)^0$

Any nonzero expression raised to the 0 power is equal to 1. Thus, the answer is 1.

89. 7.12×10^4

Since the exponent is positive, the decimal point will move to the right.

7.1200. The decimal point moves right 4 places.

$7.12 \times 10^4 = 71,200$

91. 8.92×10^{-3}

Since the exponent is negative, the decimal point will move to the left.

.008.92 The decimal point moves left 3 places.

$8.92 \times 10^{-3} = 0.00892$

93. 9.04×10^8

Since the exponent is positive, the decimal point will move to the right.

9.04000000. 8 places

$9.04 \times 10^8 = 904,000,000$

95. 2.764×10^{-10}

Since the exponent is negative, the decimal point will move to the left.

0.0000000002.764 10 places

$2.764 \times 10^{-10} = 0.0000000002764$

97. 4.209×10^9

Since the exponent is positive, the decimal point will move to the right.

4.2090000. 7 places

$4.209 \times 10^7 = 42,090,000$

99. $490,000 = 4.9 \times 10^m$

To write 4.9 as 490,000 we move the decimal point 5 places to the right. Thus, m is 5 and

$$490,000 = 4.9 \times 10^5.$$

101. $0.00583 = 5.83 \times 10^m$

To write 5.83 as 0.00583 we move the decimal point 3 places to the left. Thus, m is -3 and

$$0.00583 = 5.83 \times 10^{-3}.$$

103. $78,000,000,000 = 7.8 \times 10^m$

To write 7.8 as 78,000,000,000 we move the decimal point 10 places to the right. Thus, m is 10 and

$$78,000,000,000 = 7.8 \times 10^{10}.$$

105. $0.000000527 = 5.27 \times 10^m$

To write 5.27 as 0.000000527 we move the decimal point 7 places to the left. Thus, m is -7 and

$$0.000000527 = 5.27 \times 10^{-7}.$$

107. $0.000000018 = 1.8 \times 10^m$

To write 1.8 as 0.000000018 we move the decimal point 8 places to the left. Thus, m is -8 and

$$0.000000018 = 1.8 \times 10^{-8}.$$

109. $1,094,000,000,000,000 = 1.094 \times 10^m$

To write 1.094 as 1,094,000,000,000,000 we move the decimal point 15 places to the right. Thus, m is 15 and

$$1,094,000,000,000,000 = 1.094 \times 10^{15}.$$

111. $(4\times 10^7)(2\times 10^5) = (4\cdot 2)\times(10^7\cdot 10^5)$

$= 8\times 10^{7+5}$ Adding exponents

$= 8\times 10^{12}$

113. $(3.8\times 10^9)(6.5\times 10^{-2}) = (3.8\cdot 6.5)\times(10^9\cdot 10^{-2})$

$= 24.7\times 10^7$

The answer is not yet in scientific notation since 24.7 is not a number between 1 and 10. We convert to scientific notation.

$$24.7\times 10^7 = (2.47\times 10)\times 10^7 = 2.47\times 10^8$$

115. $(8.7\times 10^{-12})(4.5\times 10^{-5})$

$= (8.7\cdot 4.5)\times(10^{-12}\cdot 10^{-5})$

$= 39.15\times 10^{-17}$

The answer is not yet in scientific notation since 39.15 is not a number between 1 and 10. We convert to scientific notation.

$$39.15\times 10^{-17} = (3.915\times 10)\times 10^{-17} = 3.915\times 10^{-16}$$

117. $\dfrac{8.5\times 10^8}{3.4\times 10^{-5}} = \dfrac{8.5}{3.4}\times\dfrac{10^8}{10^{-5}}$

$= 2.5\times 10^{8-(-5)}$

$= 2.5\times 10^{13}$

119. $(3.0\times 10^6)\div(6.0\times 10^9) = \dfrac{3.0\times 10^6}{6.0\times 10^9}$

$= \dfrac{3.0}{6.0}\times\dfrac{10^6}{10^9}$

$= 0.5\times 10^{6-9}$

$= 0.5\times 10^{-3}$

The answer is not yet in scientific notation because 0.5 is not between 1 and 10. We convert to scientific notation.

$$0.5\times 10^{-3} = (5\times 10^{-1})\times 10^{-3} = 5\times 10^{-4}$$

121. $\dfrac{7.5\times 10^{-9}}{2.5\times 10^{12}} = \dfrac{7.5}{2.5}\times\dfrac{10^{-9}}{10^{12}}$

$= 3.0\times 10^{-9-12}$

$= 3.0\times 10^{-21}$

123. *Writing Exercise*

125. $(3-8)(9-12)$

$= (-5)(-3)$ Subtracting

$= 15$ Multiplying

127. $7\cdot 2+8^2$

$= 7\cdot 2+64$ Evaluating the exponential expression

$= 14+64$ Multiplying

$= 78$ Adding

129. To plot $(-3,2)$, we start at the origin and move 3 units to the left and then 2 units up. To plot $(4,-1)$, we start at the origin and move 4 units to the right and then 1 unit down. To plot $(5,3)$, we start at the origin and move 5 units to the right and 3 units up. To plot $(-5,-2)$, we start at the origin and move 5 units to the left and then 2 units down.

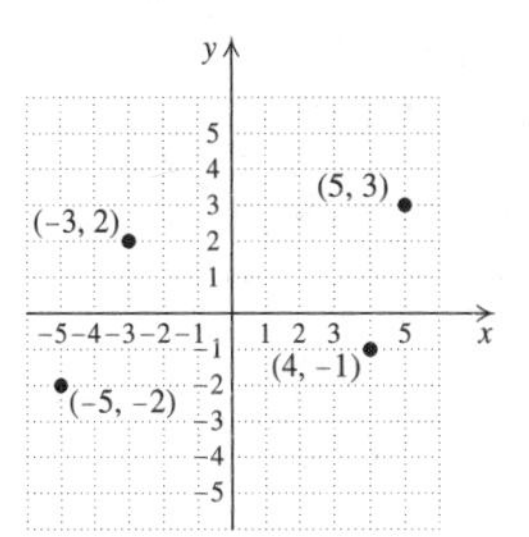

131. *Writing Exercise*

133. $\dfrac{1}{1.25\times 10^{-6}} = \dfrac{1}{1.25}\times\dfrac{1}{10^{-6}} = 0.8\times 10^6 = (8\times 10^{-1})\times 10^6 = 8\times 10^5$

135. $8^{-3}\cdot 32\div 16^2 = (2^3)^{-3}\cdot 2^5\div(2^4)^2 = 2^{-9}\cdot 2^5\div 2^8 = 2^{-4}\div 2^8 = 2^{-12}$

137. $\dfrac{125^{-4}(25^2)^4}{125} = \dfrac{(5^3)^{-4}((5^2)^2)^4}{5^3} = \dfrac{5^{-12}(5^4)^4}{5^3} = \dfrac{5^{-12}\cdot 5^{16}}{5^3} = \dfrac{5^4}{5^3} = 5^1 = 5$

139. $\dfrac{4.2\times 10^8[(2.5\times 10^{-5})\div(5.0\times 10^{-9})]}{3.0\times 10^{-12}}$

$= \dfrac{4.2\times 10^8[0.5\times 10^4]}{3.0\times 10^{-12}}$

$= \dfrac{2.1\times 10^{12}}{3.0\times 10^{-12}}$

$= 0.7\times 10^{24}$

$= (7\times 10^{-1})\times 10^{24}$

$= 7\times 10^{23}$

141. $\dfrac{7.4\times 10^{29}}{(5.4\times 10^{-6})(2.8\times 10^8)}$

$= \dfrac{7.4}{(5.4\cdot 2.8)}\times\dfrac{10^{29}}{(10^{-6}\cdot 10^8)}$

$\approx 0.4894179894\times 10^{27}$

$\approx (4.894179894\times 10^{-1})\times 10^{27}$

$\approx 4.894179894\times 10^{26}$

143. $\dfrac{(7.8\times 10^7)(8.4\times 10^{23})}{2.1\times 10^{-12}}$

$= \dfrac{(7.8\cdot 8.4)}{2.1}\times\dfrac{(10^7\cdot 10^{23})}{10^{-12}}$

$= 31.2\times 10^{42}$

$= (3.12\times 10)\times 10^{42}$

$= 3.12\times 10^{43}$

145. a) False; let $x=2$, $y=3$, $m=4$, and $n=2$:

$2^4\cdot 3^2 = 16\cdot 9 = 144$, but

$(2\cdot 3)^{4\cdot 2} = 6^8 = 1,679,616$

b) False; let $x=3$, $y=4$, and $m=2$:

$3^2\cdot 4^2 = 9\cdot 16 = 144$, but

$(3\cdot 4)^{2\cdot 2} = 12^4 = 20,736$

c) False; let $x=5$, $y=3$, and $m=2$:

$(5-3)^2 = 2^2 = 4$, but

$5^2 - 3^2 = 25 - 9 = 16$

147. ***Familiarize***. Express 1 billion and 2500 in scientific notation:

1 billion $= 1,000,000,000 = 10^9$

$2500 = 2.5\times 10^3$

Let $b =$ the number of bytes in the network.

Translate. We reword the problem.

What is 2500 times 1 gigabyte?

$b \quad = 2.5\times 10^3 \quad \times \quad 10^9$

Carry out. We do the computation.

$b = (2.5\times 10^3)\times 10^9$

$b = 2.5\times(10^3\times 10^9)$

$b = 2.5\times 10^{12}$

Check. We review the computation. Also, the answer seems reasonable since it is larger than 1 billion.

State. There are 2.5×10^{12} bytes in the network.

149. ***Familiarize***. We must express both dimensions using the same units. Let's choose centimeters. First, convert 1.5 m to centimeters and express the result in scientific notation.

$1.5\text{ m} = 1.5\times 1\text{ m} = 1.5\times 100\text{ cm} = 1.5\times 10^2\text{ cm}$

Let l represent how many times the DNA is longer than it is wide.

Translate. We reword the problem.

The length is how many times the width.

$1.5\times 10^2 \quad = \quad l \quad \cdot \quad 1.3\times 10^{-10}$

Carry out. We solve the equation.

$1.5\times 10^2 = l\cdot 1.3\times 10^{-10}$

$\dfrac{1.5\times 10^2}{1.3\times 10^{-10}} = l$

$1.15385\times 10^{12}\approx l$

Check. Since $(1.15385\times 10^{12})\times(1.3\times 10^{-10}) = 1.498705\times 10^2\approx 1.5\times 10^2$, the answer checks.

State. A strand of DNA is about 1.15385×10^{12} times longer than it is wide.

Chapter 5

Polynomials and Factoring

Exercise Set 5.1

1. Since $7a \cdot 5ab = 35a^2b$, choice (h) is most appropriate.

3. $5x + 10 = 5(x + 2)$ and $4x + 8 = 4(x + 2)$, so $x + 2$ is a common factor of $5x+10$ and $4x+8$ and choice (b) is most appropriate.

5. $3x^2(3x^2 - 1) = 9x^4 - 3x^2$, so choice (c) is most appropriate.

7. $3a + 6a^2 = 3a(1 + 2a)$, so $1 + 2a$ is a factor of $3a + 6a^2$ and choice (d) is most appropriate.

9. Answers may vary. $10x^3 = (5x)(2x^2) = (10x^2)(x) = (-2)(-5x^3)$

11. Answers may vary. $-15a^4 = (-15)(a^4) = (-5a)(3a^3) = (-3a^2)(5a^2)$

13. Answers may vary. $26x^5 = (2x^4)(13x) = (2x^3)(13x^2) = (-x^2)(-26x^3)$

15. $7x - 14 = 7 \cdot x - 7 \cdot 2$
$= 7(x - 2)$

17. $3t^2 + t = t \cdot 3t + t \cdot 1$
$= t(3t + 1)$

19. $-4a^2 - 8a = -4a \cdot a - 4a \cdot 2$
$= -4a(a + 2)$

We might also factor as follows:

$-4a^2 - 8a = 4a \cdot (-a) + 4a(-2)$
$= 4a(-a - 2)$

21. $x^3 + 6x^2 = x^2 \cdot x + x^2 \cdot 6$
$= x^2(x + 6)$

23. $8x^4 - 24x^2 = 8x^2 \cdot x^2 - 8x^2 \cdot 3$
$= 8x^2(x^2 - 3)$

25. $2x^2 + 2x - 8 = 2 \cdot x^2 + 2 \cdot x - 2 \cdot 4$
$= 2(x^2 + x - 4)$

27. $-7a^6 + 10a^4 - 14a^2 = -a^2 \cdot 7a^4 - a^2 \cdot (-10a^2) - a^2 \cdot 14$
$= -a^2(7a^4 - 10a^2 + 14)$

29. $6x^8 + 12x^6 - 24x^4 + 30x^2$
$= 6x^2 \cdot x^6 + 6x^2 \cdot 2x^4 - 6x^2 \cdot 4x^2 + 6x^2 \cdot 5$
$= 6x^2(x^6 + 2x^4 - 4x^2 + 5)$

31. $x^5y^5 + x^4y^3 + x^3y^3 - x^2y^2$
$= x^2y^2 \cdot x^3y^3 + x^2y^2 \cdot x^2y + x^2y^2 \cdot xy - x^2y^2 \cdot 1$
$= x^2y^2(x^3y^3 + x^2y + xy - 1)$

33. $-5a^3b^4 + 10a^2b^3 - 15a^3b^2$
$= -5a^2b^2 \cdot ab^2 - 5a^2b^2 \cdot (-2b) - 5a^2b^2 \cdot 3a$
$= -5a^2b^2(ab^2 - 2b + 3a)$

35. $y(y - 2) + 7(y - 2)$
$= (y - 2)(y + 7)$ Factoring out the common binomial factor $y-2$

37. $x^2(x + 3) - 7(x + 3)$
$= (x + 3)(x^2 - 7)$ Factoring out the common binomial factor $x + 3$

39. $y^2(y + 8) + (y + 8) = y^2(y + 8) + 1(y + 8)$
$= (y + 8)(y^2 + 1)$ Factoring out the common factor

41. $x^3 + 3x^2 + 4x + 12$
$= (x^3 + 3x^2) + (4x + 12)$
$= x^2(x + 3) + 4(x + 3)$ Factoring each binomial
$= (x + 3)(x^2 + 4)$ Factoring out the common factor $x + 3$

43. $5a^3 + 15a^2 + 2a + 6$
$= (5a^3 + 15a^2) + (2a + 6)$
$= 5a^2(a + 3) + 2(a + 3)$ Factoring each binomial
$= (a + 3)(5a^2 + 2)$ Factoring out the common factor $a + 3$

45. $9x^3 - 12x^2 + 3x - 4$
$= 3x^2(3x - 4) + 1(3x - 4)$
$= (3x - 4)(3x^2 + 1)$

47. $4t^3 - 20t^2 + 3t - 15$
$= 4t^2(t - 5) + 3(t - 5)$
$= (t - 5)(4t^2 + 3)$

49. $7x^3 + 2x^2 - 14x - 4$
$= x^2(7x + 2) - 2(7x + 2)$
$= (7x + 2)(x^2 - 2)$

51. $6a^3 - 7a^2 + 6a - 7$
$= a^2(6a - 7) + 1(6a - 7)$
$= (6a - 7)(a^2 + 1)$

53. $x^3 + 8x^2 - 3x - 24 = x^2(x + 8) - 3(x + 8)$
$= (x + 8)(x^2 - 3)$

55. $2x^3 + 12x^2 - 5x - 30 = 2x^2(x + 6) - 5(x + 6)$
$= (x + 6)(2x^2 - 5)$

57. We try factoring by grouping.

$p^3 + p^2 - 3p + 10 = p^2(p+1) - (3p - 10)$, or
$p^3 - 3p + p^2 + 10 = p(p^2 - 3) + p^2 + 10$

Because we cannot find a common binomial factor, this polynomial cannot be factored using factoring by grouping.

59. $y^3 + 8y^2 - 2y - 16 = y^2(y+8) - 2(y+8) = (y+8)(y^2-2)$

61. $2x^3 - 8x^2 - 9x + 36 = 2x^2(x-4) - 9(x-4)$
$= (x-4)(2x^2-9)$

63. *Writing Exercise*

65. $(x+3)(x+5)$
F O I L
$= x \cdot x + x \cdot 5 + 3 \cdot x + 3 \cdot 5$
$= x^2 + 5x + 3x + 15$
$= x^2 + 8x + 15$

67. $(a-7)(a+3)$
F O I L
$= a \cdot a + a \cdot 3 - 7 \cdot a - 7 \cdot 3$
$= a^2 + 3a - 7a - 21$
$= a^2 - 4a - 21$

69. $(2x+5)(3x-4)$
F O I L
$= 2x \cdot 3x - 2x \cdot 4 + 5 \cdot 3x - 5 \cdot 4$
$= 6x^2 - 8x + 15x - 20$
$= 6x^2 + 7x - 20$

71. $(3t-5)^2$
$= (3t)^2 - 2 \cdot 3t \cdot 5 + 5^2$
$[(A-B)^2 = A^2 - 2AB + B^2]$
$= 9t^2 - 30t + 25$

73. *Writing Exercise*

75. $4x^5 + 6x^2 + 6x^3 + 9 = 2x^2(2x^3+3) + 3(2x^3+3)$
$= (2x^3+3)(2x^2+3)$

77. $x^{12} + x^7 + x^5 + 1 = x^7(x^5+1) + (x^5+1)$
$= (x^5+1)(x^7+1)$

79. $5x^5 - 5x^4 + x^3 - x^2 + 3x - 3$
$= 5x^4(x-1) + x^2(x-1) + 3(x-1)$
$= (x-1)(5x^4 + x^2 + 3)$

We could also do this exercise as follows:
$5x^5 - 5x^4 + x^3 - x^2 + 3x - 3$
$= (5x^5 + x^3 + 3x) - (5x^4 + x^2 + 3)$
$= x(5x^4 + x^2 + 3) - 1(5x^4 + x^2 + 3)$
$= (5x^4 + x^2 + 3)(x-1)$

81. Answers may vary. $8x^4y^3 - 24x^3y^3 + 16x^2y^4$

Exercise Set 5.2

1. If c is positive, then p and q have the same sign. If both are negative, then b is negative; if both are positive then c is positive. Thus we replace each blank with "positive."

3. If p is negative and q is negative, then b is negative because it is the sum of two negative numbers and c is positive because it is the product of two negative numbers.

5. Since c is negative, it is the product of a negative and a positive number. Then because c is the product of p and q and we know that p is negative, q must be positive.

7. $x^2 + 7x + 6$

Since the constant term and the coefficient of the middle term are both positive, we look for a factorization of 6 in which both factors are positive. Their sum must be 7.

Pairs of factors	Sums of factors
1, 6	7
2, 3	5

The numbers we want are 1 and 6.

$x^2 + 7x + 6 = (x+1)(x+6)$

9. $x^2 + 7x + 12$

Since the constant term is positive and the coefficient of the middle term is positive, we look for a factorization of 12 in which both factors are positive. Their sum must be 7.

Pairs of factors	Sums of factors
1, 2	11
2, 6	7
3, 4	7

The numbers we want are 3 and 4.

$x^2 + 7x + 12 = (x+3)(x+4)$

11. $x^2 - 6x + 9$

Since the constant term is positive and the middle term is negative, we look for a factorization of 9 in which both factors are negative.

Pairs of factors	Sums of factors
−1, −9	−10
−3, −3	−6

The numbers we want are −3 and −3.

$x^2 - 6x + 9 = (x-3)(x-3)$, or $(x-3)^2$

13. $x^2 + 9x + 14$

Since the constant term is positive and the coefficient of the middle term is positive, we look for a factorization of 14 in which both factors are positive. Their sum must be 9.

Pairs of factors	Sums of factors
1, 14	15
2, 7	9

The numbers we want are 2 and 7.

$x^2 + 9x + 14 = (x+2)(x+7)$

15. $b^2 - 5b + 4$

Since the constant term is positive and the coefficient of the middle term is negative, we look for a factorization of 4 in which both factors are negative. Their sum must be -5.

Pairs of factors	Sums of factors
$-1, -4$	-5
$-2, -2$	-4

The numbers we want are -1 and -4.

$b^2 - 5b + 4 = (b - 1)(b - 4)$.

17. $a^2 + 4a - 12$

Since the constant term is negative, we look for a factorization of -12 in which one factor is positive and one factor is negative. Their sum must be 4, the coefficient of the middle term, so the positive factor must have the larger absolute value. Thus we consider only pairs of factors in which the positive factor has the larger absolute value.

Pairs of factors	Sums of factors
$-1, \quad 12$	11
$-2, \quad 6$	4
$-3, \quad 4$	1

The numbers we need are -2 and 6.

$a^2 + 4a - 12 = (a - 2)(a + 6)$.

19. $d^2 - 7d + 10$

Since the constant term is positive and the coefficient of the middle term is negative, we look for a factorization of 10 in which both factors are negative. Their sum must be -7.

Pairs of factors	Sums of factors
$-1, -10$	-11
$-2, \quad -5$	-7

The numbers we want are -2 and -5.

$d^2 - 7d + 10 = (d - 2)(d - 5)$.

21. $x^2 - 2x - 15$

The constant term, -15, must be expressed as the product of a negative number and a positive number. Since the sum of those two numbers must be negative, the negative number must have the greater absolute value.

Pairs of factors	Sums of factors
$1, \quad -15$	-14
$3, \quad -5$	-2

The numbers we need are 3 and -5.

$x^2 - 2x - 15 = (x + 3)(x - 5)$.

23. $x^2 + 2x - 15$

The constant term, -15, must be expressed as the product of a negative number and a positive number. Since the sum of those two numbers must be positive, the positive number must have the greater absolute value.

Pairs of factors	Sums of factors
$-1, \quad 15$	14
$-3, \quad 5$	2

The numbers we need are -3 and 5.

$x^2 + 2x - 15 = (x - 3)(x + 5)$.

25. $3y^2 - 9y - 84 = 3(y^2 - 3y - 28)$

After factoring out the common factor, 3, we consider $y^2-3y+28$. The constant term, -28, must be expressed as the product of a negative number and a positive number. Since the sum of those two numbers must be negative, the negative number must have the greater absolute value.

Pairs of factors	Sums of factors
$1, \quad -28$	-27
$2, \quad -14$	-12
$4, \quad -7$	-3

The numbers we need are 4 and -7. The factorization of $y^2 - 3y - 28$ is $(y + 4)(y - 7)$. We must not forget the common factor, 3. Thus, $3y^2 - 9y - 84 = 3(y^2 - 3y - 28) = 3(y + 4)(y - 7)$.

27. $-x^3 + x^2 + 42x = -x(x^2 - x - 42)$

After factoring out the common factor, $-x$, we consider $x^2 - x - 42$. The constant term, -42, must be expressed as the product of a negative number and a positive number. Since the sum of those two numbers must be negative, the negative number must have the greater absolute value.

Pairs of factors	Sums of factors
$1, \quad -42$	-41
$2, \quad -21$	-19
$3, \quad -14$	-11
$6, \quad -7$	-1

The numbers we need are 6 and -7. The factorization of x^2-x-42 is $(x+6)(x-7)$. We must not forget the common factor, $-x$. Thus, $-x^3 + x^2 + 42x = -x(x^2 - x - 42) = -x(x + 6)(x - 7)$.

29. $7x - 60 + x^2 = x^2 + 7x - 60$

The constant term, -60, must be expressed as the product of a negative number and a positive number. Since the sum of those two numbers must be positive, the positive number must have the greater absolute value.

Pairs of factors	Sums of factors
$-1, \quad 60$	59
$-2, \quad 30$	28
$-3, \quad 20$	17
$-4, \quad 15$	11
$-5, \quad 12$	7

The numbers we need are -5 and 12.

$7x - 60 + x^2 = (x - 5)(x + 12)$

31. $x^2 - 72 + 6x = x^2 + 6x - 72$

The constant term, -72, must be expressed as the product of a negative number and a positive number. Since the sum of those two numbers must be positive, the positive number must have the greater absolute value.

Pairs of factors	Sums of factors
-1, 72	71
-2, 36	34
-3, 24	21
-4, 18	14
-6, 12	6

The numbers we need are -6 and 12.

$x^2 - 72 + 6x = (x-6)(x+12)$

33. $-5b^2 - 25b + 120 = -5(b^2 + 5b - 24)$

After factoring out the common factor, -5, we consider $b^2 + 5b - 24$. The constant term, -24, must be expressed as the product of a negative number and a positive number. Since the sum of those two numbers must be positive, the positive number must have the greater absolute value.

Pairs of factors	Sums of factors
-1, 24	23
-2, 12	10
-3, 8	5

The numbers we need are -3 and 8. The factorization of $b^2+5b-24$ is $(b-3)(b+8)$. We must not forget the common factor. Thus, $-5b^2 - 25b + 120 = -5(b^2 + 5b - 24) = -5(b-3)(b+8)$.

35. $x^5 - x^4 - 2x^3 = x^3(x^2 - x - 2)$

After factoring out the common factor, x^3, we consider $x^2 - x - 2$. The constant term, -2, must be expressed as the product of a negative number and a positive number. Since the sum of those two numbers must be negative, the negative number must have the greater absolute value. The only possible factors that fill these requirements are 1 and -2. These are the numbers we need. The factorization of $x^2 - x - 2$ is $(x+1)(x-2)$. We must not forget the common factor, x^3. Thus, $x^5 - x^4 - 2x^3 = x^3(x^2 - x - 2) = x^3(x+1)(x-2)$.

37. $x^2 + 2x + 3$

Since the constant term and the coefficient of the middle term are both positive, we look for a factorization of 3 in which both factors are positive. Their sum must be 2. The only possible pair of positive factors is 1 and 3, but their sum is not 2. Thus, this polynomial is not factorable into polynomials with integer coefficients. It is prime.

39. $50 + 15t + t^2 = t^2 + 15t + 50$

Since the constant term is positive and the coefficient of the middle term is positive, we look for a factorization of 50 in which both factors are positive. Their sum must be 15.

Pairs of factors	Sums of factors
1, 50	51
2, 25	27
5, 10	15

The numbers we want are 5 and 10.

$50 + 15t + t^2 = (t+5)(t+10)$.

41. $x^2 + 20x + 99$

We look for two factors, both positive, whose product is 99 and whose sum is 20.

They are 9 and 11: $9 \cdot 11 = 99$ and $9 + 11 = 20$.

$x^2 + 20 + 99 = (x+9)(x+11)$

43. $2x^3 - 40x^2 + 192x = 2x(x^2 - 20x + 96)$

After factoring out the common factor, $2x$, we consider $x^2 - 20x + 96$. We look for two factors, both negative, whose product is 96 and whose sum is -20.

They are -8 and -12: $-8(-12) = 96$ and $-8 + (-12) = -20$.

$x^2 - 20x + 96 = (x-8)(x-12)$, so $2x^3 - 40x^2 + 192x = 2x(x-8)(x-12)$.

45. $-4x^2 - 40x - 100 = -4(x^2 + 10x + 25)$

After factoring out the common factor, -4, we consider $x^2 + 10x + 25$. We look for two factors, both positive, whose product is 25 and whose sum is 10. They are 5 and 5.

$x^2 + 10x + 25 = (x+5)(x+5)$, so $-4x^2 - 40x - 100 = -4(x+5)(x+5)$, or $-4(x+5)^2$.

47. $y^2 - 21y + 108$

We look for two factors, both negative, whose product is 108 and whose sum is -21. They are -9 and -12.

$y^2 - 21y + 108 = (y-9)(y-12)$

49. $-a^6 - 9a^5 + 90a^4 = -a^4(a^2 + 9a - 90)$

After factoring out the common factor, $-a^4$, we consider $a^2 + 9a - 90$. We look for two factors, one positive and one negative, whose product is -90 and whose sum is 9. They are -6 and 15.

$a^2 + 9a - 90 = (a-6)(a+15)$, so $-a^6 - 9a^5 + 90a^4 = -a^4(a-6)(a+15)$.

51. $t^2 + \frac{2}{3}t + \frac{1}{9}$

We look for two factors, both positive, whose product is $\frac{1}{9}$ and whose sum is $\frac{2}{3}$. They are $\frac{1}{3}$ and $\frac{1}{3}$.

$t^2 + \frac{2}{3}t + \frac{1}{9} = \left(t + \frac{1}{3}\right)\left(t + \frac{1}{3}\right)$, or $\left(t + \frac{1}{3}\right)^2$

53. $11 + w^2 - 4w = w^2 - 4w + 11$

Since the constant term is positive and the coefficient of the middle term is negative, we look for a factorization of 11 in which both factors are negative. Their sum must be -4. The only possible pair of factors is -1 and -11, but their sum is not -4. Thus, this polynomial is not factorable into polynomials with integer coefficients. It is prime.

55. $p^2 + 3pq - 10q^2 = p^2 + 3qp - 10q^2$

Think of $3q$ as a "coefficient" of p. Then we look for factors of $-10q^2$ whose sum is $3q$. They are $5q$ and $-2q$.

$p^2 + 3pq - 10q^2 = (p + 5q)(p - 2q)$.

57. $m^2 + 5mn + 5n^2 = m^2 + 5nm + 5n^2$

We look for factors of $5n^2$ whose sum is $5n$. The only reasonable possibilities are shown below.

Pairs of factors	Sums of factors
$5n,\ n$	$6n$
$-5n,\ -n$	$-6n$

There are no factors whose sum is $5n$. Thus, the polynomial is not factorable into polynomials with integer coefficients. It is prime.

59. $s^2 - 2st - 15t^2 = s^2 - 2ts - 15t^2$

We look for factors of $-15t^2$ whose sum is $-2t$. They are $-5t$ and $3t$.

$s^2 - 2st - 15t^2 = (s - 5t)(s + 3t)$

61. $6a^{10} - 30a^9 - 84a^8 = 6a^8(a^2 - 5a - 14)$

After factoring out the common factor, $6a^8$, we consider $a^2 - 5a - 14$. We look for two factors, one positive and one negative, whose product is -14 and whose sum is -5. They are 2 and -7.

$a^2 - 5a - 14 = (a + 2)(a - 7)$, so $6a^{10} - 30a^9 - 84a^8 = 6a^8(a + 2)(a - 7)$.

63. *Writing Exercise*

65. $3x - 8 = 0$

$3x = 8$ Adding 8 on both sides

$x = \frac{8}{3}$ Dividing by 3 on both sides

The solution is $\frac{8}{3}$.

67. $(x + 6)(3x + 4)$

$= 3x^2 + 4x + 18x + 24$ Using FOIL

$= 3x^2 + 22x + 24$

69. ***Familiarize.*** Let n = the number of people arrested the year before.

Translate. We reword the problem.

Number arrested the year before less 1.2% of that number is 29,090.

$$n - 1.2\% \cdot n = 29{,}090$$

Carry out. We solve the equation.

$$n - 1.2\% \cdot n = 29{,}090$$
$$1 \cdot n - 0.012n = 29{,}090$$
$$0.988n = 29{,}090$$
$$n \approx 29{,}443 \quad \text{Rounding}$$

Check. 1.2% of 29,443 is $0.012(29,443) \approx 353$ and $29,443 - 353 = 29,090$. The answer checks.

State. Approximately 29,443 people were arrested the year before.

71. *Writing Exercise*

73. $a^2 + ba - 50$

We look for all pairs of integer factors whose product is -50. The sum of each pair is represented by b.

Pairs of factors whose product is -50	Sums of factors
$-1,\ 50$	49
$1, -50$	-49
$-2,\ 25$	23
$2, -25$	-23
$-5,\ 10$	5
$5, -10$	-5

The polynomial $a^2 + ba - 50$ can be factored if b is 49, -49, 23, -23, 5, or -5.

75. $y^2 - 0.2y - 0.08$

We look for two factors, one positive and one negative, whose product is -0.08 and whose sum is -0.2. They are -0.4 and 0.2.

$y^2 - 0.2y - 0.08 = (y - 0.4)(y + 0.2)$

77. $-\frac{1}{3}a^3 + \frac{1}{3}a^2 + 2a = -\frac{1}{3}a(a^2 - a - 6)$

After factoring out the common factor, $-\frac{1}{3}a$, we consider $a^2 - a - 6$. We look for two factors, one positive and one negative, whose product is -6 and whose sum is -1. They are 2 and -3.

$a^2 - a - 6 = (a + 2)(a - 3)$, so
$-\frac{1}{3}a^3 + \frac{1}{3}a^2 + 2a = -\frac{1}{3}a(a + 2)(a - 3)$.

79. $x^{2m} + 11x^m + 28 = (x^m)^2 + 11x^m + 28$

We look for numbers p and q such that $x^{2m} + 11x^m + 28 = (x^m + p)(x^m + q)$. We find two factors, both positive, whose product is 28 and whose sum is 11. They are 4 and 7.

$x^{2m} + 11x^m + 28 = (x^m + 4)(x^m + 7)$

81. $(a + 1)x^2 + (a + 1)3x + (a + 1)2$

$= (a + 1)(x^2 + 3x + 2)$

After factoring out the common factor $a + 1$, we consider $x^2 + 3x + 2$. We look for two factors, whose product is 2 and whose sum is 3. They are 1 and 2.

$x^2 + 3x + 2 = (x + 1)(x + 2)$, so
$(a + 1)x^2 + (a + 1)3x + (a + 1)2 =$
$(a + 1)(x + 1)(x + 2)$.

83. $6x^2 + 36x + 54 = 6(x^2 + 6x + 9) = 6(x + 3)(x + 3) = 6(x + 3)^2$

Since the surface area of a cube with sides is given by $6s^2$, we know that this cube has side $x + 3$. The volume of a cube with side s is given by s^3, so the volume of this cube is $(x + 3)^3$, or $x^3 + 9x^2 + 27x + 27$.

85. Shaded area = Area of circle − Area of triangle =

$\pi x^2 - \frac{1}{2}(2x)(x) = \pi x^2 - x^2 = x^2(\pi - 1)$

87. The shaded area consists of the area of a square with side $x + x + x$, or $3x$, less the area of a semicircle with radius x. It can be expressed as follows:

$3x \cdot 3x - \frac{1}{2}\pi x^2 = 9x^2 - \frac{1}{2}\pi x^2 = x^2\left(9 - \frac{1}{2}\pi\right)$

89. $x^2 + 5x + 7x + 35 = x^2 + 12x + 35 = (x+5)(x+7)$

Exercise Set 5.3

1. Since $(6x-1)(2x+3) = 12x^2+16x-3$, choice (c) is correct.

3. Since $(7x+1)(2x-3) = 14x^2 - 19x - 3$, choice (d) is correct.

5. $2x^2 + 7x - 4$

(1) There is no common factor (other than 1 or −1).

(2) Because $2x^2$ can be factored as $2x \cdot x$, we have this possibility:

$(2x + \quad)(x + \quad)$

(3) There are 3 pairs of factors of −4 and they can be listed two ways:

$-4, 1 \quad 4, -1 \quad 2, -2$

and $\quad 1, -4 \quad -1, 4 \quad -2, 2$

(4) Look for Outer and Inner products resulting from steps (2) and (3) for which the sum is $7x$. We can immediately reject all possibilities in which a factor has a common factor, such as $(2x-4)$ or $(2x+2)$, because we determined at the outset that there is no common factor other than 1 and −1. We try some possibilities:

$(2x+1)(x-4) = 2x^2 - 7x - 4$

$(2x-1)(x+4) = 2x^2 + 7x - 4$

The factorization is $(2x-1)(x+4)$.

7. $3t^2 + 4t - 15$

(1) There is no common factor (other than 1 or −1).

(2) Because $3t^2$ can be factored as $3t \cdot t$, we have this possibility:

$(3t + \quad)(t + \quad)$

(3) There are 4 pairs of factors of −15 and they can be listed two ways:

$-15, 1 \quad 15, -1 \quad -5, 3 \quad 5, -3$

and $\quad 1, -15 \quad -1, 15 \quad 3, -5 \quad -3, 5$

(4) Look for Outer and Inner products resulting from steps (2) and (3) for which the sum is $4t$. We can immediately reject all possibilities in which either factor has a common factor, such as $(3t-15)$ or $(3t+3)$, because at the outset we determined that there is no common factor other than 1 or −1. We try some possibilities:

$(3t+1)(t-15) = 3t^2 - 44t - 15$

$(3t-5)(t+3) = 3t^2 + 4t - 15$

The factorization is $(3t-5)(t+3)$.

9. $6x^2 - 23x + 7$

(1) There is no common factor (other than 1 or −1).

(2) Because $6x^2$ can be factored as $6x \cdot x$ or $3x \cdot 2x$, we have these possibilities:

$(6x + \quad)(x + \quad)$ and $(3x + \quad)(2x + \quad)$

(3) There are 2 pairs of factors of 7 and they can be listed two ways:

$7, 1 \quad -7, -1$

and $\quad 1, 7 \quad -1, -7$

(4) Look for Outer and Inner products resulting from steps (2) and (3) for which the sum is $-23x$. Since the sign of the middle term is negative and the sign of the last term is positive, the factors of 7 must both be negative. We try some possibilities:

$(6x-7)(x-1) = 6x^2 - 13x + 7$

$(3x-7)(2x-1) = 6x^2 - 17x + 7$

$(6x-1)(x-7) = 6x^2 - 43x + 7$

$(3x-1)(2x-7) = 6x^2 - 23x + 7$

The factorization is $(3x-1)(2x-7)$.

11. $7x^2 + 15x + 2$

(1) There is no common factor (other than 1 or −1).

(2) Because $7x^2$ can be factored as $7x \cdot x$, we have this possibility:

$(7x + \quad)(x + \quad)$

(3) There are 2 pairs of factors of 2 and they can be listed two ways:

$2, 1 \quad -2, -1$

and $\quad 1, 2 \quad -1, -2$

(4) Look for Outer and Inner products resulting from steps (2) and (3) for which the sum is $15x$. Since all coefficients are positive, we need consider only positive factors of 2. We try some possibilities:

$(7x+2)(x+1) = 7x^2 + 9x + 2$

$(7x+1)(x+2) = 7x^2 + 15x + 2$

The factorization is $(7x+1)(x+2)$.

13. $9a^2 - 6a - 8$

(1) There is no common factor (other than 1 or −1).

(2) Because $9a^2$ can be factored as $9a \cdot a$ and $3a \cdot 3a$, we have these possibilities:

$(9a + \quad)(a + \quad)$ and $(3a + \quad)(3a + \quad)$

(3) There are 4 pairs of factors of −8 and they can be listed two ways:

$-8, 1 \quad 8, -1 \quad -4, 2 \quad 4, -2$

and $\quad 1, -8 \quad -1, 8 \quad 2, -4 \quad -2, 4$

(4) Look for Outer and Inner products resulting from steps (2) and (3) for which the sum is $-6a$. We try some possibilities:

$(9a-8)(a+1) = 9a^2 + a - 8$

$(9a-4)(a+2) = 9a^2 + 14a - 8$

$(3a+8)(3a-1) = 9a^2 + 21a - 8$

$(3a-4)(3a+2) = 9a^2 - 6a - 8$

The factorization is $(3a-4)(3a+2)$.

15. $6x^2 - 10x - 4$

(1) We factor out the largest common factor, 2:

$2(3x^2 - 5x - 2)$.

Then we factor the trinomial $3x^2 - 5x - 2$.

(2) Because $3x^2$ can be factored as $3x \cdot x$, we have this possibility:

$(3x + \quad)(x + \quad)$

(3) There are 2 pairs of factors of -2 and they can be listed two ways:

$-2, 1 \quad 2, -1$

and $\quad 1, -2 \quad -1, 2$

(4) Look for Outer and Inner products resulting from steps (2) and (3) for which the sum is $-5x$. We try some possibilities:

$(3x-2)(x+1) = 3x^2 + x - 2$

$(3x+2)(x-1) = 3x^2 - x - 2$

$(3x+1)(x-2) = 3x^2 - 5x - 2$

The factorization of $3x^2 - 5x - 2$ is $(3x+1)(x-2)$. We must include the common factor in order to get a factorization of the original trinomial.

$6x^2 - 10x - 4 = 2(3x+1)(x-2)$

17. $12t^2 - 6t - 6$

(1) We factor out the common factor, 6:

$6(2t^2 - t - 1)$.

Then we factor the trinomial $2t^2 - t - 1$.

(2) Because $2t^2$ can be factored as $2t \cdot t$, we have this possibility:

$(2t + \quad)(t + \quad)$

(3) There are 2 pairs of factors of -1. In this case they can be listed in only one way:

$-1, 1 \quad 1, -1$

(4) Look for Outer and Inner products resulting from steps (2) and (3) for which the sum is $-t$. We try some possibilities:

$(2t-1)(t+1) = 2t^2 + t - 1$

$(2t+1)(t-1) = 2t^2 - t - 1$

The factorization of $2t^2 - t - 1$ is $(2t+1)(t-1)$. We must include the common factor in order to get a factorization of the original trinomial.

$12t^2 - 6t - 6 = 6(2t+1)(t-1)$

19. $6 + 19x + 15x^2 = 15x^2 + 19x + 6$

(1) There is no common factor (other than 1 or -1).

(2) Because $15x^2$ can be factored as $15x \cdot x$ and $5x \cdot 3x$, we have these possibilities:

$(15x + \quad)(x + \quad)$ and $(5x + \quad)(3x + \quad)$

(3) Since all coefficients are positive, we need consider only positive pairs of factors of 6. There are 2 such pairs and they can be listed two ways:

$6, 1 \quad 3, 2$

and $\quad 1, 6 \quad 2, 3$

(4) We can immediately reject all possibilities in which either factor has a common factor, such as $(15x+6)$ or $(3x+3)$, because we determined at the outset that there is no common factor other than 1 or -1. We try some possibilities:

$(15x+2)(x+3) = 15x^2 + 47x + 6$

$(5x+6)(3x+1) = 15x^2 + 23x + 6$

$(5x+3)(3x+2) = 15x^2 + 19x + 6$

The factorization is $(5x+3)(3x+2)$.

21. $-35x^2 - 34x - 8$

(1) We factor out -1 in order to have a trinomial with a positive leading coefficient.

$-35x^2 - 34x - 8 = -1(35x^2 + 34x + 8)$

Now we factor $35x^2 + 34x + 8$.

(2) Because $35x^2$ can be factored as $35x \cdot x$ or $7x \cdot 5x$, we have these possibilities:

$(35x + \quad)(x + \quad)$ and $(7x + \quad)(5x + \quad)$

(3) Since all coefficients are positive, we need consider only positive pairs of factors of 8. There are 2 such pairs and they can be listed two ways:

$8, 1 \quad 4, 2$

and $\quad 1, 8 \quad 2, 4$

(4) We try some possibilities:

$(35x+8)(x+1) = 35x^2 + 43x + 8$

$(7x+8)(5x+1) = 35x^2 + 47x + 8$

$(7x+4)(5x+2) = 35x^2 + 34x + 8$

The factorization of $35x^2 + 34x + 8$ is $(7x+4)(5x+2)$.

We must include the factor of -1 in order to get a factorization of the original trinomial.

$-35x^2 - 34x - 8 = -1(7x+4)(5x+2)$, or $-(7x+4)(5x+2)$.

23. $4 + 6t^2 - 13t = 6t^2 - 13t + 4$

(1) There is no common factor (other than 1 or -1).

(2) Because $6t^2$ can be factored as $6t \cdot t$ or $3t \cdot 2t$, we have these possibilities:

$(6t + \quad)(t + \quad)$ and $(3t + \quad)(2t + \quad)$

(3) Since the sign of the middle term is negative but the sign of the last term is positive, we need to consider only negative factors of 4. There is only 1 such pair and it can be listed two ways:

$-4, -1$ and $-1, -4$

(4) We can immediately reject all possibilities in which either factor has a common factor, such as $(6t-4)$ or $(2t-4)$, because we determined at the outset that there is no common factor other than 1 or -1. We try some possibilities:

$(6t-1)(t-4) = 6t^2 - 25t + 4$

$(3t-4)(2t-1) = 6t^2 - 11t + 4$

These are the only possibilities that do not contain a common factor. Since neither is the desired factorization, we must conclude that $4 + 6t^2 - 13t$ is prime.

25. $25x^2 + 40x + 16$

(1) There is no common factor (other than 1 or -1).

(2) Because $25x^2$ can be factored as $25x \cdot x$ or $5x \cdot 5x$, we have these possibilities:

$(25x + \quad)(x + \quad)$ and $(5x + \quad)(5x + \quad)$

(3) Since all coefficients are positive, we need consider only positive pairs of factors of 16. There are 3 such pairs and two of them can be listed two ways:

$16, 1 \quad 8, 2 \quad 4, 4$

and $\quad 1, 16 \quad 2, 8$

(4) We try some possibilities:

$(25x+16)(x+1) = 25x^2 + 41x + 16$

$(5x+8)(5x+2) = 25x^2 + 50x + 16$

$(5x+4)(5x+4) = 25x^2 + 40x + 16$

The factorization is $(5x+4)(5x+4)$, or $(5x+4)^2$.

27. $16a^2 + 78a + 27$

(1) There is no common factor (other than 1 or -1).

(2) Because $16a^2$ can be factored as $16a \cdot a$, $8a \cdot 2a$, or $4a \cdot 4a$, we have these possibilities:

$(16a + \quad)(a + \quad)$ and $(8a + \quad)(2a + \quad)$ and $(4a + \quad)(4a + \quad)$

(3) Since all coefficients are positive, we need consider only positive pairs of factors of 27. There are 2 such pairs and two of them can be listed two ways:

$27, 1 \quad 3, 9$

and $\quad 1, 27 \quad 9, 3$

(4) We try some possibilities:

$(16a+27)(a+1) = 16a^2 + 43a + 27$

$(8a+3)(2a+9) = 16a^2 + 78a + 27$

The factorization is $(8a+3)(2a+9)$.

29. $18t^2 + 24t - 10$

(1) Factor out the common factor, 2:

$2(9t^2 + 12t - 5)$

Then we factor the trinomial $9t^2 + 12t - 5$.

(2) Because $9t^2$ can be factored as $9t \cdot t$ or $3t \cdot 3t$, we have these possibilities:

$(9t + \quad)(t + \quad)$ and $(3t + \quad)(3t + \quad)$

(3) There are 2 pairs of factors of -5 and they can be listed two ways:

$-5, 1 \quad 5, -1$

and $\quad 1, -5 \quad -1, 5$

(4) We try some possibilities:

$(9t-5)(t+1) = 9t^2 + 4t - 5$

$(9t+1)(t-5) = 9t^2 - 44t - 5$

$(3t+1)(3t-5) = 9t^2 - 12t - 5$

$(3t-1)(3t+5) = 9t^2 + 12t - 5$

The factorization of $9t^2+12t-5$ is $(3t-1)(3t+5)$. We must include the common factor in order to get a factorization of the original trinomial.

$18t^2 + 24t - 10 = 2(3t-1)(3t+5)$

31. $-2x^2 + 15 + x = -2x^2 + x + 15$

(1) We factor out -1 in order to have a trinomial with a positive leading coefficient.

$-2x^2 + x + 15 = -1(2x^2 - x - 15)$

Now we factor $2x^2 - x - 15$.

(2) Because $2x^2$ can be factored as $2x \cdot x$ we have this possibility:

$(2x + \quad)(x + \quad)$

(3) There are 4 pairs of factors of -15 and they can be listed two ways:

$-15, 1 \quad 15, -1 \quad -5, 3 \quad 5, -3$

and $\quad 1, -15 \quad -1, 15 \quad 3, -5 \quad -3, 5$

(4) We try some possibilities:

$(2x-15)(x+1) = 2x^2 - 13x - 15$

$(2x-5)(x+3) = 2x^2 + x - 15$

$(2x+5)(x-3) = 2x^2 - x - 15$

The factorization of $2x^2-x-15$ is $(2x+5)(x-3)$. We must include the factor of -1 in order to get a factorization of the original trinomial.

$-2x^2 + 15 + x = -1(2x+5)(x-3)$, or $-(2x+5)(x-3)$

33. $-6x^2 - 33x - 15$

(1) Factor out -3. This not only removes the largest common factor, 3. It also produces a trinomial with a positive leading coefficient.

$-3(2x^2 + 11x + 5)$

Then we factor the trinomial $2x^2 + 11x + 5$.

(2) Because $2x^2$ can be factored as $2x \cdot x$ we have this possibility:

$(2x + \quad)(x + \quad)$

(3) Since all coefficients are positive, we need consider only positive pairs of factors of 5. There is one such pair and it can be listed two ways:

$5, 1 \quad$ and $\quad 1, 5$

(4) We try some possibilities:

$(2x+5)(x+1) = 2x^2 + 7x + 5$

$(2x+1)(x+5) = 2x^2 + 11x + 5$

The factorization of $2x^2+11x+5$ is $(2x+1)(x+5)$. We must include the common factor in order to get a factorization of the original trinomial.

$-6x^2 - 33x - 15 = -3(2x+1)(x+5)$

35. $20x^2 - 25x + 5$

(1) Factor out the common factor, 5:

$5(4x^2 - 5x + 1)$

Then we factor the trinomial $4x^2 - 5x + 1$.

(2) Because $4x^2$ can be factored as $4x \cdot x$ or $2x \cdot 2x$, we have these possibilities:

$(4x + \quad)(x + \quad)$ and $(2x + \quad)(2x + \quad)$

(3) Since the sign of the middle term is negative but the sign of the last term is positive, we need to consider only negative factors of 1. There is only 1 such pair, $-1, -1$.

(4) We try the possibilities:

$(4x - 1)(x - 1) = 4x^2 - 5x + 1$

The factorization of $4x^2-5x+1$ is $(4x-1)(x-1)$. We must include the common factor in order to get a factorization of the original trinomial.

$20x^2 - 25x + 5 = 5(4x - 1)(x - 1)$

37. $12x^2 + 68x - 24$

(1) Factor out the common factor, 4:

$4(3x^2 + 17x - 6)$

Then we factor the trinomial $3x^2 + 17x - 6$.

(2) Because $3x^2$ can be factored as $3x \cdot x$ we have this possibility:

$(3x + \quad)(x + \quad)$

(3) There are 4 pairs of factors of -6 and they can be listed two ways:

$$\begin{array}{lllll} & 6,-1 & -6,1 & 3,-2 & -3,2 \\ \text{and} & -1,6 & 1,-6 & -2,3 & 2,-3 \end{array}$$

(4) We can immediately reject all possibilities in which either factor has a common factor, such as $(3x + 6)$ or $(3x - 3)$, because we determined at the outset that there is no common factor other than 1 or -1. We try some possibilities:

$(3x - 1)(x + 6) = 3x^2 + 17x - 6$

The factorization of $3x^2+17x-6$ is $(3x-1)(x+6)$. We must include the common factor in order to get a factorization of the original trinomial.

$12x^2 + 68x - 24 = 4(3x - 1)(x + 6)$

39. $4x + 1 + 3x^2 = 3x^2 + 4x + 1$

(1) There is no common factor (other than 1 or -1).

(2) Because $3x^2$ can be factored as $3x \cdot x$ we have this possibility:

$(3x + \quad)(x + \quad)$

(3) Since all coefficients are positive, we need consider only positive pairs of factors of 1. There is one such pair: 1,1.

(4) We try the possible factorization:

$(3x + 1)(x + 1) = 3x^2 + 4x + 1$

The factorization is $(3x + 1)(x + 1)$.

41. $y^2 + 4y - 2y - 8 = y(y + 4) - 2(y + 4)$
$= (y + 4)(y - 2)$

43. $8t^2 - 6t - 28t + 21 = 2t(4t - 3) - 7(4t - 3)$
$= (4t - 3)(2t - 7)$

45. $6x^2 + 4x + 9x + 6 = 2x(3x + 2) + 3(3x + 2)$
$= (3x + 2)(2x + 3)$

47. $2t^2 + 6t - t - 3 = 2t(t + 3) - 1(t + 3)$
$= (t + 3)(2t - 1)$

49. $3a^2 - 12a - a + 4 = 3a(a - 4) - 1(a - 4)$
$= (a - 4)(3a - 1)$

51. $9t^2 + 14t + 5$

(1) First note that there is no common factor (other than 1 or -1).

(2) Multiply the leading coefficient, 9, and the constant, 5:

$9 \cdot 5 = 45$

(3) We look for factors of 45 that add to 14. Since all coefficients are positive, we need to consider only positive factors.

Pairs of factors	Sums of factors
1, 45	46
3, 15	18
5, 9	14

The numbers we need are 5 and 9.

(4) Rewrite the middle term:

$14t = 5t + 9t$

(5) Factor by grouping:

$$\begin{aligned} 9t^2 + 14t + 5 &= 9t^2 + 5t + 9t + 5 \\ &= t(9t + 5) + 1(9t + 5) \\ &= (9t + 5)(t + 1) \end{aligned}$$

53. $-16x^2 - 32x - 7$

(1) We factor out -1 in order to have a trinomial with a positive leading coefficient.

$-16x^2 - 32x - 7 = -1(16x^2 + 32x + 7)$

Now we factor $16x^2 + 32x + 7$.

(2) Multiply the leading coefficient, 16, and the constant, 7:

$16 \cdot 7 = 112$

(3) We look for factors of 112 that add to 32. Since all coefficients are positive, we need to consider only positive factors.

Pairs of factors	Sums of factors
1, 112	113
2, 56	58
4, 28	32
7, 16	23
8, 14	22

The numbers we need are 4 and 28.

(4) Rewrite the middle term:

$32x = 4x + 28x$

(5) Factor by grouping:

$$\begin{aligned}16x^2+32x+7 &= 16x^2+4x+28x+7\\ &= 4x(4x+1)+7(4x+1)\\ &= (4x+1)(4x+7)\end{aligned}$$

We must include the factor of -1 in order to get a factorization of the original trinomial.

$-16x^2-32x-7=-1(4x+1)(4x+7)$, or $-(4x+1)(4x+7)$

55. $10a^2+25a-15$

(1) Factor out the largest common factor, 5:

$$10a^2+25a-15=5(2a^2+5a-3)$$

(2) To factor $2a^2+5a-3$ by grouping we first multiply the leading coefficient, 2, and the constant, -3:

$$2(-3)=-6$$

(3) We look for factors of -6 that add to 5.

Pairs of factors	Sums of factors
$-1, 6$	5
$-6, 1$	-5
$-2, 3$	1
$2, -3$	-1

The numbers we need are -1 and 6.

(4) Rewrite the middle term:

$$5a=-a+6a$$

(5) Factor by grouping:

$$\begin{aligned}2a^2+5a-3 &= 2a^2-a+6a-3\\ &= a(2a-1)+3(2a-1)\\ &= (2a-1)(a+3)\end{aligned}$$

The factorization of $2a^2+5a-3$ is $(2a-1)(a+3)$. We must include the common factor in order to get a factorization of the original trinomial:

$$10a^2+25a-15=5(2a-1)(a+3)$$

57. $18x^3+21x^2-9x$

(1) Factor out the largest common factor, $3x$:

$$18x^3+21x^2-9x=3x(6x^2+7x-3)$$

(2) To factor $6x^2+7x-3$ by grouping we first multiply the leading coefficient, 6, and the constant, -3:

$$6(-3)=-18$$

(3) We look for factors of -18 that add to 7.

Pairs of factors	Sums of factors
$-1, 18$	17
$1, -18$	-17
$-2, 9$	7
$2, -9$	-7
$-3, 6$	3
$3, -6$	-3

The numbers we need are -2 and 9.

(4) Rewrite the middle term:

$$7x=-2x+9x$$

(5) Factor by grouping:

$$\begin{aligned}6x^2+7x-3 &= 6x^2-2x+9x-3\\ &= 2x(3x-1)+3(3x-1)\\ &= (3x-1)(2x+3)\end{aligned}$$

The factorization of $6x^2+7x-3$ is $(3x-1)(2x+3)$. We must include the common factor in order to get a factorization of the original trinomial:

$$18x^3+21x^2-9x=3x(3x-1)(2x+3)$$

59. $89x+64+25x^2=25x^2+89x+64$

(1) First note that there is no common factor (other than 1 or -1).

(2) Multiply the leading coefficient, 25, and the constant, 64:

$$25\cdot 64=1600$$

(3) We look for factors of 1600 that add to 89. Since all coefficients are positive, we need to consider only positive factors. The numbers we need are 25 and 64.

(4) Rewrite the middle term:

$$89x=25x+64x$$

(5) Factor by grouping:

$$\begin{aligned}25x^2+89x+64 &= 25x^2+25x+64x+64\\ &= 25x(x+1)+64(x+1)\\ &= (x+1)(25x+64)\end{aligned}$$

61. $168x^3+45x^2+3x$

(1) Factor out the largest common factor, $3x$:

$$168x^3+45x^2+3x=3x(56x^2+15x+1)$$

(2) To factor $56x^2+15x+1$ we first multiply the leading coefficient, 56, and the constant, 1:

$$56\cdot 1=56$$

(3) We look for factors of 56 that add to 15. Since all coefficients are positive, we need to consider only positive factors. The numbers we need are 7 and 8.

(4) Rewrite the middle term:

$$15x=7x+8x$$

(5) Factor by grouping:

$$\begin{aligned}56x^2+15x+1 &= 56x^2+7x+8x+1\\ &= 7x(8x+1)+1(8x+1)\\ &= (8x+1)(7x+1)\end{aligned}$$

The factorization of $56x^2+15x+1$ is $(8x+1)(7x+1)$. We must include the common factor in order to get a factorization of the original trinomial:

$$168x^3+45x^2+3x=3x(8x+1)(7x+1)$$

63. $-14t^4+19t^3+3t^2$

(1) Factor out $-t^2$. This not only removes the largest common factor, t^2. It also produces a trinomial with a positive leading coefficient.

$$-14t^4+19t^3+3t^2=-t^2(14t^2-19t-3)$$

(2) To factor $14t^2-19t-3$ we first multiply the leading coefficient, 14, and the constant, -3:

$14(-3) = -42$

(3) We look for factors of -42 that add to -19. The numbers we need are -21 and 2.

(4) Rewrite the middle term:

$-19t = -21t + 2t$

(5) Factor by grouping:

$$\begin{aligned} 14t^2 - 19t - 3 &= 14t^2 - 21t + 2t - 3 \\ &= 7t(2t-3) + 1(2t-3) \\ &= (2t-3)(7t+1) \end{aligned}$$

The factorization of $14t^2 - 19t - 3$ is $(2t-3)(7t+1)$. We must include the common factor in order to get a factorization of the original trinomial:

$-14t^4 + 19t^3 + 3t^2 = -t^2(2t-3)(7t+1)$

65. $3x + 45x^2 - 18 = 45x^2 + 3x - 18$

(1) Factor out the largest common factor, 3:

$45x^2 + 3x - 18 = 3(15x^2 + x - 6)$

(2) To factor $15x^2 + x - 6$ we first multiply the leading coefficient, 15, and the constant, -6:

$15(-6) = -90$

(3) We look for factors of -90 that add to 1. The numbers we need are 10 and -9.

(4) Rewrite the middle term:

$x = 10x - 9x$

(5) Factor by grouping:

$$\begin{aligned} 15x^2 + x - 6 &= 15x^2 + 10x - 9x - 6 \\ &= 5x(3x+2) - 3(3x+2) \\ &= (3x+2)(5x-3) \end{aligned}$$

The factorization of $15x^2+x-6$ is $(3x+2)(5x-3)$. We must include the common factor in order to get a factorization of the original trinomial:

$3x + 45x^2 - 18 = 3(3x+2)(5x-3)$

67. $2a^2 + 5ab + 2b^2$

(1) There is no common factor (other than 1 or -1).

(2) Multiply the leading coefficient, 2, and the constant, 2:

$2 \cdot 2 = 4$

(3) We look for factors of 4 that add to 5. The numbers we need are 1 and 4.

(4) Rewrite the middle term:

$5ab = ab + 4ab$

(5) Factor by grouping:

$$\begin{aligned} 2a^2 + 5ab + 2b^2 &= 2a^2 + ab + 4ab + 2b^2 \\ &= a(2a+b) + 2b(2a+b) \\ &= (2a+b)(a+2b) \end{aligned}$$

69. $8s^2 + 18st + 9t^2$

(1) There is no common factor (other than 1 or -1).

(2) Multiply the leading coefficient, 8, and the constant, 9:

$8 \cdot 9 = 72$

(3) We look for factors of 72 that add to 18. The numbers we need are 6 and 12.

(4) Rewrite the middle term:

$18st = 6st + 12st$

(5) Factor by grouping:

$$\begin{aligned} 8s^2 + 18st + 9t^2 &= 8s^2 + 6st + 12st + 9t^2 \\ &= 2s(4s+3t) + 3t(4s+3t) \\ &= (4s+3t)(2s+3t) \end{aligned}$$

71. $18x^2 - 6xy - 24y^2$

(1) Factor out the largest common factor, 6:

$18x^2 - 6xy - 24y^2 = 6(3x^2 - xy - 4y^2)$

(2) To factor $3x^2 - xy - 4y^2$, we first multiply the leading coefficient, 3, and the constant, -4:

$3(-4) = -12$

(3) We look for factors of -12 that add to -1. The numbers we need are -4 and 3.

(4) Rewrite the middle term:

$-xy = -4xy + 3xy$

(5) Factor by grouping:

$$\begin{aligned} 3x^2 - xy - 4y^2 &= 3x^2 - 4xy + 3xy - 4y^2 \\ &= x(3x-4y) + y(3x-4y) \\ &= (3x-4y)(x+y) \end{aligned}$$

The factorization of $3x^2 - xy - 4y^2$ is $(3x-4y)(x+y)$. We must include the common factor in order to get a factorization of the original trinomial:

$18x^2 - 6xy - 24y^2 = 6(3x-4y)(x+y)$

73. $-24a^2 + 34ab - 12b^2$

(1) Factor out -2. This not only removes the largest common factor, 2. It also produces a trinomial with a positive leading coefficient.

$-24a^2 + 34ab - 12b^2 = -2(12a^2 - 17ab + 6b^2)$

(2) To factor $12a^2-17ab+6b^2$, we first multiply the leading coefficient, 12, and the constant, 6:

$12 \cdot 6 = 72$

(3) We look for factors of 72 that add to -17. The numbers we need are -8 and -9.

(4) Rewrite the middle term:

$-17ab = -8ab - 9ab$

(5) Factor by grouping:

$$\begin{aligned} 12a^2 - 17ab + 6b^2 &= 12a^2 - 8ab - 9ab + 6b^2 \\ &= 4a(3a-2b) - 3b(3a-2b) \\ &= (3a-2b)(4a-3b) \end{aligned}$$

The factorization of $12a^2 - 17ab + 6b^2$ is $(3a-2b)(4a-3b)$. We must include the common factor in order to get a factorization of the original trinomial:

$-24a^2 + 34ab - 12b^2 = -2(3a-2b)(4a-3b)$

75. $35x^2 + 34x^3 + 8x^4 = 8x^4 + 34x^3 + 35x^2$

(1) Factor out the largest common factor, x^2:

$$x^2(8x^2 + 34x + 35)$$

(2) To factor $8x^2 + 34x + 35$ by grouping we first multiply the leading coefficient, 8, and the constant, 35:

$$8 \cdot 35 = 280$$

(3) We look for factors of 280 that add to 34. The numbers we need are 14 and 20.

(4) Rewrite the middle term:

$$34x = 14x + 20x$$

(5) Factor by grouping:

$$\begin{aligned} 8x^2 + 34x + 35 &= 8x^2 + 14x + 20x + 35 \\ &= 2x(4x+7) + 5(4x+7) \\ &= (4x+7)(2x+5) \end{aligned}$$

The factorization of $8x^2 + 34x + 35$ is $(4x+7)(2x+5)$. We must include the common factor in order to get a factorization of the original trinomial:

$$35x^2 + 34x^3 + 8x^4 = x^2(4x+7)(2x+5)$$

77. $18a^7 + 8a^6 + 9a^8 = 9a^8 + 18a^7 + 8a^6$

(1) Factor out the largest common factor, a^6:

$$9a^8 + 18a^7 + 8a^6 = a^6(9a^2 + 18a + 8)$$

(2) To factor $9a^2 + 18a + 8$ we first multiply the leading coefficient, 9, and the constant, 8:

$$9 \cdot 8 = 72$$

(3) Look for factors of 72 that add to 18. The numbers we need are 6 and 12.

(4) Rewrite the middle term:

$$18a = 6a + 12a$$

(5) Factor by grouping:

$$\begin{aligned} 9a^2 + 18a + 8 &= 9a^2 + 6a + 12a + 8 \\ &= 3a(3a+2) + 4(3a+2) \\ &= (3a+2)(3a+4) \end{aligned}$$

The factorization of $9a^2 + 18a + 8$ is $(3a+2)(3a+4)$. We must include the common factor in order to get a factorization of the original trinomial:

$$18a^7 + 8a^6 + 9a^8 = a^6(3a+2)(3a+4)$$

79. *Writing Exercise*

81. ***Familiarize.*** We will use the formula $C = 2\pi r$, where C is circumference and r is radius, to find the radius in kilometers. Then we will multiply that number by 0.62 to find the radius in miles.

Translate.

$$\underbrace{\text{Circumference}} = \underbrace{2 \cdot \pi \cdot \text{radius}}$$
$$\downarrow \qquad \downarrow \qquad \downarrow$$
$$40,000 \quad \approx \quad 2(3.14)r$$

Carry out. First we solve the equation.

$$40,000 \approx 2(3.14)r$$
$$40,000 \approx 6.28r$$
$$6369 \approx r$$

Then we multiply to find the radius in miles:

$$6369(0.62) \approx 3949$$

Check. If $r = 6369$, then $2\pi r = 2(3.14)(6369) \approx 40,000$. We should also recheck the multiplication we did to find the radius in miles. Both values check.

State. The radius of the earth is about 6369 km or 3949 mi. (These values may differ slightly if a different approximation is used for π.)

83. $$\begin{aligned} (3x+1)^2 &= (3x)^2 + 2 \cdot 3x \cdot 1 + 1^2 \\ &\qquad [(A+B)^2 = A^2 + 2AB + B^2] \\ &= 9x^2 + 6x + 1 \end{aligned}$$

85. $$\begin{aligned} (4t-5)^2 &= (4t)^2 - 2 \cdot 4t \cdot 5 + 5^2 \\ &\qquad [(A-B)^2 = A^2 - 2AB + B^2] \\ &= 16t^2 - 40t + 25 \end{aligned}$$

87. $$\begin{aligned} (5x-2)(5x+2) &= (5x)^2 - 2^2 \\ &\qquad [(A+B)(A-B) = A^2 - B^2] \\ &= 25x^2 - 4 \end{aligned}$$

89. $$\begin{aligned} (2t+7)(2t-7) &= (2t)^2 - 7^2 \\ &\qquad [(A+B)(A-B) = A^2 - B^2] \\ &= 4t^2 - 49 \end{aligned}$$

91. *Writing Exercise*

93. $18x^2y^2 - 3xy - 10$

We will factor by grouping.

(1) There is no common factor (other than 1 or -1).

(2) Multiply the leading coefficient, 18, and the constant, -10:

$$18(-10) = -180$$

(3) We look for factors of -180 that add to -3. The numbers we want are -15 and 12.

(4) Rewrite the middle term:

$$-3xy = -15xy + 12xy$$

(5) Factor by grouping:

$$\begin{aligned} 18x^2y^2 - 3xy - 10 &= 18x^2y^2 - 15xy + 12xy - 10 \\ &= 3xy(6xy-5) + 2(6xy-5) \\ &= (6xy-5)(3xy+2) \end{aligned}$$

95. $9a^2b^3 + 25ab^2 + 16$

We cannot factor the leading term, $9a^2b^3$, in a way that will produce a middle term with variable factors ab^2, so this trinomial is prime.

97. $16t^{10} - 8t^5 + 1 = 16(t^5)^2 - 8t^5 + 1$

(1) There is no common factor (other than 1 or -1).

(2) Because $16t^{10}$ can be factored as $16t^5 \cdot t^5$ or $8t^5 \cdot 2t^5$ or $4t^5 \cdot 4t^5$, we have these possibilities:

$$(16t^5 + \quad)(t^5 + \quad) \text{ and } (8t^5 + \quad)(2t^5 + \quad)$$

and $(4t^5 + \quad)(4t^5 + \quad)$

(3) Since the last term is positive and the middle term is negative we need consider only negative factors of 1. The only negative pair of factors is -1, -1.

(4) We try some possibilities:

$(16t^5 - 1)(t^5 - 1) = 16t^{10} - 17t^5 + 1$

$(8t^5 - 1)(2t^5 - 1) = 16t^{10} - 10t^5 + 1$

$(4t^5 - 1)(4t^5 - 1) = 16t^{10} - 8t^5 + 1$

The factorization is $(4t^5 - 1)(4t^5 - 1)$, or $(4t^5 - 1)^2$.

99. $-20x^{2n} - 16x^n - 3 = -20(x^n)^2 - 16x^n - 3$

(1) Factor out -1 in order to have a trinomial with a positive leading coëfficient.

$-20x^{2n} - 16x^n - 3 = -1(20x^{2n} + 16x^n + 3)$

(2) Because $20x^{2n}$ can be factored as $20x^n \cdot x^n$, $10x^n \cdot 2x^n$, or $5x^n \cdot 4x^n$, we have these possibilities:

$(20x^n + \quad)(x^n + \quad)$ and $(10x^n + \quad)(2x^n + \quad)$ and $(5x^n + \quad)(4x^n + \quad)$

(3) Since all the signs are positive, we need consider only the positive factor pair 3,1 when factoring 3. This pair can also be listed as 1,3.

(4) We try some possibilities:

$(20x^n + 3)(x^n + 1) = 20x^{2n} + 23x^n + 3$

$(10x^n + 3)(2x^n + 1) = 20x^{2n} + 16x^n + 3$

The factorization of $20x^{2n} + 16x^n + 3$ is $(10x^n + 3)(2x^n + 1)$. We must include the common factor to get a factorization of the original trinomial.

$-20x^{2n} - 16x^n - 3 = -1(10x^n + 3)(2x^n + 1)$, or $-(10x^n + 3)(2x^n + 1)$

101. $a^{2n+1} - 2a^{n+1} + a$

(1) Factor out the largest common factor, a:

$a^{2n+1} - 2a^{n+1} + a = a(a^{2n} - 2a^n + 1)$

(2) Multiply the leading coefficient, 1, and the constant, 1:

$1 \cdot 1 = 1$

(3) Look for factors of 1 that add to -2. The numbers we need are -1 and -1.

(4) Rewrite the middle term.

$-2a^n = -a^n - a^n$

(5) Factor by grouping:

$$\begin{aligned} a^{2n} - 2a^n + 1 &= a^{2n} - a^n - a^n + 1 \\ &= a^n(a^n - 1) - 1(a^n - 1) \\ &= (a^n - 1)(a^n - 1), \text{ or } (a^n - 1)^2 \end{aligned}$$

The factorization of $a^{2n} - 2a^n + 1$ is $(a^n - 1)^2$. We must include the common factor in order to get a factorization of the original trinomial:

$a^{2n+1} - 2a^{n+1} + a = a(a^n - 1)^2$

103. $3(a+1)^{n+1}(a+3)^2 - 5(a+1)^n(a+3)^3$

$= (a+1)^n(a+3)^2[3(a+1) - 5(a+3)]$ Removing the common factors

$= (a+1)^n(a+3)^2[3a + 3 - 5a - 15]$ Simplifying inside the brackets

$= (a+1)^n(a+3)^2(-2a - 12)$

$= (a+1)^n(a+3)^2(-2)(a+6)$ Removing the common factor

$= -2(a+1)^n(a+3)^2(a+6)$ Rearranging

Exercise Set 5.4

1. $x^2 - 64 = x^2 - 8^2$, so $x^2 - 64$ is a difference of squares.

3. Two terms of $x^2 - 5x + 4$ are squares (x^2 and 4) and neither is being subtracted. However, the remaining term is neither $2 \cdot x \cdot 2$ nor $-2 \cdot x \cdot 2$, so the polynomial is not a perfect-square trinomial. It is not a binomial so it cannot be a difference of squares. It is not prime since it can be factored: $x^2 - 5x + 4 = (x-1)(x-4)$. Thus it is none of the given possibilities.

5. $a^2 - 8a + 16 = a^2 - 2 \cdot a \cdot 4 + 4^2$, so this is a perfect-square trinomial.

7. $-25x^2 - 9$ is not a trinomial. It is not a difference of squares because the terms do not have different signs. There is no common factor (other than 1 or -1), so $-25x^2 - 9$ is a prime polynomial.

9. $4t^2 + 20t + 25 = (2t)^2 + 2 \cdot 2t \cdot 5 + 5^2$, so this is a perfect square trinomial.

11. $x^2 + 18x + 81$

(1) Two terms, x^2 and 81, are squares.

(2) Neither x^2 nor 81 is being subtracted.

(3) Twice the product of the square roots, $2 \cdot x \cdot 9$, is $18x$, the remaining term.

Thus, $x^2 + 18x + 81$ is a perfect-square trinomial.

13. $x^2 - 16x - 64$

(1) Two terms, x^2 and 64, are squares.

(2) There is a minus sign before 64, so $x^2 + 16x - 64$ is not a perfect-square trinomial.

15. $x^2 - 3x + 9$

(1) Two terms, x^2 and 9, are squares.

(2) There is no minus sign before x^2 or 9.

(3) Twice the product of the square roots, $2 \cdot x \cdot 3$, is $6x$. This is neither the remaining term nor its opposite, so $x^2 - 3x + 9$ is not a perfect-square trinomial.

17. $9x^2 - 36x + 24$

(1) Only one term, $9x^2$, is a square. Thus, $9x^2 - 36x + 24$ is not a perfect-square trinomial.

19.

$$\begin{aligned} & x^2 + 16x + 64 \\ &= x^2 + 2 \cdot x \cdot 8 + 8^2 = (x+8)^2 \\ & \uparrow \quad \uparrow \quad \uparrow \quad \uparrow \quad \uparrow \\ &= A^2 + 2 \ \ A \ \ B + B^2 = (A+B)^2 \end{aligned}$$

21.

$$\begin{aligned} & x^2 - 14x + 49 \\ &= x^2 - 2 \cdot x \cdot 7 + 7^2 = (x-7)^2 \\ & \uparrow \quad \uparrow \quad \uparrow \quad \uparrow \quad \uparrow \\ &= A^2 - 2 \ \ A \ \ B + B^2 = (A-B)^2 \end{aligned}$$

23.

$$\begin{aligned} 3x^2 + 6x + 3 &= 3(x^2 + 2x + 1) \\ &= 3(x^2 + 2 \cdot x \cdot 1 + 1^2) \\ &= 3(x+1)^2 \end{aligned}$$

25. $4-4x+x^2 = 2^2-2\cdot 2\cdot x+x^2$
$= (2-x)^2$

We could also factor as follows:
$4-4x+x^2 = x^2-4x+4$
$= x^2-2\cdot x\cdot 2+2^2$
$= (x-2)^2$

27. $18x^2+12x+2 = 2(9x^2+6x+1)$
$= 2[(3x)^2+2\cdot 3x\cdot 1+1^2]$
$= 2(3x+1)^2$

29. $49-56y+16y^2 = 16y^2-56y+49$
$= (4y)^2-2\cdot 4y\cdot 7+7^2$
$= (4y-7)^2$

We could also factor as follows:
$49-56y+16y^2 = 7^2-2\cdot 7\cdot 4y+(4y)^2$
$= (7-4y)^2$

31. $-x^5+18x^4-81x^3 = -x^3(x^2-18x+81)$
$= -x^3(x^2-2\cdot x\cdot 9+9^2)$
$= -x^3(x-9)^2$

33. $2x^3-4x^2+2x = 2x(x^2-2x+1)$
$= 2x(x^2-2\cdot x\cdot 1+1^2)$
$= 2x(x-1)^2$

35. $20x^2+100x+125 = 5(4x^2+20x+25)$
$= 5[(2x)^2+2\cdot 2x\cdot 5+5^2]$
$= 5(2x+5)^2$

37. $49-42x+9x^2 = 7^2-2\cdot 7\cdot 3x+(3x)^2 = (7-3x)^2$, or $(3x-7)^2$

39. $16x^2+24x+9 = (4x)^2+2\cdot 4x\cdot 3+3^2 =$ $(4x+3)^2$

41. $2+20x+50x^2 = 2(1+10x+25x^2)$
$= 2[1^2+2\cdot 1\cdot 5x+(5x)^2]$
$= 2(1+5x)^2$, or $2(5x+1)^2$

43. $4p^2+12pq+9q^2 = (2p)^2+2\cdot 2p\cdot 3q+(3q)^2$
$= (2p+3q)^2$

45. $a^2-12ab+49b^2$

This is not a perfect square trinomial because $-2\cdot a\cdot 7b = -14ab \neq -12ab$. Nor can it be factored using the methods of Sections 5.2 and 5.3. Thus, it is prime.

47. $-64m^2-16mn-n^2 = -1(64m^2+16mn+n^2)$
$= -1[(8m)^2+2\cdot 8m\cdot n+n^2]$
$= -1(8m+n)^2$, or $-(8m+n)^2$

49. $-32s^2+80st-50t^2 = -2(16s^2-40st+25t^2)$
$= -2[(4s)^2-2\cdot 4s\cdot 5t+(5t)^2]$
$= -2(4s-5t)^2$

51. x^2+100

(1) The first expression is a square: x^2
The second expression is a square: $100 = 10^2$
(2) The terms do not have different signs.
Thus, x^2+100 is not a difference of squares.

53. x^2-81

(1) The first expression is a square: x^2
The second expression is a square: $81 = 9^2$
(2) The terms have different signs.
Thus, x^2-81 is a difference of squares, x^2-9^2.

55. $-26+4t^2$

(1) The expression 26 is not a square. Thus, $-26+4t^2$ is not a difference of squares.

57. $y^2-4 = y^2-2^2 = (y+2)(y-2)$

59. $p^2-9 = p^2-3^2 = (p+3)(p-3)$

61. $-49+t^2 = t^2-49 = t^2-7^2 = (t+7)(t-7)$, or $(7+t)(-7+t)$

63. $6a^2-54 = 6(a^2-9) = 6(a^2-3^2) = 6(a+3)(a-3)$

65. $49x^2-14x+1 = (7x)^2-2\cdot 7x\cdot 1+1^2 = (7x-1)^2$

67. $200-2t^2 = 2(100-t^2) = 2(10^2-t^2) =$ $2(10+t)(10-t)$

69. $-80a^2+45 = -5(16a^2-9) = -5[(4a^2)-3^2] =$ $-5(4a+3)(4a-3)$

71. $5t^2-80 = 5(t^2-16) = 5(t^2-4^2) =$ $5(t+4)(t-4)$

73. $8x^2-98 = 2(4x^2-49) = 2[(2x)^2-7^2] =$ $2(2x+7)(2x-7)$

75. $36x-49x^3 = x(36-49x^2) = x[6^2-(7x)^2] =$ $x(6+7x)(6-7x)$

77. $49a^4-20$

There is no common factor (other than 1 or -1). Since 20 is not a square, this is not a difference of squares. Thus, the polynomial is prime.

79. t^4-1
$= (t^2)^2-1^2$
$= (t^2+1)(t^2-1)$
$= (t^2+1)(t+1)(t-1)$ Factoring further; t^2-1 is a difference of squares

81. $-3x^3+24x^2-48x = -3x(x^2-8x+16)$
$= -3x(x^2-2\cdot x\cdot 4+4^2)$
$= -3x(x-4)^2$

83. $48t^2-27 = 3(16t^2-9)$
$= 3[(4t)^2-3^2]$
$= 3(4t+3)(4t-3)$

85. $a^8 - 2a^7 + a^6 = a^6(a^2 - 2a + 1)$
$= a^6(a^2 - 2 \cdot a \cdot 1 + 1^2)$
$= a^6(a - 1)^2$

87. $7a^2 - 7b^2 = 7(a^2 - b^2)$
$= 7(a + b)(a - b)$

89. $25x^2 - 4y^2 = (5x)^2 - (2y)^2$
$= (5x + 2y)(5x - 2y)$

91. $18t^2 - 8s^2 = 2(9t^2 - 4s^2)$
$= 2[(3t)^2 - (2s)^2]$
$= 2(3t + 2s)(3t - 2s)$

93. *Writing Exercise*

95. ***Familiarize***. Let $a =$ the amount of oxygen, in liters, that can be dissolved in 100 L of water at 20° C.

Translate. We reword the problem.

$\underbrace{5 \text{ L}}$ is 1.6 times $\underbrace{\text{amount } a}$.

$\downarrow \quad \downarrow \quad \downarrow \quad \downarrow \quad \downarrow$

$5 \quad = 1.6 \quad \cdot \quad a$

Carry out. We solve the equation.

$5 = 1.6a$

$3.125 = a$ Dividing both sides by 1.6

Check. Since 1.6 times 3.125 is 5, the answer checks.

State. 3.125 L of oxygen can be dissolved in 100 L of water at 20° C.

97. $(x^3y^5)(x^9y^7) = x^{3+9}y^{5+7} = x^{12}y^{12}$

99. Graph: $y = \frac{3}{2}x - 3$

Because the equation is in the form $y = mx + b$, we know the y-intercept is $(0, -3)$. We find two other points on the line, substituting multiples of 2 for x to avoid fractions.

When $x = -2$, $y = \frac{3}{2}(-2) - 3 = -3 - 3 = -6$.

When $x = 4$, $y = \frac{3}{2} \cdot 4 - 3 = 6 - 3 = 3$.

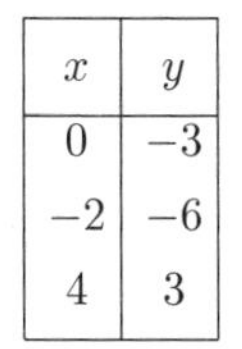

x	y
0	−3
−2	−6
4	3

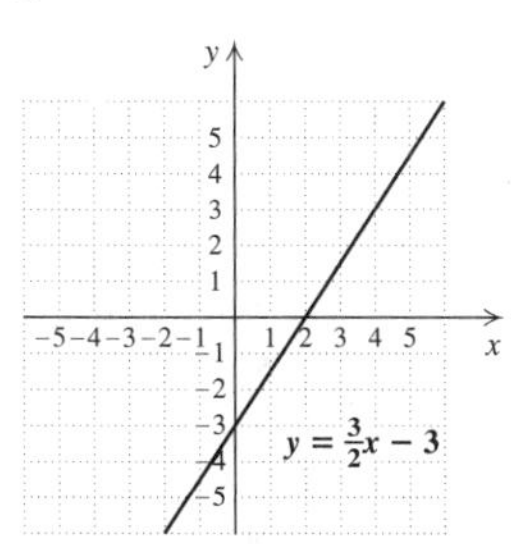

101. *Writing Exercise*

103. $x^8 - 2^8 = (x^4 + 2^4)(x^4 - 2^4)$
$= (x^4 + 2^4)(x^2 + 2^2)(x^2 - 2^2)$
$= (x^4 + 2^4)(x^2 + 2^2)(x + 2)(x - 2)$, or
$(x^4 + 16)(x^2 + 4)(x + 2)(x - 2)$

105. $18x^3 - \frac{8}{25}x = 2x\left(9x^2 - \frac{4}{25}\right) =$
$2x\left(3x + \frac{2}{5}\right)\left(3x - \frac{2}{5}\right)$

107. $(y - 5)^4 - z^8$
$= [(y - 5)^2 + z^4][(y - 5)^2 - z^4]$
$= [(y - 5)^2 + z^4][y - 5 + z^2][y - 5 - z^2]$
$= (y^2 - 10y + 25 + z^4)(y - 5 + z^2)(y - 5 - z^2)$

109. $-x^4 + 8x^2 + 9 = -1(x^4 - 8x^2 - 9)$
$= -1(x^2 - 9)(x^2 + 1)$
$= -1(x + 3)(x - 3)(x^2 + 1)$

111. $(y + 3)^2 + 2(y + 3) + 1$
$= (y + 3)^2 + 2 \cdot (y + 3) \cdot 1 + 1^2$
$= [(y + 3) + 1]^2$
$= (y + 4)^2$

113. $27x^3 - 63x^2 - 147x + 343$
$= 9x^2(3x - 7) - 49(3x - 7)$
$= (3x - 7)(9x^2 - 49)$
$= (3x - 7)(3x + 7)(3x - 7)$, or
$(3x - 7)^2(3x + 7)$

115. $81 - b^{4k} = 9^2 - (b^{2k})^2$
$= (9 + b^{2k})(9 - b^{2k})$
$= (9 + b^{2k})[3^2 - (b^k)^2]$
$= (9 + b^{2k})(3 + b^k)(3 - b^k)$

117. $x^2(x + 1)^2 - (x^2 + 1)^2$
$= x^2(x^2 + 2x + 1) - (x^4 + 2x^2 + 1)$
$= x^4 + 2x^3 + x^2 - x^4 - 2x^2 - 1$
$= 2x^3 + x^2 - 2x^2 - 1$
$= (2x^3 - 2x^2) + (x^2 - 1)$
$= 2x^3 - x^2 - 1$

119. $y^2 + 6y + 9 - x^2 - 8x - 16$
$= (y^2 + 6y + 9) - (x^2 + 8x + 16)$
$= (y + 3)^2 - (x + 4)^2$
$= [(y + 3) + (x + 4)][(y + 3) - (x + 4)]$
$= (y + 3 + x + 4)(y + 3 - x - 4)$
$= (y + x + 7)(y - x - 1)$

121. For $c = a^2$, $2 \cdot a \cdot 3 = 24$. Then $a = 4$, so $c = 4^2 = 16$.

123. $(x + 1)^2 - x^2$
$= [(x + 1) + x][(x + 1) - x]$
$= 2x + 1$
$= (x + 1) + x$

Exercise Set 5.5

1. common factor

3. grouping

5. $10a^2 - 640$
$= 10(a^2 - 64)$ 10 is a common factor.
$= 10(a+8)(a-8)$ Factoring the difference of squares

7. $y^2 + 49 - 14y$
$= y^2 - 14y + 49$ Perfect-square trinomial
$= (y-7)^2$

9. $2t^2 + 11t + 12$

There is no common factor (other than 1 of -1). This trinomial has three terms, but it is not a perfect-square trinomial. Multiply the leading coefficient and the constant, 2 and 12: $2 \cdot 12 = 24$. Try to factor 24 so that the sum of the factors is 11. The numbers we want are 3 and 8: $3 \cdot 8 = 24$ and $3 + 8 = 11$. Split the middle term and factor by grouping.

$$2t^2 + 11t + 12 = 2t^2 + 3t + 8t + 12$$
$$= t(2t+3) + 4(2t+3)$$
$$= (2t+3)(t+4)$$

11. $x^3 - 18x^2 + 81x$
$= x(x^2 - 18x + 81)$ x is a common factor.
$= x(x^2 - 2 \cdot x \cdot 9 + 9^2)$ Perfect-square trinomial
$= x(x-9)^2$

13. $x^3 - 5x^2 - 25x + 125$
$= x^2(x-5) - 25(x-5)$ Factoring by grouping
$= (x-5)(x^2-25)$
$= (x-5)(x+5)(x-5)$ Factoring the difference of squares
$= (x-5)^2(x+5)$

15. $27t^3 - 3t$
$= 3t(9t^2 - 1)$ $3t$ is a common factor.
$= 3t[(3t)^2 - 1^2]$ Difference of squares
$= 3t(3t+1)(3t-1)$

17. $9x^3 + 12x^2 - 45x$
$= 3x(3x^2 + 4x - 15)$ $3x$ is a common factor.
$= 3x(x+3)(3x-5)$ Factoring the trinomial

19. $t^2 + 25$

The polynomial has no common factor and is not a difference of squares. It is prime.

21. $6x^2 + 3x - 45$
$= 3(2x^2 + x - 15)$ 3 is a common factor.
$= 3(2x-5)(x+3)$ Factoring the trinomial

23. $-2a^6 + 8a^5 - 8a^4$
$= -2a^4(a^2 - 4a + 4)$ Factoring out $-2a^4$
$= -2a^4(a-2)^2$ Factoring the perfect-square trinomial

25. $5x^5 - 80x$
$= 5x(x^4 - 16)$ $5x$ is a common factor.
$= 5x[(x^2)^2 - 4^2]$ Difference of squares
$= 5x(x^2+4)(x^2-4)$ Difference of squares
$= 5x(x^2+4)(x+2)(x-2)$

27. $t^4 - 9$ Difference of squares
$= (t^2+3)(t^2-3)$

29. $-x^6 + 2x^5 - 7x^4$
$= -x^4(x^2 - 2x + 7)$

The trinomial is prime, so this is the complete factorization.

31. $p^2q^2 - r^2$ Difference of squares
$= (pq+r)(pq-r)$

33. $ax^2 + ay^2 = a(x^2 + y^2)$

35. $36mn - 9m^2n^2 = 9mn(4 - mn)$

37. $2\pi rh + 2\pi r^2 = 2\pi r(h+r)$

39. $(a+b)(5a) + (a+b)(3b)$
$= (a+b)(5a+3b)$ $(a+b)$ is a common factor.

41. $x^2 + x + xy + y$
$= x(x+1) + y(x+1)$ Factoring by grouping
$= (x+1)(x+y)$

43. $a^2 - 3a + ay - 3y$
$= a(a-3) + y(a-3)$ Factoring by grouping
$= (a-3)(a+y)$

45. $3x^2 + 13xy - 10y^2 = (3x-2y)(x+5y)$

47. $4b^2 + a^2 - 4ab$
$= 4b^2 - 4ab + a^2$
$= (2b)^2 - 2 \cdot 2b \cdot a + a^2$ Perfect-square trinomial
$= (2b-a)^2$

This result can also be expressed as $(a-2b)^2$.

49. $16x^2 + 24xy + 9y^2$
$= (4x)^2 + 2 \cdot 4x \cdot 3y + (3y)^2$ Perfect-square trinomial
$= (4x+3y)^2$

51. $t^2 - 8t + 10$

There is no common factor, this is not a perfect-square trinomial, and we cannot find a pair of factors whose product is 10 and whose sum is -8, so this polynomial is prime.

53. $a^4b^4 - 16$
$= (a^2b^2)^2 - 4^2$ Difference of squares
$= (a^2b^2 + 4)(a^2b^2 - 4)$ Difference of squares
$= (a^2b^2 + 4)(ab + 2)(ab - 2)$

55. $4p^2q - pq^2 + 4p^3$
$= p(4pq - q^2 + 4p^2)$
$= p(4p^2 + 4pq - q^2)$
The trinomial cannot be factored further, so this is the complete factorization.

57. $3b^2 + 17ab - 6a^2 = (3b - a)(b + 6a)$

59. $-12 - x^2y^2 - 8xy$
$= -x^2y^2 - 8xy - 12$
$= -1(x^2y^2 + 8xy + 12)$
$= -1(xy + 2)(xy + 6)$, or $-(xy + 2)(xy + 6)$

61. $p^2q^2 + 7pq + 6 = (pq + 1)(pq + 6)$

63. $4ab^5 - 32b^4 + a^2b^6$
$= b^4(4ab - 32 + a^2b^2)$ b^2 is a common factor.
$= b^4(a^2b^2 + 4ab - 32)$
$= b^4(ab + 8)(ab - 4)$ Factoring the trinomial

65. $x^6 + x^5y - 2x^4y^2$
$= x^4(x^2 + xy - 2y^2)$ x^4 is a common factor.
$= x^4(x + 2y)(x - y)$ Factoring the trinomial

67. $36a^2 - 15a + \frac{25}{16}$
$= (6a)^2 - 2 \cdot 6a \cdot \frac{5}{4} + \left(\frac{5}{4}\right)^2$ Perfect-square trinomial
$= \left(6a - \frac{5}{4}\right)^2$

69. $\frac{1}{81}x^2 - \frac{8}{27}x + \frac{16}{9}$
$= \left(\frac{1}{9}x\right)^2 - 2 \cdot \frac{1}{9}x \cdot \frac{4}{3} + \left(\frac{4}{3}\right)^2$ Perfect-square trinomial
$= \left(\frac{1}{9}x - \frac{4}{3}\right)^2$
If we had factored out $\frac{1}{9}$ at the outset, the final result would have been $\frac{1}{9}\left(\frac{1}{3}x - 4\right)^2$.

71. $1 - 16x^{12}y^{12}$
$= (1 + 4x^6y^6)(1 - 4x^6y^6)$ Difference of squares
$= (1 + 4x^6y^6)(1 + 2x^3y^3)(1 - 2x^3y^3)$ Difference of squares

73. $4a^2b^2 + 12ab + 9$
$= (2ab)^2 + 2 \cdot 2ab \cdot 3 + 3^2$ Perfect-square trinomial
$= (2ab + 3)^2$

75. $a^4 + 8a^2 + 8a^3 + 64a$
$= a(a^3 + 8a + 8a^2 + 64)$ a is a common factor.
$= a[a(a^2 + 8) + 8(a^2 + 8)]$ Factoring by grouping
$= a(a^2 + 8)(a + 8)$

77. *Writing Exercise*

79.
$y = -4x + 7$	
11	$-4(-1) + 7$
	$4 + 7$

$11 \stackrel{?}{=} 11$ TRUE
Since $11 = 11$ is true, $(-1, 11)$ is a solution.

$y = -4x + 7$	
7	$-4 \cdot 0 + 7$
	$0 + 7$

$7 \stackrel{?}{=} 7$ TRUE
Since $7 = 7$ is true, $(0, 7)$ is a solution.

$y = -4x + 7$	
-5	$-4 \cdot 3 + 7$
	$-12 + 7$

$-5 \stackrel{?}{=} -5$ TRUE
Since $-5 = -5$ is true, $(3, -5)$ is a solution.

81. $3x + 7 = 0$
$3x = -7$ Subtracting 7 from both sides
$x = -\frac{7}{3}$ Dividing both sides by 3
The solution is $-\frac{7}{3}$.

83. $4x - 9 = 0$
$4x = 9$ Adding 9 to both sides
$x = \frac{9}{4}$ Dividing both sides by 4
The solution is $\frac{9}{4}$.

85. *Writing Exercise*

87. $-(x^5 + 7x^3 - 18x)$
$= -x(x^4 + 7x^2 - 18)$
$= -x(x^2 + 9)(x^2 - 2)$

89. $-3a^4 + 15a^2 - 12$
$= -3(a^4 - 5a^2 + 4)$
$= -3(a^2 - 1)(a^2 - 4)$
$= -3(a + 1)(a - 1)(a + 2)(a - 2)$

91. $y^2(y + 1) - 4y(y + 1) - 21(y + 1)$
$= (y + 1)(y^2 - 4y - 21)$
$= (y + 1)(y - 7)(y + 3)$

93. $6(x-1)^2 + 7y(x-1) - 3y^2$
$= [2(x-1)+3y][3(x-1)-y]$
$= (2x+3y-2)(3x-y-3)$

95. $2(a+3)^4 - (a+3)^3(b-2) - (a+3)^2(b-2)^2$
$= (a+3)^2[2(a+3)^2 - (a+3)(b-2) - (b-2)^2]$
$= (a+3)^2[2(a+3)+(b-2)][(a+3)-(b-2)]$
$= (a+3)^2(2a+6+b-2)(a+3-b+2)$
$= (a+3)^2(2a+b+4)(a-b+5)$

Exercise Set 5.6

1. Equations of the type $ax^2+bx+c=0$, with $a \neq 0$, are quadratic, so choice (c) is correct.

3. Most quadratic equations have 2 solutions, so choice (a) is correct.

5. $(x+5)(x+7)=0$

We use the principle of zero products.

$x+5=0$ *or* $x+7=0$
$x=-5$ *or* $x=-7$

Check:

For -5:

$(x+5)(x+7)=0$	
$(-5+5)(-5+7)$	0
$0\cdot 2$	
$0 \stackrel{?}{=} 0$	TRUE

For -7:

$(x+5)(x+7)=0$	
$(-7+5)(-7+7)$	0
$-2\cdot 0$	
$0 \stackrel{?}{=} 0$	TRUE

The solutions are -5 and -7.

7. $(x-3)(x+6)=0$

$x-3=0$ *or* $x+6=0$
$x=3$ *or* $x=-6$

Check:

For 3:

$(x-3)(x+6)=0$	
$(3-3)(3+6)$	0
$0\cdot 9$	
$0 \stackrel{?}{=} 0$	TRUE

For -6:

$(x-3)(x+6)=0$	
$(-6-3)(-6+6)$	0
$-9\cdot 0$	
$0 \stackrel{?}{=} 0$	TRUE

The solutions are 3 and -6.

9. $(2x-9)(x+4)=0$

$2x-9=0$ *or* $x+4=0$
$2x=9$ *or* $x=-4$
$x=\frac{9}{2}$ *or* $x=-4$

The solutions are $\frac{9}{2}$ and -4.

11. $(10x-9)(4x+7)=0$

$10x-9=0$ *or* $4x+7=0$
$10x=9$ *or* $4x=-7$
$x=\frac{9}{10}$ *or* $x=-\frac{7}{4}$

The solutions are $\frac{9}{10}$ and $-\frac{7}{4}$.

13. $x(x+2)=0$

$x=0$ *or* $x+2=0$
$x=0$ *or* $x=-2$

The solutions are 0 and -2.

15. $\left(\frac{2}{3}x-\frac{12}{11}\right)\left(\frac{7}{4}x-\frac{1}{12}\right)=0$

$\frac{2}{3}x-\frac{12}{11}=0$ *or* $\frac{7}{4}x-\frac{1}{12}=0$
$\frac{2}{3}x=\frac{12}{11}$ *or* $\frac{7}{4}x=\frac{1}{12}$
$x=\frac{3}{2}\cdot\frac{12}{11}$ *or* $x=\frac{4}{7}\cdot\frac{1}{12}$
$x=\frac{18}{11}$ *or* $x=\frac{1}{21}$

The solutions are $\frac{18}{11}$ and $\frac{1}{21}$.

17. $5x(2x+9)=0$

$5x=0$ *or* $2x+9=0$
$x=0$ *or* $2x=-9$
$x=0$ *or* $x=-\frac{9}{2}$

The solutions are 0 and $-\frac{9}{2}$.

19. $(20x-0.4x)(7-0.1x)=0$

$20-0.4x=0$ *or* $7-0.1x=0$
$-0.4x=-20$ *or* $-0.1x=-7$
$x=50$ *or* $x=70$

The solutions are 50 and 70.

21. $x^2-7x+6=0$

$(x-6)(x-1)=0$ Factoring
$x-6=0$ *or* $x-1=0$ Using the principle of zero products
$x=6$ *or* $x=1$

The solutions are 6 and 1.

23. $x^2 + 4x - 21 = 0$

$(x-3)(x+7) = 0$ Factoring

$x - 3 = 0$ *or* $x + 7 = 0$ Using the principle of zero products

$x = 3$ *or* $x = -7$

The solutions are 3 and -7.

25. $x^2 + 9x + 14 = 0$

$(x+7)(x+2) = 0$

$x + 7 = 0$ *or* $x + 2 = 0$

$x = -7$ *or* $x = -2$

The solutions are -7 and -2.

27. $x^2 - 6x = 0$

$x(x-6) = 0$

$x = 0$ *or* $x - 6 = 0$

$x = 0$ *or* $x = 6$

The solutions are 0 and 6.

29. $6t + t^2 = 0$

$t(6+t) = 0$

$t = 0$ *or* $6 + t = 0$

$t = 0$ *or* $t = -6$

The solutions are 0 and -6.

31. $9x^2 = 4$

$9x^2 - 4 = 0$

$(3x+2)(3x-2) = 0$

$3x + 2 = 0$ *or* $3x - 2 = 0$

$3x = -2$ *or* $3x = 2$

$x = -\frac{2}{3}$ *or* $x = \frac{2}{3}$

The solutions are $-\frac{2}{3}$ and $\frac{2}{3}$.

33. $0 = 25 + x^2 + 10x$

$0 = x^2 + 10x + 25$ Writing in descending order

$0 = (x+5)(x+5)$

$x + 5 = 0$ *or* $x + 5 = 0$

$x = -5$ *or* $x = -5$

The solution is -5.

35. $1 + x^2 = 2x$

$x^2 - 2x + 1 = 0$

$(x-1)(x-1) = 0$

$x - 1 = 0$ *or* $x - 1 = 0$

$x = 1$ *or* $x = 1$

The solution is 1.

37. $4t^2 = 8t$

$4t^2 - 8t = 0$

$4t(t-2) = 0$

$t = 0$ *or* $t - 2 = 0$

$t = 0$ *or* $t = 2$

The solutions are 0 and 2.

39. $3x^2 - 7x = 20$

$3x^2 - 7x - 20 = 0$

$(3x+5)(x-4) = 0$

$3x + 5 = 0$ *or* $x - 4 = 0$

$3x = -5$ *or* $x = 4$

$x = -\frac{5}{3}$ *or* $x = 4$

The solutions are $-\frac{5}{3}$ and 4.

41. $2y^2 + 12y = -10$

$2y^2 + 12y + 10 = 0$

$2(y^2 + 6y + 5) = 0$

$2(y+5)(y+1) = 0$

$y + 5 = 0$ *or* $y + 1 = 0$

$y = -5$ *or* $y = -1$

The solutions are -5 and -1.

43. $(x-7)(x+1) = -16$

$x^2 - 6x - 7 = -16$

$x^2 - 6x + 9 = 0$

$(x-3)(x-3) = 0$

$x - 3 = 0$ *or* $x - 3 = 0$

$x = 3$ *or* $x = 3$

The solution is 3.

45. $14z^2 - 3 = 21z - 3$

$14z^2 - 21z - 3 = -3$

$14z^2 - 21z = 0$

$7z(2z-3) = 0$

$z = 0$ *or* $2z - 3 = 0$

$z = 0$ *or* $2z = 3$

$z = 0$ *or* $z = \frac{3}{2}$

The solutions are 0 and $\frac{3}{2}$.

47. $81x^2 - 5 = 20$

$81x^2 - 25 = 0$

$(9x+5)(9x-5) = 0$

$9x + 5 = 0$ *or* $9x - 5 = 0$

$9x = -5$ *or* $9x = 5$

$x = -\frac{5}{9}$ *or* $x = \frac{5}{9}$

The solutions are $-\frac{5}{9}$ and $\frac{5}{9}$.

49. $(x-1)(5x+4)=2$

$5x^2-x-4=2$

$5x^2-x-6=0$

$(5x-6)(x+1)=0$

$5x-6=0 \quad or \quad x+1=0$

$5x=6 \quad or \quad x=-1$

$x=\frac{6}{5} \quad or \quad x=-1$

The solutions are $\frac{6}{5}$ and -1.

51. $x^2-2x=18+5x$

$x^2-7x-18=0$ Subtracting 18 and $5x$

$(x-9)(x+2)=0$

$x-9=0 \quad or \quad x+2=0$

$x=9 \quad or \quad x=-2$

The solutions are 9 and -2.

53. $(6a+1)(a+1)=21$

$6a^2+7a+1=21$

$6a^2+7a-20=0$

$(3a-4)(2a+5)=0$

$3a-4=0 \quad or \quad 2a+5=0$

$3a=4 \quad or \quad 2a=-5$

$a=\frac{4}{3} \quad or \quad a=-\frac{5}{2}$

The solutions are $\frac{4}{3}$ and $-\frac{5}{2}$.

55. The solutions of the equation are the first coordinates of the x-intercepts of the graph. From the graph we see that the x-intercepts are $(-3,0)$ and $(2,0)$, so the solutions of the equation are -3 and 2.

57. The solutions of the equation are the first coordinates of the x-intercepts of the graph. From the graph we see that the x-intercepts are $(-1,0)$ and $(3,0)$, so the solutions of the equation are -1 and 3.

59. We let $y=0$ and solve for x.

$0=x^2+3x-4$

$0=(x+4)(x-1)$

$x+4=0 \quad or \quad x-1=0$

$x=-4 \quad or \quad x=1$

The x-intercepts are $(-4,0)$ and $(1,0)$.

61. We let $y=0$ and solve for x.

$0=x^2-2x-15$

$0=(x-5)(x+3)$

$x-5=0 \quad or \quad x+3=0$

$x=5 \quad or \quad x=-3$

The x-intercepts are $(5,0)$ and $(-3,0)$.

63. We let $y=0$ and solve for x

$0=2x^2+x-10$

$0=(2x+5)(x-2)$

$2x+5=0 \quad or \quad x-2=0$

$2x=-5 \quad or \quad x=2$

$x=-\frac{5}{2} \quad or \quad x=2$

The x-intercepts are $\left(-\frac{5}{2},0\right)$ and $(2,0)$.

65. *Writing Exercise*

67. $(a+b)^2$

69. Let x represent the smaller integer; $x+(x+1)$

71. Let x represent the number; $\frac{1}{2}x-7>24$

73. *Writing Exercise*

75. $(2x-5)(3x^2+29x+56)=0$

$(2x-5)(x+7)(3x+8)=0$

$2x-5=0 \quad or \quad x+7=0 \quad or \quad 3x+8=0$

$2x=5 \quad or \quad x=-7 \quad or \quad 3x=-8$

$x=\frac{5}{2} \quad or \quad x=-7 \quad or \quad x=-\frac{8}{3}$

The solutions are $\frac{5}{2}$, -7, and $-\frac{8}{3}$.

77. a) $x=-4 \quad or \quad x=5$

$x+4=0 \quad or \quad x-5=0$

$(x+4)(x-5)=0$ Principle of zero products

$x^2-x-20=0$ Multiplying

b) $x=-1 \quad or \quad x=7$

$x+1=0 \quad or \quad x-7=0$

$(x+1)(x-7)=0$

$x^2-6x+-7=0$

c) $x=\frac{1}{4} \quad or \quad x=3$

$x-\frac{1}{4}=0 \quad or \quad x-3=0$

$\left(x-\frac{1}{4}\right)(x-3)=0$

$x^2-\frac{13}{4}x+\frac{3}{4}=0$

$4\left(x^2-\frac{13}{4}x+\frac{3}{4}\right)=4\cdot 0$ Multiplying both sides by 4

$4x^2-13x+3=0$

d) $x=\frac{1}{2} \quad or \quad x=\frac{1}{3}$

$x-\frac{1}{2}=0 \quad or \quad x-\frac{1}{3}=0$

$$\left(x-\frac{1}{2}\right)\left(x-\frac{1}{3}\right)=0$$
$$x^2-\frac{5}{6}x+\frac{1}{6}=0$$
$$6x^2-5x+1=0 \quad \text{Multiplying by 6}$$

e) $x=\frac{2}{3}$ *or* $x=\frac{3}{4}$

$x-\frac{2}{3}=0$ *or* $x-\frac{3}{4}=0$

$$\left(x-\frac{2}{3}\right)\left(x-\frac{3}{4}\right)=0$$
$$x^2-\frac{17}{12}x+\frac{1}{2}=0$$
$$12x^2-17x+6=0 \quad \text{Multiplying by 12}$$

f) $x=-1$ *or* $x=2$ *or* $x=3$

$x+1=0$ *or* $x-2=0$ *or* $x-3=0$

$$(x+1)(x-2)(x-3)=0$$
$$(x^2-x-2)(x-3)=0$$
$$x^3-4x^2+x+6=0$$

79. $a(9+a)=4(2a+5)$

$9a+a^2=8a+20$

$a^2+a-20=0$ Subtracting $8a$ and 20

$(a+5)(a-4)=0$

$a+5=0$ *or* $a-4=0$

$a=-5$ *or* $a=4$

The solutions are -5 and 4.

81. $-x^2+\frac{9}{25}=0$

$x^2-\frac{9}{25}=0$ Multiplying by -1

$\left(x-\frac{3}{5}\right)\left(x+\frac{3}{5}\right)=0$

$x-\frac{3}{5}=0$ *or* $x+\frac{3}{5}=0$

$x=\frac{3}{5}$ *or* $x=-\frac{3}{5}$

The solutions are $\frac{3}{5}$ and $-\frac{3}{5}$.

83. $(t+1)^2=9$

Observe that $t+1$ is a number which yields 9 when it is squared. Thus, we have

$t+1=-3$ *or* $t+1=3$

$t=-4$ *or* $t=2$

The solutions are -4 and 2.

We could also do this exercise as follows:

$$(t+1)^2=9$$
$$t^2+2t+1=9$$
$$t^2+2t-8=0$$
$$(t+4)(t-2)=0$$

$t+4=0$ *or* $t-2=0$

$t=-4$ *or* $t=2$

Again we see that the solutions are -4 and 2.

85. a) $2(x^2+10x-2)=2\cdot 0$ Multiplying (a) by 2

$2x^2+20x-4=0$

(a) and $2x^2+20x-4=0$ are equivalent.

b) $(x-6)(x+3)=x^2-3x-18$ Multiplying

(b) and $x^2-3x-18=0$ are equivalent.

c) $5x^2-5=5(x^2-1)=5(x+1)(x-1)=$
$(x+1)5(x-1)=(x+1)(5x-5)$

(c) and $(x+1)(5x-5)=0$ are equivalent.

d) $2(2x-5)(x+4)=2\cdot 0$ Multiplying (d) by 2

$2(x+4)(2x-5)=0$

$(2x+8)(2x-5)=0$

(d) and $(2x+8)(2x-5)=0$ are equivalent.

e) $4(x^2+2x+9)=4\cdot 0$ Multiplying (e) by 4

$4x^2+8x+36=0$

(e) and $4x^2+8x+36=0$ are equivalent.

f) $3(3x^2-4x+8)=3\cdot 0$ Multiplying (f) by 3

$9x^2-12x+24=0$

(f) and $9x^2-12x+24=0$ are equivalent.

87. *Writing Exercise*

89. $-0.25, 0.88$

91. $4.55, -3.23$

93. -3.76, 0

Exercise Set 5.7

1. ***Familiarize***. Let x = the number (or numbers).

Translate. We reword the problem.

The square of a number minus the number is 20.

$$x^2 \quad - \quad x \quad = 20$$

Carry out. We solve the equation.

$$x^2-x=20$$
$$x^2-x-20=0$$
$$(x-5)(x+4)=0$$

$x-5=0$ *or* $x+4=0$

$x=5$ *or* $x=-4$

Check. For 5: The square of 5 is 5^2, or 25, and $25-5=20$.

For -4: The square of -4 is $(-4)^2$, or 16, and $16-(-4)=16+4=20$. Both numbers check.

State. The numbers are 5 and -4.

3. ***Familiarize.*** Let x = the length of the shorter leg, in cm. Then $x+3$ = the length of the longer leg.

Translate. we use the Pythagorean theorem.

$$a^2+b^2=c^2$$
$$x^2+(x+3)^2=15^2$$

Carry out. We solve the equation.

$$\begin{aligned} x^2+(x+3)^2&=15^2\\ x^2+x^2+6x+9&=225\\ 2x^2+6x+9&=225\\ 2x^2+6x-216&=0\\ 2(x^2+3x-108)&=0\\ 2(x+12)(x-9)&=0 \end{aligned}$$

$$x+12=0 \quad \textit{or} \quad x-9=0$$
$$x=-12 \quad \textit{or} \quad x=9$$

Check. The number -12 cannot be the length of a side because it is negative. When $x=9$, then $x+3=12$, and $9^2+12^2=81+144=225=15^2$, so the number 9 checks.

State. The lengths of the sides are 9 cm, 12 cm, and 15 cm.

5. ***Familiarize.*** The locker numbers are consecutive integers. Let x = the smaller integer. Then $x+1$ = the larger integer.

Translate. We reword the problem.

$$\begin{array}{ccccc} \underbrace{\text{Smaller integer}} & \text{times} & \underbrace{\text{larger integer}} & \text{is} & 90. \\ \downarrow & \downarrow & \downarrow & \downarrow & \downarrow \\ x & \cdot & (x+1) & = & 90 \end{array}$$

Carry out. We solve the equation.

$$\begin{aligned} x(x+1)&=90\\ x^2+x&=90\\ x^2+x-90&=0\\ (x+10)(x-9)&=0 \end{aligned}$$

$$x+10=0 \quad \textit{or} \quad x-9=0$$
$$x=-10 \quad \textit{or} \quad x=9$$

Check. The solutions of the equation are -10 and 9. Since a locker number cannot be negative, -10 cannot be a solution of the original problem. We only need to check 9. When $x=9$, then $x+1=10$, and $9\cdot 10=90$. This checks.

State. The locker numbers are 9 and 10.

7. ***Familiarize.*** Let x = the smaller odd integer. Then $x+2$ = the larger odd integer.

Translate. We reword the problem.

$$\begin{array}{ccccc} \underbrace{\text{Smaller odd integer}} & \text{times} & \underbrace{\text{larger odd integer}} & \text{is} & 255. \\ \downarrow & \downarrow & \downarrow & \downarrow & \downarrow \\ x & \cdot & (x+2) & = & 255 \end{array}$$

Carry out.

$$\begin{aligned} x(x+2)&=255\\ x^2+2x&=255\\ x^2+2x-255&=0\\ (x+17)(x-15)&=0 \end{aligned}$$

$$x+17=0 \quad \textit{or} \quad x-15=0$$
$$x=-17 \quad \textit{or} \quad x=15$$

Check. The solutions of the equation are -17 and 15. When x is -17, then $x+2$ is -15 and $-17(-15)=225$. The numbers -17 and -15 are consecutive odd integers which are solutions of the problem. When x is 15, then $x+2=17$ and $15\cdot 17=255$. The numbers 15 and 17 are also consecutive odd integers which are solutions of the problem.

State. We have two solutions, each of which consists of a pair of numbers: -17 and -15 or 15 and 17.

9. ***Familiarize.*** Let w = the width of the table, in feet. Then $6w$ = the width. Recall that the area of a rectangle is Length $\cdot$ Width.

Translate.

$$\begin{array}{ccc} \underbrace{\text{The area of the rectangle}} & \text{is} & \underbrace{24\text{ ft}^2}. \\ \downarrow & \downarrow & \downarrow \\ 6w\cdot w & = & 24 \end{array}$$

Carry out. We solve the equation.

$$\begin{aligned} 6w\cdot w&=24\\ 6w^2&=24\\ 6w^2-24&=0\\ 6(w^2-4)&=0\\ 2(w+2)(w-2)&=0 \end{aligned}$$

$$w+2=0 \quad \textit{or} \quad w-2=0$$
$$w=-2 \quad \textit{or} \quad w=2$$

Check. Since the width must be positive, -2 cannot be a solution. If the width is 2 ft, then the length is $6\cdot 2$ ft, or 12 ft, and the area is 2 ft $\cdot$ 12 ft $=24$ ft^2. Thus, 2 checks.

State. The table is 12 ft long and 2 ft wide.

11. ***Familiarize.*** We make a drawing. Let w = the width, in cm. Then $w+2$ = the length, in cm.

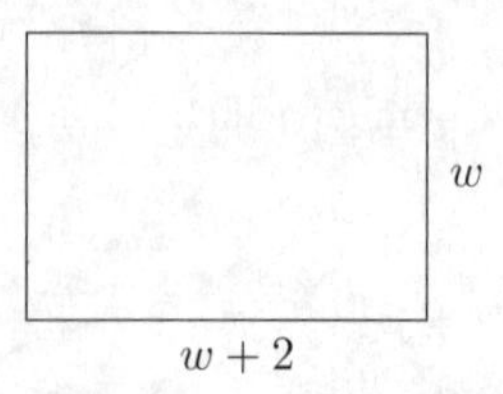

Recall that the area of a rectangle is length times width.

Translate. We reword the problem.

$$\begin{array}{ccccc} \text{Length} & \text{times} & \text{width} & \text{is} & \underbrace{24\text{ cm}^2}. \\ \downarrow & \downarrow & \downarrow & \downarrow & \downarrow \\ (w+2) & \cdot & w & = & 24 \end{array}$$

Carry out. We solve the equation.

$$(w+2)w = 24$$
$$w^2+2w = 24$$
$$w^2+2w-24 = 0$$
$$(w+6)(w-4) = 0$$
$$w+6=0 \quad or \quad w-4=0$$
$$w=-6 \quad or \quad w=4$$

Check. Since the width must be positive, -6 cannot be a solution. If the width is 4 cm, then the length is $4+2$, or 6 cm, and the area is $6\cdot 4$, or 24 cm^2. Thus, 4 checks.

State. The width is 4 cm, and the length is 6 cm.

13. ***Familiarize.*** Using the labels shown on the drawing in the text, we let h = the height, in cm, and $h+10$ = the base, in cm. Recall that the formula for the area of a triangle is $\frac{1}{2}\cdot(\text{base})\cdot(\text{height})$.

Translate.

$$\underbrace{\tfrac{1}{2}}\ \text{times}\ \ \text{base}\ \ \text{times}\ \ \text{height}\ \ \text{is}\ \ \underbrace{28\ \text{cm}^2}.$$
$$\frac{1}{2}\cdot(h+10)\cdot h = 28$$

Carry out.

$$\frac{1}{2}(h+10)h = 28$$
$$(h+10)h = 56 \quad \text{Multiplying by 2}$$
$$h^2+10h = 56$$
$$h^2+10h-56 = 0$$
$$(h+14)(h-4) = 0$$
$$h+14=0 \quad or \quad h-4=0$$
$$h=-14 \quad or \quad h=4$$

Check. Since the height of the triangle must be positive, -14 cannot be a solution. If the height is 4 cm, then the base is $4+10$, or 14 cm, and the area is $\frac{1}{2}\cdot 14\cdot 4$, or 28 cm^2. Thus, 4 checks.

State. The height of the triangle is 4 cm, and the base is 14 cm.

15. ***Familiarize.*** Using the labels show on the drawing in the text, we let x = the length of the foot of the sail, in ft, and $x+5$ = the height of the sail, in ft. Recall that the formula for the area of a triangle is $\frac{1}{2}\cdot(\text{base})\cdot(\text{height})$.

Translate.

$$\tfrac{1}{2}\ \text{times}\ \text{base}\ \text{times}\ \text{height}\ \text{is}\ \underbrace{42\ \text{ft}^2}.$$
$$\frac{1}{2}\cdot x\cdot(x+5) = 42$$

Carry out.

$$\frac{1}{2}x(x+5) = 42$$
$$x(x+5) = 84 \quad \text{Multiplying by 2}$$
$$x^2+5x = 84$$
$$x^2+5x-84 = 0$$
$$(x+12)(x-7) = 0$$
$$x+12=0 \quad or \quad x-7=0$$
$$x=-12 \quad or \quad x=7$$

Check. The solutions of the equation are -12 and 7. The length of the base of a triangle cannot be negative, so -12 cannot be a solution. Suppose the length of the foot of the sail is 7 ft. Then the height is $7+5$, or 12 ft, and the area is $\frac{1}{2}\cdot 7\cdot 12$, or 42 ft^2. These numbers check.

State. The length of the foot of the sail is 7 ft, and the height is 12 ft.

17. ***Familiarize and Translate.*** We substitute 150 for A in the formula.

$$A = -50t^2+200t$$
$$150 = -50t^2+200t$$

Carry out. We solve the equation.

$$150 = -50t^2+200t$$
$$0 = -50t^2+200t-150$$
$$0 = -50(t^2-4t+3)$$
$$0 = -50(t-1)(t-3)$$
$$t-1=0 \quad or \quad t-3=0$$
$$t=1 \quad or \quad t=3$$

Check. Since $-50\cdot 1^2+200\cdot 1 = -50+200 = 150$, the number 1 checks. Since $-50\cdot 3^2+200\cdot 3 = -450+600 = 150$, the number 3 checks also.

State. There will be about 150 micrograms of Albuterol in the bloodstream 1 minute and 3 minutes after an inhalation.

19. ***Familiarize.*** We will use the formula $x^2-x=N$.

Translate. Substitute 240 for N.

$$x^2-x = 240$$

Carry out.

$$x^2-x = 240$$
$$x^2-x-240 = 0$$
$$(x-16)(x+15) = 0$$
$$x-16=0 \quad \text{or} \quad x+15=0$$
$$x=16 \quad \text{or} \quad x=-15$$

Check. The solutions of the equation are 16 and -15. Since the number of teams cannot be negative, -15 cannot be a solution. But 16 checks since $16^2-16 = 256-16 = 240$.

State. There are 16 teams in the league.

21. ***Familiarize.*** We will use the formula

$$H = \frac{1}{2}(n^2-n).$$

Translate. Substitute 15 for n.

$$H = \frac{1}{2}(15^2-15)$$

Carry out. We do the computation on the right.

$$H = \frac{1}{2}(15^2 - 15)$$
$$H = \frac{1}{2}(225 - 15)$$
$$H = \frac{1}{2}(210)$$
$$H = 105$$

Check. We can recheck the computation, or we can solve the equation $105 = \frac{1}{2}(n^2 - n)$. The answer checks.

State. 105 handshakes are possible.

23. ***Familiarize***. We will use the formula $H = \frac{1}{2}(n^2 - n)$, since "high fives" can be substituted for handshakes.

Translate. Substitute 66 for H.

$$66 = \frac{1}{2}(n^2 - n)$$

Carry out.

$$66 = \frac{1}{2}(n^2 - n)$$
$$132 = n^2 - n \quad \text{Multiplying by 2}$$
$$0 = n^2 - n - 132$$
$$0 = (n - 12)(n + 11)$$
$$n - 12 = 0 \quad or \quad n + 11 = 0$$
$$n = 12 \quad or \quad n = -11$$

Check. The solutions of the equation are 12 and -11. Since the number of players cannot be negative, -11 cannot be a solution. However, 12 checks since $\frac{1}{2}(12^2 - 12) = \frac{1}{2}(144 - 12) = \frac{1}{2}(132) = 66$.

State. 12 players were on the team.

25. ***Familiarize***. Let h = the vertical height to which each brace reaches, in feet. We have a right triangle with hypotenuse 15 ft and legs 12 ft and h.

Translate. We use the Pythagorean theorem.

$$a^2 + b^2 = c^2$$
$$12^2 + h^2 = 15^2$$

Carry out. We solve the equation.

$$12^2 + h^2 = 15^2$$
$$144 + h^2 = 225$$
$$h^2 - 81 = 0$$
$$(h + 9)(h - 9) = 0$$
$$h + 9 = 0 \quad \text{or} \quad h - 9 = 0$$
$$h = -9 \quad \text{or} \quad h = 9$$

Check. Since the vertical height must be positive, -9 cannot be a solution. If the height is 9 ft, then we have $12^2 + 9^2 = 144 + 81 = 225 = 15^2$. The number 9 checks.

State. Each brace reaches 9 ft vertically.

27. ***Familiarize***. We make a drawing. Let l = the length of the cable, in ft.

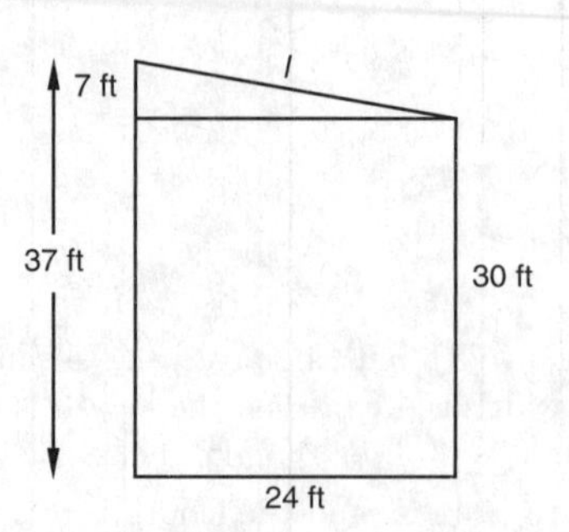

Note that we have a right triangle with hypotenuse l and legs of 24 ft and $37 - 30$, or 7 ft.

Translate. We use the Pythagorean theorem.

$$a^2 + b^2 = c^2$$
$$7^2 + 24^2 = l^2 \quad \text{Substituting}$$

Carry out.

$$7^2 + 24^2 = l^2$$
$$49 + 576 = l^2$$
$$625 = l^2$$
$$0 = l^2 - 625$$
$$0 = (l + 25)(l - 25)$$
$$l + 25 = 0 \quad or \quad l - 25 = 0$$
$$l = -25 \quad or \quad l = 25$$

Check. The integer -25 cannot be the length of the cable, because it is negative. When $l = 25$, we have $7^2 + 24^2 = 25^2$. This checks.

State. The cable is 25 ft long.

29. ***Familiarize***. We label the drawing. Let x = the length of a side of the dining room, in ft. Then the dining room has dimensions x by x and the kitchen has dimensions x by 10. The entire rectangular space has dimension x by $x + 10$. Recall that we multiply these dimensions to find the area of the rectangle.

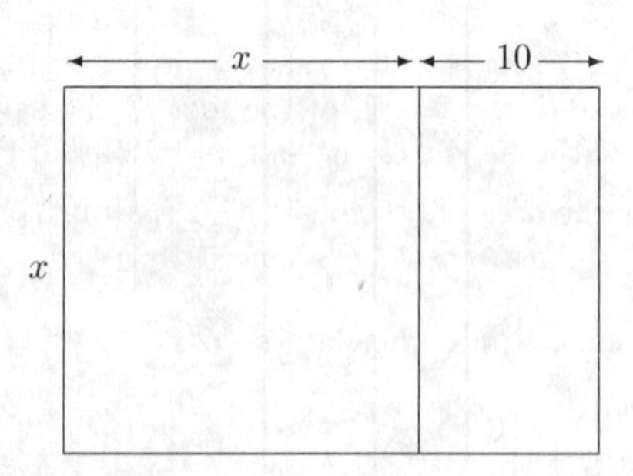

Translate.

$$\underbrace{\text{The area of the rectangular space}}_{x(x+10)} \quad \text{is} \quad \underbrace{264 \text{ ft}^2}_{264}.$$

$$x(x + 10) = 264$$

Carry out. We solve the equation.

$$x(x+10) = 264$$
$$x^2 + 10x = 264$$
$$x^2 + 10x - 264 = 0$$
$$(x+22)(x-12) = 0$$
$$x + 22 = 0 \quad or \quad x - 12 = 0$$
$$x = -22 \quad or \quad x = 12$$

Check. Since the length of a side of the dining room must be positive, -22 cannot be a solution. If x is 12 ft, then $x+10$ is 22 ft, and the area of the space is $12 \cdot 22$, or 264 ft^2. The number 12 checks.

State. The dining room is 12 ft by 12 ft, and the kitchen is 12 ft by 10 ft.

31. ***Familiarize.*** We will use the formula $h = 48t - 16t^2$.

Translate. Substitute $\frac{1}{2}$ for t.

$$h = 48 \cdot \frac{1}{2} - 16\left(\frac{1}{2}\right)^2$$

Carry out. We do the computation on the right.

$$h = 48 \cdot \frac{1}{2} - 16\left(\frac{1}{2}\right)^2$$
$$h = 48 \cdot \frac{1}{2} - 16 \cdot \frac{1}{4}$$
$$h = 24 - 4$$
$$h = 20$$

Check. We can recheck the computation, or we can solve the equation $20 = 48t - 16t^2$. The answer checks.

State. The rocket is 20 ft high $\frac{1}{2}$ sec after it is launched.

33. ***Familiarize.*** We will use the formula $h = 48t - 16t^2$.

Translate. Substitute 32 for h.

$$32 = 48t - 16t^2$$

Carry out. We solve the equation.

$$32 = 48t - 16t^2$$
$$0 = -16t^2 + 48t - 32$$
$$0 = -16(t^2 - 3t + 2)$$
$$0 = -16(t-1)(t-2)$$
$$t - 1 = 0 \quad or \quad t - 2 = 0$$
$$t = 1 \quad or \quad t = 2$$

Check. When $t = 1$, $h = 48 \cdot 1 - 16 \cdot 1^2 = 48 - 16 = 32$. When $t = 2$, $h = 48 \cdot 2 - 16 \cdot 2^2 = 96 - 64 = 32$. Both numbers check.

State. The rocket will be exactly 32 ft above the ground at 1 sec and at 2 sec after it is launched.

35. *Writing Exercise*

37. $-\frac{2}{3} \cdot \frac{4}{7} = -\frac{2 \cdot 4}{3 \cdot 7} = -\frac{8}{21}$

39. $\frac{5}{6}\left(\frac{-7}{9}\right) = \frac{5(-7)}{6 \cdot 9} = \frac{-35}{54}$, or $-\frac{35}{54}$

41. $-\frac{2}{3} + \frac{4}{7} = -\frac{2}{3} \cdot \frac{7}{7} + \frac{4}{7} \cdot \frac{3}{3}$

$$= -\frac{14}{21} + \frac{12}{21}$$
$$= -\frac{2}{21}$$

43. $\frac{5}{6} + \frac{-7}{9} = \frac{5}{6} \cdot \frac{3}{3} + \frac{-7}{9} \cdot \frac{2}{2}$

$$= \frac{15}{18} + \frac{-14}{18}$$
$$= \frac{1}{18}$$

45. *Writing Exercise*

47. ***Familiarize.*** First we find the length of the other leg of the right triangle. Then we find the area of the triangle, and finally we multiply by the cost per square foot of the sailcloth. Let x = the length of the other leg of the right triangle, in feet.

Translate. We use the Pythagorean theorem to find x.

$$a^2 + b^2 = c^2$$
$$x^2 + 24^2 = 26^2 \quad \text{Substituting}$$

Carry out.

$$x^2 + 24^2 = 26^2$$
$$x^2 + 576 = 676$$
$$x^2 - 100 = 0$$
$$(x+10)(x-10) = 0$$
$$x + 10 = 0 \quad or \quad x - 10 = 0$$
$$x = -10 \quad or \quad x = 10$$

Since the length of the leg must be positive, -10 cannot be a solution. We use the number 10. Find the area of the triangle:

$$\frac{1}{2}bh = \frac{1}{2} \cdot 10 \text{ ft} \cdot 24 \text{ ft} = 120 \text{ ft}^2$$

Finally, we multiply the area, 120 ft^2, by the price per square foot of the sailcloth, \$10:

$$120 \cdot 10 = 1200$$

Check. Recheck the calculations. The answer checks.

State. A new main sail costs \$1200.

49. ***Familiarize.*** We add labels to the drawing in the text.

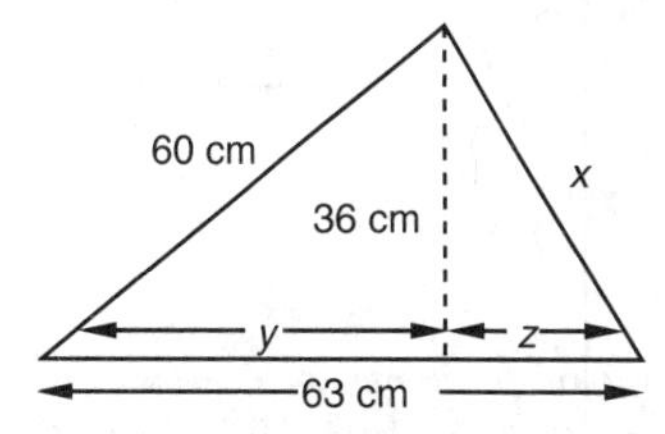

First we will use the Pythagorean theorem to find y. Then we will subtract to find z and, finally, we will use the Pythagorean theorem again to find x.

Translate. We use the Pythagorean theorem to find y.

$$a^2 + b^2 = c^2$$
$$y^2 + 36^2 = 60^2 \quad \text{Substituting}$$

Carry out.

$$y^2 + 36^2 = 60^2$$
$$y^2 + 1296 = 3600$$
$$y^2 - 2304 = 0$$
$$(y+48)(y-48) = 0$$
$$y + 48 = 0 \quad or \quad y - 48 = 0$$
$$y = -48 \quad or \quad y = 48$$

Since the length y cannot be negative, we use 48 cm. Then $z = 63 - 48 = 15$ cm.

Now we find x. We use the Pythagorean theorem again.

$$15^2 + 36^2 = x^2$$
$$225 + 1296 = x^2$$
$$1521 = x^2$$
$$0 = x^2 - 1521$$
$$0 = (x+39)(x-39)$$
$$x + 39 = 0 \quad or \quad x - 39 = 0$$
$$x = -39 \quad or \quad x = 39$$

Since the length x cannot be negative, we use 39 cm.

Check. We repeat all of the calculations. The answer checks.

State. The value of x is 39 cm.

51. ***Familiarize.*** Let w = the width of the piece of cardboard, in cm. Then $2w$ = the length, in cm. The length and width of the base of the box are $2x - 8$ and $x - 8$, respectively, and its height is 4.

Recall that the formula for the volume of a rectangular solid is given by length · width · height.

Translate.

$$\underbrace{\text{The volume}} \quad \text{is} \quad \underbrace{616\ \text{cm}^3}.$$
$$\downarrow \qquad\qquad \downarrow \qquad \downarrow$$
$$(2w-8)(w-8)(4) = 616$$

Carry out. We solve the equation.

$$(2w-8)(w-8)(4) = 616$$
$$(2w^2 - 24w + 64)(4) = 616$$
$$8w^2 - 96 + 256 = 616$$
$$8w^2 - 96w - 360 = 0$$
$$8(w^2 - 12w - 45) = 0$$
$$w^2 - 12w - 45 = 0 \quad \text{Dividing by 8}$$
$$(w-15)(w+3) = 0$$
$$w - 15 = 0 \quad or \quad w + 3 = 0$$
$$w = 15 \quad or \quad w = -3$$

Check. The width cannot be negative, so we only need to check 15. When $w = 15$, then $2w = 30$ and the dimensions of the box are $30 - 8$ by $15 - 8$ by 4, or 22 by 7 by 4. The volume is $22 \cdot 7 \cdot 4$, or 616.

State. The original dimension of the cardboard are 15 cm by 30 cm.

53. ***Familiarize.*** First we can use the Pythagorean theorem to find x, in ft. Then the height of the telephone pole is $x + 5$.

Translate. We use the Pythagorean theorem.

$$a^2 + b^2 = c^2$$
$$\left(\frac{1}{2}x + 1\right)^2 + x^2 = 34^2$$

Carry out. We solve the equation.

$$\left(\frac{1}{2}x + 1\right)^2 + x^2 = 34^2$$
$$\frac{1}{4}x^2 + x + 1 + x^2 = 1156$$
$$x^2 + 4x + 4 + 4x^2 = 4624 \quad \text{Multiplying by 4}$$
$$5x^2 + 4 + 4 = 4624$$
$$5x^2 + 4x - 4620 = 0$$
$$(5x + 154)(x - 30) = 0$$
$$5x + 154 = 0 \quad or \quad x - 30 = 0$$
$$5x = -154 \quad or \quad x = 30$$
$$x = -30.8 \quad or \quad x = 30$$

Check. Since the length x must be positive, -30.8 cannot be a solution. If x is 30 ft, then $\frac{1}{2}x + 1$ is $\frac{1}{2} \cdot 30 + 1$, or 16 ft. Since $16^2 + 30^2 = 1156 = 34^2$, the number 30 checks. When x is 30 ft, then $x + 5$ is 35 ft.

State. The height of the telephone pole is 35 ft.

55. First substitute 18 for N in the given formula.

$$18 = -0.009t(t - 12)^3$$

Graph $y_1 = 18$ and $y_2 = -0.009x(x - 12)^3$ in the given window and use the TRACE feature to find the first coordinates of the points of intersection of the graphs. We find $x \approx 2$ hr and $x \approx 4.2$ hr.

57. Graph $y = -0.009x(x - 12)^3$ and use the TRACE feature to find the first coordinate of the highest point on the graph. We find $x = 3$ hr.

Chapter 6

Rational Expressions and Equations

Exercise Set 6.1

1. $x - 2 = 0$ when $x = 2$ and $x + 3 = 0$ when $x = -3$, so choice (e) is correct.

3. $a^2 - a - 12 = (a - 4)(a + 3)$; $a - 4 = 0$ when $a = 4$ and $a + 3 = 0$ when $a = -3$, so choice (d) is correct.

5. $2t - 1 = 0$ when $t = \frac{1}{2}$ and $3t + 4 = 0$ when $t = -\frac{4}{3}$, so choice (c) is correct.

7. $\frac{12}{-7x}$

We find the real number(s) that make the denominator 0. To do so we set the denominator equal to 0 and solve for x:

$$-7x = 0$$
$$x = 0$$

The expression is undefined for $x = 0$.

9. $\frac{t - 6}{t + 8}$

Set the denominator equal to 0 and solve for t:

$$t + 8 = 0$$
$$t = -8$$

The expression is undefined for $t = -8$.

11. $\frac{a - 4}{3a - 12}$

Set the denominator equal to 0 and solve for a:

$$3a - 12 = 0$$
$$3a = 12$$
$$a = 4$$

The expression is undefined for $a = 4$.

13. $\frac{x^2 - 25}{x^2 - 3x - 28}$

Set the denominator equal to 0 and solve for x:

$$x^2 - 3x - 28 = 0$$
$$(x - 7)(x + 4) = 0$$
$$x - 7 = 0 \quad or \quad x + 4 = 0$$
$$x = 7 \quad or \quad x = -4$$

The expression is undefined for $x = 7$ and $x = -4$.

15. $\frac{t^2 - 4}{2t^2 + 11t - 6}$

Set the denominator equal to 0 and solve for t:

$$2t^2 + 11t - 6 = 0$$
$$(2t - 1)(t + 6) = 0$$
$$2t - 1 = 0 \quad or \quad t + 6 = 0$$
$$2t = 1 \quad or \quad t = -6$$
$$t = \frac{1}{2} \quad or \quad t = -6$$

The expression is undefined for $t = \frac{1}{2}$ and $t = -6$.

17. $\frac{50a^2b}{40ab^3}$

$= \frac{5a \cdot 10ab}{4b^2 \cdot 10ab}$ Factoring the numerator and denominator. Note the common factor of $10ab$.

$= \frac{5a}{4b^2} \cdot \frac{10ab}{10ab}$ Rewriting as a product of two rational expressions

$= \frac{5a}{4b^2} \cdot 1$ $\frac{10ab}{10ab} = 1$

$= \frac{5a}{4b^2}$ Removing the factor 1

19.
$$\frac{28x^2y}{21x^3y^5} = \frac{4 \cdot 7x^2y}{3xy^4 \cdot 7x^2y} = \frac{4}{3xy^4} \cdot \frac{7x^2y}{7x^2y} = \frac{4}{3xy^4} \cdot 1 = \frac{4}{3xy^4}$$

21.
$$\frac{9x + 15}{12x + 20} = \frac{3(3x + 5)}{4(3x + 5)} = \frac{3}{4} \cdot \frac{3x + 5}{3x + 5} = \frac{3}{4} \cdot 1 = \frac{3}{4}$$

23.
$$\frac{a^2 - 9}{a^2 + 4a + 3} = \frac{(a + 3)(a - 3)}{(a + 3)(a + 1)} = \frac{a + 3}{a + 3} \cdot \frac{a - 3}{a + 1} = 1 \cdot \frac{a - 3}{a + 1} = \frac{a - 3}{a + 1}$$

25. $\frac{36x^6}{24x^9} = \frac{3 \cdot 12x^6}{2x^3 \cdot 12x^6}$

$= \frac{3}{2x^3} \cdot \frac{12x^6}{12x^6}$

$= \frac{3}{2x^3} \cdot 1$

$= \frac{3}{2x^3}$

Check: Let $x = 1$.

$\frac{36x^6}{24x^9} = \frac{36 \cdot 1^6}{24 \cdot 1^9} = \frac{36}{24} = \frac{3}{2}$

$\frac{3}{2x^3} = \frac{3}{2 \cdot 1^3} = \frac{3}{2}$

The answer is probably correct.

27. $\frac{-2y+6}{-8y} = \frac{-2(y-3)}{-2 \cdot 4y}$

$= \frac{-2}{-2} \cdot \frac{y-3}{4y}$

$= 1 \cdot \frac{y-3}{4y}$

$= \frac{y-3}{4y}$

Check: Let $x = 2$.

$\frac{-2y+6}{-8y} = \frac{-2 \cdot 2 + 6}{-8 \cdot 2} = \frac{2}{-16} = -\frac{1}{8}$

$\frac{y-3}{4y} = \frac{2-3}{4 \cdot 2} = \frac{-1}{8} = -\frac{1}{8}$

The answer is probably correct.

29. $\frac{6a^2-3a}{7a^2-7a} = \frac{3a(2a-1)}{7a(a-1)}$

$= \frac{a}{a} \cdot \frac{3(2a-1)}{7(a-1)}$

$= 1 \cdot \frac{3(2a-1)}{7(a-1)}$

$= \frac{3(2a-1)}{7(a-1)}$

Check: Let $a = 2$.

$\frac{6a^2-3a}{7a^2-7a} = \frac{6 \cdot 2^2 - 3 \cdot 2}{7 \cdot 2^2 - 7 \cdot 2} = \frac{18}{14} = \frac{9}{7}$

$\frac{3(2a-1)}{7(a-1)} = \frac{3(2 \cdot 2 - 1)}{7(2-1)} = \frac{3 \cdot 3}{7 \cdot 1} = \frac{9}{7}$

The answer is probably correct.

31. $\frac{t^2-16}{t^2+t-20} = \frac{(t+4)(t-4)}{(t+5)(t-4)}$

$= \frac{t+4}{t+5} \cdot \frac{t-4}{t-4}$

$= \frac{t+4}{t+5} \cdot 1$

$= \frac{t+4}{t+5}$

Check: Let $t = 1$.

$\frac{t^2-16}{t^2+t-20} = \frac{1^2-16}{1^2+1-20} = \frac{-15}{-18} = \frac{5}{6}$

$\frac{t+4}{t+5} = \frac{1+4}{1+5} = \frac{5}{6}$

The answer is probably correct.

33. $\frac{3a^2+9a-12}{6a^2-30a+24} = \frac{3(a^2+3a-4)}{6(a^2-5a+4)}$

$= \frac{3(a+4)(a-1)}{3 \cdot 2(a-4)(a-1)}$

$= \frac{3(a-1)}{3(a-1)} \cdot \frac{a+4}{2(a-4)}$

$= 1 \cdot \frac{a+4}{2(a-4)}$

$= \frac{a+4}{2(a-4)}$

Check: Let $a = 2$.

$\frac{3a^2+9a-12}{6a^2-30a+24} = \frac{3 \cdot 2^2 + 9 \cdot 2 - 12}{6 \cdot 2^2 - 30 \cdot 2 + 24} = \frac{18}{-12} = -\frac{3}{2}$

$\frac{a+4}{2(a-4)} = \frac{2+4}{2(2-4)} = \frac{6}{-4} = -\frac{3}{2}$

The answer is probably correct.

35. $\frac{x^2+8x+16}{x^2-16} = \frac{(x+4)(x+4)}{(x+4)(x-4)}$

$= \frac{x+4}{x+4} \cdot \frac{x+4}{x-4}$

$= 1 \cdot \frac{x+4}{x-4}$

$= \frac{x+4}{x-4}$

Check: Let $x = 1$.

$\frac{x^2+8x+16}{x^2-16} = \frac{1^2+8 \cdot 1+16}{1^2-16} = \frac{25}{-15} = -\frac{5}{3}$

$\frac{x+4}{x-4} = \frac{1+4}{1-4} = \frac{5}{-3} = -\frac{5}{3}$

The answer is probably correct.

37. $\frac{t^2-1}{t+1} = \frac{(t+1)(t-1)}{t+1}$

$= \frac{t+1}{t+1} \cdot \frac{t-1}{1}$

$= 1 \cdot \frac{t-1}{1}$

$= t-1$

Check: Let $t = 2$.

$\frac{t^2-1}{t+1} = \frac{2^2-1}{2+1} = \frac{3}{3} = 1$

$t-1 = 2-1 = 1$

The answer is probably correct.

39. $\frac{y^2+4}{y+2}$ cannot be simplified.

Neither the numerator nor the denominator can be factored.

41. $\frac{5x^2+20}{10x^2+40} = \frac{5(x^2+4)}{10(x^2+4)}$

$= \frac{1 \cdot \cancel{5} \cdot \cancel{(x^2+4)}}{2 \cdot \cancel{5} \cdot \cancel{(x^2+4)}}$

$= \frac{1}{2}$

Check: Let $x = 1$.

$\dfrac{5x^2+20}{10x^2+40} = \dfrac{5\cdot 1^2+20}{10\cdot 1^2+40} = \dfrac{25}{50} = \dfrac{1}{2}$

$\dfrac{1}{2} = \dfrac{1}{2}$

The answer is probably correct.

43. $\dfrac{y^2+6y}{2y^2+13y+6} = \dfrac{y(y+6)}{(2y+1)(y+6)}$

$= \dfrac{y}{2y+1}\cdot\dfrac{y+6}{y+6}$

$= \dfrac{y}{2y+1}\cdot 1$

$= \dfrac{y}{2y+1}$

Check: Let $y = 1$.

$\dfrac{y^2+6y}{2y^2+13y+6} = \dfrac{1^2+6\cdot 1}{2\cdot 1^2+13\cdot 1+6} = \dfrac{7}{21} = \dfrac{1}{3}$

$\dfrac{y}{2y+1} = \dfrac{1}{2\cdot 1+1} = \dfrac{1}{3}$

The answer is probably correct.

45. $\dfrac{4x^2-12x+9}{10x^2-11x-6} = \dfrac{(2x-3)(2x-3)}{(2x-3)(5x+2)}$

$= \dfrac{2x-3}{2x-3}\cdot\dfrac{2x-3}{5x+2}$

$= 1\cdot\dfrac{2x-3}{5x+2}$

$= \dfrac{2x-3}{5x+2}$

Check: Let $t = 1$.

$\dfrac{4x^2-12x+9}{10x^2-11x-6} = \dfrac{4\cdot 1^2-12\cdot 1+9}{10\cdot 1^2-11\cdot 1-6} = \dfrac{1}{-7} = -\dfrac{1}{7}$

$\dfrac{2x-3}{5x+2} = \dfrac{2\cdot 1-3}{5\cdot 1+2} = \dfrac{-1}{7} = -\dfrac{1}{7}$

The answer is probably correct.

47. $\dfrac{x-9}{9-x} = \dfrac{x-9}{-(x-9)}$

$= \dfrac{1}{-1}\cdot\dfrac{x-9}{x-9}$

$= \dfrac{1}{-1}\cdot 1$

$= -1$

Check: Let $x = 2$.

$\dfrac{x-9}{9-x} = \dfrac{2-9}{9-2} = \dfrac{-7}{7} = -1$

The answer is probably correct.

49. $\dfrac{7t-14}{2-t} = \dfrac{7(t-2)}{-(t-2)}$

$= \dfrac{7}{-1}\cdot\dfrac{t-2}{t-2}$

$= \dfrac{7}{-1}\cdot 1$

$= -7$

Check: Let $t = 1$.

$\dfrac{7t-14}{2-t} = \dfrac{7\cdot 1-14}{2-1} = \dfrac{-7}{1} = -7$

The answer is probably correct.

51. $\dfrac{a-b}{3b-3a} = \dfrac{a-b}{-3(a-b)}$

$= \dfrac{1}{-3}\cdot\dfrac{a-b}{a-b}$

$= \dfrac{1}{-3}\cdot 1$

$= -\dfrac{1}{3}$

Check: Let $a = 2$ and $b = 1$.

$\dfrac{a-b}{3b-3a} = \dfrac{2-1}{3\cdot 1-3\cdot 2} = \dfrac{1}{-3} = -\dfrac{1}{3}$

The answer is probably correct.

53. $\dfrac{3x^2-3y^2}{2y^2-2x^2} = \dfrac{3(x^2-y^2)}{2(y^2-x^2)}$

$= \dfrac{3(x^2-y^2)}{2(-1)(x^2-y^2)}$

$= \dfrac{3}{2(-1)}\cdot\dfrac{x^2-y^2}{x^2-y^2}$

$= \dfrac{3}{2(-1)}\cdot 1$

$= -\dfrac{3}{2}$

Check: Let $x = 1$ and $y = 2$.

$\dfrac{3x^2-3y^2}{2y^2-2x^2} = \dfrac{3\cdot 1^2-3\cdot 2^2}{2\cdot 2^2-2\cdot 1^2} = \dfrac{-9}{6} = -\dfrac{3}{2}$

$-\dfrac{3}{2} = -\dfrac{3}{2}$

The answer is probably correct.

55. $\dfrac{7s^2-28t^2}{28t^2-7s^2}$

Note that the numerator and denominator are opposites. Thus, we have an expression divided by its opposite, so the result is -1.

57. *Writing Exercise*

59. $-\dfrac{2}{3}\cdot\dfrac{6}{7} = -\dfrac{2\cdot 6}{3\cdot 7}$

$= -\dfrac{2\cdot 2\cdot \cancel{3}}{\cancel{3}\cdot 7}$

$= -\dfrac{4}{7}$

61. $\dfrac{5}{8}\div\left(-\dfrac{1}{6}\right) = \dfrac{5}{8}\cdot(-6)$

$= -\dfrac{5\cdot 6}{8}$

$= -\dfrac{5\cdot \cancel{2}\cdot 3}{\cancel{2}\cdot 4}$

$= -\dfrac{15}{4}$

63. $\dfrac{7}{9}-\dfrac{2}{3}\cdot\dfrac{6}{7} = \dfrac{7}{9}-\dfrac{4}{7} = \dfrac{7}{9}\cdot\dfrac{7}{7}-\dfrac{4}{7}\cdot\dfrac{9}{2} =$

$\dfrac{49}{63}-\dfrac{36}{63} = \dfrac{13}{63}$

65. *Writing Exercise*

67.
$$\frac{16y^4-x^4}{(x^2+4y^2)(x-2y)}$$
$$=\frac{(4y^2+x^2)(4y^2-x^2)}{(x^2+4y^2)(x-2y)}$$
$$=\frac{(4y^2+x^2)(2y+x)(2y-x)}{(x^2+4y^2)(x-2y)}$$
$$=\frac{(x^2+4y^2)(2y+x)(-1)(x-2y)}{(x^2+4y^2)(x-2y)}$$
$$=\frac{(x^2+4y^2)(x-2y)}{(x^2+4y^2)(x-2y)}\cdot\frac{(2y+x)(-1)}{1}$$
$$=-2y-x,\ \text{or}\ -x-2y,\ \text{or}\ -(2y+x)$$

69.
$$\frac{x^5-2x^3+4x^2-8}{x^7+2x^4-4x^3-8}$$
$$=\frac{x^3(x^2-2)+4(x^2-2)}{x^4(x^3+2)-4(x^3+2)}$$
$$=\frac{(x^2-2)(x^3+4)}{(x^3+2)(x^4-4)}$$
$$=\frac{(x^2-2)(x^3+4)}{(x^3+2)(x^2+2)(x^2-2)}$$
$$=\frac{\cancel{(x^2-2)}(x^3+4)}{(x^3+2)(x^2+2)\cancel{(x^2-2)}}$$
$$=\frac{x^3+4}{(x^3+2)(x^2+2)}$$

71.
$$\frac{(t^4-1)(t^2-9)(t-9)^2}{(t^4-81)(t^2+1)(t+1)^2}$$
$$=\frac{(t^2+1)(t+1)(t-1)(t+3)(t-3)(t-9)(t-9)}{(t^2+9)(t+3)(t-3)(t^2+1)(t+1)(t+1)}$$
$$=\frac{\cancel{(t^2+1)}\cancel{(t+1)}(t-1)\cancel{(t+3)}\cancel{(t-3)}(t-9)(t-9)}{(t^2+9)\cancel{(t+3)}\cancel{(t-3)}\cancel{(t^2+1)}\cancel{(t+1)}(t+1)}$$
$$=\frac{(t-1)(t-9)(t-9)}{(t^2+9)(t+1)},\ \text{or}\ \frac{(t-1)(t-9)^2}{(t^2+9)(t+1)}$$

73.
$$\frac{(x^2-y^2)(x^2-2xy+y^2)}{(x+y)^2(x^2-4xy-5y^2)}$$
$$=\frac{(x+y)(x-y)(x-y)(x-y)}{(x+y)(x+y)(x-5y)(x+y)}$$
$$=\frac{\cancel{(x+y)}(x-y)(x-y)(x-y)}{\cancel{(x+y)}(x+y)(x-5y)(x+y)}$$
$$=\frac{(x-y)^3}{(x+y)^2(x-5y)}$$

75. *Writing Exercise*

Exercise Set 6.2

1. $\frac{7x}{5}\cdot\frac{x-5}{2x+1}=\frac{7x(x-5)}{5(2x+1)}$

3. $\frac{a-4}{a+6}\cdot\frac{a+2}{a+6}=\frac{(a-4)(a+2)}{(a+6)(a+6)}$, or $\frac{(a-4)(a+2)}{(a+6)^2}$

5. $\frac{2x+3}{4}\cdot\frac{x+1}{x-5}=\frac{(2x+3)(x+1)}{4(x-5)}$

7. $\frac{a-4}{a^2+4}\cdot\frac{a+4}{a^2-4}=\frac{(a-4)(a+4)}{(a^2+4)(a^2-4)}$

9. $\frac{x+6}{3+x}\cdot\frac{x-1}{x+1}=\frac{(x+6)(x-1)}{(3+x)(x+1)}$

11. $\frac{5a^4}{6a}\cdot\frac{2}{a}$

$=\frac{5a^4\cdot 2}{6a\cdot a}$ Multiplying the numerators and the denominators

$=\frac{5\cdot a\cdot a\cdot a\cdot a\cdot 2}{2\cdot 3\cdot a\cdot a}$ Factoring the numerator and the denominator

$=\frac{5\cdot\cancel{a}\cdot\cancel{a}\cdot a\cdot a\cdot\cancel{2}}{\cancel{2}\cdot 3\cdot\cancel{a}\cdot\cancel{a}}$ Removing a factor equal to 1

$=\frac{5a^2}{3}$ Simplifying

13. $\frac{3c}{d^2}\cdot\frac{8d}{6c^3}$

$=\frac{3c\cdot 8d}{d^2\cdot 6c^3}$ Multiplying the numerators and the denominators

$=\frac{3\cdot c\cdot 2\cdot 4\cdot d}{d\cdot d\cdot 3\cdot 2\cdot c\cdot c\cdot c}$ Factoring the numerator and the denominator

$=\frac{\cancel{3}\cdot\cancel{c}\cdot\cancel{2}\cdot 4\cdot\cancel{d}}{\cancel{d}\cdot d\cdot\cancel{3}\cdot\cancel{2}\cdot\cancel{c}\cdot c\cdot c}$

$=\frac{4}{dc^2}$

15.
$$\frac{x^2-3x-10}{(x-2)^2}\cdot\frac{x-2}{x-5}=\frac{(x^2-3x-10)(x-2)}{(x-2)^2(x-5)}$$
$$=\frac{(x-5)(x+2)(x-2)}{(x-2)(x-2)(x-5)}$$
$$=\frac{\cancel{(x-5)}(x+2)\cancel{(x-2)}}{\cancel{(x-2)}(x-2)\cancel{(x-5)}}$$
$$=\frac{x+2}{x-2}$$

17.
$$\frac{a^2+25}{a^2-4a+3}\cdot\frac{a-5}{a+5}=\frac{(a^2+25)(a-5)}{(a^2-4a+3)(a+5)}$$
$$=\frac{(a^2+25)(a-5)}{(a-3)(a-1)(a+5)}$$

(No simplification is possible.)

19.
$$\frac{a^2-9}{a^2}\cdot\frac{7a}{a^2+a-12}=\frac{(a+3)(a-3)\cdot 7\cdot a}{a\cdot a(a+4)(a-3)}$$
$$=\frac{(a+3)\cancel{(a-3)}\cdot 7\cdot\cancel{a}}{\cancel{a}\cdot a(a+4)\cancel{(a-3)}}$$
$$=\frac{7(a+3)}{a(a+4)}$$

21. $\dfrac{4a^2}{3a^2-12a+12}\cdot\dfrac{3a-6}{2a}$

$=\dfrac{4a^2(3a-6)}{(3a^2-12a+12)2a}$

$=\dfrac{2\cdot 2\cdot a\cdot a\cdot 3\cdot(a-2)}{3\cdot(a-2)\cdot(a-2)\cdot 2\cdot a}$

$=\dfrac{\cancel{2}\cdot 2\cdot\cancel{a}\cdot a\cdot\cancel{3}\cdot\cancel{(a-2)}}{\cancel{3}\cdot\cancel{(a-2)}\cdot(a-2)\cdot\cancel{2}\cdot\cancel{a}}$

$=\dfrac{2a}{a-2}$

23. $\dfrac{t^2+2t-3}{t^2+4t-5}\cdot\dfrac{t^2-3t-10}{t^2+5t+6}$

$=\dfrac{(t^2+2t-3)(t^2-3t-10)}{(t^2+4t-5)(t^2+5t+6)}$

$=\dfrac{(t+3)(t-1)(t-5)(t+2)}{(t+5)(t-1)(t+3)(t+2)}$

$=\dfrac{\cancel{(t+3)}\cancel{(t-1)}(t-5)\cancel{(t+2)}}{(t+5)\cancel{(t-1)}\cancel{(t+3)}\cancel{(t+2)}}$

$=\dfrac{t-5}{t+5}$

25. $\dfrac{5a^2-180}{10a^2-10}\cdot\dfrac{20a+20}{2a-12}$

$=\dfrac{(5a^2-180)(20a+20)}{(10a^2-10)(2a-12)}$

$=\dfrac{5(a+6)(a-6)(2)(10)(a+1)}{10(a+1)(a-1)(2)(a-6)}$

$=\dfrac{5(a+6)\cancel{(a-6)}\cancel{(2)}\cancel{(10)}\cancel{(a+1)}}{\cancel{10}\cancel{(a+1)}(a-1)\cancel{(2)}\cancel{(a-6)}}$

$=\dfrac{5(a+6)}{a-1}$

27. $\dfrac{x^2+4x+4}{(x-1)^2}\cdot\dfrac{x^2-2x+1}{(x+2)^2}=\dfrac{(x+2)^2(x-1)^2}{(x-1)^2(x+2)^2}=1$

29. $\dfrac{5t^2+12t+4}{t^2+4t+4}\cdot\dfrac{t^2+8t+16}{5t^2+22t+8}$

$=\dfrac{(5t^2+12t+4)(t^2+8t+16)}{(t^2+4t+4)(5t^2+22t+8)}$

$=\dfrac{(5t+2)(t+2)(t+4)(t+4)}{(t+2)(t+2)(5t+2)(t+4)}$

$=\dfrac{\cancel{(5t+2)}\cancel{(t+2)}\cancel{(t+4)}(t+4)}{\cancel{(t+2)}(t+2)\cancel{(5t+2)}\cancel{(t+4)}}$

$=\dfrac{t+4}{t+2}$

31. $\dfrac{2x^2-5x+3}{6x^2-5x-1}\cdot\dfrac{6x^2+13x+2}{2x^2+3x-9}$

$=\dfrac{(2x^2-5x+3)(6x^2+13x+2)}{(6x^2-5x-1)(2x^2+3x-9)}$

$=\dfrac{(2x-3)(x-1)(6x+1)(x+2)}{(6x+1)(x-1)(2x-3)(x+3)}$

$=\dfrac{\cancel{(2x-3)}\cancel{(x-1)}\cancel{(6x+1)}(x+2)}{\cancel{(6x+1)}\cancel{(x-1)}\cancel{(2x-3)}(x+3)}$

$=\dfrac{x+2}{x+3}$

33. The reciprocal of $\dfrac{3x}{7}$ is $\dfrac{7}{3x}$ because $\dfrac{3x}{7}\cdot\dfrac{7}{3x}=1$.

35. The reciprocal of a^3-8a is $\dfrac{1}{a^3-8a}$ because $\dfrac{a^3-8a}{1}\cdot\dfrac{1}{a^3-8a}=1$.

37. The reciprocal of $\dfrac{x^2+2x-5}{x^2-4x+7}$ is $\dfrac{x^2-4x+7}{x^2+2x-5}$ because $\dfrac{x^2+2x-5}{x^2-4x+7}\cdot\dfrac{x^2-4x+7}{x^2+2x-5}=1$.

39. $\dfrac{5}{8}\div\dfrac{3}{7}$

$=\dfrac{5}{8}\cdot\dfrac{7}{3}$ Multiplying by the reciprocal of the divisor

$=\dfrac{5\cdot 7}{8\cdot 3}$

$=\dfrac{35}{24}$

No simplification is possible.

41. $\dfrac{x}{4}\div\dfrac{5}{x}$

$=\dfrac{x}{4}\cdot\dfrac{x}{5}$ Multiplying by the reciprocal of the divisor

$=\dfrac{x\cdot x}{4\cdot 5}$

$=\dfrac{x^2}{20}$

43. $\dfrac{a^5}{b^4}\div\dfrac{a^2}{b}=\dfrac{a^5}{b^4}\cdot\dfrac{b}{a^2}$

$=\dfrac{a^5\cdot b}{b^4\cdot a^2}$

$=\dfrac{a^2\cdot a^3\cdot b}{b\cdot b^3\cdot a^2}$

$=\dfrac{a^2b}{a^2b}\cdot\dfrac{a^3}{b^3}$

$=\dfrac{a^3}{b^3}$

45. $\dfrac{y+5}{4}\div\dfrac{y}{2}=\dfrac{y+5}{4}\cdot\dfrac{2}{y}$

$=\dfrac{(y+5)(2)}{4\cdot y}$

$=\dfrac{(y+5)(\cancel{2})}{\cancel{2}\cdot 2y}$

$=\dfrac{y+5}{2y}$

47. $\dfrac{4y-8}{y+2}\div\dfrac{y-2}{y^2-4}=\dfrac{4y-8}{y+2}\cdot\dfrac{y^2-4}{y-2}$

$=\dfrac{(4y-8)(y^2-4)}{(y+2)(y-2)}$

$=\dfrac{4\cancel{(y-2)}\cancel{(y+2)}(y-2)}{\cancel{(y+2)}\cancel{(y-2)}(1)}$

$=4(y-2)$

49. $$\frac{a}{a-b} \div \frac{b}{b-a} = \frac{a}{a-b} \cdot \frac{b-a}{b}$$
$$= \frac{a(b-a)}{(a-b)(b)}$$
$$= \frac{a(-1)\cancel{(a-b)}}{\cancel{(a-b)}(b)}$$
$$= \frac{-a}{b} = -\frac{a}{b}$$

51. $$(y^2-9) \div \frac{y^2-2y-3}{y^2+1} = \frac{(y^2-9)}{1} \cdot \frac{y^2+1}{y^2-2y-3}$$
$$= \frac{(y^2-9)(y^2+1)}{y^2-2y-3}$$
$$= \frac{(y+3)(y-3)(y^2+1)}{(y-3)(y+1)}$$
$$= \frac{(y+3)\cancel{(y-3)}(y^2+1)}{\cancel{(y-3)}(y+1)}$$
$$= \frac{(y+3)(y^2+1)}{y+1}$$

53. $$\frac{-3+3x}{16} \div \frac{x-1}{5} = \frac{3x-3}{16} \cdot \frac{5}{x-1}$$
$$= \frac{(3x-3)\cdot 5}{16(x-1)}$$
$$= \frac{3(x-1)\cdot 5}{16(x-1)}$$
$$= \frac{3\cancel{(x-1)}\cdot 5}{16\cancel{(x-1)}}$$
$$= \frac{15}{16}$$

55. $$\frac{-6+3x}{5} \div \frac{4x-8}{25} = \frac{-6+3x}{5} \cdot \frac{25}{4x-8}$$
$$= \frac{(-6+3x)\cdot 25}{5(4x-8)}$$
$$= \frac{3(x-2)\cdot 5\cdot 5}{5\cdot 4(x-2)}$$
$$= \frac{3\cancel{(x-2)}\cdot \cancel{5}\cdot 5}{\cancel{5}\cdot 4\cancel{(x-2)}}$$
$$= \frac{15}{4}$$

57. $$\frac{a+2}{a-1} \div \frac{3a+6}{a-5} = \frac{a+2}{a-1} \cdot \frac{a-5}{3a+6}$$
$$= \frac{(a+2)(a-5)}{(a-1)(3a+6)}$$
$$= \frac{(a+2)(a-5)}{(a-1)\cdot 3\cdot (a+2)}$$
$$= \frac{\cancel{(a+2)}(a-5)}{(a-1)\cdot 3\cdot \cancel{(a+2)}}$$
$$= \frac{a-5}{3(a-1)}$$

59. $$(2x-1) \div \frac{2x^2-11x+5}{4x^2-1}$$
$$= \frac{2x-1}{1} \cdot \frac{4x^2-1}{2x^2-11x+5}$$
$$= \frac{(2x-1)(4x^2-1)}{1\cdot(2x^2-11x+5)}$$
$$= \frac{(2x-1)(2x+1)(2x-1)}{1\cdot(2x-1)(x-5)}$$
$$= \frac{\cancel{(2x-1)}(2x+1)(2x-1)}{1\cdot\cancel{(2x-1)}(x-5)}$$
$$= \frac{(2x-1)(2x+1)}{x-5}$$

61. $$\frac{a^2-10a+25}{2a^2-a-21} \div \frac{a^2-a-20}{3a^2+5a-12}$$
$$= \frac{a^2-10a+25}{2a^2-a-21} \cdot \frac{3a^2+5a-12}{a^2-a-20}$$
$$= \frac{\cancel{(a-5)}(a-5)\cancel{(a+3)}(3a-4)}{\cancel{(a+3)}(2a-7)\cancel{(a-5)}(a+4)}$$
$$= \frac{(a-5)(3a-4)}{(2a-7)(a+4)}$$

63. $$\frac{c^2+10c+21}{c^2-2c-15} \div (5c^2+32c-21)$$
$$= \frac{c^2+10c+21}{c^2-2c-15} \cdot \frac{1}{5c^2+32c-21}$$
$$= \frac{(c^2+10c+21)\cdot 1}{(c^2-2c-15)(5c^2+32c-21)}$$
$$= \frac{(c+7)(c+3)}{(c-5)(c+3)(5c-3)(c+7)}$$
$$= \frac{(c+7)(c+3)}{(c+7)(c+3)} \cdot \frac{1}{(c-5)(5c-3)}$$
$$= \frac{1}{(c-5)(5c-3)}$$

65. $$\frac{x-y}{x^2+2xy+y^2} \div \frac{x^2-y^2}{x^2-5xy+4y^2}$$
$$= \frac{x-y}{x^2+2xy+y^2} \cdot \frac{x^2-5xy+4y^2}{x^2-y^2}$$
$$= \frac{\cancel{(x-y)}(x-y)(x-4y)}{(x+y)(x+y)(x+y)\cancel{(x-y)}}$$
$$= \frac{(x-y)(x-4y)}{(x+y)^3}$$

67. *Writing Exercise*

69. $$\frac{3}{4} + \frac{5}{6} = \frac{3}{4}\cdot\frac{3}{3} + \frac{5}{6}\cdot\frac{2}{2}$$
$$= \frac{9}{12} + \frac{10}{12}$$
$$= \frac{19}{12}$$

71. $$\frac{2}{9} - \frac{1}{6} = \frac{2}{9}\cdot\frac{2}{2} - \frac{1}{6}\cdot\frac{3}{3}$$
$$= \frac{4}{18} - \frac{3}{18}$$
$$= \frac{1}{18}$$

73. $\frac{2}{5} - \left(\frac{3}{2}\right)^2 = \frac{2}{5} - \frac{9}{4} = \frac{8}{20} - \frac{45}{20} = -\frac{37}{20}$

75. *Writing Exercise*

77. $\frac{2a^2 - 5ab}{c - 3d} \div (4a^2 - 25b^2)$

$= \frac{2a^2 - 5ab}{c - 3d} \cdot \frac{1}{4a^2 - 25b^2}$

$= \frac{a(2a - 5b)}{(c - 3d)(2a + 5b)(2a - 5b)}$

$= \frac{2a - 5b}{2a - 5b} \cdot \frac{a}{(c - 3d)(2a + 5b)}$

$= \frac{a}{(c - 3d)(2a + 5b)}$

79. $\frac{3a^2 - 5ab - 12b^2}{3ab + 4b^2} \div (3b^2 - ab)^2$

$= \frac{3a^2 - 5ab - 12b^2}{3ab + 4b^2} \cdot \frac{1}{(3b^2 - ab)^2}$

$= \frac{(3a + 4b)(a - 3b)}{b(3a + 4b) \cdot [b(3b - a)]^2}$

$= \frac{(3a + 4b)(-1)(3b - a)}{b(3a + 4b)(b^2)(3b - a)(3b - a)}$

$= \frac{(3a + 4b)(-1)(3b - a)}{b(3a + 4b)(b^2)(3b - a)(3b - a)}$

$= -\frac{1}{b^3(3b - a)}$, or $\frac{1}{b^3(a - 3b)}$

81. $\frac{a^2 - 3b}{a^2 + 2b} \cdot \frac{a^2 - 2b}{a^2 + 3b} \cdot \frac{a^2 + 2b}{a^2 - 3b}$

Note that $\frac{a^2 - 3b}{a^2 + 2b} \cdot \frac{a^2 + 2b}{a^2 - 3b}$ is the product of reciprocals and thus is equal to 1. Then the product in the original exercise is the remaining factor, $\frac{a^2 - 2b}{a^2 + 3b}$.

83. $\frac{z^2 - 8z + 16}{z^2 + 8z + 16} \div \frac{(z-4)^5}{(z+4)^5} \div \frac{3z+12}{z^2-16}$

$= \frac{(z - 4)^2}{(z + 4)^2} \cdot \frac{(z + 4)^5}{(z - 4)^5} \cdot \frac{(z + 4)(z - 4)}{3(z + 4)}$

$= \frac{(z - 4)^2(z + 4)^2(z + 4)^3(z + 4)(z - 4)}{(z + 4)^2(z - 4)^2(z - 4)(z - 4)^2(3)(z + 4)}$

$= \frac{(z + 4)^3}{3(z - 4)^2}$

85. $\frac{3x + 3y + 3}{9x} \div \frac{x^2 + 2xy + y^2 - 1}{x^4 + x^2}$

$= \frac{3x + 3y + 3}{9x} \cdot \frac{x^4 + x^2}{x^2 + 2xy + y^2 - 1}$

$= \frac{3(x + y + 1)(x^2)(x^2 + 1)}{9x[(x + y) + 1][(x + y) - 1]}$

$= \frac{3(x + y + 1)(x)(x)(x^2 + 1)}{3 \cdot 3 \cdot x(x + y + 1)(x + y - 1)}$

$= \frac{3x(x + y + 1)}{3x(x + y + 1)} \cdot \frac{x(x^2 + 1)}{3(x + y - 1)}$

$= \frac{x(x^2 + 1)}{3(x + y - 1)}$

87. $\frac{3y^3 + 6y^2}{y^2 - y - 12} \div \frac{y^2 - y}{y^2 - 2y - 8} \cdot \frac{y^2 + 5y + 6}{y^2}$

$= \frac{3y^3 + 6y^2}{y^2 - y - 12} \cdot \frac{y^2 - 2y - 8}{y^2 - y} \cdot \frac{y^2 + 5y + 6}{y^2}$

$= \frac{3 \cdot y^2(y + 2)(y - 4)(y + 2)(y + 3)(y + 2)}{(y - 4)(y + 3)(y)(y - 1)(y^2)}$

$= \frac{3(y + 2)^3}{y(y - 1)}$

89. Enter $y_1 = \frac{x - 1}{x^2+2x+1} \div \frac{x^2 - 1}{x^2-5x+4}$ and $y_2 = \frac{x^2 - 5x + 4}{(x + 1)^3}$, display the values of y_1 and y_2 in a table, and compare the values. (See the Technology Connection on page 374 in the text.)

Exercise Set 6.3

1. To add two rational expressions when the denominators are the same, add numerators and keep the common denominator. (See page 378 in the text.)

3. The least common multiple of two denominators is usually referred to as the least common denominator and is abbreviated LCD. (See page 380 in the text.)

5. $\frac{6}{x} + \frac{4}{x} = \frac{10}{x}$ Adding numerators

7. $\frac{x}{12} + \frac{2x + 5}{12} = \frac{3x + 5}{12}$ Adding numerators

9. $\frac{4}{a + 3} + \frac{5}{a + 3} = \frac{9}{a + 3}$

11. $\frac{8}{a + 2} - \frac{2}{a + 2} = \frac{6}{a + 2}$ Subtracting numerators

13. $\frac{3y + 8}{2y} - \frac{y + 1}{2y}$

$= \frac{3y + 8 - (y + 1)}{2y}$

$= \frac{3y + 8 - y - 1}{2y}$ Removing parentheses

$= \frac{2y + 7}{2y}$

15. $\dfrac{7x+8}{x+1}+\dfrac{4x+3}{x+1}$

$=\dfrac{11x+11}{x+1}$ Adding numerators

$=\dfrac{11(x+1)}{x+1}$ Factoring

$=\dfrac{11\cancel{(x+1)}}{\cancel{x+1}}$ Removing a factor equal to 1

$=11$

17. $\dfrac{7x+8}{x+1}-\dfrac{4x+3}{x+1}=\dfrac{7x+8-(4x+3)}{x+1}$

$=\dfrac{7x+8-4x-3}{x+1}$

$=\dfrac{3x+5}{x+1}$

19. $\dfrac{a^2}{a-4}+\dfrac{a-20}{a-4}=\dfrac{a^2+a-20}{a-4}$

$=\dfrac{(a+5)(a-4)}{a-4}$

$=\dfrac{(a+5)\cancel{(a-4)}}{\cancel{a-4}}$

$=a+5$

21. $\dfrac{x^2}{x-2}-\dfrac{6x-8}{x-2}=\dfrac{x^2-(6x-8)}{x-2}$

$=\dfrac{x^2-6x+8}{x-2}$

$=\dfrac{(x-4)(x-2)}{x-2}$

$=\dfrac{(x-4)\cancel{(x-2)}}{\cancel{x-2}}$

$=x-4$

23. $\dfrac{t^2-5t}{t-1}+\dfrac{5t-t^2}{t-1}$

Note that the numerators are opposites, so their sum is 0. Then we have $\dfrac{0}{t-1}$, or 0.

25. $\dfrac{x-6}{x^2+5x+6}+\dfrac{9}{x^2+5x+6}=\dfrac{x+3}{x^2+5x+6}$

$=\dfrac{x+3}{(x+3)(x+2)}$

$=\dfrac{\cancel{x+3}}{\cancel{(x+3)}(x+2)}$

$=\dfrac{1}{x+2}$

27. $\dfrac{t^2-5t}{t^2+6t+9}+\dfrac{4t-12}{t^2+6t+9}=\dfrac{t^2-t-12}{t^2+6t+9}$

$=\dfrac{(t-4)(t+3)}{(t+3)^2}$

$=\dfrac{(t-4)\cancel{(t+3)}}{(t+3)\cancel{(t+3)}}$

$=\dfrac{t-4}{t+3}$

29. $\dfrac{2x^2+x}{x^2-8x+12}-\dfrac{x^2-2x+10}{x^2-8x+12}$

$=\dfrac{2x^2+x-(x^2-2x+10)}{x^2-8x+12}$

$=\dfrac{2x^2+x-x^2+2x-10}{x^2-8x+12}$

$=\dfrac{x^2+3x-10}{x^2-8x+12}$

$=\dfrac{(x+5)(x-2)}{(x-6)(x-2)}$

$=\dfrac{(x+5)\cancel{(x-2)}}{(x-6)\cancel{(x-2)}}$

$=\dfrac{x+5}{x-6}$

31. $\dfrac{3-2x}{x^2-6x+8}+\dfrac{7-3x}{x^2-6x+8}$

$=\dfrac{10-5x}{x^2-6x+8}$

$=\dfrac{5(2-x)}{(x-4)(x-2)}$

$=\dfrac{5(-1)(x-2)}{(x-4)(x-2)}$

$=\dfrac{5(-1)\cancel{(x-2)}}{(x-4)\cancel{(x-2)}}$

$=\dfrac{-5}{x-4}$, or $-\dfrac{5}{x-4}$, or $\dfrac{5}{4-x}$

33. $\dfrac{x-9}{x^2+3x-4}-\dfrac{2x-5}{x^2+3x-4}$

$=\dfrac{x-9-(2x-5)}{x^2+3x-4}$

$=\dfrac{x-9-2x+5}{x^2+3x-4}$

$=\dfrac{-x-4}{x^2+3x-4}$

$=\dfrac{-(x+4)}{(x+4)(x-1)}$

$=\dfrac{-1\cancel{(x+4)}}{\cancel{(x+4)}(x-1)}$

$=\dfrac{-1}{x-1}$, or $-\dfrac{1}{x-1}$, or $\dfrac{1}{1-x}$

35. $15=3\cdot 5$

$27=3\cdot 3\cdot 3$

$\text{LCM}=3\cdot 3\cdot 3\cdot 5$, or 135

37. $8=2\cdot 2\cdot 2$

$9=3\cdot 3$

$\text{LCM}=2\cdot 2\cdot 2\cdot 3\cdot 3$, or 72

39. $6=2\cdot 3$

$9=3\cdot 3$

$21=3\cdot 7$

$\text{LCM}=2\cdot 3\cdot 3\cdot 7$, or 126

41. $12x^2 = 2 \cdot 2 \cdot 3 \cdot x \cdot x$

$6x^3 = 2 \cdot 3 \cdot x \cdot x \cdot x$

$\text{LCM} = 2 \cdot 2 \cdot 3 \cdot x \cdot x \cdot x$, or $12x^3$

43. $15a^4b^7 = 3 \cdot 5 \cdot a \cdot a \cdot a \cdot a \cdot b \cdot b \cdot b \cdot b \cdot b \cdot b \cdot b$

$10a^2b^8 = 2 \cdot 5 \cdot a \cdot a \cdot b \cdot b \cdot b \cdot b \cdot b \cdot b \cdot b \cdot b$

$\text{LCM} = 2 \cdot 3 \cdot 5 \cdot a \cdot a \cdot a \cdot a \cdot b \cdot b \cdot b \cdot b \cdot b \cdot b \cdot b \cdot b$, or $30a^4b^8$

45. $2(y-3) = 2 \cdot (y-3)$

$6(y-3) = 2 \cdot 3 \cdot (y-3)$

$\text{LCM} = 2 \cdot 3 \cdot (y-3)$, or $6(y-3)$

47. $x^2 - 4 = (x+2)(x-2)$

$x^2 + 5x + 6 = (x+3)(x+2)$

$\text{LCM} = (x+2)(x-2)(x+3)$

49. $t^3 + 4t^2 + 4t = t(t^2 + 4t + 4) = t(t+2)(t+2)$

$t^2 - 4t = t(t-4)$

$\text{LCM} = t(t+2)(t+2)(t-4) = t(t+2)^2(t-4)$

51. $10x^2y = 2 \cdot 5 \cdot x \cdot x \cdot y$

$6y^2z = 2 \cdot 3 \cdot y \cdot y \cdot z$

$5xz^3 = 5 \cdot x \cdot z \cdot z \cdot z$

$\text{LCM} = 2 \cdot 3 \cdot 5 \cdot x \cdot x \cdot y \cdot y \cdot z \cdot z \cdot z = 30x^2y^2z^3$

53. $a + 1 = a + 1$

$(a-1)^2 = (a-1)(a-1)$

$a^2 - 1 = (a+1)(a-1)$

$\text{LCM} = (a+1)(a-1)(a-1) = (a+1)(a-1)^2$

55. $m^2 - 5m + 6 = (m-3)(m-2)$

$m^2 - 4m + 4 = (m-2)(m-2)$

$\text{LCM} = (m-3)(m-2)(m-2) = (m-3)(m-2)^2$

57. $6x^3 - 24x^2 + 18x = 6x(x^2 - 4x + 3) =$

$2 \cdot 3 \cdot x(x-1)(x-3)$

$4x^5 - 24x^4 + 20x^3 = 4x^3(x^2 - 6x + 5) =$

$2 \cdot 2 \cdot x \cdot x \cdot x(x-1)(x-5)$

$\text{LCM} = 2 \cdot 2 \cdot 3 \cdot x \cdot x \cdot x(x-1)(x-3)(x-5) =$

$12x^3(x-1)(x-3)(x-5)$

59. $10a^3 = 2 \cdot 5 \cdot a \cdot a \cdot a$

$5a^6 = 5 \cdot a \cdot a \cdot a \cdot a \cdot a \cdot a$

The LCD is $2 \cdot 5 \cdot a \cdot a \cdot a \cdot a \cdot a \cdot a$, or $10a^6$.

$\frac{3}{10a^3} \cdot \frac{a^3}{a^3} = \frac{3a^3}{10a^6}$ and

$\frac{b}{5a^6} \cdot \frac{2}{2} = \frac{2b}{10a^6}$

61. $3x^4y^2 = 3 \cdot x \cdot x \cdot x \cdot x \cdot y \cdot y$

$9xy^3 = 3 \cdot 3 \cdot x \cdot y \cdot y \cdot y$

The LCD is $3 \cdot 3 \cdot x \cdot x \cdot x \cdot x \cdot y \cdot y \cdot y$, or $9x^4y^3$.

$\frac{7}{3x^4y^2} \cdot \frac{3y}{3y} = \frac{21y}{9x^4y^3}$ and

$\frac{4}{9xy^3} \cdot \frac{x^3}{x^3} = \frac{4x^3}{9x^4y^3}$

63. The LCD is $(x+2)(x-2)(x+3)$. (See Exercise 47.)

$$\frac{2x}{x^2-4} = \frac{2x}{(x+2)(x-2)} \cdot \frac{x+3}{x+3} = \frac{2x(x+3)}{(x+2)(x-2)(x+3)}$$

$$\frac{4x}{x^2+5x+6} = \frac{4x}{(x+3)(x+2)} \cdot \frac{x-2}{x-2} = \frac{4x(x-2)}{(x+3)(x+2)(x-2)}$$

65. *Writing Exercise*

67. $\frac{7}{-9} = -\frac{7}{9} = \frac{-7}{9}$

69. $\frac{5}{18} - \frac{7}{12} = \frac{5}{18} \cdot \frac{2}{2} - \frac{7}{12} \cdot \frac{3}{3} = \frac{10}{36} - \frac{21}{36} = -\frac{11}{36}$

71. The shaded area has dimensions $x-6$ by $x-3$. Then the area is $(x-6)(x-3)$, or $x^2 - 9x + 18$.

73. *Writing Exercise*

75. $$\frac{6x-1}{x-1} + \frac{3(2x+5)}{x-1} + \frac{3(2x-3)}{x-1} = \frac{6x-1+6x+15+6x-9}{x-1} = \frac{18x+5}{x-1}$$

77. $$\frac{x^2}{3x^2-5x-2} - \frac{2x}{3x+1} \cdot \frac{1}{x-2} = \frac{x^2}{(3x+1)(x-2)} - \frac{2x}{(3x+1)(x-2)} = \frac{x^2-2x}{(3x+1)(x-2)} = \frac{x(x-2)}{(3x+1)(x-2)} = \frac{x}{3x+1}$$

79. The smallest number of strands that can be used is the LCM of 10 and 3.

$10 = 2 \cdot 5$

$3 = 3$

$\text{LCM} = 2 \cdot 5 \cdot 3 = 30$

The smallest number of stands that can be used is 30.

81. If the number of strands must also be a multiple of 4, we find the smallest multiple of 30 that is also a multiple of 4.

$1 \cdot 30 = 30$, not a multiple of 4

$2 \cdot 30 = 60 = 15 \cdot 4$, a multiple of 4

The smallest number of strands that can be used is 60.

83. $6t^2-6=6(t^2-1)=2\cdot 3(t+1)(t-1)$

$(3t^2-6t+3)^2=[3(t^2-2t+1)]^2=[3(t-1)^2]^2=$

$3^2(t-1)^4=3\cdot 3(t-1)(t-1)(t-1)(t-1)$

$8t-8=8(t-1)=2\cdot 2\cdot 2(t-1)$

$\text{LCM}=2\cdot 2\cdot 2\cdot 3\cdot 3(t+1)(t-1)^4=72(t+1)(t-1)^4$

85. The time it takes until the machines begin copying a page at exactly the same time again is the LCM of their copying rates.

$10=2\cdot 5$

$14=2\cdot 7$

$\text{LCM}=2\cdot 5\cdot 7=70$

It takes 70 min.

87. The number of minutes after 5:00 A.M. when the shuttles will first leave at the same time again is the LCM of their departure intervals, 25 minutes and 35 minutes.

$25=5\cdot 5$

$35=5\cdot 7$

$\text{LCM}\ =5\cdot 5\cdot 7, \text{ or } 175$

Thus, the shuttles will leave at the same time 175 minutes after 5:00 A.M., or at 7:55 A.M.

89. *Writing Exercise*

Exercise Set 6.4

1. To add or subtract when denominators are different, first find the LCD.

3. Add or subtract the numerators, as indicated. Write the sum or difference over the LCD.

5. $\frac{4}{x}+\frac{9}{x^2}=\frac{4}{x}+\frac{9}{x\cdot x}$ LCD $=x\cdot x$, or x^2

$$=\frac{4}{x}\cdot\frac{x}{x}+\frac{9}{x\cdot x}$$

$$=\frac{4x+9}{x^2}$$

7. $\left.\begin{array}{l}6r=2\cdot 3\cdot r\\ 8r=2\cdot 2\cdot 2\cdot r\end{array}\right\}\text{LCD}=2\cdot 2\cdot 2\cdot 3\cdot r, \text{ or } 24r$

$$\frac{1}{6r}-\frac{3}{8r}=\frac{1}{6r}\cdot\frac{4}{4}-\frac{3}{8r}\cdot\frac{3}{3}$$

$$=\frac{4-9}{24r}$$

$$=\frac{-5}{24r}, \text{ or } -\frac{5}{24r}$$

9. $\left.\begin{array}{l}c^2d=c\cdot c\cdot d\\ cd^3=c\cdot d\cdot d\cdot d\end{array}\right\}\text{LCD}=c\cdot c\cdot d\cdot d\cdot d, \text{ or } c^2d^3$

$$\frac{2}{c^2d}+\frac{7}{cd^3}=\frac{2}{c^2d}\cdot\frac{d^2}{d^2}+\frac{7}{cd^3}\cdot\frac{c}{c}=\frac{2d^2+7c}{c^2d^3}$$

11. $\left.\begin{array}{l}3xy^2=3\cdot x\cdot y\cdot y\\ x^2y^3=x\cdot x\cdot y\cdot y\cdot y\end{array}\right\}\text{LCD}=3\cdot x\cdot x\cdot y\cdot y\cdot y, \text{ or } 3x^2y^3$

$$\frac{-2}{3xy^2}-\frac{6}{x^2y^3}=\frac{-2}{3xy^2}\cdot\frac{xy}{xy}-\frac{6}{x^2y^3}\cdot\frac{3}{3}=\frac{-2xy-18}{3x^2y^3}$$

13. $\left.\begin{array}{l}9=3\cdot 3\\ 6=2\cdot 3\end{array}\right\}\text{LCD}=3\cdot 3\cdot 2, \text{ or } 18$

$$\frac{x-4}{9}+\frac{x+5}{6}=\frac{x-4}{9}\cdot\frac{2}{2}+\frac{x+5}{6}\cdot\frac{3}{3}$$

$$=\frac{2(x-4)+3(x+5)}{18}$$

$$=\frac{2x-8+3x+15}{18}$$

$$=\frac{5x+7}{18}$$

15. $\left.\begin{array}{l}6=2\cdot 3\\ 3=3\end{array}\right\}\text{LCD}=2\cdot 3, \text{ or } 6$

$$\frac{x-2}{6}-\frac{x+1}{3}=\frac{x-2}{6}-\frac{x+1}{3}\cdot\frac{2}{2}$$

$$=\frac{x-2}{6}-\frac{2x+2}{6}$$

$$=\frac{x-2-(2x+2)}{6}$$

$$=\frac{x-2-2x-2}{6}$$

$$=\frac{-x-4}{6}, \text{ or } \frac{-(x+4)}{6}$$

17. $\left.\begin{array}{l}16a=2\cdot 2\cdot 2\cdot 2\cdot a\\ 4a^2=2\cdot 2\cdot a\cdot a\end{array}\right\}\text{LCD}=2\cdot 2\cdot 2\cdot 2\cdot a\cdot a, \text{ or } 16a^2$

$$\frac{a+4}{16a}+\frac{3a+4}{4a^2}=\frac{a+4}{16a}\cdot\frac{a}{a}+\frac{3a+4}{4a^2}\cdot\frac{4}{4}$$

$$=\frac{a^2+4a}{16a^2}+\frac{12a+16}{16a^2}$$

$$=\frac{a^2+16a+16}{16a^2}$$

19. $\left.\begin{array}{l}3z=3\cdot z\\ 4z=2\cdot 2\cdot z\end{array}\right\}\text{LCD}=2\cdot 2\cdot 3\cdot z, \text{ or } 12z$

$$\frac{4z-9}{3z}-\frac{3z-8}{4z}=\frac{4z-9}{3z}\cdot\frac{4}{4}-\frac{3z-8}{4z}\cdot\frac{3}{3}$$

$$=\frac{16z-36}{12z}-\frac{9z-24}{12z}$$

$$=\frac{16z-36-(9z-24)}{12z}$$

$$=\frac{16z-36-9z+24}{12z}$$

$$=\frac{7z-12}{12z}$$

21. $\left.\begin{array}{l} xy^2 = x\cdot y\cdot y \\ x^2y = x\cdot x\cdot y\end{array}\right\}$LCD $= x\cdot x\cdot y\cdot y$, or x^2y^2

$$\frac{x+y}{xy^2}+\frac{3x+y}{x^2y} = \frac{x+y}{xy^2}\cdot\frac{x}{x}+\frac{3x+y}{x^2y}\cdot\frac{y}{y}$$
$$= \frac{x(x+y)+y(3x+y)}{x^2y^2}$$
$$= \frac{x^2+xy+3xy+y^2}{x^2y^2}$$
$$= \frac{x^2+4xy+y^2}{x^2y^2}$$

23. $\left.\begin{array}{l} 3xt^2 = 3\cdot x\cdot t\cdot t \\ x^2t = x\cdot x\cdot t\end{array}\right\}$LCD $= 3\cdot x\cdot x\cdot t\cdot t$, or $3x^2t^2$

$$\frac{4x+2t}{3xt^2}-\frac{5x-3t}{x^2t}$$
$$= \frac{4x+2t}{3xt^2}\cdot\frac{x}{x}-\frac{5x-3t}{x^2t}\cdot\frac{3t}{3t}$$
$$= \frac{4x^2+2tx}{3x^2t^2}-\frac{15xt-9t^2}{3x^2t^2}$$
$$= \frac{4x^2+2tx-(15xt-9t^2)}{3x^2t^2}$$
$$= \frac{4x^2+2tx-15xt+9t^2}{3x^2t^2}$$
$$= \frac{4x^2-13xt+9t^2}{3x^2t^2}$$

(Although $4x^2-13xt+9t^2$ can be factored, doing so will not enable us to simplify the result further.)

25. The denominators cannot be factored, so the LCD is their product, $(x-2)(x+2)$.

$$\frac{3}{x-2}+\frac{3}{x+2} = \frac{3}{x-2}\cdot\frac{x+2}{x+2}+\frac{3}{x+2}\cdot\frac{x-2}{x-2}$$
$$= \frac{3(x+2)+3(x-2)}{(x-2)(x+2)}$$
$$= \frac{3x+6+3x-6}{(x-2)(x+2)}$$
$$= \frac{6x}{(x-2)(x+2)}$$

27. $\dfrac{5}{x+5}-\dfrac{3}{x-5}$ LCD $= (x+5)(x-5)$

$$= \frac{5}{x+5}\cdot\frac{x-5}{x-5}-\frac{3}{x-5}\cdot\frac{x+5}{x+5}$$
$$= \frac{5x-25}{(x+5)(x-5)}-\frac{3x+15}{(x+5)(x-5)}$$
$$= \frac{5x-25-(3x+15)}{(x+5)(x-5)}$$
$$= \frac{5x-25-3x-15}{(x+5)(x-5)}$$
$$= \frac{2x-40}{(x+5)(x-5)}$$

(Although $2x-40$ can be factored, doing so will not enable us to simplify the result further.)

29. $\left.\begin{array}{l} 3x = 3\cdot x \\ x+1 = x+1\end{array}\right\}$LCD $= 3x(x+1)$

$$\frac{3}{x+1}+\frac{2}{3x} = \frac{3}{x+1}\cdot\frac{3x}{3x}+\frac{2}{3x}\cdot\frac{x+1}{x+1}$$
$$= \frac{9x+2(x+1)}{3x(x+1)}$$
$$= \frac{9x+2x+2}{3x(x+1)}$$
$$= \frac{11x+2}{3x(x+1)}$$

31. $\dfrac{3}{2t^2-2t}-\dfrac{5}{2t-2}$

$$= \frac{3}{2t(t-1)}-\frac{5}{2(t-1)} \quad \text{LCD} = 2t(t-1)$$
$$= \frac{3}{2t(t-1)}-\frac{5}{2(t-1)}\cdot\frac{t}{t}$$
$$= \frac{3-5t}{2t(t-1)}$$

33. $\dfrac{2x}{x^2-16}+\dfrac{x}{x-4}$

$$= \frac{2x}{(x+4)(x-4)}+\frac{x}{x-4}$$

LCD $= (x+4)(x-4)$

$$= \frac{2x}{(x+4)(x-4)}+\frac{x}{x-4}\cdot\frac{x+4}{x+4}$$
$$= \frac{2x+x(x+4)}{(x+4)(x-4)}$$
$$= \frac{2x+x^2+4x}{(x+4)(x-4)}$$
$$= \frac{x^2+6x}{(x+4)(x-4)}$$

(Although x^2+6x can be factored, doing so will not enable us to simplify the result further.)

35. $\dfrac{6}{z+4}-\dfrac{2}{3z+12} = \dfrac{6}{z+4}-\dfrac{2}{3(z+4)}$

LCD $= 3(z+4)$

$$= \frac{6}{z+4}\cdot\frac{3}{3}-\frac{2}{3(z+4)}$$
$$= \frac{18}{3(z+4)}-\frac{2}{3(z+4)}$$
$$= \frac{16}{3(z+4)}$$

37. $\dfrac{3}{x-1}+\dfrac{2}{(x-1)^2}$ LCD $= (x-1)^2$

$$= \frac{3}{x-1}\cdot\frac{x-1}{x-1}+\frac{2}{(x-1)^2}$$
$$= \frac{3(x-1)+2}{(x-1)^2}$$
$$= \frac{3x-3+2}{(x-1)^2}$$
$$= \frac{3x-1}{(x-1)^2}$$

39. $\dfrac{3}{x+2} - \dfrac{8}{x^2-4}$

$= \dfrac{3}{x+2} - \dfrac{8}{(x+2)(x-2)}$ LCD $= (x+2)(x-2)$

$= \dfrac{3}{x+2} \cdot \dfrac{x-2}{x-2} - \dfrac{8}{(x+2)(x-2)}$

$= \dfrac{3(x-2) - 8}{(x+2)(x-2)}$

$= \dfrac{3x - 6 - 8}{(x+2)(x-2)}$

$= \dfrac{3x - 14}{(x+2)(x-2)}$

41. $\dfrac{3a}{4a-20} + \dfrac{9a}{6a-30}$

$= \dfrac{3a}{2 \cdot 2(a-5)} + \dfrac{9a}{2 \cdot 3(a-5)}$

LCD $= 2 \cdot 2 \cdot 3(a-5)$

$= \dfrac{3a}{2 \cdot 2(a-5)} \cdot \dfrac{3}{3} + \dfrac{9a}{2 \cdot 3(a-5)} \cdot \dfrac{2}{2}$

$= \dfrac{9a + 18a}{2 \cdot 2 \cdot 3(a-5)}$

$= \dfrac{27a}{2 \cdot 2 \cdot 3(a-5)}$

$= \dfrac{\cancel{3} \cdot 9 \cdot a}{2 \cdot 3 \cdot \cancel{3}(a-5)}$

$= \dfrac{9a}{4(a-5)}$

43. $\dfrac{x}{x-5} + \dfrac{x}{5-x} = \dfrac{x}{x-5} + \dfrac{x}{5-x} \cdot \dfrac{-1}{-1}$

$= \dfrac{x}{x-5} + \dfrac{-x}{x-5}$

$= 0$

45. $\dfrac{6}{a^2+a-2} + \dfrac{4}{a^2-4a+3}$

$= \dfrac{6}{(a+2)(a-1)} + \dfrac{4}{(a-3)(a-1)}$

LCD $= (a+2)(a-1)(a-3)$

$= \dfrac{6}{(a+2)(a-1)} \cdot \dfrac{a-3}{a-3} + \dfrac{4}{(a-3)(a-1)} \cdot \dfrac{a+2}{a+2}$

$= \dfrac{6(a-3) + 4(a+2)}{(a+2)(a-1)(a-3)}$

$= \dfrac{6a - 18 + 4a + 8}{(a+2)(a-1)(a-3)}$

$= \dfrac{10a - 10}{(a+2)(a-1)(a-3)}$

$= \dfrac{10\cancel{(a-1)}}{(a+2)\cancel{(a-1)}(a-3)}$

$= \dfrac{10}{(a+2)(a-3)}$

47. $\dfrac{x}{x^2+9x+20} - \dfrac{4}{x^2+7x+12}$

$= \dfrac{x}{(x+4)(x+5)} - \dfrac{4}{(x+3)(x+4)}$

LCD $= (x+3)(x+4)(x+5)$

$= \dfrac{x}{(x+4)(x+5)} \cdot \dfrac{x+3}{x+3} - \dfrac{4}{(x+3)(x+4)} \cdot \dfrac{x+5}{x+5}$

$= \dfrac{x(x+3) - 4(x+5)}{(x+3)(x+4)(x+5)}$

$= \dfrac{x^2 + 3x - 4x - 20}{(x+3)(x+4)(x+5)}$

$= \dfrac{x^2 - x - 20}{(x+3)(x+4)(x+5)}$

$= \dfrac{\cancel{(x+4)}(x-5)}{(x+3)\cancel{(x+4)}(x+5)}$

$= \dfrac{x-5}{(x+3)(x+5)}$

49. $\dfrac{3z}{z^2-4x+4} + \dfrac{10}{z^2+z-6}$

$= \dfrac{3z}{(z-2)^2} + \dfrac{10}{(z-2)(z+3)}$,

LCD $= (z-2)^2(z+3)$

$= \dfrac{3z}{(z-2)^2} \cdot \dfrac{z+3}{z+3} + \dfrac{10}{(z-2)(z+3)} \cdot \dfrac{z-2}{z-2}$

$= \dfrac{3z(z+3) + 10(z-2)}{(x-2)^2(z+3)}$

$= \dfrac{3z^2 + 9z + 10z - 20}{(z-2)^2(z+3)}$

$= \dfrac{3z^2 + 19z - 20}{(z-2)^2(z+3)}$

51. $\dfrac{-5}{x^2+17x+16} - \dfrac{0}{x^2+9x+8}$

Note that $\dfrac{0}{x^2+9x+8} = 0$, so the difference is $\dfrac{-5}{x^2+17x+16}$.

53. $\dfrac{4x}{5} - \dfrac{x-3}{-5} = \dfrac{4x}{5} - \dfrac{x-3}{-5} \cdot \dfrac{-1}{-1}$

$= \dfrac{4x}{5} - \dfrac{3-x}{5}$

$= \dfrac{4x - (3-x)}{5}$

$= \dfrac{4x - 3 + x}{5}$

$= \dfrac{5x - 3}{5}$

55. $\dfrac{y^2}{y-3}+\dfrac{9}{3-y}=\dfrac{y^2}{y-3}+\dfrac{9}{3-y}\cdot\dfrac{-1}{-1}$

$=\dfrac{y^2}{y-3}+\dfrac{-9}{-3+y}$

$=\dfrac{y^2-9}{y-3}$

$=\dfrac{(y+3)\cancel{(y-3)}}{\cancel{y-3}}$

$=y+3$

57. $\dfrac{b-7}{b^2-16}+\dfrac{7-b}{16-b^2}=\dfrac{b-7}{b^2-16}+\dfrac{7-b}{16-b^2}\cdot\dfrac{-1}{-1}$

$=\dfrac{b-7}{b^2-16}+\dfrac{b-7}{b^2-16}$

$=\dfrac{2b-14}{b^2-16}$

(Although both $2b-14$ and b^2-16 can be factored, doing so will not enable us to simplify the result further.)

59. $\dfrac{4-p}{25-p^2}+\dfrac{p+1}{p-5}$

$=\dfrac{4-p}{(5+p)(5-p)}+\dfrac{p+1}{p-5}$

$=\dfrac{4-p}{(5+p)(5-p)}\cdot\dfrac{-1}{-1}+\dfrac{p+1}{p-5}$

$=\dfrac{p-4}{(p+5)(p-5)}+\dfrac{p+1}{p-5}$ LCD $=(p+5)(p-5)$

$=\dfrac{p-4}{(p+5)(p-5)}+\dfrac{p+1}{p-5}\cdot\dfrac{p+5}{p+5}$

$=\dfrac{p-4+p^2+6p+5}{(p+5)(p-5)}$

$=\dfrac{p^2+7p+1}{(p+5)(p-5)}$

61. $\dfrac{8x}{16-x^2}-\dfrac{5}{x-4}$

$=\dfrac{8x}{(4+x)(4-x)}-\dfrac{5}{x-4}$

$=\dfrac{8x}{(4+x)(4-x)}-\dfrac{5}{x-4}\cdot\dfrac{-1}{-1}$

$=\dfrac{8x}{(4+x)(4-x)}-\dfrac{-5}{4-x}$ LCD $=(4+x)(4-x)$

$=\dfrac{8x}{(4+x)(4-x)}-\dfrac{-5}{4-x}\cdot\dfrac{4+x}{4+x}$

$=\dfrac{8x-(-5)(4+x)}{(4+x)(4-x)}$

$=\dfrac{8x+20+5x}{(4+x)(4-x)}$

$=\dfrac{13x+20}{(4+x)(4-x)}$, or $\dfrac{-13x-20}{(x+4)(x-4)}$

63. $\dfrac{a}{a^2-1}+\dfrac{2a}{a-a^2}=\dfrac{a}{a^2-1}+\dfrac{2\cdot\cancel{a}}{\cancel{a}(1-a)}$

$=\dfrac{a}{(a+1)(a-1)}+\dfrac{2}{1-a}$

$=\dfrac{a}{(a+1)(a-1)}+\dfrac{2}{1-a}\cdot\dfrac{-1}{-1}$

$=\dfrac{a}{(a+1)(a-1)}+\dfrac{-2}{a-1}$

LCD $=(a+1)(a-1)$

$=\dfrac{a}{(a+1)(a-1)}+\dfrac{-2}{a-1}\cdot\dfrac{a+1}{a+1}$

$=\dfrac{a-2a-2}{(a+1)(a-1)}$

$=\dfrac{-a-2}{(a+1)(a-1)}$, or

$=\dfrac{a+2}{(1+a)(1-a)}$

65. $\dfrac{4x}{x^2-y^2}-\dfrac{6}{y-x}$

$=\dfrac{4x}{(x+y)(x-y)}-\dfrac{6}{y-x}$

$=\dfrac{4x}{(x+y)(x-y)}-\dfrac{6}{y-x}\cdot\dfrac{-1}{-1}$

$=\dfrac{4x}{(x+y)(x-y)}-\dfrac{-6}{x-y}$ LCD $=(x+y)(x-y)$

$=\dfrac{4x}{(x+y)(x-y)}-\dfrac{-6}{x-y}\cdot\dfrac{x+y}{x+y}$

$=\dfrac{4x-(-6)(x+y)}{(x+y)(x-y)}$

$=\dfrac{4x+6x+6y}{(x+y)(x-y)}$

$=\dfrac{10x+6y}{(x+y)(x-y)}$

(Although $10x+6y$ can be factored, doing so will not enable us to simplify the result further.)

67. $\dfrac{x-3}{2-x}-\dfrac{x+3}{x+2}+\dfrac{x+6}{4-x^2}$

$=\dfrac{x-3}{2-x}-\dfrac{x+3}{x+2}+\dfrac{x+6}{(2+x)(2-x)}$

LCD $=(2+x)(2-x)$

$=\dfrac{x-3}{2-x}\cdot\dfrac{2+x}{2+x}-\dfrac{x+3}{x+2}\cdot\dfrac{2-x}{2-x}+\dfrac{x+6}{(2+x)(2-x)}$

$=\dfrac{(x-3)(2+x)-(x+3)(2-x)+(x+6)}{(2+x)(2-x)}$

$=\dfrac{x^2-x-6-(-x^2-x+6)+x+6}{(2+x)(2-x)}$

$=\dfrac{x^2-x-6+x^2+x-6+x+6}{(2+x)(2-x)}$

$=\dfrac{2x^2+x-6}{(2+x)(2-x)}$

$=\dfrac{(2x-3)\cancel{(x+2)}}{\cancel{(2+x)}(2-x)}$

$=\dfrac{2x-3}{2-x}$

69. $\dfrac{x+5}{x+3}+\dfrac{x+7}{x+2}-\dfrac{7x+19}{(x+3)(x+2)}$

LCD is $(x+3)(x+2)$

$= \dfrac{x+5}{x+3}\cdot\dfrac{x+2}{x+2}+\dfrac{x+7}{x+2}\cdot\dfrac{x+3}{x+3}-\dfrac{7x+19}{(x+3)(x+2)}$

$= \dfrac{(x+5)(x+2)+(x+7)(x+3)-(7x+19)}{(x+3)(x+2)}$

$= \dfrac{x^2+7x+10+x^2+10x+21-7x-19}{(x+3)(x+2)}$

$= \dfrac{2x^2+10x+12}{(x+3)(x+2)}$

$= \dfrac{2(x^2+5x+6)}{(x+3)(x+2)}$

$= \dfrac{2\cancel{(x+3)}\cancel{(x+2)}}{\cancel{(x+3)}\cancel{(x+2)}}$

$= 2$

71. $\dfrac{1}{x+y}+\dfrac{1}{x-y}-\dfrac{2x}{x^2-y^2}$

LCD $=(x+y)(x-y)$

$= \dfrac{1}{x+y}\cdot\dfrac{x-y}{x-y}+\dfrac{1}{x-y}\cdot\dfrac{x+y}{x+y}-\dfrac{2x}{(x+y)(x-y)}$

$= \dfrac{(x-y)+(x+y)-2x}{(x+y)(x-y)}$

$= 0$

73. *Writing Exercise*

75. $-\dfrac{3}{7}\div\dfrac{6}{13}=-\dfrac{3}{7}\cdot\dfrac{13}{6}$

$= -\dfrac{\cancel{3}\cdot 13}{7\cdot 2\cdot\cancel{3}}$

$= -\dfrac{13}{14}$

77. $\dfrac{\frac{2}{9}}{\frac{5}{3}}=\dfrac{2}{9}\div\dfrac{5}{3}$

$= \dfrac{2}{9}\cdot\dfrac{3}{5}$

$= \dfrac{2\cdot\cancel{3}}{\cancel{3}\cdot 3\cdot 5}$

$= \dfrac{2}{15}$

79. Graph: $y=-\dfrac{1}{2}x-5$

Since the equation is in the form $y=mx+b$, we know the y–intercept is $(0,-5)$. We find two other solutions, substituting multiples of 2 for x to avoid fractions.

When $x=-2$, $y=-\dfrac{1}{2}(-2)-5=1-5=-4$.

When $x=-4$, $y=-\dfrac{1}{2}(-4)-5=2-5=-3$.

x	y
0	−5
−2	−4
−4	−3

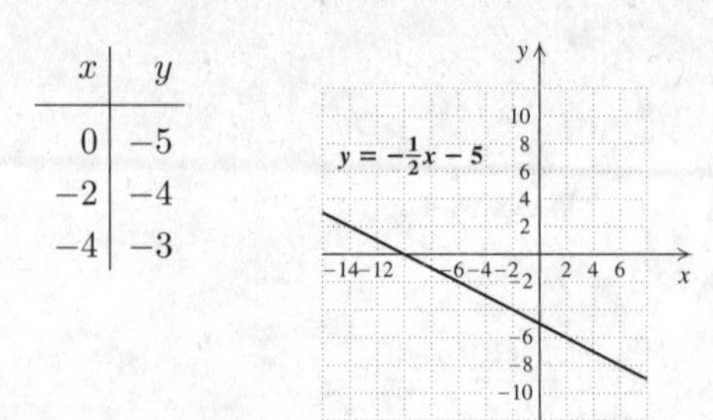

81. *Writing Exercise*

83. $P=2\left(\dfrac{3}{x+4}\right)+2\left(\dfrac{2}{x-5}\right)$

$= \dfrac{6}{x+4}+\dfrac{4}{x-5}$ LCD $=(x+4)(x-5)$

$= \dfrac{6}{x+4}\cdot\dfrac{x-5}{x-5}+\dfrac{4}{x-5}\cdot\dfrac{x+4}{x+4}$

$= \dfrac{6x-30+4x+16}{(x+4)(x-5)}$

$= \dfrac{10x-14}{(x+4)(x-5)}$, or $\dfrac{10x-14}{x^2-x-20}$

$A=\left(\dfrac{3}{x+4}\right)\left(\dfrac{2}{x-5}\right)=\dfrac{6}{(x+4)(x-5)}$

85. $\dfrac{2x+11}{x-3}\cdot\dfrac{3}{x+4}+\dfrac{2x+1}{4+x}\cdot\dfrac{3}{3-x}$

$= \dfrac{6x+33}{(x-3)(x+4)}+\dfrac{6x+3}{(4+x)(3-x)}$

$= \dfrac{6x+33}{(x-3)(x+4)}+\dfrac{6x+3}{(4+x)(3-x)}\cdot\dfrac{-1}{-1}$

$= \dfrac{6x+33}{(x-3)(x+4)}+\dfrac{-6x-3}{(x+4)(x-3)}$

$= \dfrac{6x+33-6x-3}{(x-3)(x+4)}$

$= \dfrac{30}{(x-3)(x+4)}$

87. We recognize that this is the product of the sum and difference of two terms: $(A+B)(A-B)=A^2-B^2$.

$\left(\dfrac{x}{x+7}-\dfrac{3}{x+2}\right)\left(\dfrac{x}{x+7}+\dfrac{3}{x+2}\right)$

$= \dfrac{x^2}{(x+7)^2}-\dfrac{9}{(x+2)^2}$ LCD $=(x+7)^2(x+2)^2$

$= \dfrac{x^2}{(x+7)^2}\cdot\dfrac{(x+2)^2}{(x+2)^2}-\dfrac{9}{(x+2)^2}\cdot\dfrac{(x+7)^2}{(x+7)^2}$

$= \dfrac{x^2(x+2)^2-9(x+7)^2}{(x+7)^2(x+2)^2}$

$= \dfrac{x^2(x^2+4x+4)-9(x^2+14x+49)}{(x+7)^2(x+2)^2}$

$= \dfrac{x^4+4x^3+4x^2-9x^2-126x-441}{(x+7)^2(x+2)^2}$

$= \dfrac{x^4+4x^3-5x^2-126x-441}{(x+7)^2(x+2)^2}$

89. $\frac{2x^2+5x-3}{2x^2-9x+9}+\frac{x+1}{3-2x}+\frac{4x^2+8x+3}{x-3}\cdot\frac{x+3}{9-4x^2}$

$=\frac{2x^2+5x-3}{(2x-3)(x-3)}+\frac{x+1}{3-2x}+\frac{(4x^2+8x+3)(x+3)}{(x-3)(3+2x)(3-2x)}$

$=\frac{2x^2+5x-3}{(2x-3)(x-3)}\cdot\frac{-1}{-1}+\frac{x+1}{3-2x}+\frac{4x^3+20x^2+27x+9}{(x-3)(3+2x)(3-2x)}$

$=\frac{-2x^2-5x+3}{(3-2x)(x-3)}+\frac{x+1}{3-2x}+\frac{4x^3+20x^2+27x+9}{(x-3)(3+2x)(3-2x)}$

LCD $=(x-3)(3+2x)(3-2x)$

$=\frac{-2x^2-5x+3}{(3-2x)(x-3)}\cdot\frac{3+2x}{3+2x}+\frac{x+1}{3-2x}\cdot\frac{(x-3)(3+2x)}{(x-3)(3+2x)}+\frac{4x^3+20x^2+27x+9}{(x-3)(3+2x)(3-2x)}$

$=[(-4x^3-16x^2-9x+9+2x^3-x^2-12x-9+4x^3+20x^2+27x+9)]/[(x-3)(3+2x)(3-2x)]$

$=\frac{2x^3+3x^2+6x+9}{(x-3)(3+2x)(3-2x)}$

$=\frac{x^2(2x+3)+3(2x+3)}{(x-3)(3+2x)(3-2x)}$

$=\frac{\cancel{(2x+3)}(x^2+3)}{(x-3)\cancel{(3+2x)}(3-2x)}$

$=\frac{x^2+3}{(x-3)(3-2x)}$, or $\frac{-x^2-3}{(x-3)(2x-3)}$

91. Answers may vary. $\frac{a}{a-b}+\frac{3b}{b-a}$

93. *Writing Exercise*

Exercise Set 6.5

1. The LCD is the LCM of x^2, x, 2, and $4x$. It is $4x^2$.

$$\frac{\frac{5}{x^2}+\frac{1}{x}}{\frac{7}{2}-\frac{3}{4x}}\cdot\frac{4x^2}{4x^2}=\frac{\frac{5}{x^2}\cdot 4x^2+\frac{1}{x}\cdot 4x^2}{\frac{7}{2}\cdot 4x^2-\frac{3}{4x}\cdot 4x^2}$$

Choice (d) is correct.

3. We subtract to get a single rational expression in the numerator and add to get a single rational expression in the denominator.

$$\frac{\frac{4}{5x}-\frac{1}{10}}{\frac{8}{x^2}+\frac{7}{2}}=\frac{\frac{4}{5x}\cdot\frac{2}{2}-\frac{1}{10}\cdot\frac{x}{x}}{\frac{8}{x^2}\cdot\frac{2}{2}+\frac{7}{2}\cdot\frac{x^2}{x^2}}$$

$$=\frac{\frac{8}{10x}-\frac{x}{10x}}{\frac{16}{2x^2}+\frac{7x^2}{2x^2}}$$

$$=\frac{\frac{8-x}{10x}}{\frac{16+7x^2}{2x^2}}$$

Choice (b) is correct.

5. $\frac{1+\frac{1}{2}}{1+\frac{1}{4}}$ LCD is 4

$=\frac{1+\frac{1}{2}}{1+\frac{1}{4}}\cdot\frac{4}{4}$ Multiplying by $\frac{4}{4}$

$=\frac{\left(1+\frac{1}{2}\right)4}{\left(1+\frac{1}{4}\right)4}$ Multiplying numerator and denominator by 4

$=\frac{1\cdot 4+\frac{1}{2}\cdot 4}{1\cdot 4+\frac{1}{4}\cdot 4}$

$=\frac{4+2}{4+1}$

$=\frac{6}{5}$

7. $\frac{4+\frac{1}{3}}{1-\frac{5}{27}}$

$=\frac{4\cdot\frac{3}{3}+\frac{1}{3}}{1\cdot\frac{27}{27}-\frac{5}{27}}$ Getting a common denominator in numerator and in denominator

$=\frac{\frac{12}{3}+\frac{1}{3}}{\frac{27}{27}-\frac{5}{27}}$

$=\frac{\frac{13}{3}}{\frac{22}{27}}$ Adding in the numerator; subtracting in the denominator

$=\frac{13}{3}\cdot\frac{27}{22}$ Multiplying by the reciprocal of the divisor

$=\frac{13\cdot 3\cdot 9}{3\cdot 22}$

$=\frac{13\cdot\cancel{3}\cdot 9}{\cancel{3}\cdot 22}$

$=\frac{117}{22}$

9. $\dfrac{\dfrac{s}{3}+s}{\dfrac{3}{s}+s}$ LCD is $3s$

$= \dfrac{\dfrac{s}{3}+s}{\dfrac{3}{s}+s} \cdot \dfrac{3s}{3s}$

$= \dfrac{\left(\dfrac{s}{3}+s\right)(3s)}{\left(\dfrac{3}{s}+s\right)(3s)}$

$= \dfrac{\dfrac{5}{3}\cdot 3s + s\cdot 3s}{\dfrac{3}{s}\cdot 3s + s\cdot 3s}$

$= \dfrac{s^2+3s^2}{9+3s^2}$

$= \dfrac{4s^2}{9+3s^2}$

11. $\dfrac{\dfrac{4}{x}}{\dfrac{3}{x}+\dfrac{2}{x^2}}$ LCD is x^2

$= \dfrac{\dfrac{4}{x}}{\dfrac{3}{x}+\dfrac{2}{x^2}} \cdot \dfrac{x^2}{x^2}$

$= \dfrac{\dfrac{4}{x}\cdot x^2}{\left(\dfrac{3}{x}+\dfrac{2}{x^2}\right)x^2}$

$= \dfrac{4x}{\dfrac{3}{x}\cdot x^2 + \dfrac{2}{x^2}\cdot x^2}$

$= \dfrac{4x}{3x+2}$

13. $\dfrac{\dfrac{2a-5}{3a}}{\dfrac{a-7}{6a}}$

$= \dfrac{2a-5}{3a}\cdot\dfrac{6a}{a-7}$ Multiplying by the reciprocal of the divisor

$= \dfrac{(2a-5)\cdot 2\cdot 3a}{3a\cdot(a-7)}$

$= \dfrac{(2a-5)\cdot 2\cdot \cancel{3a}}{\cancel{3a}\cdot(a-7)}$

$= \dfrac{2(2a-5)}{a-7}$

$= \dfrac{4a-10}{a-7}$

15. $\dfrac{\dfrac{x}{4}-\dfrac{4}{x}}{\dfrac{1}{4}+\dfrac{1}{x}}$ LCD is $4x$

$= \dfrac{\dfrac{x}{4}-\dfrac{4}{x}}{\dfrac{1}{4}+\dfrac{1}{x}} \cdot \dfrac{4x}{4x}$

$= \dfrac{\dfrac{x}{4}\cdot 4x - \dfrac{4}{x}\cdot 4x}{\dfrac{1}{4}\cdot 4x + \dfrac{1}{x}\cdot 4x}$

$= \dfrac{x^2-16}{x+4}$

$= \dfrac{(x+4)(x-4)}{x+4}$

$= \dfrac{\cancel{(x+4)}(x-4)}{\cancel{(x+4)}\cdot 1}$

$= x-4$

17. $\dfrac{\dfrac{1}{6}-\dfrac{1}{x}}{\dfrac{6-x}{6}}$ LCD is $6x$

$= \dfrac{\dfrac{1}{6}-\dfrac{1}{x}}{\dfrac{6-x}{6}} \cdot \dfrac{6x}{6x}$

$= \dfrac{\dfrac{1}{6}\cdot 6x - \dfrac{1}{x}\cdot 6x}{\left(\dfrac{6-x}{6}\right)(6x)}$

$= \dfrac{x-6}{(6-x)(x)}$

$= \dfrac{x-6}{-(x-6)(x)}$ $(6-x=-1(-6+x)=-(x-6))$

$= \dfrac{\cancel{(x-6)}\cdot 1}{-\cancel{(x-6)}(x)}$

$= \dfrac{1}{-x} = -\dfrac{1}{x}$

19. $\dfrac{\dfrac{1}{t^2}+1}{\dfrac{1}{t}-1}$ LCD is t^2

$= \dfrac{\dfrac{1}{t^2}+1}{\dfrac{1}{t}-1} \cdot \dfrac{t^2}{t^2}$

$= \dfrac{\dfrac{1}{t^2}\cdot t^2 + 1\cdot t^2}{\dfrac{1}{t}\cdot t^2 - 1\cdot t^2}$

$= \dfrac{1+t^2}{t-t^2}$

(Although the denominator can be factored, doing so will not enable us to simplify further.)

21. $\dfrac{\dfrac{x^2}{x^2-y^2}}{\dfrac{x}{x+y}}$

$= \dfrac{x^2}{x^2-y^2}\cdot\dfrac{x+y}{x}$ Multiplying by the reciprocal of the divisor

$= \dfrac{x^2(x+y)}{(x^2-y^2)(x)}$

$= \dfrac{x\cdot x\cdot(x+y)}{(x+y)(x-y)(x)}$

$= \dfrac{\cancel{x}\cdot x\cdot\cancel{(x+y)}}{\cancel{(x+y)}(x-y)(\cancel{x})}$

$= \dfrac{x}{x-y}$

23. $\dfrac{\dfrac{7}{a^2}+\dfrac{2}{a}}{\dfrac{5}{a^3}-\dfrac{3}{a}}$ LCD is a^3

$= \dfrac{\dfrac{7}{a^2}+\dfrac{2}{a}}{\dfrac{5}{a^3}-\dfrac{3}{a}}\cdot\dfrac{a^3}{a^3}$

$= \dfrac{\dfrac{7}{a^2}\cdot a^3+\dfrac{2}{a}\cdot a^3}{\dfrac{5}{a^3}\cdot a^3-\dfrac{3}{a}\cdot a^3}$

$= \dfrac{7a+2a^2}{5-3a^2}$

(Although the numerator can be factored, doing so will not enable us to simplify further.)

25. $\dfrac{\dfrac{2}{7a^4}-\dfrac{1}{14a}}{\dfrac{3}{5a^2}+\dfrac{2}{15a}} = \dfrac{\dfrac{2}{7a^4}\cdot\dfrac{2}{2}-\dfrac{1}{14a}\cdot\dfrac{a^3}{a^3}}{\dfrac{3}{5a^2}\cdot\dfrac{3}{3}+\dfrac{2}{15a}\cdot\dfrac{a}{a}}$

$= \dfrac{\dfrac{4-a^3}{14a^4}}{\dfrac{9+2a}{15a^2}}$

$= \dfrac{4-a^3}{14a^4}\cdot\dfrac{15a^2}{9+2a}$

$= \dfrac{15\cdot\cancel{a^2}(4-a^3)}{14a^2\cdot\cancel{a^2}(9+2a)}$

$= \dfrac{15(4-a^3)}{14a^2(9+2a)}$, or $\dfrac{60-15a^3}{126a^2+28a^3}$

27. $\dfrac{\dfrac{x}{5y^3}+\dfrac{3}{10y}}{\dfrac{3}{10y}+\dfrac{x}{5y^3}}$

Observe that, by the commutative law of addition, the numerator and denominator are equivalent, so the result is 1.

29. $\dfrac{\dfrac{3}{ab^4}+\dfrac{4}{a^3b}}{\dfrac{5}{a^3b}-\dfrac{3}{ab}} = \dfrac{\dfrac{3}{ab^4}\cdot\dfrac{a^2}{a^2}+\dfrac{4}{a^3b}\cdot\dfrac{b^3}{b^3}}{\dfrac{5}{a^3b}-\dfrac{3}{ab}\cdot\dfrac{a^2}{a^2}}$

$= \dfrac{\dfrac{3a^2+4b^3}{a^3b^4}}{\dfrac{5-3a^2}{a^3b}}$

$= \dfrac{3a^2+4b^3}{a^3b^4}\cdot\dfrac{a^3b}{5-3a^2}$

$= \dfrac{\cancel{a^3b}(3a^2+4b^3)}{\cancel{a^3b}\cdot b^3(5-3a^2)}$

$= \dfrac{3a^2+4b^3}{b^3(5-3a^2)}$, or $\dfrac{3a^2+4b^3}{5b^3-3a^2b^3}$

31. $\dfrac{2-\dfrac{3}{x^2}}{2+\dfrac{3}{x^4}} = \dfrac{2-\dfrac{3}{x^2}}{2+\dfrac{3}{x^4}}\cdot\dfrac{x^4}{x^4}$

$= \dfrac{2\cdot x^4-\dfrac{3}{x^2}\cdot x^4}{2\cdot x^4+\dfrac{3}{x^4}\cdot x^4}$

$= \dfrac{2x^4-3x^2}{2x^4+3}$

33. $\dfrac{t-\dfrac{2}{t}}{t+\dfrac{5}{t}} = \dfrac{t\cdot\dfrac{t}{t}-\dfrac{2}{t}}{t\cdot\dfrac{t}{t}+\dfrac{5}{t}}$

$= \dfrac{\dfrac{t^2-2}{t}}{\dfrac{t^2+5}{t}}$

$= \dfrac{t^2-2}{t}\cdot\dfrac{t}{t^2+5}$

$= \dfrac{\cancel{t}(t^2-2)}{\cancel{t}(t^2+5)}$

$= \dfrac{t^2-2}{t^2+5}$

35. $\dfrac{3+\dfrac{4}{ab^3}}{\dfrac{3+a}{a^2b}} = \dfrac{3+\dfrac{4}{ab^3}}{\dfrac{3+a}{a^2b}}\cdot\dfrac{a^2b^3}{a^2b^3}$

$= \dfrac{3\cdot a^2b^3+\dfrac{4}{ab^3}\cdot a^2b^3}{\dfrac{3+a}{a^2b}\cdot a^2b^3}$

$= \dfrac{3a^2b^3+4a}{b^2(3+a)}$, or $\dfrac{3a^2b^3+4a}{3b^2+ab^2}$

37. $\dfrac{t+5+\frac{3}{t}}{t+2+\frac{1}{t}}$ LCD is t

$= \dfrac{t+5+\frac{3}{t}}{t+2+\frac{1}{t}} \cdot \dfrac{t}{t}$

$= \dfrac{t\cdot t+5\cdot t+\frac{3}{t}\cdot t}{t\cdot t+2\cdot t+\frac{1}{t}\cdot t}$

$= \dfrac{t^2+5t+3}{t^2+2t+1}$

$= \dfrac{t^2+5t+3}{(t+1)^2}$

39. $\dfrac{x-2-\frac{1}{x}}{x-5-\frac{4}{x}} = \dfrac{x-2-\frac{1}{x}}{x-5-\frac{4}{x}} \cdot \dfrac{x}{x}$

$= \dfrac{x\cdot x-2\cdot x-\frac{1}{x}\cdot x}{x\cdot x-5\cdot x-\frac{4}{x}\cdot x}$

$= \dfrac{x^2-2x-1}{x^2-5x-4}$

41. *Writing Exercise*

43. $3x-5+2(4x-1) = 12x-3$

$3x-5+8x-2 = 12x-3$

$11x-7 = 12x-3$

$-7 = x-3$

$-4 = x$

The solution is -4.

45. $\frac{3}{4}x - \frac{5}{8} = \frac{3}{8}x + \frac{7}{4}$ LCD is 8

$8\left(\frac{3}{4}x - \frac{5}{8}\right) = 8\left(\frac{3}{8}x + \frac{7}{4}\right)$

$8\cdot\frac{3}{4}x - 8\cdot\frac{5}{8} = 8\cdot\frac{3}{8}x + 8\cdot\frac{7}{4}$

$6x-5 = 3x+14$

$3x-5 = 14$

$3x = 19$

$x = \frac{19}{3}$

The solution is $\frac{19}{3}$.

47. $x^2-7x-30 = 0$

$(x-10)(x+3) = 0$

$x-10 = 0$ *or* $x+3 = 0$

$x = 10$ *or* $x = -3$

The solutions are 10 and -3.

49. *Writing Exercise*

51. $\dfrac{\frac{x-5}{x-6}}{\frac{x-7}{x-8}}$

This expression is undefined for any value of x that makes a denominator 0. We see that $x-6=0$ when $x=6$, $x-7=0$ when $x=7$, and $x-8=0$ when $x=8$, so the expression is undefined for the x-values 6, 7, and 8.

53. $\dfrac{\frac{2x+3}{5x+4}}{\frac{3}{7}-\frac{2x}{21}}$

This expression is undefined for any value of x that makes a denominator 0. First we find the value of x for which $5x+4=0$.

$5x+4 = 0$

$5x = -4$

$x = -\frac{4}{5}$

Then we find the value of x for which $\frac{3}{7}-\frac{2x}{21}=0$:

$\frac{3}{7}-\frac{2x}{21} = 0$

$21\left(\frac{3}{7}-\frac{2x}{21}\right) = 21\cdot 0$

$21\cdot\frac{3}{7} - 21\cdot\frac{2x}{21} = 0$

$9-2x = 0$

$9 = 2x$

$\frac{9}{2} = x$

The expression is undefined for the x-values $-\frac{4}{5}$ and $\frac{9}{2}$.

55. $$\frac{\frac{P\left(1+\frac{i}{12}\right)^2}{\left(1+\frac{1}{12}\right)^2-1}}{\frac{i}{12}} = \frac{\frac{P\left(1+\frac{i}{6}+\frac{i^2}{144}\right)}{\left(1+\frac{i}{6}+\frac{i^2}{144}\right)-1}}{\frac{i}{12}}$$

$$= \frac{P\left(1+\frac{i}{6}+\frac{i^2}{144}\right)}{\frac{\frac{i}{6}+\frac{i^2}{144}}{\frac{i}{12}}}$$

$$= \frac{P\left(1+\frac{i}{6}+\frac{i^2}{144}\right)}{\left(\frac{i}{6}+\frac{i^2}{144}\right)\left(\frac{12}{i}\right)}$$

$$= \frac{P\left(1+\frac{i}{6}+\frac{i^2}{144}\right)}{2+\frac{i}{12}}$$

$$= \frac{P\left(1+\frac{i}{6}+\frac{i^2}{144}\right)}{2+\frac{i}{12}} \cdot \frac{144}{144}$$

$$= \frac{144P\left(1+\frac{i}{6}+\frac{i^2}{144}\right)}{144\left(2+\frac{i}{12}\right)}$$

$$= \frac{P(144+24i+i^2)}{288+12i}$$

$$= \frac{P(12+i)^2}{12(24+i)}, \text{ or}$$

$$\frac{P(i+12)^2}{12(i+24)}$$

57. $$\frac{\frac{5}{x+2}-\frac{3}{x-2}}{\frac{x}{x-1}+\frac{x}{x+1}} = \frac{\frac{5}{x+2}\cdot\frac{x-2}{x-2}-\frac{3}{x-2}\cdot\frac{x+2}{x+2}}{\frac{x}{x-1}\cdot\frac{x+1}{x+1}+\frac{x}{x+1}\cdot\frac{x-1}{x-1}}$$

$$= \frac{\frac{5(x-2)-3(x+2)}{(x+2)(x-2)}}{\frac{x(x+1)+x(x-1)}{(x+1)(x-1)}}$$

$$= \frac{\frac{5x-10-3x-6}{(x+2)(x-2)}}{\frac{x^2+x+x^2-x}{(x+1)(x-1)}}$$

$$= \frac{\frac{2x-16}{(x+2)(x-2)}}{\frac{2x^2}{(x+1)(x-1)}}$$

$$= \frac{2x-16}{(x+2)(x-2)} \cdot \frac{(x+1)(x-1)}{2x^2}$$

$$= \frac{\not{2}(x-8)(x+1)(x-1)}{\not{2}\cdot x^2(x+2)(x-2)}$$

$$= \frac{(x-8)(x+1)(x-1)}{x^2(x+2)(x-2)}$$

59. $$\left[\frac{\frac{x-1}{x-1}-1}{\frac{x+1}{x-1}+1}\right]^5$$

Consider the numerator of the complex rational expression:

$$\frac{x-1}{x-1}-1 = 1-1 = 0$$

Since the denominator, $\frac{x+1}{x-1}+1$ is not equal to 0, the simplified form of the original expression is 0.

61. $$\frac{\dfrac{z}{1-\dfrac{z}{2+2z}}-2z}{\dfrac{2z}{5z-2}-3} = \frac{\dfrac{z}{\dfrac{2+2z-z}{2+2z}}-2z}{\dfrac{2z-15z+6}{5z-2}}$$

$$= \frac{\dfrac{z}{\dfrac{2+z}{2+2z}}-2z}{\dfrac{-13z+6}{5z-2}}$$

$$= \frac{z\cdot\dfrac{2+2z}{2+z}-2z}{\dfrac{-13z+6}{5z-2}}$$

$$= \frac{\dfrac{z(2+2z)-2z(2+z)}{2+z}}{\dfrac{-13z+6}{5z-2}}$$

$$= \frac{\dfrac{2z+2z^2-4z-2z^2}{2+z}}{\dfrac{-13z+6}{5z-2}}$$

$$= \frac{\dfrac{-2z}{2+z}}{\dfrac{-13z+6}{5z-2}}$$

$$= \frac{-2z}{2+z}\cdot\frac{5z-2}{-13z+6}$$

$$= \frac{-2z(5z-2)}{(2+z)(-13z+6)}, \text{ or }$$

$$\frac{2z(5z-2)}{(2+z)(13z-6)}$$

63. 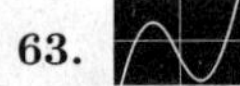

Exercise Set 6.6

1. The statement is false. See Example 2(c).

3. The statement is true. See page 404 in the text.

5. Because no variable appears in a denominator, no restrictions exist.

$$\frac{3}{5}-\frac{5}{8}=\frac{x}{20},\ \text{LCD} = 40$$

$$40\left(\frac{3}{5}-\frac{5}{8}\right) = 40\cdot\frac{x}{20}$$

$$40\cdot\frac{3}{5}-40\cdot\frac{5}{8} = 40\cdot\frac{x}{20}$$

$$24-25 = 2x$$

$$-1 = 2x$$

$$-\frac{1}{2} = x$$

Check:

$$\frac{3}{5}-\frac{5}{8}=\frac{x}{20}$$

$\frac{3}{5}-\frac{5}{8}$	$\dfrac{-\frac{1}{2}}{20}$
$\frac{24}{40}-\frac{25}{40}$	$-\frac{1}{2}\cdot\frac{1}{20}$
$-\frac{1}{40} \stackrel{?}{=} -\frac{1}{40}$	TRUE

This checks, so the solution is $-\frac{1}{2}$.

7. Note that x cannot be 0.

$$\frac{1}{3}+\frac{5}{6}=\frac{1}{x},\ \text{LCD} = 6x$$

$$6x\left(\frac{1}{3}+\frac{5}{6}\right) = 6x\cdot\frac{1}{x}$$

$$6x\cdot\frac{1}{3}+6x\cdot\frac{5}{6} = 6x\cdot\frac{1}{x}$$

$$2x+5x = 6$$

$$7x = 6$$

$$x = \frac{6}{7}$$

Check:

$$\frac{1}{3}+\frac{5}{6}=\frac{1}{x}$$

$\frac{1}{3}+\frac{5}{6}$	$\dfrac{1}{\frac{6}{7}}$
$\frac{2}{6}+\frac{5}{6}$	$1\cdot\frac{7}{6}$
$\frac{7}{6} \stackrel{?}{=} \frac{7}{6}$	TRUE

This checks, so the solution is $\frac{6}{7}$.

9. Note that t cannot be 0.

$$\frac{1}{6}+\frac{1}{8}=\frac{1}{t},\ \text{LCD} = 24t$$

$$24t\left(\frac{1}{6}+\frac{1}{8}\right) = 24t\cdot\frac{1}{t}$$

$$24t\cdot\frac{1}{6}+24t\cdot\frac{1}{8} = 24t\cdot\frac{1}{t}$$

$$4t+3t = 24$$

$$7t = 24$$

$$t = \frac{24}{7}$$

Check:

$$\frac{1}{6}+\frac{1}{8}=\frac{1}{t}$$

$\frac{1}{6}+\frac{1}{8}$	$\frac{1}{\frac{24}{7}}$
$\frac{4}{24}+\frac{3}{24}$	$1\cdot\frac{7}{24}$

$$\frac{7}{24} \stackrel{?}{=} \frac{7}{24} \quad \text{TRUE}$$

This checks, so the solution is $\frac{24}{7}$.

11. Note that x cannot be 0.

$$x+\frac{5}{x}=-6,\ \text{LCD}=x$$
$$x\left(x+\frac{5}{x}\right)=-6\cdot x$$
$$x\cdot x+x\cdot\frac{5}{x}=-6\cdot x$$
$$x^2+5=-6x$$
$$x^2+6x+5=0$$
$$(x+5)(x+1)=0$$
$$x+5=0 \quad or \quad x+1=0$$
$$x=-5 \quad or \quad x=-1$$

Check:

$$x+\frac{5}{x}=-6$$

$-5+\frac{5}{-5}$	-6
$-5-1$	

$$-6 \stackrel{?}{=} -6 \quad \text{TRUE}$$

$$x+\frac{5}{x}=-6$$

$-1+\frac{5}{-1}$	-6
$-1-5$	

$$-6 \stackrel{?}{=} -6 \quad \text{TRUE}$$

Both of these check, so the two solutions are -5 and -1.

13. Note that x cannot be 0.

$$\frac{x}{6}-\frac{6}{x}=0,\ \text{LCD}=6x$$
$$6x\left(\frac{x}{6}-\frac{6}{x}\right)=6x\cdot 0$$
$$6x\cdot\frac{x}{6}-6x\cdot\frac{6}{x}=6x\cdot 0$$
$$x^2-36=0$$
$$(x+6)(x-6)=0$$
$$x+6=0 \quad or \quad x-6=0$$
$$x=-6 \quad or \quad x=6$$

Check:

$$\frac{x}{6}-\frac{6}{x}=0$$

$\frac{-6}{6}-\frac{6}{-6}$	0
$-1+1$	

$$0 \stackrel{?}{=} 0 \quad \text{TRUE}$$

$$\frac{x}{6}-\frac{6}{x}=0$$

$\frac{6}{6}-\frac{6}{6}$	0
$1-1$	

$$0 \stackrel{?}{=} 0 \quad \text{TRUE}$$

Both of these check, so the two solutions are -6 and 6.

15. Note that x cannot be 0.

$$\frac{5}{x}=\frac{6}{x}-\frac{1}{3},\ \text{LCD}=3x$$
$$3x\cdot\frac{5}{x}=3x\left(\frac{6}{x}-\frac{1}{3}\right)$$
$$3x\cdot\frac{5}{x}=3x\cdot\frac{6}{x}-3x\cdot\frac{1}{3}$$
$$15=18-x$$
$$-3=-x$$
$$3=x$$

Check:

$$\frac{5}{x}=\frac{6}{x}-\frac{1}{3}$$

$\frac{5}{3}$	$\frac{6}{3}-\frac{1}{3}$

$$\frac{5}{3} \stackrel{?}{=} \frac{5}{3} \quad \text{TRUE}$$

This checks, so the solution is 3.

17. Note that t cannot be 0.

$$\frac{5}{3t}+\frac{3}{t}=1,\ \text{LCD}=3t$$
$$3t\left(\frac{5}{3t}+\frac{3}{t}\right)=3t\cdot 1$$
$$3t\cdot\frac{5}{3t}+3t\cdot\frac{3}{t}=3t\cdot 1$$
$$5+9=3t$$
$$14=3t$$
$$\frac{14}{3}=t$$

Check:

$$\frac{5}{3t}+\frac{3}{t}=1$$

$\frac{5}{3\cdot\frac{14}{3}}+\frac{3}{\frac{14}{3}}$	1
$\frac{5}{14}+\frac{9}{14}$	
$\frac{14}{14}$	

$$1 \stackrel{?}{=} 1 \quad \text{TRUE}$$

This checks, so the solution is $\frac{14}{3}$.

19. To avoid division by 0, we must have $x+3 \neq 0$, or $x \neq -3$.

$$\frac{x-8}{x+3} = \frac{1}{4}, \text{ LCD} = 4(x+3)$$
$$4(x+3) \cdot \frac{x-8}{x+3} = 4(x+3) \cdot \frac{1}{4}$$
$$4(x-8) = x+3$$
$$4x - 32 = x+3$$
$$3x = 35$$
$$x = \frac{35}{3}$$

Check:

$$\frac{x-8}{x+3} = \frac{1}{4}$$
$$\frac{\frac{35}{3}-8}{\frac{35}{3}+3} \;\Big|\; \frac{1}{4}$$
$$\frac{\frac{35}{3}-\frac{24}{3}}{\frac{35}{3}+\frac{9}{3}}$$
$$\frac{\frac{11}{3}}{\frac{44}{3}}$$
$$\frac{11}{3} \cdot \frac{3}{44}$$
$$\frac{1}{4} \stackrel{?}{=} \frac{1}{4} \quad \text{TRUE}$$

This checks, so the solution is $\frac{35}{3}$.

21. Note that x cannot be 0.

$$x + \frac{12}{x} = -7, \text{ LCD is } x$$
$$x\left(x + \frac{12}{x}\right) = x \cdot (-7)$$
$$x \cdot x + x \cdot \frac{12}{x} = -7x$$
$$x^2 + 12 = -7x$$
$$x^2 + 7x + 12 = 0$$
$$(x+3)(x+4) = 0$$
$$x+3 = 0 \quad or \quad x+4 = 0$$
$$x = -3 \quad or \quad x = -4$$

Both numbers check, so the solutions are -3 and -4.

23. To avoid division by 0, we must have $x+1 \neq 0$ and $x-2 \neq 0$, or $x \neq -1$ and $x \neq 2$.

$$\frac{2}{x+1} = \frac{1}{x-2},$$
$$\text{LCD} = (x+1)(x-2)$$
$$(x+1)(x-2) \cdot \frac{2}{x+1} = (x+1)(x-2) \cdot \frac{1}{x-2}$$
$$2(x-2) = x+1$$
$$2x - 4 = x+1$$
$$x = 5$$

This checks, so the solution is 5.

25. Because no variable appears in a denominator, no restrictions exist.

$$\frac{a}{6} - \frac{a}{10} = \frac{1}{6}, \text{ LCD} = 30$$
$$30\left(\frac{a}{6} - \frac{a}{10}\right) = 30 \cdot \frac{1}{6}$$
$$30 \cdot \frac{a}{6} - 30 \cdot \frac{a}{10} = 30 \cdot \frac{1}{6}$$
$$5a - 3a = 5$$
$$2a = 5$$
$$a = \frac{5}{2}$$

This checks, so the solution is $\frac{5}{2}$.

27. Because no variable appears in a denominator, no restrictions exist.

$$\frac{x+1}{3} - 1 = \frac{x-1}{2}, \text{ LCD} = 6$$
$$6\left(\frac{x+1}{3} - 1\right) = 6 \cdot \frac{x-1}{2}$$
$$6 \cdot \frac{x+1}{3} - 6 \cdot 1 = 6 \cdot \frac{x-1}{2}$$
$$2(x+1) - 6 = 3(x-1)$$
$$2x + 2 - 6 = 3x - 3$$
$$2x - 4 = 3x - 3$$
$$-1 = x$$

This checks, so the solution is -1.

29. To avoid division by 0, we must have $t-5 \neq 0$, or $t \neq 5$.

$$\frac{4}{t-5} = \frac{t-1}{t-5}, \text{ LCD} = t-5$$
$$(t-5) \cdot \frac{4}{t-5} = (t-5) \cdot \frac{t-1}{t-5}$$
$$4 = t-1$$
$$5 = t$$

Because of the restriction $t \neq 5$, the number 5 must be rejected as a solution. The equation has no solution.

31. To avoid division by 0, we must have $x+4\neq 0$ and $x\neq 0$, or $x\neq -4$ and $x\neq 0$.

$$\frac{3}{x+4}=\frac{5}{x},\quad \text{LCD}=x(x+4)$$
$$x(x+4)\cdot\frac{3}{x+4}=x(x+4)\cdot\frac{5}{x}$$
$$3x=5(x+4)$$
$$3x=5x+20$$
$$-2x=20$$
$$x=-10$$

This checks, so the solution is -10.

33. To avoid division by 0, we must have $a-1\neq 0$ and $a-2\neq 0$, or $a\neq 1$ and $a\neq 2$.

$$\frac{a-4}{a-1}=\frac{a+2}{a-2},\quad \text{LCD}=(a-1)(a-2)$$
$$(a-1)(a-2)\cdot\frac{a-4}{a-1}=(a-1)(a-2)\cdot\frac{a+2}{a-2}$$
$$(a-2)(a-4)=(a-1)(a+2)$$
$$a^2-6a+8=a^2+a-2$$
$$-6a+8=a-2$$
$$10=7a$$
$$\frac{10}{7}=a$$

This checks, so the solution is $\frac{10}{7}$.

35. To avoid division by 0, we must have $t-2\neq 0$, or $t\neq 2$.

$$\frac{5}{t-2}+\frac{3t}{t-2}=\frac{4}{t^2-4t+4},\quad \text{LCD is }(t-2)^2$$
$$(t-2)^2\left(\frac{5}{t-2}+\frac{3t}{t-2}\right)=(t-2)^2\cdot\frac{4}{(t-2)^2}$$
$$5(t-2)+3t(t-2)=4$$
$$5t-10+3t^2-6t=4$$
$$3t^2-t-10=4$$
$$3t^2-t-14=0$$
$$(3t-7)(t+2)=0$$
$$3t-7=0 \quad \text{or} \quad t+2=0$$
$$3t=7 \quad \text{or} \quad t=-2$$
$$t=\frac{7}{3} \quad \text{or} \quad t=-2$$

Both numbers check. The solutions are $\frac{7}{3}$ and -2.

37. To avoid division by 0, we must have $x-3\neq 0$ and $x+3\neq 0$, or $x\neq 3$ and $x\neq -3$.

$$\frac{4}{x-3}+\frac{2x}{x^2-9}=\frac{1}{x+3},$$
$$\text{LCD}=(x-3)(x+3)$$
$$(x-3)(x+3)\left(\frac{4}{x-3}+\frac{2x}{(x+3)(x-3)}\right)=$$
$$(x-3)(x+3)\cdot\frac{1}{x+3}$$
$$4(x+3)+2x=x-3$$
$$4x+12+2x=x-3$$
$$6x+12=x-3$$
$$5x=-15$$
$$x=-3$$

Because of the restriction of $x\neq -3$, we must reject the number -3 as a solution. The equation has no solution.

39. To avoid division by 0, we must have $y-3\neq 0$ and $y+3\neq 0$, or $y\neq 3$ and $y\neq -3$.

$$\frac{5}{y-3}-\frac{30}{y^2-9}=1$$
$$\frac{5}{y-3}-\frac{30}{(y+3)(y-3)}=1,$$
$$\text{LCD}=(y-3)(y+3)$$
$$(y-3)(y+3)\left(\frac{5}{y-3}-\frac{30}{(y+3)(y-3)}\right)=$$
$$(y-3)(y+3)\cdot 1$$
$$5(y+3)-30=(y+3)(y-3)$$
$$5y+15-30=y^2-9$$
$$0=y^2-5y+6$$
$$0=(y-3)(y-2)$$
$$y-3=0 \quad \text{or} \quad y-2=0$$
$$y=3 \quad \text{or} \quad y=2$$

Because of the restriction $y\neq 3$, we must reject the number 3 as a solution. The number 2 checks, so it is the solution.

41. To avoid division by 0, we must have $8-a\neq 0$ (or equivalently $a-8\neq 0$), or $a\neq 8$.

$$\frac{4}{8-a}=\frac{4-a}{a-8}$$
$$\frac{-1}{-1}\cdot\frac{4}{8-a}=\frac{4-a}{a-8}$$
$$\frac{-4}{a-8}=\frac{4-a}{a-8},\quad \text{LCD}=a-8$$
$$(a-8)\cdot\frac{-4}{a-8}=(a-8)\cdot\frac{4-a}{a-8}$$
$$-4=4-a$$
$$-8=-a$$
$$8=a$$

Because of the restriction $a\neq 8$, we must reject the number 8 as a solution. The equation has no solution.

43. $\dfrac{-2}{x+2}=\dfrac{x}{x+2}$

To avoid division by 0, we must have $x+2\neq 0$, or $x\neq -2$. Now observe that the denominators are the same, so the numerators must be the same. Thus, we have $-2=x$, but because of the restriction $x \neq -2$ this cannot be a solution. The equation has no solution.

45. *Writing Exercise*

47. ***Familiarize.*** Let $x =$ the first odd integer. Then $x+2 =$ the next odd integer.

Translate.

$$\underbrace{\text{The sum of two consecutive odd integers}}_{x+(x+2)} \quad \underset{=}{\text{is}} \quad \underset{276}{276.}$$

Carry out. We solve the equation.

$$x+(x+2) = 276$$
$$2x+2 = 276$$
$$2x = 274$$
$$x = 137$$

When $x = 137$, then $x+2 = 137+2 = 139$.

Check. The numbers 137 and 139 are consecutive odd integers and $137+139 = 276$. These numbers check.

State. The integers are 137 and 139.

49. ***Familiarize.*** Let $b =$ the base of the triangle, in cm. Then $b+3 =$ the height. Recall that the area of a triangle is given by $\frac{1}{2} \times$ base $\times$ height.

Translate.

$$\underbrace{\text{The area of the triangle}}_{\frac{1}{2}\cdot b\cdot(b+3)} \quad \underset{=}{\text{is}} \quad \underset{54}{54\text{ cm}^2.}$$

Carry out. We solve the equation.

$$\frac{1}{2}b(b+3) = 54$$
$$2\cdot\frac{1}{2}b(b+3) = 2\cdot 54$$
$$b(b+3) = 108$$
$$b^2+3b = 108$$
$$b^2+3b-108 = 0$$
$$(b-9)(b+12) = 0$$
$$b-9 = 0 \quad or \quad b+12 = 0$$
$$b = 9 \quad or \quad b = -12$$

Check. The length of the base cannot be negative so we need to check only 9. If the base is 9 cm, then the height is 9+3, or 12 cm, and the area is $\frac{1}{2}\cdot 9\cdot 12$, or 54 cm^2. The answer checks.

State. The base measures 9 cm, and the height measures 12 cm.

51. To find the rate, in centimeters per day, we divide the amount of growth by the number of days. From June 9 to June 24 is $24-9 = 15$ days.

$$\text{Rate, in cm per day} = \frac{0.9\text{ cm}}{15\text{ days}}$$
$$= 0.06\text{ cm/day}$$
$$= 0.06\text{ cm per day}$$

53. *Writing Exercise*

55. To avoid division by 0, we must have $x-3 \neq 0$, or $x \neq 3$.

$$1+\frac{x-1}{x-3} = \frac{2}{x-3}-x,\ \text{LCD}=x-3$$
$$(x-3)\left(1+\frac{x-1}{x-3}\right) = (x-3)\left(\frac{2}{x-3}-x\right)$$
$$(x-3)\cdot 1+(x-3)\cdot\frac{x-1}{x-3} = (x-3)\cdot\frac{2}{x-3}-(x-3)x$$
$$x-3+x-1 = 2-x^2+3x$$
$$2x-4 = 2-x^2+3x$$
$$x^2-x-6 = 0$$
$$(x-3)(x+2) = 0$$
$$x-3 = 0 \quad or \quad x+2 = 0$$
$$x = 3 \quad or \quad x = -2$$

Because of the restriction $x \neq 3$, we must reject the number 3 as a solution. The number -2 checks, so it is the solution.

57. To avoid division by 0, we must have $x+4 \neq 0$ and $x-1 \neq 0$ and $x+2 \neq 0$, or $x \neq -4$ and $x \neq 1$ and $x \neq -2$.

$$\frac{x}{x^2+3x-4}+\frac{x+1}{x^2+6x+8} = \frac{2x}{x^2+x-2}$$
$$\frac{x}{(x+4)(x-1)}+\frac{x+1}{(x+2)(x+4)} = \frac{2x}{(x+2)(x-1)},$$
$$\text{LCD} = (x+4)(x-1)(x+2)$$
$$(x+4)(x-1)(x+2)\left(\frac{x}{(x+4)(x-1)}+\frac{x+1}{(x+2)(x+4)}\right) =$$
$$(x+4)(x-1)(x+2)\left(\frac{2x}{(x+2)(x-1)}\right)$$
$$x(x+2)+(x-1)(x+1) = 2x(x+4)$$
$$x^2+2x+x^2-1 = 2x^2+8x$$
$$2x^2+2x-1 = 2x^2+8x$$
$$-1 = 6x$$
$$-\frac{1}{6} = x$$

This checks, so the solution is $-\frac{1}{6}$.

59. To avoid division by 0, we must have $x+2 \neq 0$ and $x-2 \neq 0$, or $x \neq -2$ and $x \neq 2$.

$$\frac{x^2}{x^2-4} = \frac{x}{x+2}-\frac{2x}{2-x}$$
$$\frac{x^2}{x^2-4} = \frac{x}{x+2}-\frac{2x}{2-x}\cdot\frac{-1}{-1}$$
$$\frac{x^2}{(x+2)(x-2)} = \frac{x}{x+2}-\frac{-2x}{x-2},$$
$$\text{LCD} = (x+2)(x-2)$$
$$(x+2)(x-2)\cdot\frac{x^2}{(x+2)(x-2)} =$$
$$(x+2)(x-2)\left(\frac{x}{x+2}-\frac{-2x}{x-2}\right)$$
$$x^2 = x(x-2)-(-2x)(x+2)$$
$$x^2 = x^2-2x+2x^2+4x$$
$$x^2 = 3x^2+2x$$
$$0 = 2x^2+2x$$
$$0 = 2x(x+1)$$

$$2x = 0 \quad or \quad x + 1 = 0$$
$$x = 0 \quad or \quad x = -1$$

Both of these check, so the solutions are -1 and 0.

61. To avoid division by 0, we must have $x - 1 \neq 0$, or $x \neq 1$.

$$\frac{1}{x-1} + x - 5 = \frac{5x-4}{x-1} - 6, \text{ LCD} = x - 1$$
$$(x-1)\left(\frac{1}{x-1} + x - 5\right) = (x-1)\left(\frac{5x-4}{x-1} - 6\right)$$
$$1 + x(x-1) - 5(x-1) = 5x - 4 - 6(x-1)$$
$$1 + x^2 - x - 5x + 5 = 5x - 4 - 6x + 6$$
$$x^2 - 6x + 6 = -x + 2$$
$$x^2 - 5x + 4 = 0$$
$$(x-1)(x-4) = 0$$
$$x - 1 = 0 \quad or \quad x - 4 = 0$$
$$x = 1 \quad or \quad x = 4$$

Because of the restriction $x \neq 1$, we must reject the number 1 as a solution. The number 4 checks, so it is the solution.

63. 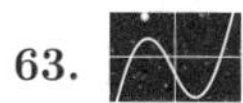

Exercise Set 6.7

1. ***Familiarize.*** The job takes Ned 20 min working alone and Linda 30 min working alone. Then in 1 min Ned does $\frac{1}{20}$ of the job and Linda does $\frac{1}{30}$ of the job. Working together they can do $\frac{1}{20} + \frac{1}{30}$, or $\frac{5}{60}$, or $\frac{1}{12}$ of the job in 1 min. In 10 min, Ned does $10 \cdot \frac{1}{20}$ or the job and Linda does $10 \cdot \frac{1}{30}$ of the job. Working together they can do $10 \cdot \frac{1}{20} + 10 \cdot \frac{1}{30}$, or $\frac{5}{6}$, of the job in 10 min. In 15 min, Ned does $15 \cdot \frac{1}{20}$ of the job and Linda does $15 \cdot \frac{1}{30}$ of the job. Working together they can do $15 \cdot \frac{1}{20} + 15 \cdot \frac{1}{30}$, or $1\frac{1}{4}$ of the job which is more of the job than needs to be done. The answer is somewhere between 10 min and 15 min. (When we determined that Ned and Linda could do $\frac{1}{12}$ of the job working together for 1 min, we could have observed that it would take them 12 min to do the entire job, but we will continue with the full solution here.)

Translate. If they work together t minutes, then Ned does $t \cdot \frac{1}{20}$ of the job and Linda does $t \cdot \frac{1}{30}$ of the job. We want a number t such that

$$\left(\frac{1}{20} + \frac{1}{30}\right)t = 1, \text{ or } \frac{1}{12} \cdot t = 1.$$

Carry out. We solve the equation.

$$\frac{1}{12} \cdot t = 1$$
$$12 \cdot \frac{1}{12} \cdot t = 12 \cdot 1$$
$$t = 12$$

Check. We can repeat the computations. We also expected the result to be between 10 min and 15 min as it is.

State. Working together, it takes Ned and Linda 12 min to do the job.

3. ***Familiarize.*** The job takes Juanita 12 hours working alone and Anton 16 hours working alone. Then in 1 hour Juanita does $\frac{1}{12}$ of the job and Anton does $\frac{1}{16}$ of the job. Working together, they can do $\frac{1}{12} + \frac{1}{16}$, or $\frac{7}{48}$ of the job in 1 hour. In four hours, Juanita does $4 \cdot \frac{1}{12}$ of the job and Anton does $4 \cdot \frac{1}{16}$ of the job. Working together they can do $4 \cdot \frac{1}{12} + 4 \cdot \frac{1}{16}$, or $\frac{7}{12}$ of the job in 4 hours. In 7 hours they can do $7 \cdot \frac{1}{12} + 7 \cdot \frac{1}{16}$, or $\frac{49}{48}$ or $1\frac{1}{48}$ of the job which is more of the job than needs to be done. The answer is somewhere between 4 hr and 7 hr.

Translate. If they work together t hours, then Juanita does $t\left(\frac{1}{12}\right)$ of the job and Anton does $t\left(\frac{1}{16}\right)$ of the job. We want some number t such that

$$\left(\frac{1}{12} + \frac{1}{16}\right)t = 1, \text{ or } \frac{7}{48} \cdot t = 1.$$

Carry out. We solve the equation.

$$\frac{7}{48} \cdot t = 1$$
$$\frac{48}{7} \cdot \frac{7}{48} \cdot t = \frac{48}{7} \cdot 1$$
$$t = \frac{48}{7}, \text{ or } 6\frac{6}{7}$$

Check. The check can be done by repeating the computations. We also have a partial check in that we expected from our familiarization step that the answer would be between 4 hr and 7 hr.

State. Working together, it takes them $\frac{48}{7}$ hr, or 6 hr, $51\frac{3}{7}$ min, to build the shed.

5. ***Familiarize.*** The job takes Zoe 4 hr working alone and Steffi 3 hr working alone. Then in 1 hr Zoe does $\frac{1}{4}$ of the job and Steffi does $\frac{1}{3}$ of the job. Working together they can do $\frac{1}{4} + \frac{1}{3}$, or $\frac{7}{12}$ of the job in 1 hr. In 2 hr, Zoe does $2 \cdot \frac{1}{4}$ of the job and Steffi does $2 \cdot \frac{1}{3}$ of the job. Working together they can do $2 \cdot \frac{1}{4} + 2 \cdot \frac{1}{3}$, or $1\frac{1}{6}$ of the job which is more of the job than needs to be done. The answer is between 1 hr and 2 hr.

Translate. If they work together t hr, then Zoe does $t \cdot \frac{1}{4}$ of the job and Steffi does $t \cdot \frac{1}{3}$ of the job. We want a number t such that

$$\left(\frac{1}{4} + \frac{1}{3}\right)t = 1, \text{ or } \frac{7}{12} \cdot t = 1.$$

Carry out. We solve the equation.

$$\frac{7}{12}\cdot t = 1$$
$$\frac{12}{7}\cdot\frac{7}{12}\cdot t = \frac{12}{7}\cdot 1$$
$$t = \frac{12}{7}$$

Check. We can repeat the computations. We also expected the result to be between 1 hr and 2 hr as it is.

State. Working together, it takes Zoe and Steffi $\frac{12}{7}$ hr, or 1 hr, $42\frac{6}{7}$ min, to do the job.

7. ***Familiarize.*** The job takes Raul 48 hours working alone and Mira 36 hours working alone. Then in 1 hour Raul does $\frac{1}{48}$ of the job and Mira does $\frac{1}{36}$ of the job. Working together they can do $\frac{1}{48}+\frac{1}{36}$, or $\frac{7}{144}$ of the job in 1 hour. In two hours, Raul does $2\left(\frac{1}{48}\right)$ of the job and Mira does $2\left(\frac{1}{36}\right)$ of the job. Working together they can do $2\left(\frac{1}{48}\right)+2\left(\frac{1}{36}\right)$, or $\frac{7}{72}$ of the job in two hours. In 10 hours, they can do $\frac{35}{72}$ of the job. In 25 hours, they can do $\frac{175}{144}$, or $1\frac{31}{144}$ of the job which is more of the job than needs to be done. The answer is somewhere between 10 hr and 25 hr.

Translate. If they work together t hours, then Raul does $t\left(\frac{1}{48}\right)$ of the job and Mira does $t\left(\frac{1}{36}\right)$ of the job. We want some number t such that

$$\left(\frac{1}{48}+\frac{1}{36}\right)t = 1, \text{ or } \frac{7}{144}\cdot t = 1.$$

Carry out. We solve the equation.

$$\frac{7}{144}\cdot t = 1$$
$$\frac{144}{7}\cdot\frac{7}{144}\cdot t = \frac{144}{7}\cdot 1$$
$$t = \frac{144}{7}, \text{ or } 20\frac{4}{7}$$

Check. The check can be done by repeating the computations. We also have a partial check in that we expected from our familiarization step that the answer would be between 10 hr and 25 hr.

State. Working together, it takes them $\frac{144}{7}$ hr, or 20 hr, $34\frac{2}{7}$ min, to complete the job.

9. ***Familiarize.*** The job takes Jorge 4 days working alone and Carla 5 days working alone. Then in 1 day Jorge does $\frac{1}{4}$ of the job and Carla does $\frac{1}{5}$ of the job. Working together, they can do $\frac{1}{4}+\frac{1}{5}$, or $\frac{9}{20}$ of the job in 1 day. In two days, Jorge does $2\left(\frac{1}{4}\right)$ of the job and Carla does $2\left(\frac{1}{5}\right)$ of the job. Working together they can do $2\left(\frac{1}{4}\right)+2\left(\frac{1}{5}\right)$, or $\frac{9}{10}$ of the job in 2 days. In 3 days they can do $3\left(\frac{1}{4}\right)+3\left(\frac{1}{5}\right)$, or $1\frac{7}{20}$ of the job which is more of the job then needs to be done. The answer is somewhere between 2 days and 3 days.

Translate. If they work together t days, then Jorge does $t\left(\frac{1}{4}\right)$ of the job and Carla does $t\left(\frac{1}{5}\right)$ of the job. We want some number t such that

$$\left(\frac{1}{4}+\frac{1}{5}\right)t = 1, \text{ or } \frac{9}{20}\cdot t = 1.$$

Carry out. We solve the equation.

$$\frac{9}{20}\cdot t = 1$$
$$\frac{20}{9}\cdot\frac{9}{20}\cdot t = \frac{20}{9}\cdot 1$$
$$t = \frac{20}{9}, \text{ or } 2\frac{2}{9}$$

Check. The check can be done by repeating the computations. We also have a partial check in that we expected from our familiarization step that the answer would be between 2 days and 3 days.

State. Working together, it takes them $\frac{20}{9}$ days, or $2\frac{2}{9}$ days, to complete the job.

11. ***Familiarize.*** Let t = the number of minutes it would take the two machines to copy the dissertation, working together.

Translate. We use the work principle.

$$\left(\frac{1}{12}+\frac{1}{18}\right)t = 1, \text{ or } \frac{5}{36}\cdot t = 1$$

Carry out. We solve the equation.

$$\frac{5}{36}\cdot 1 = 1$$
$$\frac{36}{5}\cdot\frac{5}{36}\cdot t = \frac{36}{5}\cdot 1$$
$$t = \frac{36}{5}$$

Check. In $\frac{36}{5}$ min, the portion of the job done is $\frac{1}{12}\cdot\frac{36}{5}+\frac{1}{18}\cdot\frac{36}{5} = \frac{3}{5}+\frac{2}{5} = 1$. The answer checks.

State. It would take the two machines $\frac{36}{5}$ min, or $7\frac{1}{5}$ min to copy the dissertation, working together.

13. ***Familiarize.*** We complete the table shown in the text.

	d = Distance	r · Speed	t Time
B & M	330	$r-14$	$\frac{330}{r-14}$
AMTRAK	400	r	$\frac{400}{r}$

Translate. Since the time must be the same for both trains, we have the equation

$$\frac{330}{r-14}=\frac{400}{r}.$$

Carry out. We first multiply by the LCD, $r(r-14)$.

$$r(r-14)\cdot\frac{330}{r-14}=r(r-14)\cdot\frac{400}{r}$$
$$330r=400(r-14)$$
$$330r=400r-5600$$
$$-70r=-5600$$
$$r=80$$

If the speed of the AMTRAK train is 80 km/h, then the speed of the B & M train is 80 – 14, or 66 km/h.

Check. The speed of the B&M train is 14 km/h slower than the speed of the AMTRAK train. At 66 km/h the B&M train travels 330 km in 330/66, or 5 hr. At 80 km/h the AMTRAK train travels 400 km in 400/80, or 5 hr. The times are the same, so the answer checks.

State. The speed of the AMTRAK train is 80 km/h, and the speed of the B&M freight train is 66 km/h.

15. ***Familiarize***. Let r = the speed of Bill's Harley, in mph. Then $r+30$ = the speed of Hillary's Lexus. We organize the information in a table using the formula time = distance/rate to fill in the last column.

	Distance	Speed	Time
Harley	75	r	$\frac{75}{r}$
Lexus	120	$r+30$	$\frac{120}{r+30}$

Translate. Since the times must be the same, we have the equation

$$\frac{75}{r}=\frac{120}{r+30}.$$

Carry out. We first multiply by the LCD, $r(r+30)$.

$$r(r+30)\cdot\frac{75}{r}=r(r+30)\cdot\frac{120}{r+30}$$
$$75(r+30)=120r$$
$$75r+2250=120r$$
$$2250=45r$$
$$50=r$$

Then $r+30=50+30=80$.

Check. The speed of the Lexus is 30 mph faster than the speed of the Harley. At 50 mph, the Harley travels 75 mi in 75/50, or 1.5 hr. At 80 mph, the Lexus travels 120 mi in 120/80, or 1.5 hr. The times are the same, so the answer checks.

State. The speed of Bill's Harley is 50 mph, and the speed of Hillary's Lexus is 80 mph.

17. ***Familiarize***. Let r = Ralph's speed, in km/h. Then Bonnie's speed is $r+3$. Also set t = the time, in hours, that Ralph and Bonnie walk. We organize the information in a table.

	Distance	Speed	Time
Ralph	7.5	r	t
Bonnie	12	$r+3$	t

Translate. We can replace the t's in the table shown above using the formula $r=d/t$.

	Distance	Speed	Time
Ralph	7.5	r	$\frac{7.5}{r}$
Bonnie	12	$r+3$	$\frac{12}{r+3}$

Since the times are the same for both walkers, we have the equation

$$\frac{7.5}{r}=\frac{12}{r+3}.$$

Carry out. We first multiply by the LCD, $r(r+3)$.

$$r(r+3)\cdot\frac{7.5}{r}=r(r+3)\cdot\frac{12}{r+3}$$
$$7.5(r+3)=12r$$
$$7.5r+22.5=12r$$
$$22.5=4.5r$$
$$5=r$$

If $r=5$, then $r+3=8$.

Check. If Ralph's speed is 5 km/h and Bonnie's speed is 8 km/h, then Bonnie walks 3 km/h faster than Ralph. Ralph's time is 7.5/5, or 1.5 hr. Bonnie's time is 12/8, or 1.5 hr. Since the times are the same, the answer checks.

State. Ralph's speed is 5 km/h, and Bonnie's speed is 8 km/h.

19. ***Familiarize***. Let t = the time it takes Caledonia to drive to town and organize the given information in a table.

	Distance	Speed	Time
Caledonia	15	r	t
Manley	20	r	$t+1$

Translate. We can replace the r's in the table above using the formula $r=d/t$.

	Distance	Speed	Time
Caledonia	15	$\frac{15}{t}$	t
Manley	20	$\frac{20}{t+1}$	$t+1$

Since the speeds are the same for both riders, we have the equation

$$\frac{15}{t}=\frac{20}{t+1}.$$

Carry out. We multiply by the LCD, $t(t+1)$.

$$t(t+1)\cdot\frac{15}{t}=t(t+1)\cdot\frac{20}{t+1}$$
$$15(t+1)=20t$$
$$15t+15=20t$$
$$15=5t$$
$$3=t$$

If $t=3$, then $t+1=3+1$, or 4.

Check. If Caledonia's time is 3 hr and Manley's time is 4 hr, then Manley's time is 1 hr more than Caledonia's. Caledonia's speed is 15/3, or 5 mph. Manley's speed is 20/4, or 5 mph. Since the speeds are the same, the answer checks.

State. It takes Caledonia 3 hr to drive to town.

21. We write a proportion and then solve it.

$$\frac{b}{6}=\frac{7}{4}$$
$$b=\frac{7}{4}\cdot 6$$
$$b=\frac{42}{4},\ \text{or } 10.5$$

$\left(\text{Note that the proportions } \frac{6}{b}=\frac{4}{7}, \frac{b}{7}=\frac{6}{4}, \text{ or } \frac{7}{b}=\frac{4}{6} \text{ could also be used.}\right)$

23. We write a proportion and then solve it.

$$\frac{4}{f}=\frac{6}{4}$$
$$4f\cdot\frac{4}{f}=4f\cdot\frac{6}{4}$$
$$16=6f$$
$$\frac{8}{3}=f \qquad \text{Simplifying}$$

$\left(\text{One of the following proportions could also be used: } \frac{f}{4}=\frac{4}{6}, \frac{4}{f}=\frac{9}{6}, \frac{f}{4}=\frac{6}{9}, \frac{4}{9}=\frac{f}{6}, \frac{9}{4}=\frac{6}{f}\right)$

25. From the blueprint we see that 9 in. represents 36 ft and that p in. represent 15 ft. We use a proportion to find p.

$$\frac{9}{36}=\frac{p}{15}$$
$$180\cdot\frac{9}{36}=180\cdot\frac{p}{15}$$
$$45=12p$$
$$\frac{15}{4}=p,\ \text{or}$$
$$3\frac{3}{4}=p$$

The length of p is $3\frac{3}{4}$ in.

27. From the blueprint we see that 9 in. represents 36 ft and that 5 in. represents r ft. We use a proportion to find r.

$$\frac{9}{36}=\frac{5}{r}$$
$$36r\cdot\frac{9}{36}=36r\cdot\frac{5}{r}$$
$$9r=180$$
$$r=20$$

The length of r is 20 ft.

29. Consider the two similar right triangles in the drawing. One has legs 1.5 ft and 18 ft. The other has legs h ft and 32 ft. We use a proportion to find h.

$$\frac{1.5}{18}=\frac{h}{32}$$
$$288\cdot\frac{1.5}{18}=288\cdot\frac{h}{32}$$
$$24=9h$$
$$\frac{8}{3}=h,\ \text{or}$$
$$2\frac{2}{3}=h$$

The length of h is $2\frac{2}{3}$ ft.

31.
$$\frac{a}{b}=\frac{c}{d}$$
$$\frac{8}{5}=\frac{6}{d}$$
$$5d\cdot\frac{8}{5}=5d\cdot\frac{6}{d}$$
$$8d=30$$
$$d=\frac{30}{8}=\frac{15}{4}\ \text{cm, or } 3.75\ \text{cm}$$

33. Consider the two similar right triangles in the drawing. One has legs 5 and 7. The other has legs 9 and r. We use a proportion to find r.

$$\frac{5}{7}=\frac{9}{r}$$
$$7r\cdot\frac{5}{7}=7r\cdot\frac{9}{r}$$
$$5r=63$$
$$r=\frac{63}{5},\ \text{or } 12.6$$

35. ***Familiarize***. A rate of 140 steps per minute corresponds to a speed of 4 mph, and we wish to find the number of steps per minute S that correspond to a speed of 3 mph. We can use a proportion.

Translate.

$$\begin{array}{r} \text{Steps per min}\rightarrow \\ \text{Speed}\rightarrow \end{array}\frac{140}{4}=\frac{S}{3}\begin{array}{l}\leftarrow\text{Steps per min} \\ \leftarrow\text{Speed}\end{array}$$

Carry out. We solve the proportion.

$$12\cdot\frac{140}{4}=12\cdot\frac{S}{3}$$
$$420=4S$$
$$105=S$$

Check. $\dfrac{140}{4} = 35,\ \dfrac{105}{3} = 35$

The ratios are the same so the answer checks.

State. 105 steps per minute corresponds to a speed of 3 mph.

37. ***Familiarize.*** The coffee beans from 14 trees are required to produce 7.7 kilograms of coffee, and we wish to find how many trees T are required to produce 308 kilograms of coffee. We can use a proportion.

Translate.

$$\begin{array}{r} \text{Trees} \rightarrow \\ \text{Kilograms} \rightarrow \end{array} \dfrac{T}{308} = \dfrac{14}{7.7} \begin{array}{l} \leftarrow \text{Trees} \\ \leftarrow \text{Kilograms} \end{array}$$

Carry out. We solve the proportion.

$$308 \cdot \frac{T}{308} = 308 \cdot \frac{14}{7.7}$$

$$T = \frac{4312}{7.7}$$

$$T = 560$$

Check. $\dfrac{560}{308} = 1.8\overline{1},\ \dfrac{14}{7.7} = 1.8\overline{1}$

The ratios are the same, so the answer checks.

State. 560 trees are required to produce 308 kg of coffee.

39. ***Familiarize.*** 10 cm^3 of human blood contains 1.2 grams of hemoglobin, and we wish to find how many grams of hemoglobin H are contained in 16 cm^3 of the same blood. We can use a proportion.

Translate.

$$\begin{array}{r} \text{Grams} \rightarrow \\ \text{cm}^3 \rightarrow \end{array} \dfrac{H}{16} = \dfrac{1.2}{10} \begin{array}{l} \leftarrow \text{Grams} \\ \leftarrow \text{cm}^3 \end{array}$$

Carry out. We solve the proportion.

We multiply by 16 to get H alone.

$$16 \cdot \frac{H}{16} = 16 \cdot \frac{1.2}{10}$$

$$H = \frac{19.2}{10}$$

$$H = 1.92$$

Check.

$\dfrac{1.92}{16} = 0.12,\ \dfrac{1.2}{10} = 0.12$

The ratios are the same, so the answer checks.

State. 16 cm^3 of the same blood would contain 1.92 grams of hemoglobin.

41. ***Familiarize.*** U.S. women earn 77 cents for each dollar earned by a man. This gives us one ratio, expressed in dollars: $\dfrac{0.77}{1}$. If a male sales manager earns $42,000, we want to find how much a female would earn for comparable work. This gives us a second ratio, also expressed in dollars: $\dfrac{F}{42,000}$.

Translate. We translate to a proportion.

$$\begin{array}{r} \text{Female's earnings} \rightarrow \\ \text{Male's earnings} \rightarrow \end{array} \dfrac{0.77}{1} = \dfrac{F}{42,000} \begin{array}{l} \leftarrow \text{Female's earnings} \\ \leftarrow \text{Male's earnings} \end{array}$$

Carry out. We solve the proportion.

$$42,000 \cdot \frac{0.77}{1} = 42,000 \cdot \frac{F}{42,000}$$

$$32,340 = F$$

Check.

$\dfrac{0.77}{1} = 0.77,\ \dfrac{32,340}{42,000} = 0.77$

The ratios are the same, so the answer checks.

State. If a male sales manager earns $42,000, a female would earn $32,340 for comparable work.

43. ***Familiarize.*** A helmet with a circumference of 53.8 cm corresponds to size $6\frac{3}{4}$, or 6.75, and we wish to find the size S to which a 59.8-cm helmet corresponds. We can use a proportion.

Translate.

$$\begin{array}{r} \text{Centimeters} \rightarrow \\ \text{Size} \rightarrow \end{array} \dfrac{53.8}{6.75} = \dfrac{59.8}{S} \begin{array}{l} \leftarrow \text{Centimeters} \\ \leftarrow \text{Size} \end{array}$$

Carry out. We solve the proportion. We multiply by $6.75S$ to clear the fractions.

$$6.75S \cdot \frac{53.8}{6.75} = \frac{59.8}{S} \cdot 6.75S$$

$$53.85 = 403.65$$

$$S \approx 7.5, \text{ or } 7\frac{1}{2}$$

Check. $\dfrac{53.8}{6.75} \approx 7.97,\ \dfrac{59.8}{7.5} \approx 7.97$

Since the ratios are the same, the answer checks.

State. The U.S. size $7\frac{1}{2}$ corresponds to a 59.8-cm helmet.

45. ***Familiarize.*** Let D = the number of duds you would expect in a sample of 320 firecrackers. We can use a proportion to find D.

Translate.

$$\begin{array}{r} \text{Duds} \rightarrow \\ \text{Firecrackers} \rightarrow \end{array} \dfrac{9}{144} = \dfrac{D}{320} \begin{array}{l} \leftarrow \text{Duds} \\ \leftarrow \text{Firecrackers} \end{array}$$

Carry out. We solve the proportion. We multiply by the LCD, 2880.

$$2880 \cdot \frac{9}{144} = 2880 \cdot \frac{D}{320}$$

$$180 = 9D$$

$$20 = D$$

Check. $\dfrac{9}{144} = 0.0625,\ \dfrac{20}{320} = 0.0625$

The ratios are the same, so the answer checks.

State. You would expect 20 duds in a sample of 320 fireworks.

47. ***Familiarize.*** The ratio of trout tagged to the total number of trout in the lake, T, is $\frac{112}{T}$. Of the 82 trout caught later, there were 32 tagged trout. The ratio of tagged trout to trout caught is $\frac{32}{82}$.

Translate. We translate to a proposition.

Trout tagged originally $\longrightarrow$ $\frac{112}{T} = \frac{32}{82}$ $\longleftarrow$ Tagged trout caught later
Trout in lake $\longrightarrow$ $\longleftarrow$ Trout caught later

Carry out. We solve the proportion.

$$82T \cdot \frac{112}{T} = 82T \cdot \frac{32}{82}$$
$$82 \cdot 112 = T \cdot 32$$
$$9184 = 32T$$
$$\frac{9184}{32} = T$$
$$287 = T$$

Check.
$\frac{112}{287} \approx 0.39,\ \frac{32}{82} \approx 0.39$

The ratios are the same, so the answer checks.

State. We estimate that there are 287 trout in the lake.

49. ***Familiarize.*** The ratio of moose tagged to the total number of moose in the park, M, is $\frac{69}{M}$. Of the 40 moose caught later, 15 are tagged. The ratio of tagged moose to moose caught is $\frac{15}{40}$.

Translate. We translate to a proportion.

Moose originally tagged $\rightarrow$ $\frac{69}{M} = \frac{15}{40}$ $\leftarrow$ Tagged moose caught later
Moose in forest $\rightarrow$ $\leftarrow$ Moose caught later

Carry out. We solve the proportion. We multiply by the LCD, $40M$.

$$40M \cdot \frac{69}{M} = 40M \cdot \frac{15}{40}$$
$$40 \cdot 69 = M \cdot 15$$
$$2760 = 15M$$
$$\frac{2760}{15} = M$$
$$184 = M$$

Check.
$\frac{69}{184} = 0.375,\ \frac{15}{40} = 0.375$

The ratios are the same, so the answer checks.

State. We estimate that there are 184 moose in the park.

51. ***Familiarize.*** The ratio of bears tagged to the total number of bears in the forest, B, is $\frac{16}{B}$. Of the 20 bears caught later, 8 are tagged. The ratio of tagged bears to bears caught is $\frac{8}{20}$.

Translate. We translate to a proportion.

Bears originally tagged $\rightarrow$ $\frac{16}{B} = \frac{8}{20}$ $\leftarrow$ Tagged bears caught later
Bears in forest $\rightarrow$ $\leftarrow$ Bears caught later

Carry out. We solve the proportion. We multiply by the LCD, $20B$.

$$20B \cdot \frac{16}{B} = 20B \cdot \frac{8}{20}$$
$$320 = 8B$$
$$40 = B$$

Check. $\frac{16}{40} = 0.4,\ \frac{8}{20} = 0.4$

The ratios are the same, so the answer checks.

State. We estimate that there are 40 bears in the forest.

53. ***Familiarize.*** The ratio of the weight of an object on the moon to the weight of an object on the earth is 0.16 to 1.

a) We wish to find how much a 12-ton rocket would weigh on the moon.

b) We wish to find how much a 180-lb astronaut would weigh on the moon.

Translate. We translate to proportions.

a) Weight on the moon $\rightarrow$ $\frac{0.16}{1} = \frac{T}{12}$ $\leftarrow$ Weight on the moon
Weight on earth $\rightarrow$ $\leftarrow$ Weight on earth

b) Weight on the moon $\rightarrow$ $\frac{0.16}{1} = \frac{P}{180}$ $\leftarrow$ Weight on the moon
Weight on earth $\rightarrow$ $\leftarrow$ Weight on earth

Carry out. We solve each proportion.

a) $\frac{0.16}{1} = \frac{T}{12}$ b) $\frac{0.16}{1} = \frac{P}{180}$

$12(0.16) = T$ $180(0.16) = P$

$1.92 = T$ $28.8 = P$

Check. $\frac{0.16}{1} = 0.16,\ \frac{1.92}{12} = 0.16,\ \frac{28.8}{180} = 0.16$

The ratios are the same, so the answer checks.

State.

a) A 12-ton rocket would weigh 1.92 tons on the moon.

b) A 180-lb astronaut would weigh 28.8 lb on the moon.

55. *Writing Exercise*

57. Graph: $y = 2x - 6$.

We select some x-values and compute y-values.

If $x = 1$, then $y = 2 \cdot 1 - 6 = -4$.

If $x = 3$, then $y = 2 \cdot 3 - 6 = 0$.

If $x = 5$, then $y = 2 \cdot 5 - 6 = 4$.

x	y	(x, y)
1	-4	$(1, -4)$
3	0	$(3, 0)$
5	4	$(5, 4)$

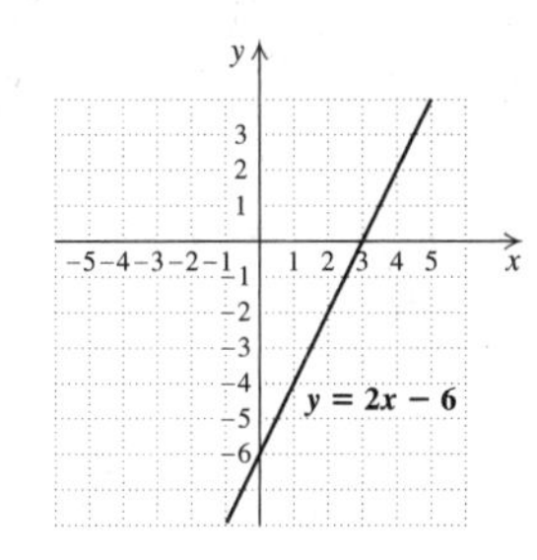

59. Graph: $3x + 2y = 12$.

We can replace either variable with a number and then calculate the other coordinate. We will find the intercepts and one other point.

If $y = 0$, we have:

$$3x + 2 \cdot 0 = 12$$
$$3x = 12$$
$$x = 4$$

The x-intercept is $(4, 0)$.

If $x = 0$, we have:

$$3 \cdot 0 + 2y = 12$$
$$2y = 12$$
$$y = 6$$

The y-intercept is $(0, 6)$.

If $y = -3$, we have:

$$3x + 2(-3) = 12$$
$$3x - 6 = 12$$
$$3x = 18$$
$$x = 6$$

The point $(6, -3)$ is on the graph.

We plot these points and draw a line through them.

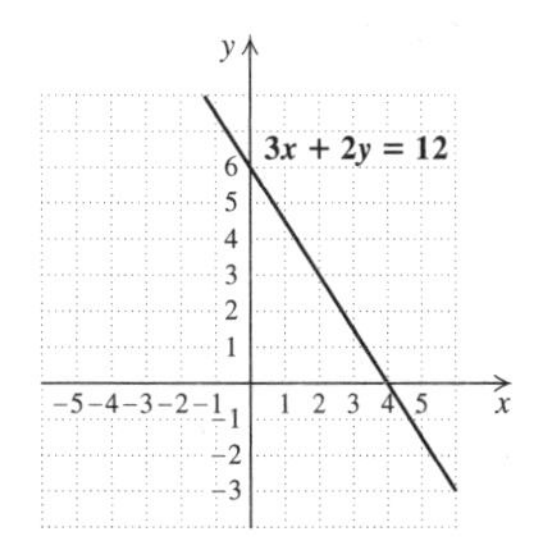

61. Graph: $y = -\frac{3}{4}x + 2$

We select some x-values and compute y-values. We use multiples of 4 to avoid fractions.

If $x = -4$, then $y = -\frac{3}{4}(-4) + 2 = 5$.

If $x = 0$, then $y = -\frac{3}{4} \cdot 0 + 2 = 2$.

If $x = 4$, then $y = -\frac{3}{4} \cdot 4 + 2 = -1$.

x	y	(x, y)
-4	5	$(-4, 5)$
0	2	$(0, 2)$
4	-1	$(4, -1)$

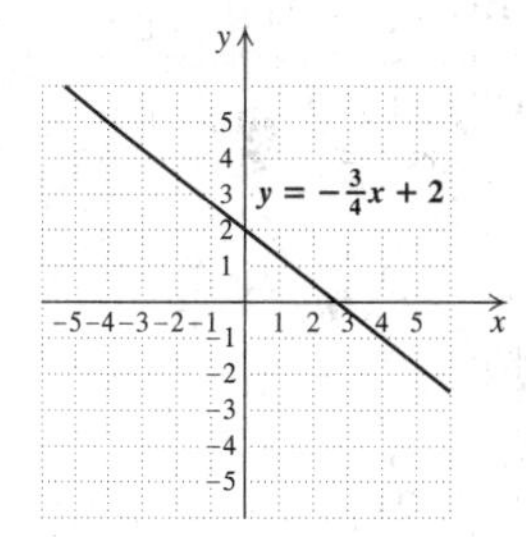

63. *Writing Exercise*

65. ***Familiarize***. Let t = the number of hours it would take Ricki to sew the quilt, working alone. Then $t + 6$ = the number of hours it would take Maura, working alone. In 4 hr Ricki does $4 \cdot \frac{1}{t}$ of the job and Maura does $4 \cdot \frac{1}{t+6}$ of the job, and together they do 1 complete job.

Translate. We use the information above to write an equation.

$$\frac{4}{t} + \frac{4}{t+6} = 1$$

Carry out. We solve the equation.

$$\frac{4}{t} + \frac{4}{t+6} = 1, \text{ LCD} = t(t+6)$$
$$t(t+6)\left(\frac{4}{t} + \frac{4}{t+6}\right) = t(t+6) \cdot 1$$
$$4(t+6) + 4t = t(t+6)$$
$$4t + 24 + 4t = t^2 + 6t$$
$$0 = t^2 - 2t - 24$$
$$0 = (t-6)(t+4)$$

$$t - 6 = 0 \quad or \quad t + 4 = 0$$
$$t = 6 \quad or \quad t = -4$$

Check. Time cannot be negative in this application, so we check only 6. If Ricki can sew the quilt working alone in 6 hr, then it would take Maura $6 + 6$, or 12 hr, working alone. In 4 hr they would do

$$\frac{4}{6} + \frac{4}{12} = \frac{2}{3} + \frac{1}{3} = 1 \text{ complete job.}$$

The answer checks.

State. It would take Ricki 6 hr to sew the quilt working alone, and it would take Maura 12 hr working alone.

67. ***Familiarize***. The job takes Rosina 8 days working alone and Ng 10 days working alone. Let x represent the number of days it would take Oscar working alone. Then in 1 day Rosina does $\frac{1}{8}$ of the job, Ng does $\frac{1}{10}$ of the job, and Oscar does $\frac{1}{x}$ of the job. In 1 day they would complete $\frac{1}{8} + \frac{1}{10} + \frac{1}{x}$ of the job, and in 3 days they would complete $3\left(\frac{1}{8} + \frac{1}{10} + \frac{1}{x}\right)$, or $\frac{3}{8} + \frac{3}{10} + \frac{3}{x}$.

Translate. The amount done in 3 days is one entire job, so we have

$$\frac{3}{8} + \frac{3}{10} + \frac{3}{x} = 1.$$

Carry out. We solve the equation.

$$\frac{3}{8}+\frac{3}{10}+\frac{3}{x}=1,\ \text{LCD}=40x$$
$$40x\left(\frac{3}{8}+\frac{3}{10}+\frac{3}{x}\right)=40x\cdot 1$$
$$40x\cdot\frac{3}{8}+40x\cdot\frac{3}{10}+40x\cdot\frac{3}{x}=40x$$
$$15x+12x+120=40x$$
$$120=13x$$
$$\frac{120}{13}=x$$

Check. If it takes Oscar $\frac{120}{13}$, or $9\frac{3}{13}$ days, to complete the job, then in one day Oscar does $\dfrac{1}{\frac{120}{13}}$, or $\frac{13}{120}$, of the job, and in 3 days he does $3\left(\frac{13}{120}\right)$, or $\frac{13}{40}$, of the job. The portion of the job done by Rosina, Ng, and Oscar in 3 days is

$\frac{3}{8}+\frac{3}{10}+\frac{13}{40}=\frac{15}{40}+\frac{12}{40}+\frac{13}{40}=\frac{40}{40}=1$ entire job.

The answer checks.

State. It will take Oscar $9\frac{3}{13}$ days to write the program working alone.

69. Russ can reshingle the roof in 12 hr and it takes him half that time when Joan works with him. Thus, Russ and Joan have each done half of the job in 6 hr, so it would take Joan 12 hr to reshingle the roof, working alone.

71. ***Familiarize.*** The correct ratio of oil to gasoline is 3.2/160, or 0.02. The ratio in Gus' original mixture is 5.6/200, or 0.028. Since this is a larger number than 0.02, Gus needs to add more gasoline to make the ratio lower. Let g = the number of ounces of gasoline Gus should add.

Translate. We translate to an equation.

$$\begin{array}{r}\text{Oil}\rightarrow\\ \text{Gasoline}\rightarrow\end{array}\frac{5.6}{200+g}=0.02$$

Carry out. We solve the equation.

$$(200+g)\cdot\frac{5.6}{200+g}=(200+g)(0.02)$$
$$5.6=4+0.02g$$
$$1.6=0.02g$$
$$80=g$$

Check. If Gus adds an additional 80 oz of gasoline, the ratio of oil to gasoline is $\frac{5.6}{200+80}=\frac{5.6}{280}=0.02$, the correct ratio. The answer checks.

State. Gus should add 80 oz of gasoline.

73. ***Familiarize.*** We organize the information in a table. Let r = the speed on the first part of the trip and t = the time driven at that speed.

	Distance	Speed	Time
First part	30	r	t
Second part	30	$r+15$	$1-t$

Translate. From the rows of the table we obtain two equations:

$$30=rt$$
$$30=(r+15)(1-t)$$

We solve each equation for t and set the results equal:

Solving $30=rt$ for t: $t=\frac{30}{r}$

Solving $20=(r+15)(1-t)$ for t: $t=1-\frac{20}{r+15}$

Then $\frac{30}{r}=1-\frac{20}{r+15}$.

Carry out. We first multiply the equation by the LCD, $r(r+15)$.

$$r(r+15)\cdot\frac{30}{r}=r(r+15)\left(1-\frac{20}{r+15}\right)$$
$$30(r+15)=r(r+15)-20r$$
$$30r+450=r^2+15r-20r$$
$$0=r^2-35r-450$$
$$0=(r-45)(r+10)$$
$$r-45=0\quad or\quad r+10=0$$
$$r=45\quad or\quad r=-10$$

Check. Since the speed cannot be negative, we only check 45. If $r=45$, then the time for the first part is $\frac{30}{45}$, or $\frac{2}{3}$ hr. If $r=45$, then $r+15=60$ and the time for the second part is $\frac{20}{60}$, or $\frac{1}{3}$ hr. The total time is $\frac{2}{3}+\frac{1}{3}$, or 1 hour. The value checks.

State. The speed for the first 30 miles was 45 mph.

75. ***Familiarize.*** Let x = the numerator in the equivalent ratio. Then $104-x$ = the denominator.

Translate. The ratios $\frac{9}{17}$ and $\frac{x}{104-x}$ are equivalent, so we write a proportion.

$$\frac{9}{17}=\frac{x}{104-x}$$

Carry out. We solve the proportion. We multiply by the LCD, $17(104-x)$.

$$17(104-x)\cdot\frac{9}{17}=17(104-x)\cdot\frac{x}{104-x}$$
$$9(104-x)=17x$$
$$936-9x=17x$$
$$936=26x$$
$$36=x$$

Then $104-x=104-36=68$.

Check. $\frac{36}{68}=\frac{4\cdot 9}{4\cdot 17}=\frac{9}{17}$, so the ratios are equivalent.

State. The numerator will be 36, and the denominator will be 68.

77. Find a second proportion:

$$\frac{A}{B} = \frac{C}{D} \quad \text{Given}$$

$$\frac{D}{A} \cdot \frac{A}{B} = \frac{D}{A} \cdot \frac{C}{D} \quad \text{Multiplying by } \frac{D}{A}$$

$$\frac{D}{B} = \frac{C}{A}$$

Find a third proportion:

$$\frac{A}{B} = \frac{C}{D} \quad \text{Given}$$

$$\frac{B}{C} \cdot \frac{A}{B} = \frac{B}{C} \cdot \frac{C}{D} \quad \text{Multiplying by } \frac{B}{C}$$

$$\frac{A}{C} = \frac{B}{D}$$

Find a fourth proportion:

$$\frac{A}{B} = \frac{C}{D} \quad \text{Given}$$

$$\frac{DB}{AC} \cdot \frac{A}{B} = \frac{DB}{AC} \cdot \frac{C}{D} \quad \text{Multiplying by } \frac{DB}{AC}$$

$$\frac{D}{C} = \frac{B}{A}$$

79. *Writing Exercise*

Chapter 7

Systems and More Graphing

Exercise Set 7.1

1. both; see page 430 in the text.

3. consistent; see page 433 in the text.

5. We check by substituting alphabetically 3 for x and 2 for y.

$$\begin{array}{c|c} \multicolumn{2}{c}{2x+3y=12} \\ \hline 2\cdot 3+3\cdot 2 & 12 \\ 6+6 & \\ \end{array} \qquad \begin{array}{c|c} \multicolumn{2}{c}{x-4y=-5} \\ \hline 3-4\cdot 2 & -5 \\ 3-8 & \\ \end{array}$$

$$12 \stackrel{?}{=} 12 \ \text{TRUE} \qquad -5 \stackrel{?}{=} -5 \ \text{TRUE}$$

The ordered pair $(3,2)$ is a solution of each equation. Therefore it is a solution of the system of equations.

7. We check by substituting alphabetically 3 for a and 2 for b.

$$\begin{array}{c|c} \multicolumn{2}{c}{3b-2a=0} \\ \hline 3\cdot 2-2\cdot 3 & 0 \\ 6-6 & \\ \end{array} \qquad \begin{array}{c|c} \multicolumn{2}{c}{b+2a=15} \\ \hline 2+2\cdot 3 & 15 \\ 2+6 & \\ \end{array}$$

$$0 \stackrel{?}{=} 0 \ \text{TRUE} \qquad 8 \stackrel{?}{=} 15 \ \text{FALSE}$$

The ordered pair $(3,2)$ is not a solution of $b+2a=15$. Therefore it is not a solution of the system of equations.

9. We check by substituting alphabetically 15 for x and 20 for y.

$$\begin{array}{c|c} \multicolumn{2}{c}{3x-2y=5} \\ \hline 3\cdot 15-2\cdot 20 & 5 \\ 45-40 & \\ \end{array}$$

$$5 \stackrel{?}{=} 5 \ \text{TRUE}$$

$$\begin{array}{c|c} \multicolumn{2}{c}{6x-5y=-10} \\ \hline 6\cdot 15-5\cdot 20 & -10 \\ 90-100 & \\ \end{array}$$

$$-10 \stackrel{?}{=} -10 \ \text{TRUE}$$

The ordered pair $(15,20)$ is a solution of each equation. Therefore it is a solution of the system of equations.

11. We graph the equations.

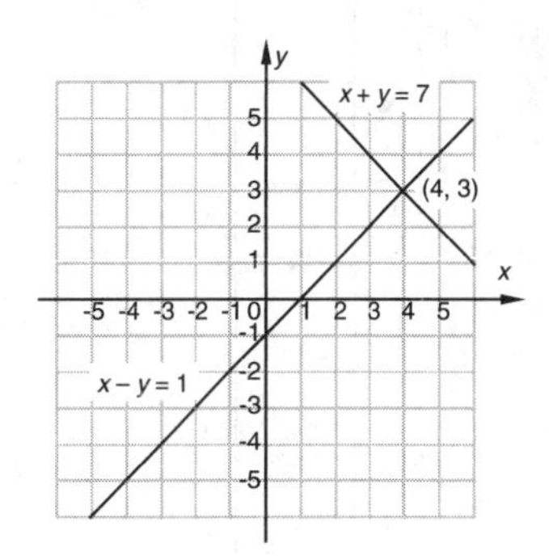

The "apparent" solution of the system, $(4,3)$, should be checked in both equations.

Check:

$$\begin{array}{c|c} \multicolumn{2}{c}{x+y=7} \\ \hline 4+3 & 7 \\ \end{array} \qquad \begin{array}{c|c} \multicolumn{2}{c}{x-y=1} \\ \hline 4-3 & 1 \\ \end{array}$$

$$7 \stackrel{?}{=} 7 \ \text{TRUE} \qquad 1 \stackrel{?}{=} 1 \ \text{TRUE}$$

The solution is $(4,3)$.

13. We graph the equations.

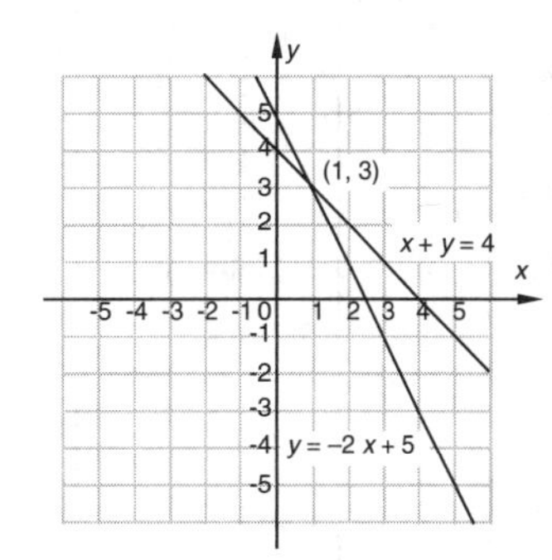

The "apparent" solution of the system, $(1,3)$, should be checked in both equations.

Check:

$$\begin{array}{c|c} \multicolumn{2}{c}{y=-2x+5} \\ \hline 3 & -2\cdot 1+5 \\ & -2+5 \\ \end{array} \qquad \begin{array}{c|c} \multicolumn{2}{c}{x+y=4} \\ \hline 1+3 & 4 \\ \end{array}$$

$$3 \stackrel{?}{=} 3 \ \text{TRUE} \qquad 4 \stackrel{?}{=} 4 \ \text{TRUE}$$

The solution is $(1,3)$.

15. We graph the equations.

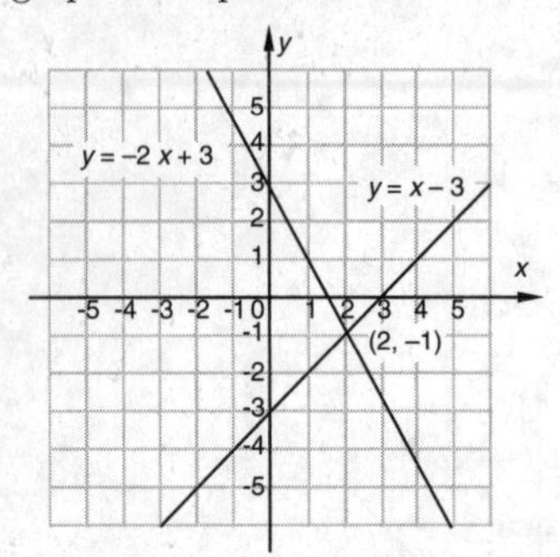

The "apparent" solution of the system, $(2,-1)$, should be checked in both equations.

Check:

$$\begin{array}{c|c} \multicolumn{2}{c}{y = x - 3} \\ \hline -1 & 2-3 \\ \multicolumn{2}{c}{-1 \stackrel{?}{=} -1 \quad \text{TRUE}} \end{array} \qquad \begin{array}{c|c} \multicolumn{2}{c}{y = -2x + 3} \\ \hline -1 & -2 \cdot 2 + 3 \\ & -4+3 \\ \multicolumn{2}{c}{-1 \stackrel{?}{=} -1 \quad \text{TRUE}} \end{array}$$

The solution is $(2,-1)$.

17. We graph the equations.

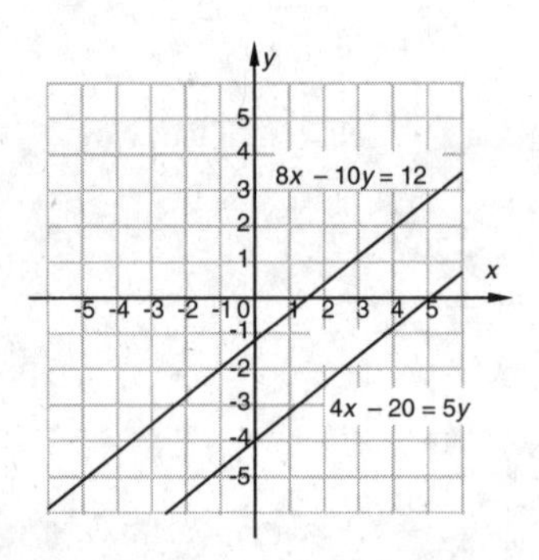

The lines are parallel. There is no solution.

19. We graph the equations.

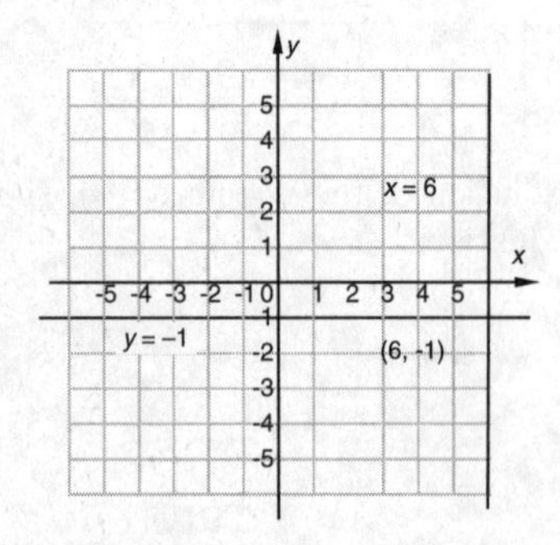

The "apparent" solution of the system, $(6,-1)$, should be checked in both equations.

Check:

$$\begin{array}{c} x = 6 \\ \hline 6 \stackrel{?}{=} 6 \quad \text{TRUE} \end{array} \qquad \begin{array}{c} y = -1 \\ \hline -1 \stackrel{?}{=} -1 \quad \text{TRUE} \end{array}$$

The solution is $(6,-1)$.

21. We graph the equations.

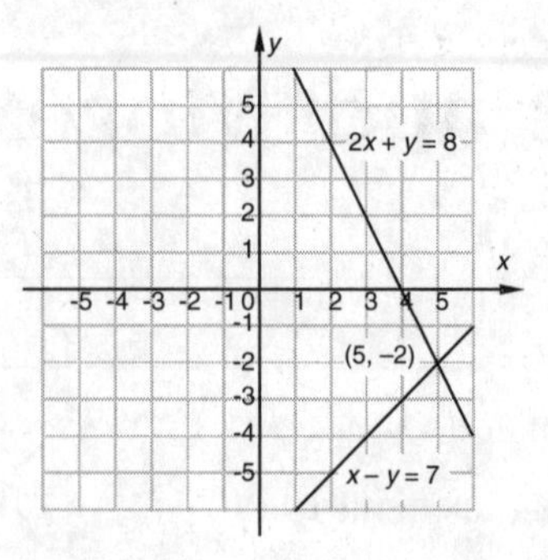

The "apparent" solution of the system, $(5,-2)$, should be checked in both equations.

Check:

$$\begin{array}{c|c} \multicolumn{2}{c}{2x + y = 8} \\ \hline 2 \cdot 5 + (-2) & 8 \\ 10 - 2 & \\ \multicolumn{2}{c}{8 \stackrel{?}{=} 8 \quad \text{TRUE}} \end{array}$$

$$\begin{array}{c|c} \multicolumn{2}{c}{x - y = 7} \\ \hline 5 - (-2) & 7 \\ 5 + 2 & \\ \multicolumn{2}{c}{7 \stackrel{?}{=} 7 \quad \text{TRUE}} \end{array}$$

The solution is $(5,-2)$.

23. We graph the equations.

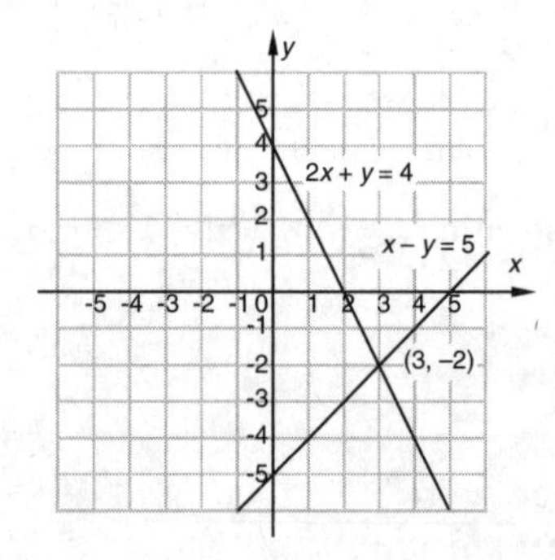

The "apparent" solution of the system, $(3,-2)$, should be checked in both equations.

Check:

$$\begin{array}{c|c} \multicolumn{2}{c}{x - y = 5} \\ \hline 3 - (-2) & 5 \\ 3 + 2 & \\ \multicolumn{2}{c}{5 \stackrel{?}{=} 5 \quad \text{TRUE}} \end{array}$$

$$\begin{array}{c|c} \multicolumn{2}{c}{2x + y = 4} \\ \hline 2 \cdot 3 + (-2) & 4 \\ 6 - 2 & \\ \multicolumn{2}{c}{4 \stackrel{?}{=} 4 \quad \text{TRUE}} \end{array}$$

The solution is $(3,-2)$.

25. We graph the equations.

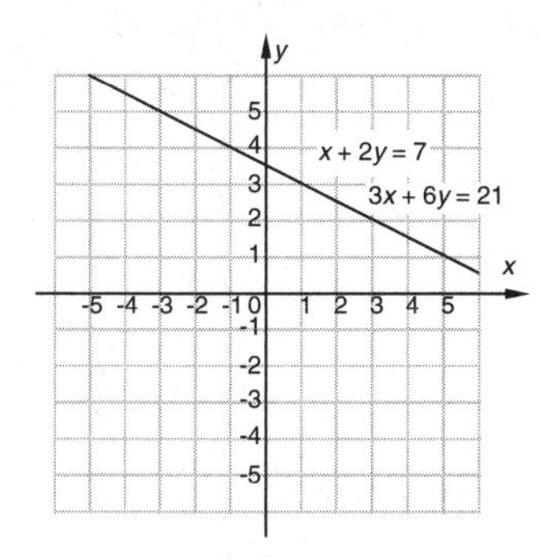

We see that the equations represent the same line. This means that any solution of one equation is a solution of the other equation as well. Thus, there is an infinite number of solutions.

27. We graph the equations.

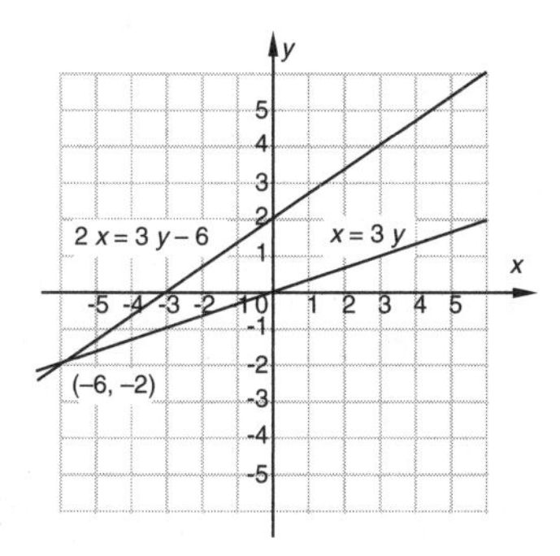

The "apparent" solution of the system, $(-6,-2)$, should be checked in both equations.

Check:

$2x = 3y - 6$	
$2(-6)$	$3(-2) - 6$
-12	$-6 - 6$

$-12 \stackrel{?}{=} -12$ TRUE

$x = 3y$	
-6	$3(-2)$

$-6 \stackrel{?}{=} -6$ TRUE

The solution is $(-6,-2)$.

29. $y = \frac{1}{5}x + 4,$

$2y = \frac{2}{5}x + 8$

Observe that we can obtain the second equation by multiplying both sides of the first equation by 2. Thus, the equations are dependent and there is an infinite number of solutions.

31. We graph the equations.

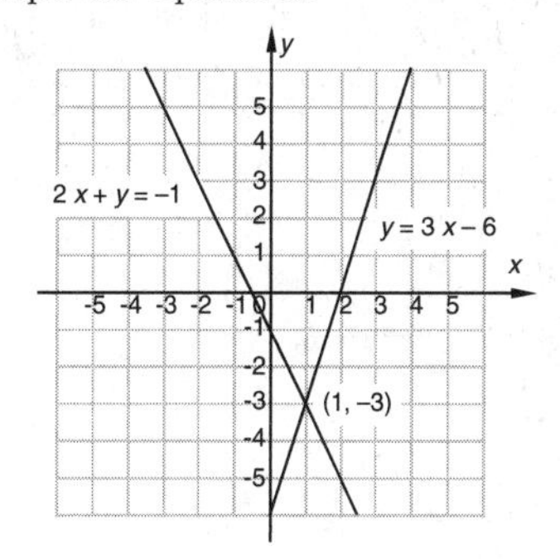

The "apparent" solution of the system, $(1,-3)$, should be checked in both equations.

$2x + y = -1$	
$2 \cdot 1 + (-3)$	-1
$2 - 3$	

$-1 \stackrel{?}{=} -1$ TRUE

$y = 3x - 6$	
-3	$3 \cdot 1 - 6$
	$3 - 6$

$-3 \stackrel{?}{=} -3$ TRUE

The solution is $(1,-3)$.

33. We graph the equations.

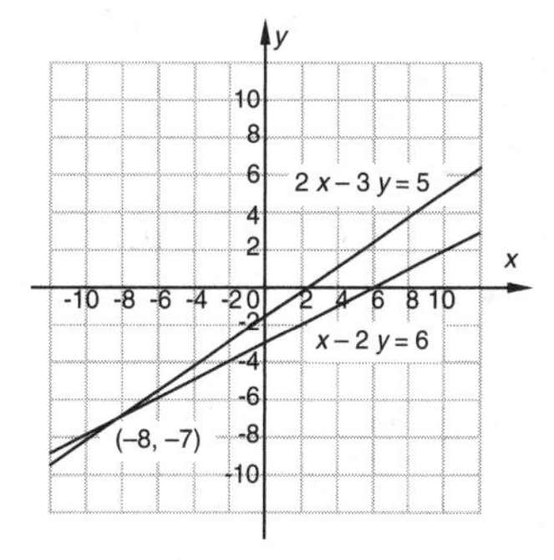

The "apparent" solution of the system, $(-8,-7)$, should be checked in both equations.

Check:

$2x - 3y = 5$	
$2(-8) - 3(-7)$	5
$-16 + 21$	

$5 \stackrel{?}{=} 5$ TRUE

$x - 2y = 6$	
$-8 - 2(-7)$	6
$-8 + 14$	

$6 \stackrel{?}{=} 6$ TRUE

The solution is $(-8,-7)$.

35. We graph the equations.

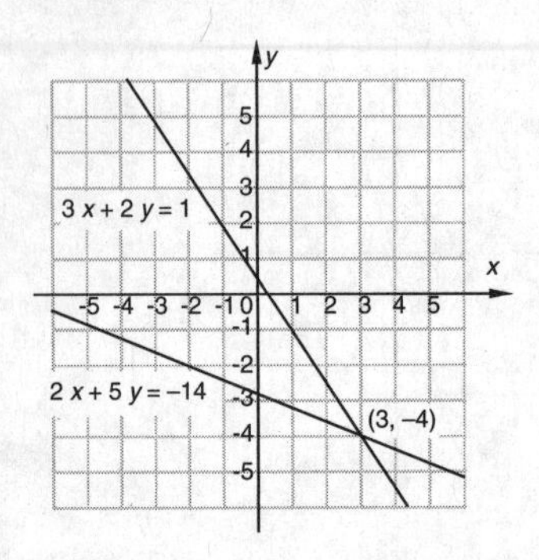

The "apparent" solution of the system, $(3, -4)$, should be checked in both equations.

Check:

$$\begin{array}{c|c} \multicolumn{2}{c}{3x + 2y = 1} \\ \hline 3 \cdot 3 + 2(-4) & 1 \\ 9 - 8 & \\ \end{array}$$
$$1 \stackrel{?}{=} 1 \quad \text{TRUE}$$

$$\begin{array}{c|c} \multicolumn{2}{c}{2x + 5y = -14} \\ \hline 2 \cdot 3 + 5(-4) & -14 \\ 6 - 20 & \\ \end{array}$$
$$-14 \stackrel{?}{=} -14 \quad \text{TRUE}$$

The solution is $(3, -4)$.

37. We graph the equations.

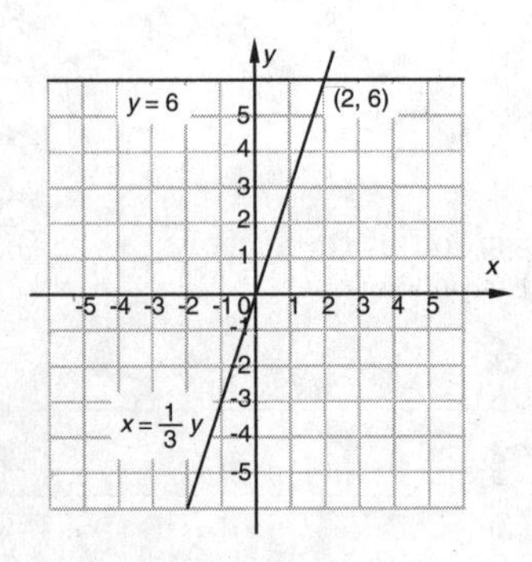

The "apparent" solution of the system, $(2, 6)$, should be checked in both equations.

Check:

$$\begin{array}{c|c} \multicolumn{2}{c}{x = \frac{1}{3}y} \\ \hline 2 & \frac{1}{3} \cdot 6 \\ \end{array}$$
$$2 \stackrel{?}{=} 2 \quad \text{TRUE}$$

$$\begin{array}{c} y = 6 \\ \hline \end{array}$$
$$6 \stackrel{?}{=} 6 \quad \text{TRUE}$$

The solution is $(2, 6)$.

39. Graph $y = 2x - 1$ and $y = 3$.

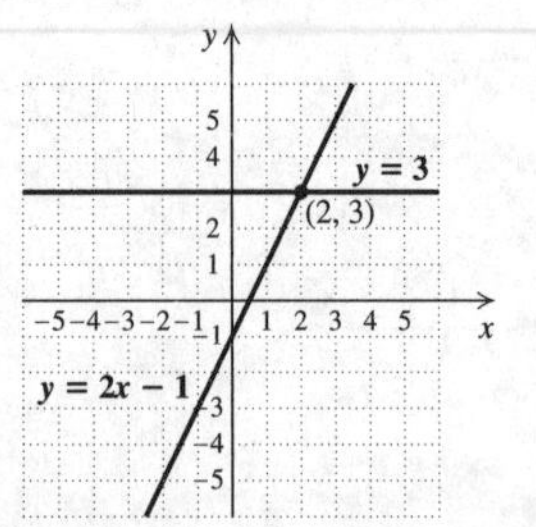

The graphs intersect at $(2, 3)$, indicating that when $x = 2$ we have $2x - 1 = 3$. The solution is 2.

41. Graph $y = x - 4$ and $y = 6 - x$.

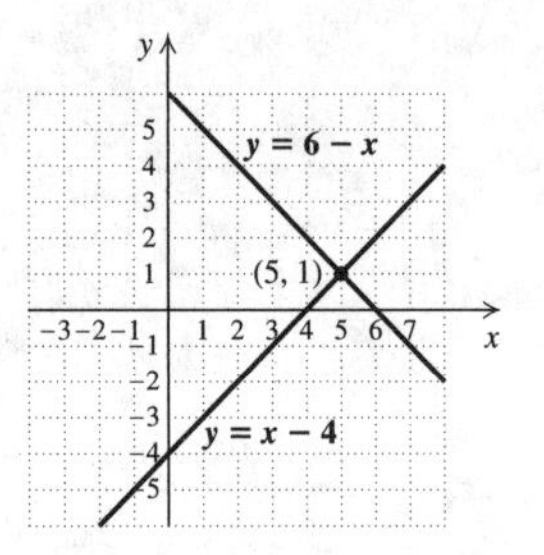

The graphs intersect at $(5, 1)$, indicating that when $x = 5$ we have $x - 4 = 6 - x$. The solution is 5.

43. Graph $y = 2x - 1$ and $y = -x + 5$.

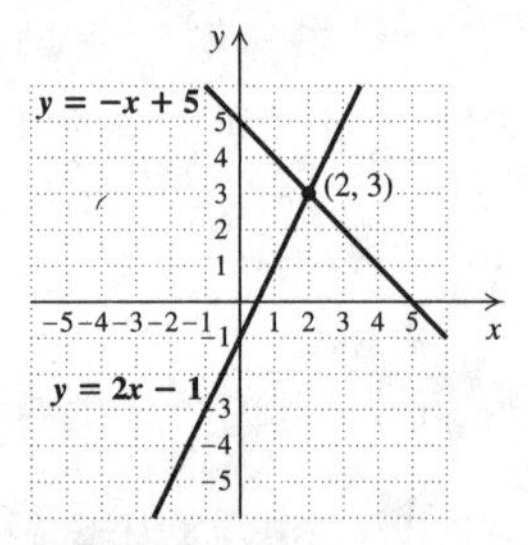

The graphs intersect at $(2, 3)$, indicating that when $x = 2$ we have $2x - 1 = -x + 5$. The solution is 2.

45. Graph $y = \frac{1}{2}x + 3$ and $y = -\frac{1}{2}x - 1$.

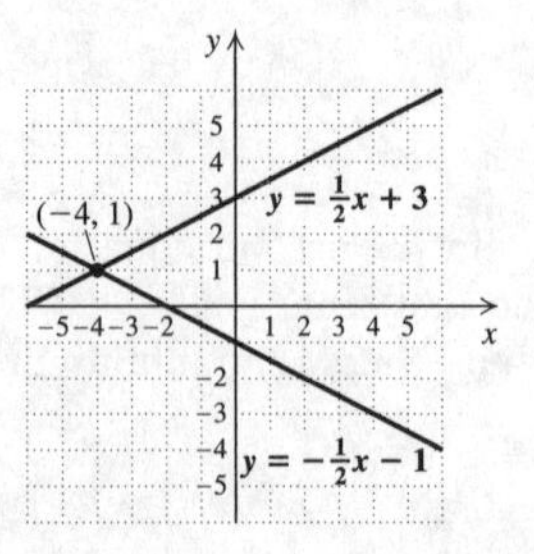

The graphs intersect at $(-4, 1)$, indicating that when $x = -4$ we have $\frac{1}{2}x + 3 = -\frac{1}{2}x - 1$. The solution is -4.

47. *Writing Exercise*

49. $4x - 5(9 - 2x) = 7$

$$4x - 45 + 10x = 7$$
$$14x - 45 = 7$$
$$14x = 52$$
$$x = \frac{52}{14} = \frac{26}{7}$$

The solution is $\frac{26}{7}$.

51. $3(4 - 2y) - 5y = 6$

$$12 - 6y - 5y = 6$$
$$12 - 11y = 6$$
$$-11y = -6$$
$$y = \frac{6}{11}$$

The solution is $\frac{6}{11}$.

53. $3a - 5b = 7$

$$-5b = -3a + 7 \quad \text{Subtracting } 3a$$
$$b = \frac{-3a+7}{-5}, \text{ or } \frac{3a-7}{5}$$

55. *Writing Exercise*

57. Systems in which the graphs of the equations coincide contain dependent equations. This is the case in Exercises 18, 25, 26, and 29.

59. Systems in which the graphs of the equations are parallel are inconsistent. This is the case in Exercises 17 and 30.

61. Answers may vary. Any equation with $(-1, 4)$ as a solution that is independent of $5x + 2y = 3$ will do. One such equation is $x + y = 3$.

63. $(2, -3)$ is a solution of $Ax - 3y = 13$. Substitute 2 for x and -3 for y and solve for A.

$$Ax - 3y = 13$$
$$A \cdot 2 - 3(-3) = 13$$
$$2A + 9 = 13$$
$$2A = 4$$
$$A = 2$$

$(2, -3)$ is a solution of $x - By = 8$. Substitute 2 for x and -3 for y and solve for B.

$$x - By = 8$$
$$2 - B(-3) = 8$$
$$2 + 3B = 8$$
$$3B = 6$$
$$B = 2$$

65. a) Let x = the number of copies, up to 500, and y = the cost. Then the cost equation for the copy card method of payment is $y = 20$. The cost equation for the method of paying per page is $y = 0.06x$.

b)

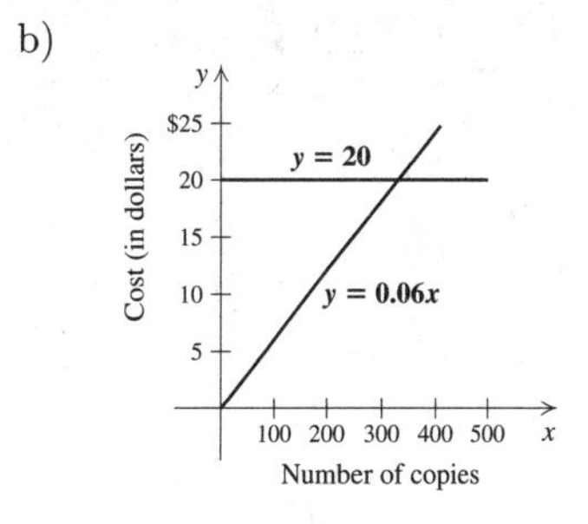

c) We see that the graphs intersect at approximately $(333, 20)$ and that for x-values greater than 333 the graph of $y = 20$ lies below the graph of $y = 0.06x$. Thus, Shelby must make more than 333 copies if the card is to be more economical.

67. a) The number of filmed prints is 33.1 billion in 2000 and decreases at a rate of 1.1 billion per year. So t years after 2000 the number of filmed prints has decreased by $1.1t$. Thus we have $n = 33.1 - 1.1t$

The number of digital prints is 2.8 billion in 2000 and is growing at a rate of 2.9 billion per year. So t years after 2000 the number of digital prints has increased by $2.9t$. Thus we have $n = 2.8 + 2.9t$.

b) Graph $y_1 = 33.1 - 1.1t$ and $y_2 = 2.8 + 2.9t$ and find the first coordinate of the point of intersection of the graphs. (See the Technology Connection on page 433 in the text.) We have $x \approx 7.6$, so the number of filmed and digital photos will be the same about 7.6 yr after 2000, or in 2007.

69. Graph $y_1 = 1.3x - 4.9$ and $y_2 = 6.3 - 3.7x$ and use the INTERSECT feature from the CALC menu to find the first coordinate of the point of intersection of the graphs. It is 2.24.

Exercise Set 7.2

1. The statement is false. See Example 2.

3. The statement is true. See Example 4(a).

5.

$$x + y = 7, \quad (1)$$
$$y = x + 3 \quad (2)$$

Substitute $x + 3$ for y in Equation (1) and solve for x.

$$x + y = 7 \quad (1)$$
$$x + (x + 3) = 7 \quad \text{Substituting}$$
$$2x + 3 = 7$$
$$2x = 4$$
$$x = 2$$

Next we substitute 2 for x in either equation of the original system and solve for x.

$$y = x + 3 \quad (2)$$
$$y = 2 + 3 \quad \text{Substituting}$$
$$y = 5$$

We check the ordered pair $(2, 5)$.

$$\begin{array}{c|c} x+y=7 & \\ \hline 2+5 & 7 \\ 7 \stackrel{?}{=} 7 & \text{TRUE} \end{array} \qquad \begin{array}{c|c} y=x+3 & \\ \hline 5 & 2+3 \\ 5 \stackrel{?}{=} 5 & \text{TRUE} \end{array}$$

Since $(2,5)$ checks in both equations, it is the solution.

7. $x = y+1,\quad (1)$

$x+2y = 4 \quad (2)$

Substitute $y+1$ for x in Equation (1) and solve for y.

$$\begin{aligned} x+2y &= 4 \quad (2) \\ (y+1)+2y &= 4 \quad \text{Substituting} \\ 3y+1 &= 4 \\ 3y &= 3 \\ y &= 1 \end{aligned}$$

Next we substitute 1 for y in either equation of the original system and solve for x.

$$\begin{aligned} x &= y+1 \quad (1) \\ x &= 1+1 \quad \text{Substituting} \\ x &= 2 \end{aligned}$$

We check the ordered pair $(2,1)$.

$$\begin{array}{c|c} x=y+1 & \\ \hline 2 & 1+1 \\ 2 \stackrel{?}{=} 2 & \text{TRUE} \end{array} \qquad \begin{array}{c|c} x+2y=4 & \\ \hline 2+2\cdot 1 & 4 \\ 2+2 & \\ 4 \stackrel{?}{=} 4 & \text{TRUE} \end{array}$$

Since $(2,1)$ checks in both equations, it is the solution.

9. $y = 2x-5,\quad (1)$

$3y-x = 5 \quad (2)$

Substitute $2x-5$ for y in Equation (2) and solve for x.

$$\begin{aligned} 3y-x &= 5 \quad (2) \\ 3(2x-5)-x &= 5 \quad \text{Substituting} \\ 6x-15-x &= 5 \\ 5x-15 &= 5 \\ 5x &= 20 \\ x &= 4 \end{aligned}$$

Next we substitute 4 for x in either equation of the original system and solve for y.

$$\begin{aligned} y &= 2x-5 \quad (1) \\ y &= 2\cdot 4-5 \quad \text{Substituting} \\ y &= 8-5 \\ y &= 3 \end{aligned}$$

We check the ordered pair $(4,3)$.

$$\begin{array}{c|c} y=2x-5 & \\ \hline 3 & 2\cdot 4-5 \\ & 8-5 \\ 3 \stackrel{?}{=} 3 & \text{TRUE} \end{array} \qquad \begin{array}{c|c} 3y-x=5 & \\ \hline 3\cdot 3-4 & 5 \\ 9-4 & \\ 5 \stackrel{?}{=} 5 & \text{TRUE} \end{array}$$

Since $(4,3)$ checks in both equations, it is the solution.

11. $3x+y = 2,\quad (1)$

$y = -2x \quad (2)$

Substitute $-2x$ for y in Equation (1) and solve for x.

$$\begin{aligned} 3x+y &= 2 \quad (1) \\ 3x+(-2x) &= 2 \quad \text{Substituting} \\ x &= 2 \end{aligned}$$

Next we substitute 2 for x in either equation of the original system and solve for y.

$$\begin{aligned} y &= -2x \quad (2) \\ y &= -2\cdot 2 \quad \text{Substituting} \\ y &= -4 \end{aligned}$$

We check the ordered pair $(2,-4)$.

$$\begin{array}{c|c} 3x+y=2 & \\ \hline 3\cdot 2+(-4) & 2 \\ 6-4 & \\ 2 \stackrel{?}{=} 2 & \text{TRUE} \end{array} \qquad \begin{array}{c|c} y=-2x & \\ \hline -4 & -2\cdot 2 \\ -4 \stackrel{?}{=} -4 & \text{TRUE} \end{array}$$

Since $(2,-4)$ checks in both equations, it is the solution.

13. $2x+3y = 8,\quad (1)$

$x = y-6 \quad (2)$

Substitute $y-6$ for x in Equation (1) and solve for y.

$$\begin{aligned} 2x+3y &= 8 \quad (1) \\ 2(y-6)+3y &= 8 \quad \text{Substituting} \\ 2y-12+3y &= 8 \\ 5y-12 &= 8 \\ 5y &= 20 \\ y &= 4 \end{aligned}$$

Next we substitute 4 for y in either equation of the original system and solve for x.

$$\begin{aligned} x &= y-6 \quad (2) \\ x &= 4-6 \quad \text{Substituting} \\ x &= -2 \end{aligned}$$

We check the ordered pair $(-2,4)$.

$$\begin{array}{c|c} 2x+3y=8 & \\ \hline 2(-2)+3\cdot 4 & 8 \\ -4+12 & \\ 8 \stackrel{?}{=} 8 & \text{TRUE} \end{array} \qquad \begin{array}{c|c} x=y-6 & \\ \hline -2 & 4-6 \\ -2 \stackrel{?}{=} -2 & \text{TRUE} \end{array}$$

Since $(-2,4)$ checks in both equations, it is the solution.

15. $x = 2y+1,\quad (1)$

$3x-6y = 2 \quad (2)$

We substitute $2y+1$ for x in Equation (2) and solve for y.

$$\begin{aligned} 3x-6y &= 2 \quad (2) \\ 3(2y+1)-6y &= 2 \\ 6y+3-6y &= 2 \\ 3 &= 2 \end{aligned}$$

We obtain a false equation, so the system has no solution.

17. $s+t=-4,\quad (1)$
$s-t=2\quad (2)$

We solve Equation (2) for s.

$$s-t=2\quad (2)$$
$$s=t+2\quad (3)$$

We substitute $t+2$ for s in Equation (1) and solve for t.

$$\begin{aligned} s+t&=-4 && (1)\\ (t+2)+t&=-4 && \text{Substituting}\\ 2t+2&=-4\\ 2t&=-6\\ t&=-3 \end{aligned}$$

Now we substitute -3 for t in either of the original equations or in Equation (3) and solve for s. It is easiest to use (3).

$$s=t+2=-3+2=-1$$

We check the ordered pair $(-1,-3)$.

$$\begin{array}{c|c} \multicolumn{2}{c}{s+t=-4} \\ \hline -1+(-3) & -4 \\ \end{array}$$
$$-4 \stackrel{?}{=} -4 \quad \text{TRUE}$$

$$\begin{array}{c|c} \multicolumn{2}{c}{s-t=2} \\ \hline -1-(-3) & 2 \\ -1+3 & \\ \end{array}$$
$$2 \stackrel{?}{=} 2 \quad \text{TRUE}$$

Since $(-1,-3)$ checks in both equations, it is the solution.

19. $x-y=5,\quad (1)$
$x+2y=7\quad (2)$

Solve Equation (1) for x.

$$x-y=5\quad (1)$$
$$x=y+5\quad (3)$$

Substitute $y+5$ for x in Equation (2) and solve for y.

$$\begin{aligned} x+2y&=7 && (2)\\ (y+5)+2y&=7 && \text{Substituting}\\ 3y+5&=7\\ 3y&=2\\ y&=\frac{2}{3} \end{aligned}$$

Substitute $\frac{2}{3}$ for y in Equation (3) and compute x.

$$x=y+5=\frac{2}{3}+5=\frac{2}{3}+\frac{15}{3}=\frac{17}{3}$$

The ordered pair $\left(\frac{17}{3},\frac{2}{3}\right)$ checks in both equations. It is the solution.

21. $x-2y=7,\quad (1)$
$3x-21=6y\quad (2)$

Solve Equation (1) for x.

$$x-2y=7$$
$$x=2y+7$$

Substitute $2y+7$ for x in Equation (2) and solve for y.

$$\begin{aligned} 3x-21&=6y && (2)\\ 3(2y+7)-21&=6y && \text{Substituting}\\ 6y+21-21&=6y\\ 6y&=6y \end{aligned}$$

The last equation is true for any choice of y, so there is an infinite number of solutions.

23. $y=2x+5,\quad (1)$
$-2y=-4x-10\quad (2)$

We substitute $2x+5$ for y in Equation (2) and solve for x.

$$\begin{aligned} -2y&=-4x-10 && (2)\\ -2(2x+5)&=-4x-10 && \text{Substituting}\\ -4x-10&=-4x-10 \end{aligned}$$

The last equation is true for any choice of x, so there is an infinite number of solutions.

25. $2x+3y=-2,\quad (1)$
$2x-y=9\quad (2)$

Solve Equation (2) for y.

$$2x-y=9\quad (2)$$
$$2x-9=y\quad (3)$$

Substitute $2x-9$ for y in Equation (1) and solve for x.

$$\begin{aligned} 2x+3y&=-2 && (1)\\ 2x+3(2x-9)&=-2 && \text{Substituting}\\ 2x+6x-27&=-2\\ 8x-27&=-2\\ 8x&=25\\ x&=\frac{25}{8} \end{aligned}$$

Now substitute $\frac{25}{8}$ for x in Equation (3) and compute y.

$$y=2x-9=2\left(\frac{25}{8}\right)-9=\frac{25}{4}-\frac{36}{4}=-\frac{11}{4}$$

The ordered pair $\left(\frac{25}{8},-\frac{11}{4}\right)$ checks in both equations. It is the solution.

27. $x-y=-3,\quad (1)$
$2x+3y=-6\quad (2)$

Solve Equation (1) for x.

$$x-y=-3\quad (1)$$
$$x=y-3\quad (3)$$

Substitute $y-3$ for x in Equation (2) and solve for y.

$$\begin{aligned} 2x+3y &= -6 \quad (2) \\ 2(y-3)+3y &= -6 \quad \text{Substituting} \\ 2y-6+3y &= -6 \\ 5y-6 &= -6 \\ 5y &= 0 \\ y &= 0 \end{aligned}$$

Now substitute 0 for y in Equation (3) and compute x.

$$x = y-3 = 0-3 = -3$$

The ordered pair $(-3, 0)$ checks in both equations. It is the solution.

29. $r-2s=0,\quad (1)$

$4r-3s=15\quad (2)$

Solve Equation (1) for r.

$$\begin{aligned} r-2s &= 0 \quad (1) \\ r &= 2s \quad (3) \end{aligned}$$

Substitute $2s$ for r in Equation (2) and solve for s.

$$\begin{aligned} 4r-3s &= 15 \quad (2) \\ 4(2s)-3s &= 15 \quad \text{Substituting} \\ 8s-3s &= 15 \\ 5s &= 15 \\ s &= 3 \end{aligned}$$

Now substitute 3 for s in Equation (3) and compute r.

$$r = 2s = 2\cdot 3 = 6$$

The ordered pair $(6, 3)$ checks in both equations. It is the solution.

31. $x-3y=7,\quad (1)$

$-4x+12y=28\quad (2)$

Solve Equation (1) for x.

$$\begin{aligned} x-3y &= 7 \quad (1) \\ x &= 3y+7 \quad (3) \end{aligned}$$

Substitute $3y+7$ for x in Equation (2).

$$\begin{aligned} -4x+12y &= 28 \quad (2) \\ -4(3y+7)+12y &= 28 \quad \text{Substituting} \\ -12y-28+12y &= 28 \\ -28 &= 28 \end{aligned}$$

We obtain a false equation, so the system has no solution.

33. $x-2y=5,\quad (1)$

$2y-3x=1\quad (2)$

Solve Equation (1) for x.

$$\begin{aligned} x-2y &= 5 \\ x &= 2y+5 \quad (3) \end{aligned}$$

Substitute $2y+5$ for x in Equation (2) and solve for y.

$$\begin{aligned} 2y-3x &= 1 \quad (2) \\ 2y-3(2y+5) &= 1 \quad \text{Substituting} \\ 2y-6y-15 &= 1 \\ -4y-15 &= 1 \\ -4y &= 16 \\ y &= -4 \end{aligned}$$

Next substitute -4 for y in Equation (3) and compute x.

$$x = 2y+5 = 2(-4)+5 = -8+5 = -3$$

The ordered pair $(-3, -4)$ checks in both equations. It is the solution.

35. $2x-y=0,\quad (1)$

$2x-y=-2\quad (2)$

Solve Equation (1) for y.

$$\begin{aligned} 2x-y &= 0 \quad (1) \\ 2x &= y \quad (3) \end{aligned}$$

Substitute $2x$ for y in Equation (2) and solve for x.

$$\begin{aligned} 2x-y &= -2 \quad (2) \\ 2x-2x &= -2 \quad \text{Substituting} \\ 0 &= -2 \end{aligned}$$

We obtain a false equation, so the system has no solution.

37. ***Familiarize***. We let x = the larger number and y = the smaller number.

Translate.

$$\underbrace{\text{The sum of two numbers}}_{x+y}\ \underbrace{\text{is}}_{=}\ \underbrace{83.}_{83}$$

$$\underbrace{\text{One number}}_{x}\ \underbrace{\text{is}}_{=}\ \underbrace{5}_{5}\ \underbrace{\text{more than}}_{+}\ \underbrace{\text{the other.}}_{y}$$

The resulting system is

$$\begin{aligned} x+y &= 83, \quad (1) \\ x &= 5+y. \quad (2) \end{aligned}$$

Carry out. We solve the system of equations. We substitute $5+y$ for x in Equation (1) and solve for y.

$$\begin{aligned} x+y &= 83 \quad (1) \\ (5+y)+y &= 83 \quad \text{Substituting} \\ 5+2y &= 83 \\ 2y &= 78 \\ y &= 39 \end{aligned}$$

Next we substitute 39 for y in either equation of the original system and solve for x.

$$\begin{aligned} x+y &= 83 \quad (1) \\ x+39 &= 83 \quad \text{Substituting} \\ x &= 44 \end{aligned}$$

Check. The sum of 44 and 39 is 83. The number 44 is 5 more than 39. These numbers check.

State. The numbers are 44 and 39.

39. ***Familiarize.*** Let x = the larger number and y = the smaller number.

Translate.

The sum of two numbers is 93.

$$x + y = 93$$

The difference of two numbers is 9.

$$x - y = 9$$

The resulting system is

$$x + y = 93, \quad (1)$$
$$x - y = 9. \quad (2)$$

Carry out. We solve the system.

We solve Equation (2) for x.

$$x - y = 9 \quad (2)$$
$$x = y + 9 \quad (3)$$

We substitute $y + 9$ for x in Equation (1) and solve for y.

$$x + y = 93 \quad (1)$$
$$(y + 9) + y = 93 \quad \text{Substituting}$$
$$2y + 9 = 93$$
$$2y = 84$$
$$y = 42$$

Now we substitute 42 for y in Equation (3) and compute x.

$$x = y + 9 = 42 + 9 = 51$$

Check. The sum of 51 and 42 is 93. The difference between 51 and 42, $51 - 42$, is 9. The numbers check.

State. The numbers are 51 and 42.

41. ***Familiarize.*** Let x = the larger number and y = the smaller number.

Translate.

The difference between two numbers is 16.

$$x - y = 16$$

Three times the larger number is seven times the smaller number.

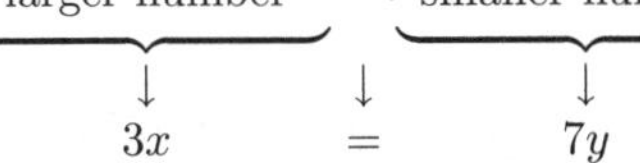

$$3x = 7y$$

The resulting system is

$$x - y = 16, \quad (1)$$
$$3x = 7y. \quad (2)$$

Carry out. We solve the system.

We solve Equation (1) for x.

$$x - y = 16 \quad (1)$$
$$x = y + 16 \quad (3)$$

We substitute $y + 16$ for x in Equation (2) and solve for y.

$$3x = 7y \quad (2)$$
$$3(y + 16) = 7y \quad \text{Substituting}$$
$$3y + 48 = 7y$$
$$48 = 4y$$
$$12 = y$$

Next we substitute 12 for y in Equation (3) and compute x.

$$x = y + 16 = 12 + 16 = 28$$

Check. The difference between 28 and 12, $28 - 12$, is 16. Three times the larger, $3 \cdot 28$ or 84, is seven times the smaller, $7 \cdot 12 = 84$. The numbers check.

State. The numbers are 28 and 12.

43. ***Familiarize.*** Let x = one angle and y = the other angle.

Translate. Since the angles are supplementary, we have one equation.

$$x + y = 180$$

The second sentence can be translated as follows:

One angle is 15° more than twice the other.

$$x = 2y + 15$$

The resulting system is

$$x + y = 180, \quad (1)$$
$$x = 2y + 15. \quad (2)$$

Carry out. We solve the system.

We substitute $2y + 15$ for x in Equation (1) and solve for y.

$$x + y = 180 \quad (1)$$
$$2y + 15 + y = 180$$
$$3y + 15 = 180$$
$$3y = 165$$
$$y = 55$$

Next we substitute 55 for y in Equation (2) and find x.

$$x = 2y + 15 = 2 \cdot 55 + 15 = 110 + 15 = 125$$

Check. The sum of the angles is $55° + 125°$, or 180°, so the angles are supplementary. If 15° is added to twice 55°, we have $2 \cdot 55° + 15°$, or 125°, which is the other angle. The answer checks.

State. One angle is 55°, and the other is 125°.

45. ***Familiarize.*** We let x = the larger angle and y = the smaller angle.

Translate. Since the angles are complementary, we have one equation.

$$x + y = 90$$

We reword and translate the second statement.

The difference of two angles is 18°.

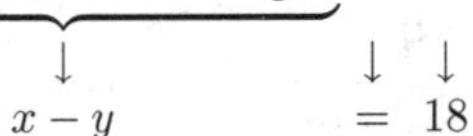

$$x - y = 18$$

The resulting system is

$$x + y = 90, \quad (1)$$
$$x - y = 18. \quad (2)$$

Carry out. We solve the system.

We first solve Equation (2) for x.

$$x - y = 18 \quad (2)$$
$$x = y + 18 \quad (3)$$

Substitute $y + 18$ for x in Equation (1) and solve for y.

$$x + y = 90 \quad (1)$$
$$y + 18 + y = 90$$
$$2y + 18 = 90$$
$$2y = 72$$
$$y = 36$$

Next we substitute 36 for y in Equation (3) and solve for x.

$$x = y + 18 = 36 + 18 = 54$$

Check. The sum of the angles is $54° + 36°$, or $90°$, so the angles are complementary. The difference of the angles is $54° - 36° = 18°$. These numbers check.

State. The angles are $54°$ and $36°$.

47. ***Familiarize***. We make a drawing. We let l = the length and w = the width.

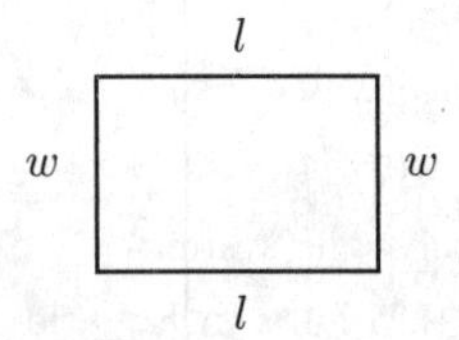

Translate. The perimeter is $2l + 2w$. We translate the first statement.

$$\underbrace{\text{The perimeter}}_{2l+2w} \ \underbrace{\text{is}}_{=} \ \underbrace{10\frac{1}{2} \text{ in.}}_{10\frac{1}{2}}$$

We translate the second statement.

$$\underbrace{\text{The length}}_{l} \ \underbrace{\text{is}}_{=} \ \underbrace{\text{twice the width.}}_{2w}$$

The resulting system is

$$2l + 2w = 10\frac{1}{2}, \quad (1)$$
$$l = 2w. \quad (2)$$

Solve. We solve the system. We substitute $2w$ for l in Equation (1) and solve for w. We also express $10\frac{1}{2}$ as $\frac{21}{2}$.

$$2l + 2w = 10\frac{1}{2} \quad \text{Equation (1)}$$
$$2(2w) + 2w = \frac{21}{2} \quad \text{Substituting}$$
$$4w + 2w = \frac{21}{2} \quad \text{Removing parentheses}$$
$$6w = \frac{21}{2} \quad \text{Collecting like terms}$$
$$w = \frac{1}{6} \cdot \frac{21}{2} \quad \text{Multiplying by } \frac{1}{6}$$
$$w = \frac{1 \cdot 21}{6 \cdot 2}$$
$$w = \frac{1 \cdot \cancel{3} \cdot 7}{\cancel{3} \cdot 2 \cdot 2}$$
$$w = \frac{7}{4}, \text{ or } 1\frac{3}{4}$$

Now we substitute $\frac{7}{4}$ for w in Equation (2) and solve for l.

$$l = 2w \quad \text{Equation (2)}$$
$$l = 2 \cdot \frac{7}{4} \quad \text{Substituting}$$
$$l = \frac{2 \cdot 7}{4}$$
$$l = \frac{\cancel{2} \cdot 7}{\cancel{2} \cdot 2}$$
$$l = \frac{7}{2}, \text{ or } 3\frac{1}{2}$$

Check. A possible solution is a length of $\frac{7}{2}$, or $3\frac{1}{2}$ in. and a width of $\frac{7}{4}$, or $1\frac{3}{4}$ in. The perimeter would be $2 \cdot \frac{7}{2} + 2 \cdot \frac{7}{4}$, or $7 + \frac{7}{2}$, or $10\frac{1}{2}$ in. Also, twice the width is $2 \cdot \frac{7}{4}$, or $\frac{7}{2}$, which is the length. These numbers check.

State. The length is $3\frac{1}{2}$ in., and the width is $1\frac{3}{4}$ in.

49. ***Familiarize***. Recall that the perimeter of a rectangle with length l and width w is given by $2l + 2w$.

Translate.

$$\underbrace{\text{The perimeter}}_{2l+2w} \ \underbrace{\text{is}}_{=} \ \underbrace{\text{1300 mi.}}_{1300}$$

$$\underbrace{\text{The length}}_{l} \ \underbrace{\text{is}}_{=} \ \underbrace{\text{110 mi more than the width.}}_{w+110}$$

The resulting system is

$$2l + 2w = 1300, \quad (1)$$
$$l = w + 110. \quad (2)$$

Carry out. We solve the system.

Substitute $w + 110$ for l in Equation (1) and solve for w.

$$2l + 2w = 1300 \quad (1)$$
$$2(w + 110) + 2w = 1300 \quad \text{Substituting}$$
$$2w + 220 + 2w = 1300$$
$$4w + 220 = 1300$$
$$4w = 1080$$
$$w = 270$$

Now substitute 270 for w in Equation (2).

$$l = w + 110 \quad (2)$$
$$l = 270 + 110 \quad \text{Substituting}$$
$$l = 380$$

Check. If the length is 380 mi and the width is 270 mi, the perimeter would be $2 \cdot 380 + 2 \cdot 270$, or $760 + 540$, or 1300 mi. Also, the length is 110 mi more than the width. These numbers check.

State. The length is 380 mi, and the width is 270 mi.

51. ***Familiarize.*** Recall that the perimeter of a rectangle with length l and width w is given by $2l + 2w$.

Translate.

$$\underbrace{\text{The perimeter}}_{2l+2w} \ \underbrace{\text{is}}_{=} \ \underbrace{280 \text{ ft.}}_{280}$$

$$\underbrace{\text{The width}}_{w} \ \underbrace{\text{is}}_{=} \ \underbrace{5}_{5} \ \underbrace{\text{more than}}_{+} \ \underbrace{\text{half the length.}}_{\frac{1}{2}l}$$

The resulting system is

$$2l + 2w = 280, \quad (1)$$
$$w = 5 + \frac{1}{2}l. \quad (2)$$

Carry out. We solve the system.

Substitute $5 + \frac{1}{2}l$ for w in Equation (1) and solve for l.

$$2l + 2\left(5 + \frac{1}{2}l\right) = 280 \quad (1)$$
$$2l + 10 + l = 280$$
$$3l + 10 = 280$$
$$3l = 270$$
$$l = 90$$

Now substitute 90 for l in Equation (2) and compute w.

$$w = 5 + \frac{1}{2}l = 5 + \frac{1}{2} \cdot 90 = 5 + 45 = 50$$

Check. If the length is 90 yd and the width is 50 yd, then the perimeter is $2 \cdot 90 + 2 \cdot 50$, or $180 + 100$, or 280 yd. Also, 5 more than half the length, $5 + \frac{1}{2} \cdot 90$, or $5 + 45$, or 50 yd, is the width. The answer checks.

State. The length is 90 yd, and the width is 50 yd.

53. ***Familiarize.*** Let h = the height of the front wall and w = the width of the service zone.

Translate.

$$\underbrace{\text{The height}}_{h} \ \underbrace{\text{is}}_{=} \ \underbrace{4}_{4} \ \underbrace{\text{times}}_{\cdot} \ \underbrace{\text{the width.}}_{w}$$

$$\underbrace{\text{The height}}_{h} \ \underbrace{\text{plus}}_{+} \ \underbrace{\text{the width}}_{w} \ \underbrace{\text{is}}_{=} \ \underbrace{25 \text{ ft.}}_{25}$$

The resulting system is

$$h = 4w, \quad (1)$$
$$h + w = 25. \quad (2)$$

Carry out. We solve the system.

Substitute $4w$ for h in Equation (2) and solve for w.

$$h + w = 25 \quad (2)$$
$$4w + w = 25 \quad \text{Substituting}$$
$$5w = 25$$
$$w = 5$$

Now substitute 5 for w in Equation (1) to find h.

$$h = 4w = 4 \cdot 5 = 20$$

Check. If the height is 20 ft and the width of the service zone is 5 ft, then the height is 4 times the width and the sum of the height and the width is 25. The answer checks.

State. The height of the court is 20 ft, and the width of the service zone is 5 ft.

55. *Writing Exercise*

57.
$$2(5x + 3y) - 3(5x + 3y)$$
$$= 10x + 6y - 15x - 9y$$
$$= -5x - 3y$$

We could also simplify this expression as follows:

$$2(5x + 3y) - 3(5x + 3y)$$
$$= -1(5x + 3y)$$
$$= -5x - 3y$$

59.
$$4(5x + 6y) - 5(4x + 7y)$$
$$= 20x + 24y - 20x - 35y$$
$$= -11y$$

61.
$$2(5x - 3y) - 5(2x + y)$$
$$= 10x - 6y - 10x - 5y$$
$$= -11y$$

63. *Writing Exercise*

65.
$$\frac{1}{6}(a + b) = 1, \quad (1)$$
$$\frac{1}{4}(a - b) = 2 \quad (2)$$

Observe that $\frac{1}{6}(a + b) = 1$, so $a + b = 6$. Also, $\frac{1}{4}(a - b) = 2$, so $a - b = 8$. We need to find two numbers whose sum is 6 and whose difference is 8. The numbers are 7 and -1, so the solution of the system of equations is $(7, -1)$.

We could also solve this system of equations using the substitution method. We first clear the fractions.

$a + b = 6 \quad (1a)$

$a - b = 8 \quad (2a)$

We solve Equation (2a) for a.

$a - b = 8 \quad (2a)$

$a = b + 8$

We substitute $b + 8$ for a in Equation (1a) and solve for b.

$(b + 8) + b = 6$

$2b + 8 = 6$

$2b = -2$

$b = -1$

Next we substitute -1 for b in Equation (2a) and solve for a.

$a - b = 8$

$a - (-1) = 8$

$a + 1 = 8$

$a = 7$

Since $(7, -1)$ checks in both equations, it is the solution.

67. Graph the equations and use the INTERSECT feature from the CALC menu to find the coordinates of the point of intersection. (It might be necessary to solve each equation for y before entering them on a graphing calculator.) The solution is approximately $(4.38, 4.33)$.

69. ***Familiarize***. Let t and d represent the ages Trudy and Dennis will be when the age requirement is met.

Translate. Dennis will be 20 years older than Trudy, so we have one equation:

$d = t + 20.$

Trudy's age will be 7 more than half of Dennis' age, so we have a second equation:

$t = \frac{1}{2}d + 7.$

The resulting system is

$d = t + 20, \quad (1)$

$t = \frac{1}{2}d + 7. \quad (2)$

Carry out. We solve the system.

First we substitute $t + 20$ for d in Equation (2) and solve for t.

$t = \frac{1}{2}d + 7 \quad (2)$

$t = \frac{1}{2}(t + 20) + 7$

$t = \frac{1}{2}t + 10 + 7$

$t = \frac{1}{2}t + 17$

$\frac{1}{2}t = 17$

$t = 34$

We are asked to find only Trudy's age but we will find Dennis' as well so that we can check the answer. Substitute 34 for t in Equation (1) and find d.

$d = t + 20 = 34 + 20 = 54$

Check. If Trudy is 34 and Dennis is 54, then Trudy is 20 yr younger than Dennis. Since $\frac{1}{2} \cdot 54 + 7 = 27 + 7 = 34$, we see that Trudy's age is 7 more than half of Dennis' age. The answer checks.

State. The youngest age at which Trudy can marry Dennis is 34 yr.

71. $x + y + z = 180, \quad (1)$

$x = z - 70, \quad (2)$

$2y - z = 0 \quad (3)$

Substitute $z - 70$ for x in Equation (1).

$(z - 70) + y + z = 180$

$y + 2z = 250 \quad (4)$

We now have a system of two equations in two variables.

$2y - z = 0 \quad (3)$

$y + 2z = 250 \quad (4)$

Solve Equation (3) for z.

$2y - z = 0 \quad (3)$

$2y = z \quad (5)$

Substitute $2y$ for z in Equation (4).

$y + 2(2y) = 250$

$5y = 250$

$y = 50$

Substitute 50 for y in Equation (5).

$z = 2y = 2 \cdot 50 = 100$

Substitute 100 for z in Equation (2).

$x = z - 70 = 100 - 70 = 30$

The triple $(30, 50, 100)$ checks in all three equations. It is the solution.

73. *Writing Exercise*

Exercise Set 7.3

1. The statement is false. See the introductory paragraph for this section on pages 444 and 445 in the text.

3. The statement is true. See Example 6.

5.
$$\begin{array}{rl} x - y = 6 & (1) \\ x + y = 12 & (2) \\ \hline 2x \quad = 18 & \text{Adding} \\ x = 9 & \end{array}$$

Substitute 9 for x in one of the original equations and solve for y.

$x + y = 12 \quad (2)$

$9 + y = 12 \quad \text{Substituting}$

$y = 3$

Check:

$$\begin{array}{c|c} x - y = 6 & \\ \hline 9-3 & 6 \\ 6 \overset{?}{=} 6 & \text{TRUE} \end{array} \qquad \begin{array}{c|c} x + y = 12 & \\ \hline 9+3 & 12 \\ 12 \overset{?}{=} 12 & \text{TRUE} \end{array}$$

Since $(9,3)$ checks, it is the solution.

7.
$$\begin{aligned} x + y &= 6 \quad (1) \\ -x + 3y &= -2 \quad (2) \\ \hline 4y &= 4 \quad \text{Adding} \\ y &= 1 \end{aligned}$$

Substitute 1 for y in one of the original equations and solve for x.

$$\begin{aligned} x + y &= 6 \quad (1) \\ x + 1 &= 6 \quad \text{Substituting} \\ x &= 5 \end{aligned}$$

Check:

$$\begin{array}{c|c} x + y = 6 & \\ \hline 5+1 & 6 \\ 6 \overset{?}{=} 6 & \text{TRUE} \end{array}$$

$$\begin{array}{c|c} -x + 3y = -2 & \\ \hline -5+3\cdot 1 & -2 \\ -5+3 & \\ -2 \overset{?}{=} -2 & \text{TRUE} \end{array}$$

Since $(5,1)$ checks, it is the solution.

9.
$$\begin{aligned} 4x - y &= 1 \quad (1) \\ 3x + y &= 13 \quad (2) \\ \hline 7x &= 14 \quad \text{Adding} \\ x &= 2 \end{aligned}$$

Substitute 2 for x in one of the original equations and solve for y.

$$\begin{aligned} 3x + y &= 13 \quad (2) \\ 3\cdot 2 + y &= 13 \quad \text{Substituting} \\ 6 + y &= 13 \\ y &= 7 \end{aligned}$$

Check:

$$\begin{array}{c|c} 4x - y = 1 & \\ \hline 4\cdot 2 - 7 & 7 \\ 8-7 & \\ 1 \overset{?}{=} 1 & \text{TRUE} \end{array}$$

$$\begin{array}{c|c} 3x + y = 13 & \\ \hline 3\cdot 2 + 7 & 8 \\ 6+7 & \\ 13 \overset{?}{=} 13 & \text{TRUE} \end{array}$$

Since $(2,7)$ checks, it is the solution.

11.
$$\begin{aligned} 5a + 4b &= 7 \quad (1) \\ -5a + b &= 8 \quad (2) \\ \hline 5b &= 15 \\ b &= 3 \end{aligned}$$

Substitute 3 for b in one of the original equations and solve for a.

$$\begin{aligned} 5a + 4b &= 7 \quad (1) \\ 5a + 4\cdot 3 &= 7 \\ 5a + 12 &= 7 \\ 5a &= -5 \\ a &= -1 \end{aligned}$$

Check:

$$\begin{array}{c|c} 5a + 4b = 7 & \\ \hline 5(-1) + 4\cdot 3 & 7 \\ -5+12 & \\ 7 \overset{?}{=} 7 & \text{TRUE} \end{array}$$

$$\begin{array}{c|c} -5a + b = 8 & \\ \hline -5(-1) + 3 & 8 \\ 5+3 & \\ 8 \overset{?}{=} 8 & \text{TRUE} \end{array}$$

Since $(-1,3)$ checks, it is the solution.

13.
$$\begin{aligned} 8x - 5y &= -9 \quad (1) \\ 3x + 5y &= -2 \quad (2) \\ \hline 11x &= -11 \quad \text{Adding} \\ x &= -1 \end{aligned}$$

Substitute -1 for x in either of the original equations and solve for y.

$$\begin{aligned} 3x + 5y &= -2 \quad \text{Equation (2)} \\ 3(-1) + 5y &= -2 \quad \text{Substituting} \\ -3 + 5y &= -2 \\ 5y &= 1 \\ y &= \frac{1}{5} \end{aligned}$$

Check:

$$\begin{array}{c|c} 8x - 5y = -9 & \\ \hline 8(-1) - 5\left(\frac{1}{5}\right) & -9 \\ -8-1 & \\ -9 \overset{?}{=} -9 & \text{TRUE} \end{array}$$

$$\begin{array}{c|c} \multicolumn{2}{c}{3x + 5y = -2} \\ \hline 3(-1) + 5\left(\frac{1}{5}\right) & -2 \\ -3 + 1 & \\ -2 \stackrel{?}{=} -2 & \text{TRUE} \end{array}$$

Since $\left(-1, \frac{1}{5}\right)$ checks, it is the solution.

15. $$\begin{aligned} 3a - 6b &= 8, \\ -3a + 6b &= -8 \\ \hline 0 &= 0 \quad \text{Adding} \end{aligned}$$

The equation $0 = 0$ is always true, so the system has an infinite number of solutions.

17. $-x - y = 8$, (1)

$2x - y = -1$ (2)

We multiply by -1 on both sides of Equation (1) and then add.

$$\begin{aligned} x + y &= -8 \quad \text{Multiplying by } -1 \\ 2x - y &= -1 \\ \hline 3x \quad &= -9 \quad \text{Adding} \\ x &= -3 \end{aligned}$$

Substitute -3 for x in one of the original equations and solve for y.

$$\begin{aligned} 2x - y &= -1 \quad (2) \\ 2(-3) - y &= -1 \quad \text{Substituting} \\ -6 - y &= -1 \\ -y &= 5 \\ y &= -5 \end{aligned}$$

Check:

$$\begin{array}{c|c} \multicolumn{2}{c}{-x - y = 8} \\ \hline -(-3) - (-5) & 8 \\ 3 + 5 & \\ 8 \stackrel{?}{=} 8 & \text{TRUE} \end{array}$$

$$\begin{array}{c|c} \multicolumn{2}{c}{2x - y = -1} \\ \hline 2(-3) - (-5) & -1 \\ -6 + 5 & \\ -1 \stackrel{?}{=} -1 & \text{TRUE} \end{array}$$

Since $(-3, -5)$ checks, it is the solution.

19. $x + 3y = 19$,

$x - y = -1$

We multiply by -1 on both sides of Equation (2) and then add.

$$\begin{aligned} x + 3y &= 19 \\ -x + y &= 1 \quad \text{Multiplying by } -1 \\ \hline 4y &= 20 \quad \text{Adding} \\ y &= 5 \end{aligned}$$

Substitute 5 for y in one of the original equations and solve for x.

$$\begin{aligned} x - y &= -1 \quad (2) \\ x - 5 &= -1 \quad \text{Substituting} \\ x &= 4 \end{aligned}$$

Check:

$$\begin{array}{c|c} \multicolumn{2}{c}{x + 3y = 19} \\ \hline 4 + 3 \cdot 5 & 19 \\ 4 + 15 & \\ 19 \stackrel{?}{=} 19 & \text{TRUE} \end{array} \qquad \begin{array}{c|c} \multicolumn{2}{c}{x - y = -1} \\ \hline 4 - 5 & -1 \\ -1 \stackrel{?}{=} -1 & \text{TRUE} \end{array}$$

Since $(4, 5)$ checks, it is the solution.

21. $x + y = 5$, (1)

$4x - 3y = 13$ (2)

We multiply by 3 on both sides of Equation (1) and then add.

$$\begin{aligned} 3x + 3y &= 15 \quad \text{Multiplying by 3} \\ 4x - 3y &= 13 \\ \hline 7x \quad &= 28 \\ x &= 4 \end{aligned}$$

Substitute 4 for x in one of the original equations and solve for y.

$$\begin{aligned} x + y &= 5 \quad (1) \\ 4 + y &= 5 \quad \text{Substituting} \\ y &= 1 \end{aligned}$$

Check:

$$\begin{array}{c|c} \multicolumn{2}{c}{x + y = 5} \\ \hline 4 + 1 & 5 \\ 5 \stackrel{?}{=} 5 & \text{TRUE} \end{array} \qquad \begin{array}{c|c} \multicolumn{2}{c}{4x - 3y = 13} \\ \hline 4 \cdot 4 - 3 \cdot 1 & 13 \\ 16 - 3 & \\ 13 \stackrel{?}{=} 13 & \text{TRUE} \end{array}$$

Since $(4, 1)$ checks, it is the solution.

23. $2w - 3z = -1$, (1)

$-4w + 6z = 5$ (2)

We multiply by 2 on both sides of Equation (1) and then add.

$$\begin{aligned} 4w - 6z &= -2 \\ -4w + 6z &= 5 \\ \hline 0 &= 3 \end{aligned}$$

We get a false equation, so there is no solution.

25. $2a + 3b = -1$, (1)

$3a + 5b = -2$ (2)

We use the multiplication principle with both equations and then add.

$$\begin{aligned} -10a - 15b &= 5 \quad \text{Multiplying (1) by } -5 \\ 9a + 15b &= -6 \quad \text{Multiplying (2) by 3} \\ \hline -a \quad &= -1 \quad \text{Adding} \\ a &= 1 \end{aligned}$$

Substitute 1 for a in one of the original equations and solve for b.

$$\begin{aligned} 2a+3b &= -1 && \text{Equation (1)} \\ 2\cdot 1+3b &= -1 && \text{Substituting} \\ 2+3b &= -1 \\ 3b &= -3 \\ b &= -1 \end{aligned}$$

Check:

$$\begin{array}{c|c} \multicolumn{2}{c}{2a+3b=-1} \\ \hline 2\cdot 1+3(-1) & -1 \\ 2-3 & \\ \end{array}$$

$-1 \stackrel{?}{=} -1$ TRUE

$$\begin{array}{c|c} \multicolumn{2}{c}{3a+5b=-2} \\ \hline 3\cdot 1+5(-1) & -2 \\ 3-5 & \\ \end{array}$$

$-2 \stackrel{?}{=} -2$ TRUE

Since $(1,-1)$ checks, it is the solution.

27. $3y = x,$ (1)

$5x+14 = y$ (2)

We first get each equation in the form $Ax+By=C$.

$$\begin{aligned} x-3y &= 0, && \text{(1a)} && \text{Adding } -3y \\ 5x-y &= -14 && \text{(2a)} && \text{Adding } -y-14 \end{aligned}$$

We multiply by -5 on both sides of Equation (1a) and add.

$$\begin{aligned} -5x+15y &= 0 && \text{Multiplying by } -5 \\ 5x-\ \ y &= -14 \\ \hline 14y &= -14 && \text{Adding} \\ y &= -1 \end{aligned}$$

Substitute -1 for y in Equation (1a) and solve for x.

$$\begin{aligned} x-3y &= 0 \\ x-3(-1) &= 0 && \text{Substituting} \\ x+3 &= 0 \\ x &= -3 \end{aligned}$$

Check:

$$\begin{array}{c|c} \multicolumn{2}{c}{x=3y} \\ \hline -3 & 3(-1) \\ \end{array}$$

$-3 \stackrel{?}{=} -3$ TRUE

$$\begin{array}{c|c} \multicolumn{2}{c}{5x+14=y} \\ \hline 5(-3)+14 & -1 \\ -15+14 & \\ \end{array}$$

$-1 \stackrel{?}{=} -1$ TRUE

Since $(-3,-1)$ checks, it is the solution.

29. $4x-10y = 13,$ (1)

$-2x+5y = 8$ (2)

We multiply by 2 on both sides of Equation (2) and then add.

$$\begin{aligned} 4x-10y &= 13 \\ -4x+10y &= 16 && \text{Multiplying by 2} \\ \hline 0 &= 29 \end{aligned}$$

The equation $0=29$ is false for any pair (x,y), so there is no solution.

31. $8n+6-3m = 0,$

$32 = m-n$

We first get each equation in the form $Am+Bn=C$.

$$\begin{aligned} -3m+8n &= -6, && \text{(1)} && \text{Subtracting 6} \\ m-n &= 32 && \text{(2)} \end{aligned}$$

We multiply by 3 on both sides of Equation (2) and add.

$$\begin{aligned} -3m+8n &= -6 \\ 3m-3n &= 96 \\ \hline 5n &= 90 \\ n &= 18 \end{aligned}$$

Substitute 18 for n in Equation (2) and solve for m.

$$\begin{aligned} m-n &= 32 \\ m-18 &= 32 \\ m &= 50 \end{aligned}$$

Check:

$$\begin{array}{c|c} \multicolumn{2}{c}{8n+6-3m=0} \\ \hline 8\cdot 18+6-3\cdot 50 & 0 \\ 144+6-150 & \\ \end{array}$$

$0 \stackrel{?}{=} 0$ TRUE

$$\begin{array}{c|c} \multicolumn{2}{c}{32=m-n} \\ \hline 32 & 50-18 \\ \end{array}$$

$32 \stackrel{?}{=} 32$ TRUE

Since $(50,18)$ checks, it is the solution.

33. $3x+5y = 4,$ (1)

$-2x+3y = 10$ (2)

We use the multiplication principle with both equations and then add.

$$\begin{aligned} 6x+10y &= 8 && \text{Multiplying (1) by 2} \\ -6x+\ 9y &= 30 && \text{Multiplying (2) by 3} \\ \hline 19y &= 38 && \text{Adding} \\ y &= 2 \end{aligned}$$

Substitute 2 for y in one of the original equations and solve for x.

$$\begin{aligned} 3x+5y &= 4 && \text{(1)} \\ 3x+5\cdot 2 &= 4 \\ 3x+10 &= 4 \\ 3x &= -6 \\ x &= -2 \end{aligned}$$

Check:

$$\begin{array}{c|c} \multicolumn{2}{c}{3x+5y=4} \\ \hline 3(-2)+5\cdot 2 & 4 \\ -6+10 & \\ \end{array}$$

$4 \stackrel{?}{=} 4$ TRUE

$$\begin{array}{c|c} \multicolumn{2}{c}{-2x+3y=10} \\ \hline -2(-2)+3\cdot 2 & 10 \\ 4+6 & \\ \end{array}$$
$$10 \stackrel{?}{=} 10 \quad \text{TRUE}$$

Since $(-2,2)$ checks, it is the solution.

35. $0.06x + 0.05y = 0.07,$

$0.04x - 0.03y = 0.11$

We first multiply each equation by 100 to clear the decimals.

$$6x + 5y = 7, \quad (1)$$
$$4x - 3y = 11 \quad (2)$$

We use the multiplication principle with both equations of the resulting system.

$$\begin{aligned} 18x + 15y &= 21 \quad \text{Multiplying (1) by 3} \\ 20x - 15y &= 55 \quad \text{Multiplying (2) by 5} \\ \hline 38x \qquad &= 76 \quad \text{Adding} \\ x &= 2 \end{aligned}$$

Substitute 2 for x in Equation (1) and solve for y.

$$\begin{aligned} 6x+5y &= 7 \\ 6\cdot 2+5y &= 7 \\ 12+5y &= 7 \\ 5y &= -5 \\ y &= -1 \end{aligned}$$

Check:

$$\begin{array}{c|c} \multicolumn{2}{c}{0.06x+0.05y=0.07} \\ \hline 0.06(2)+0.05(-1) & 0.07 \\ 0.12-0.05 & \\ \end{array}$$
$$0.07 \stackrel{?}{=} 0.07 \quad \text{TRUE}$$

$$\begin{array}{c|c} \multicolumn{2}{c}{0.04x-0.03y=0.11} \\ \hline 0.04(2)-0.03(-1) & 0.11 \\ 0.08+0.03 & \\ \end{array}$$
$$0.11 \stackrel{?}{=} 0.11 \quad \text{TRUE}$$

Since $(2,-1)$ checks, it is the solution.

37. $x + \frac{9}{2}y = \frac{15}{4},$

$\frac{9}{10}x - y = \frac{9}{20}$

First we clear fractions. We multiply both sides of the first equation by 4 and both sides of the second equation by 20.

$$\begin{aligned} 4\left(x+\frac{9}{2}y\right) &= 4\cdot\frac{15}{4} \\ 4x+4\cdot\frac{9}{2}y &= 15 \\ 4x+18 &= 15 \end{aligned}$$

$$\begin{aligned} 20\left(\frac{9}{10}x-y\right) &= 20\cdot\frac{9}{20} \\ 20\cdot\frac{9}{10}x-20y &= 9 \\ 18x-20y &= 9 \end{aligned}$$

The resulting system is

$$4x+18y = 15, \quad (1)$$
$$18x-20y = 9. \quad (2)$$

We use the multiplication principle with both equations.

$$\begin{aligned} 72x + 324y &= 270 \quad \text{Multiplying (1) by 18} \\ -72x + 80y &= -36 \quad \text{Multiplying (2) by } -4 \\ \hline 404y &= 234 \\ y &= \frac{234}{404}, \text{ or } \frac{117}{202} \end{aligned}$$

Substitute $\frac{117}{202}$ for y in (1) and solve for x.

$$\begin{aligned} 4x+18\left(\frac{117}{202}\right) &= 15 \\ 4x+\frac{1053}{101} &= 15 \\ 4x &= \frac{462}{101} \\ x &= \frac{1}{4}\cdot\frac{462}{101} \\ x &= \frac{231}{202} \end{aligned}$$

The ordered pair $\left(\frac{231}{202}, \frac{117}{202}\right)$ checks in both equations. It is the solution.

39. ***Familiarize***. We let m = the number of miles driven and c = the total cost of the truck rental.

Translate. We reword and translate the first statement, using \$0.79 for 79¢.

\$39.95	plus	79¢	times	the number of miles driven	is	cost.
↓	↓	↓	↓	↓	↓	↓
39.95	+	0.79	·	m	=	c

We reword and translate the second statement using \$0.59 for 59¢.

\$49.95	plus	59¢	times	the number of miles driven	is	cost.
↓	↓	↓	↓	↓	↓	↓
49.95	+	0.59	·	m	=	c

We have a system of equations:

$$39.95+0.79m = c,$$
$$49.95+0.59m = c$$

Carry out. To solve the system of equations, we multiply the second equation by -1 and add to eliminate c.

$$\begin{array}{r} 39.95 + 0.79m = c \\ -49.95 - 0.59m = -c \\ \hline -10 + 0.2m = 0 \\ 0.2m = 10 \\ m = 50 \end{array}$$

Check. For 50 mi, the cost of the Budget truck is $\$39.95 + \$0.79(50)$, or $\$79.45$. For 50 mi, the cost of the Penske truck is $\$49.95 + \$0.59(50)$, or $\$79.45$. The cost is the same for 50 mi.

State. When the trucks are driven 50 mi, the cost is the same.

41. ***Familiarize***. We let x = the larger angle and y = the smaller angle.

Translate. We reword and translate the first statement.

The sum of two angles is 90°.

$$x + y = 90$$

We reword and translate the second statement.

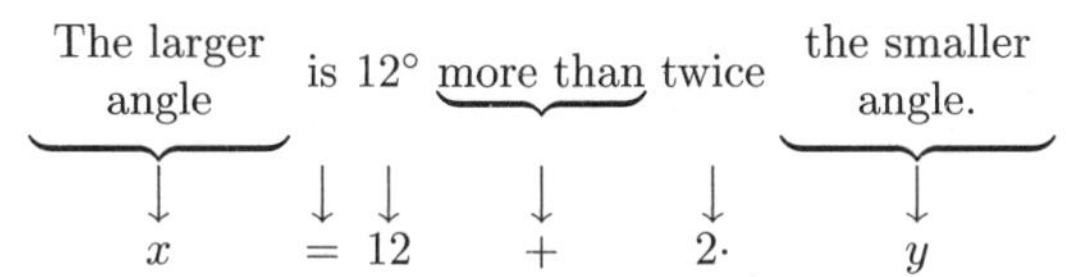

We have a system of equations:

$$x + y = 90,$$
$$x = 12 + 2y$$

Carry out. We solve the system. We will use the elimination method, although we could also easily use the substitution method. First we get the second equation in the form $Ax + By = C$.

$$x + y = 90 \quad (1)$$
$$x - 2y = 12 \quad (2) \quad \text{Adding } -2y$$

Now we multiply Equation (2) by 2 and add.

$$\begin{array}{r} 2x + 2y = 180 \\ x - 2y = 12 \\ \hline 3x \quad\;\; = 192 \\ x = 64 \end{array}$$

Then we substitute 64 for x in Equation (1) and solve for y.

$$x + y = 90 \quad (1)$$
$$64 + y = 90 \quad \text{Substituting}$$
$$y = 26$$

Check. The sum of the angles is $64° + 26°$, or $90°$, so the angles are complementary. The larger angle, 64°, is 12° more than twice the smaller angle, 26°. These numbers check.

State. The angles are 64° and 26°.

43. ***Familiarize***. Let m = the number of long distance minutes used in a month and c = the cost of the calls, in cents.

Translate. We reword the problem and translate. We will express \$4.95 as 495¢.

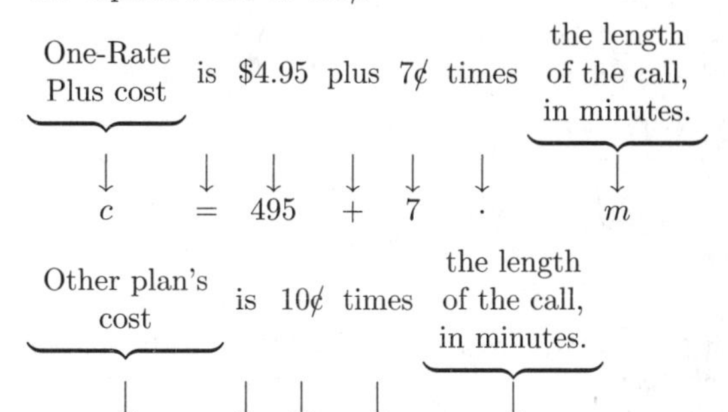

We have a system of equations:

$$c = 495 + 7m,$$
$$c = 10m$$

Carry out. To solve the system, we multiply the second equation by -1 and add to eliminate c.

$$\begin{array}{r} c = 495 + 7m \\ -c = \quad\;\; - 10m \\ \hline 0 = 495 - 3m \\ 3m = 495 \\ m = 165 \end{array}$$

Check. For 165 min, the cost of the One-Rate Plus plan is $495 + 7(165)$, or $495 + 1155$, or 1650¢ and the cost of the other plan is $10(165)$, or 1650¢. Thus the costs are the same for 165 long distance minutes.

State. For 165 minutes of long distance calls, the monthly costs are the same.

45. ***Familiarize***. Let x = the measure of one angle and y = the measure of the other angle.

Translate. We reword the problem.

The measure of one angle plus the measure of the other angle is 180°.

$$x + y = 180$$

One angle is 4 times the other angle minus 5°.

$$x = 4 \cdot y - 5$$

The resulting system is

$$x + y = 180,$$
$$x = 4y - 5.$$

Carry out. We solve the system. We will use the elimination method although we could also easily use the substitution method. First we get the second equation in the form $Ax + By = C$.

$$x + y = 180 \quad (1)$$
$$x - 4y = -5 \quad (2) \text{ Adding } -4y$$

Now we multiply Equation (2) by -1 and add.

$$\begin{aligned} x + y &= 180 \\ -x + 4y &= 5 \\ \hline 5y &= 185 \\ y &= 37 \end{aligned}$$

Then we substitute 37 for y in Equation (1) and solve for x.

$$\begin{aligned} x + y &= 180 \\ x + 37 &= 180 \\ x &= 143 \end{aligned}$$

Check. The sum of the angle measures is $37° + 143°$, or $180°$, so the angles are supplementary. Also, $5°$ less than four times the $37°$ angle is $4 \cdot 37° - 5°$, or $148° - 5°$, or $143°$, the measure of the other angle. These numbers check.

State. The measures of the angles are $37°$ and $143°$.

47. ***Familiarize.*** Let $x =$ the number of loaves of white bread baked and $y =$ the number of loaves of whole-wheat bread.

Translate.

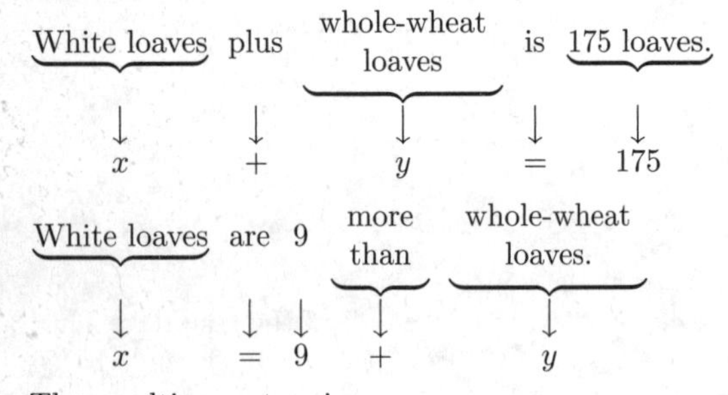

The resulting system is

$$x + y = 175,$$
$$x = 9 + y.$$

Carry out. We solve the system. We will use the elimination method although we could also easily use the substitution method. First we get the second equation in the form $Ax + By = C$.

$$\begin{aligned} x + y &= 175 \quad (1) \\ x - y &= 9 \quad (2) \text{ Adding } -y \\ \hline 2x &= 184 \quad \text{Adding (1) and (2)} \\ x &= 92 \end{aligned}$$

Substitute 92 for x in Equation (1) and solve for y.

$$\begin{aligned} x + y &= 175 \\ 92 + y &= 175 \\ y &= 83 \end{aligned}$$

Check. The total number of loaves is $92+83$, or 175. The number of loaves of white bread, 92, is 9 more than 83, the number of loaves of whole-wheat bread. The answer checks.

State. Maple Branch Bakers bake 92 loaves of white bread and 83 loaves of whole-wheat bread each day.

49. ***Familiarize.*** Let $l =$ the length of the frame and $w =$ the width, in feet.

Translate.

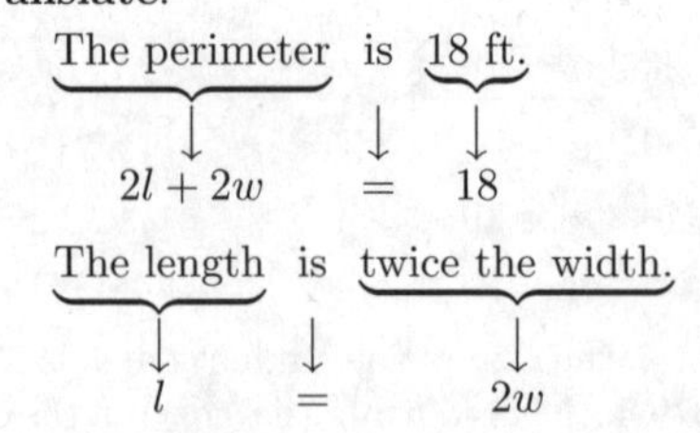

The resulting system is

$$2l + 2w = 18,$$
$$l = 2w.$$

Carry out. We solve the system. We will use the elimination method, although we could also easily use the substitution method. First we get the second equation in the form $Al + Bw = C$. Then we add the equations.

$$\begin{aligned} 2l + 2w &= 18 \quad (1) \\ l - 2w &= 0 \quad (2) \\ \hline 3l \quad &= 18 \\ l &= 6 \end{aligned}$$

Substitute 6 for l in Equation (2) and solve for w.

$$\begin{aligned} l - 2w &= 0 \\ 6 - 2w &= 0 \\ 6 &= 2w \\ 3 &= w \end{aligned}$$

Check. The perimeter is $2 \cdot 6 + 2 \cdot 3$, or 18 ft. Twice the width is $2 \cdot 3$, or 6 ft, which is the length. These numbers check.

State. The length of the mirror is 6 ft, and the width is 3 ft.

51. *Writing Exercise*

53. $8\% = 8 \times 0.01$ Replacing % by $\times 0.01$

$= 0.08$

This is equivalent to moving the decimal point two places to the left and dropping the percent symbol.

55. $0.4\% = 0.4 \times 0.01$ Replacing % by $\times 0.01$

$= 0.004$

This is equivalent to moving the decimal point two places to the left and dropping the percent symbol.

57. ***Translate.***

What is 9% of 350?

$$x = 9\% \cdot 350$$

We solve the equation.

$x = 0.09(350)$ $(9\% = 0.09)$

$x = 31.5$ Multiplying

The answer is 31.5.

59. *Writing Exercise*

61. $x + y = 7,\quad (1)$
$3(y - x) = 9\quad (2)$

Multiply Equation (1) by 3 and remove parentheses in Equation (2) and then rewrite this equation in the form $Ax + By = C$. Then add.

$$\begin{aligned} 3x + 3y &= 21 \\ -3x + 3y &= 9 \\ \hline 6y &= 30 \\ y &= 5 \end{aligned}$$

Substitute 5 for y in Equation (1).

$$\begin{aligned} x + 5 &= 7 \\ x &= 2 \end{aligned}$$

The ordered pair $(2, 5)$ checks, so it is the solution.

63. $2(5a - 5b) = 10,\quad (1)$
$-5(2a + 6b) = 10\quad (2)$

Remove parentheses and add.

$$\begin{aligned} 10a - 10b &= 10 \\ -10a - 30b &= 10 \\ \hline -40b &= 20 \\ b &= -\frac{1}{2} \end{aligned}$$

Substitute $-\frac{1}{2}$ for b in Equation (2).

$$\begin{aligned} -5\left(2a + 6\left(-\frac{1}{2}\right)\right) &= 10 \\ -5(2a - 3) &= 10 \\ -10 + 15 &= 10 \\ -10a &= -5 \\ a &= \frac{1}{2} \end{aligned}$$

The ordered pair $\left(\frac{1}{2}, -\frac{1}{2}\right)$ checks, so it is the solution.

65. $y = -\frac{2}{7}x + 3,$
$y = \frac{4}{5}x + 3$

Observe that these equations represent lines with different slopes and the same y-intercept. Thus, their point of intersection is the y-intercept, $(0, 3)$ and this is the solution of the system of equations.

67. $y = ax + b,\quad (1)$
$y = x + c\quad (2)$

Substitute $x + c$ for y in Equation (1) and solve for x.

$$\begin{aligned} y &= ax + b \\ x + c &= ax + b \quad \text{Substituting} \\ x - ax &= b - c \\ (1 - a)x &= b - c \\ x &= \frac{b - c}{1 - a} \end{aligned}$$

Substitute $\frac{b - c}{1 - a}$ for x in Equation (2) and simplify to find y.

$$\begin{aligned} y &= x + c \\ y &= \frac{b - c}{1 - a} + c \\ y &= \frac{b - c}{1 - a} + c \cdot \frac{1 - a}{1 - a} \\ y &= \frac{b - c + c - ac}{1 - a} \\ y &= \frac{b - ac}{1 - a} \end{aligned}$$

The ordered pair $\left(\frac{b - c}{1 - a}, \frac{b - ac}{1 - a}\right)$ checks and is the solution. This ordered pair could also be expressed as $\left(\frac{c - b}{a - 1}, \frac{ac - b}{a - 1}\right)$.

69. ***Familiarize***. Let x represent the number of rabbits and y the number of pheasants in the cage. Each rabbit has one head and four feet. Thus, there are x rabbit heads and $4x$ rabbit feet in the cage. Each pheasant has one head and two feet. Thus, there y pheasant heads and $2y$ pheasant feet in the cage.

Translate. We reword the problem.

Rabbit heads plus pheasant heads is 35.
$\downarrow$ $\downarrow$ $\downarrow$ $\downarrow$ $\downarrow$
$x \quad + \quad y \quad = \quad 35$

Rabbit feet plus pheasant feet is 94.
$\downarrow$ $\downarrow$ $\downarrow$ $\downarrow$ $\downarrow$
$4x \quad + \quad 2y \quad = \quad 94$

The resulting system is

$x + y = 35,\quad (1)$
$4x + 2y = 94.\quad (2)$

Carry out. We solve the system of equations. We multiply Equation (1) by -2 and then add.

$$\begin{aligned} -2x - 2y &= -70 \\ 4x + 2y &= 94 \\ \hline 2x &= 24 \quad \text{Adding} \\ x &= 12 \end{aligned}$$

Substitute 12 for x in one of the original equations and solve for y.

$$\begin{aligned} x + y &= 35 \quad (1) \\ 12 + y &= 35 \quad \text{Substituting} \\ y &= 23 \end{aligned}$$

Check. If there are 12 rabbits and 23 pheasants, the total number of heads in the cage is $12 + 23$, or 35. The total number of feet in the cage is $4 \cdot 12 + 2 \cdot 23$, or $48 + 46$, or 94. The numbers check.

State. There are 12 rabbits and 23 pheasants.

71. ***Familiarize***. Let $x =$ the man's age and $y =$ his daughter's age. Five years ago their ages were $x - 5$ and $y - 5$.

Translate.

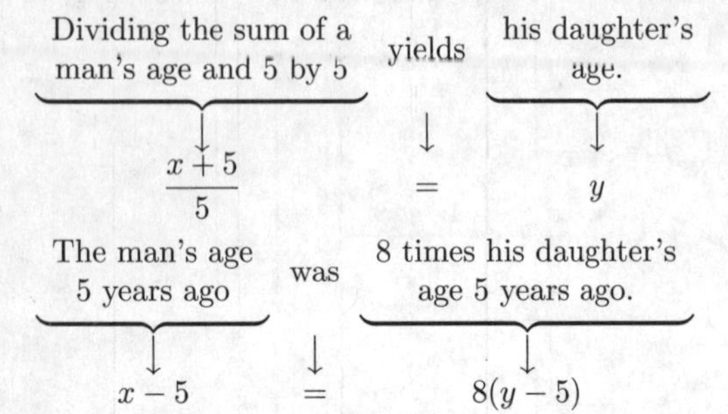

We have a system of equations:

$$\frac{x+5}{5} = y$$
$$x - 5 = 8(y-5)$$

Carry out. Solve the system.

Multiply the first Equation by 5 to clear the fraction.

$$x + 5 = 5y$$
$$x - 5y = -5$$

Simplify the second equation.

$$x - 5 = 8(y-5)$$
$$x - 5 = 8y - 40$$
$$x - 8y = -35$$

The resulting system is

$$x - 5y = -5, \quad (1)$$
$$x - 8y = -35. \quad (2)$$

Multiply Equation (2) by -1 and add.

$$\begin{aligned} x - 5y &= -5 \\ -x + 8y &= 35 \quad \text{Multiplying by } -1 \\ \hline 3y &= 30 \quad \text{Adding} \\ y &= 10 \end{aligned}$$

Substitute 10 for y in Equation (1) and solve for x.

$$x - 5y = -5$$
$$x - 5 \cdot 10 = -5 \quad \text{Substituting}$$
$$x - 50 = -5$$
$$x = 45$$

Possible solution: Man is 45, daughter is 10.

Check. If 5 is added to the man's age, $5 + 45$, the result is 50. If 50 is divided by 5, the result is 10, the daughter's age. Five years ago the father and daughter were 40 and 5, respectively, and $40 = 8 \cdot 5$. The numbers check.

State. The man is 45 years old; his daughter is 10 years old.

Exercise Set 7.4

1. ***Familiarize.*** Let $x =$ the number of two-point shots that were made and $y =$ the number of three-pointers made. Then Davis scored $2x$ points on two-point shots and $3y$ points on three-pointers.

Translate. We reword the problem and translate.

Total number of shots is 11.

$$x + y = 11$$

Total number of points is 25.

$$2x + 3y = 25$$

The resulting system is

$$x + y = 11, \quad (1)$$
$$2x + 3y = 25. \quad (2)$$

Carry out. We solve using the elimination method.

$$\begin{aligned} -2x - 2y &= -22 \quad \text{Multiplying (1) by } -2 \\ 2x + 3y &= 25 \quad (2) \\ \hline y &= 3 \quad \text{Adding} \end{aligned}$$

Substitute 3 for y in Equation (1) and solve for x.

$$x + y = 11 \quad (1)$$
$$x + 3 = 11$$
$$x = 8$$

Check. If Davis made 8 two-pointers and 3 three-pointers, then he made $8 + 3$, or 11 shots for a total of $2 \cdot 8 + 3 \cdot 3$, or $16 + 9$, or 25 points. The numbers check.

State. Davis made 8 two-point shots and 3 three-point shots.

3. ***Familiarize.*** Let $x =$ the number of two-point shots that were made and $y =$ the number of three-pointers made. Then the Nets scored $2x$ points on two-point shots and $3y$ points on three-pointers.

Translate. We reword the problem and translate.

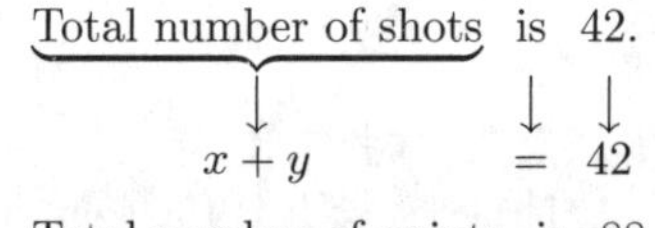

Total number of points is 88.

$$2x + 3y = 88$$

The resulting system is

$$x + y = 42, \quad (1)$$
$$2x + 3y = 88. \quad (2)$$

Carry out. We solve using the elimination method.

$$\begin{aligned} -2x - 2y &= -84 \quad \text{Multiplying (1) by } -2 \\ 2x + 3y &= 88 \quad (2) \\ \hline y &= 4 \quad \text{Adding} \end{aligned}$$

Substitute 4 for y in Equation (1) and solve for x.

$$x + y = 42 \quad (1)$$
$$x + 4 = 42$$
$$x = 38$$

Check. If the Nets made 38 two-pointers and 4 three-pointers, then they made $38 + 4$, or 42 shots and scored $2 \cdot 38 + 3 \cdot 4$, or $76 + 12$, or 88 points. The numbers check.

State. The Nets made 38 two-point shots and 4 three-point shots.

5. ***Familiarize.*** Let x = the number of 3-credit courses and y = the number of 4-credit courses. Then the 3-credit courses account for $3x$ credits and the 4-credit courses account for $4y$ credits.

Translate.

$$\underbrace{\text{Total number of courses}}_{x+y} \text{ is } 27.$$
$$x + y = 27$$

$$\underbrace{\text{Total number of credits}}_{3x+4y} \text{ is } 89.$$
$$3x + 4y = 89$$

The resulting system is

$$x + y = 27, \quad (1)$$
$$3x + 4y = 89. \quad (2)$$

Carry out. We solve using the elimination method.

$$\begin{aligned} -3x - 3y &= -81 \quad \text{Multiplying (1) by } -3 \\ 3x + 4y &= 89 \quad (2) \\ \hline y &= 8 \quad \text{Adding} \end{aligned}$$

Substitute 8 for y in Equation (1) and solve for x.

$$x + y = 27$$
$$x + 8 = 27$$
$$x = 19$$

Check. If there are 19 3-credit courses and 8 4-credit courses, the total number of courses is $19 + 8$, or 27. The total number of credits is $3 \cdot 19 + 4 \cdot 8$, or $57 + 32$, or 89. The answer checks.

State. 19 3-credit courses and 8 4-credit courses are being taken.

7. ***Familiarize.*** Let x = the number of 5-cent bottles or cans collected and y = the number of 10-cent bottles or cans collected.

Translate. We organize the given information in a table.

	\$0.05	\$0.10	Total
Number	x	y	430
Total Value	$0.05x$	$0.10y$	26.20

A system of two equations can be formed using the rows of the table.

$$x + y = 430,$$
$$0.05x + 0.10y = 26.20$$

Carry out. First we multiply on both sides of the second equation by 100 to clear the decimals.

$$x + y = 430, \quad (1)$$
$$5x + 10y = 2620 \quad (2)$$

Now multiply Equation (1) by -5 and add.

$$\begin{aligned} -5x - 5y &= -2150 \\ 5x + 10y &= 2620 \\ \hline 5y &= 470 \\ y &= 94 \end{aligned}$$

Substitute 94 for y in Equation (1) and solve for x.

$$x + y = 430$$
$$x + 94 = 430$$
$$x = 336$$

Check. If 336 5-cent bottles and cans and 94 10-cent bottles and cans were collected, then a total of $336 + 94$, or 430 bottles and cans were collected. Their total value is \$0.05(336) + \$0.10(94), or \$16.80 + \$9.40, or \$26.20. These numbers check.

State. 336 5-cent bottles and cans and 94 10-cent bottles and cans were collected.

9. ***Familiarize.*** Let c = the number of cars and m = the number of motorcycles that enter the park on a typical day. Then the cars account for payments of $20x$ and the motorcycles for $15y$.

$$\underbrace{\text{Total number of cars and motorcycles}}_{c+m} \text{ is } 5950.$$
$$c + m = 5950$$

$$\underbrace{\text{Total payment}}_{20c+15m} \text{ is } \$107,875.$$
$$20c + 15m = 107,875$$

The resulting system is

$$c + m = 5950, \quad (1)$$
$$20c + 15m = 107,875 \quad (2)$$

Carry out. We use the elimination method.

$$\begin{aligned} -15c - 15m &= -89,250 \quad \text{Multiplying (1) by } -15 \\ 20c + 15m &= 107,875 \quad (2) \\ \hline 5c &= 18,625 \quad \text{Adding} \\ c &= 3725 \end{aligned}$$

Since the problem asks only for the number of motorcycles, we could have solved for m first and stopped there, but we solve for both c and m so that we can check the solution.

Substitute 3725 for c in Equation (1) and solve for m.

$$c + m = 5950$$
$$3725 + m = 5950$$
$$m = 2225$$

Check. If there are 3725 cars and 2225 motorcycles, the total number of cars and motorcycles is $3725 + 2225$, or 5950. The total payments are $20(3725) + 15(2225)$, or $74,500 + 33,375$, or \$107,875. The answer checks.

State. On a typical day 2225 motorcycles enter the park.

11. ***Familiarize.*** Let x = the number of adult admissions and y = the number of children and senior admissions.

Translate. We organize the information in a table.

	Adult	Children and Senior	Total
Admission	\$8	\$6	
Number	x	y	394
Money Taken In	$8x$	$6y$	2610

We use the last two rows of the table to form a system of equations.

$$x + y = 394, \quad (1)$$
$$8x + 6y = 2610 \quad (2)$$

Carry out. We solve using the elimination method. We first multiply Equation (1) by -6 and add.

$$\begin{array}{rl} -6x - 6y = & -2364 \\ 8x + 6y = & 2610 \\ \hline 2x \quad = & 246 \\ x = & 123 \end{array}$$

Since the problem asks only for the number of adult admissions, this is the number that we needed to find. We will find y also, however, in order to be able to check the solution.

Substitute 123 for x in Equation (1) and solve for y.

$$x + y = 394$$
$$123 + y = 394$$
$$y = 271$$

Check. If $x = 123$ and $y = 271$, then there were $123+271$, or 394 admissions sold. The amount collected for adult admissions was $\$8 \cdot 123$, or \$984, and the amount collected for children and senior admissions was $\$6 \cdot 271$, or \$1626. The total amount collected was \$984 + \$1626, or \$2610. The numbers check.

State. There were 123 adult admissions.

13. ***Familiarize.*** Let x = the number of students receiving private lessons and y = the number of students receiving group lessons.

Translate. We present the information in a table.

	Private	Group	Total
Price	\$25	\$18	
Number	x	y	12
Money Earned	$25x$	$18y$	265

The last two rows of the table give us a system of equations.

$$x + y = 12, \quad (1)$$
$$25x + 18y = 265 \quad (2)$$

Carry out. We solve using the elimination method. First we multiply Equation (1) by -18 and add.

$$\begin{array}{rl} -18x - 18y = & -216 \\ 25x + 18y = & 265 \\ \hline 7x \quad = & 49 \\ x = & 7 \end{array}$$

Substitute 7 for x in Equation (1) and solve for y.

$$x + y = 12$$
$$7 + y = 12$$
$$y = 5$$

Check. If $x = 7$ and $y = 5$, then a total of $7 + 5$, or 12 students received lessons. Alice earned $\$25 \cdot 7$, or \$175 teaching private lessons and $\$18 \cdot 5$, or \$90 teaching group lessons. Thus, she earned a total of \$175 + \$90, or \$265. The numbers check.

State. Alice gave private lessons to 7 students and group lessons to 5 students.

15. ***Familiarize.*** Let x = the number of CDs Holly bought and y = the number of videos. Then the CDs cost a total of $5.99x$ dollars and the videos cost $7.99y$ dollars.

Translate.

Total CDs and videos purchased was 14.

$$x + y = 14$$

Total purchase price was \$93.86.

$$5.99x + 7.99y = 93.86$$

The resulting system is

$$x + y = 14, \quad (1)$$
$$5.99x + 7.99y = 93.86. \quad (2)$$

Carry out. We use the elimination method. First we multiply Equation (1) by -5.99 and add.

$$\begin{array}{rl} -5.99x - 5.99y = & 83.86 \\ 5.99x + 7.99y = & 93.86 \\ \hline 2y = & 10 \\ y = & 5 \end{array}$$

Substitute 5 for y in Equation (1) and solve for x.

$$x + y = 14$$
$$x + 5 = 14$$
$$x = 9$$

Check. If Holly bought 9 CDs and 5 videos, she bought a total of $9 + 5$, or 14 items. She paid a total of \$5.99(9) + \$7.99(5), or \$53.91 + 39.95, or \$93.86. The answer checks.

State. Holly bought 9 CDs and 5 videos.

17. ***Familiarize.*** Let x = the number of kg of Brazilian coffee to be used and y = the number of kg of Turkish coffee to be used.

Translate. Organize the given information in a table.

Type of coffee	Brazilian	Turkish	Mixture
Cost of coffee	\$19	\$22	\$20
Amount (in kg)	x	y	300
Value	$19x$	$22y$	\$20(300); or \$6000

The last two rows of the table give us two equations. Since the total amount of the mixture is 300 lb, we have

$$x + y = 300.$$

The value of the Brazilian coffee is $19x$ (x lb at \$19 per pound), the value of the Turkish coffee is $22y$ (y lb at \$22 per pound), and the value of the mixture is \$20(300) or \$6000. Thus we have

$$19x + 22y = 6000.$$

The resulting system is

$$x + y = 300, \quad (1)$$
$$19x + 22y = 6000. \quad (2)$$

Carry out. We use the elimination method. We multiply on both sides of Equation (1) by -19 and then add.

$$\begin{aligned} -19x - 19y &= -5700 \\ 19x + 22y &= 6000 \\ \hline 3y &= 300 \\ y &= 100 \end{aligned}$$

Now substitute 100 for y in Equation (1) and solve for x.

$$\begin{aligned} x + y &= 300 \\ x + 100 &= 300 \\ x &= 200 \end{aligned}$$

Check. The sum of 100 and 200 is 300. The value of the mixture is $19(200)+$22(100), or $3800+$2200, or $6000. These numbers check.

State. 200 kg of Brazilian coffee and 100 kg of Turkish coffee should be used.

19. ***Familiarize.*** Let x and y represent the number of pounds of peanuts and Brazil nuts to be used, respectively.

Translate. We organize the given information in a table.

Type of nut	Peanuts	Brazil nuts	Mixture
Cost per pound	$2.52	$3.80	$3.44
Amount	x	y	480
Value	$2.52x$	$3.80y$	3.44(480); or 1651.20

The last two rows of the table form a system of equations.

$$\begin{aligned} x + \quad y &= 480, \\ 2.52x + 3.80y &= 1651.20 \end{aligned}$$

Carry out. First we multiply the second equation by 100 to clear decimals.

$$\begin{aligned} x + \quad y &= 480, & (1) \\ 252x + 380y &= 165,120 & (2) \end{aligned}$$

Now multiply Equation (1) by -252 and add.

$$\begin{aligned} -252x - 252y &= -120,960 \\ 252x + 380y &= 165,120 \\ \hline 128y &= 44,160 \\ y &= 345 \end{aligned}$$

Substitute 345 for y in Equation (1) and solve for x.

$$\begin{aligned} x + y &= 480 \\ x + 345 &= 480 \\ x &= 135 \end{aligned}$$

Check. The sum of 135 and 345 is 480. The value of the mixture is $2.52(135) + $3.80(345), or $340.20 + $1311, or $1651.20. These numbers check.

State. 135 lb of peanuts and 345 lb of Brazil nuts should be used.

21. ***Familiarize.*** From the table in the text, note that x represents the number of milliliters of solution A to be used and y represents the number of milliliters of solution B.

Translate. We complete the table in the text.

Type of solution	50%-acid	80%-acid	68%-acid
Amount of solution	x	y	200
Percent acid	50%	80%	68%
Amount of acid in solution	$0.5x$	$0.8y$	0.68×200, or 136

Since the total amount of solution is 200 mL, we have

$$x + y = 200.$$

The amount of acid in the mixture is to be 68% of 200 mL, or 136 mL. The amounts of acid from the two solutions are $50\%x$ and $80\%y$. Thus

$$\begin{aligned} 50\%x + 80\%y &= 136, \\ \text{or} \quad 0.5x + 0.8y &= 136, \\ \text{or} \quad 5x + 8y &= 1360 \quad \text{Clearing decimals} \end{aligned}$$

Carry out. We use the elimination method.

$$\begin{aligned} x + y &= 200, & (1) \\ 5x + 8y &= 1360 & (2) \end{aligned}$$

We multiply Equation (1) by -5 and then add.

$$\begin{aligned} -5x - 5y &= -1000 \\ 5x + 8y &= 1360 \\ \hline 3y &= 360 \\ y &= 120 \end{aligned}$$

Next we substitute 120 for y in one of the original equations and solve for x.

$$\begin{aligned} x + y &= 200 & (1) \\ x + 120 &= 200 & \text{Substituting} \\ x &= 80 \end{aligned}$$

Check. The sum of 80 and 120 is 200. Now 50% of 80 is 40 and 80% of 120 is 96. These add up to 136. The numbers check.

State. 80 mL of the 50%-acid solution and 120 mL of the 80%-acid solution should be used.

23. ***Familiarize.*** Let x and y represent the number of liters of 28%-fungicide solution and 40%-fungicide solution to be used in the mixture, respectively.

Translate. We organize the given information in a table.

Type of solution	28%	40%	36%
Amount of solution	x	y	300
Percent fungicide	28%	40%	36%
Amount of fungicide in solution	$0.28x$	$0.4y$	0.36(300), or 108

We get a system of equations from the first and third rows of the table.

$$x + y = 300,$$
$$0.28x + 0.4y = 108$$

Clearing decimals we have

$$x + y = 300, \quad (1)$$
$$28x + 40y = 10,800 \quad (2)$$

Carry out. We use the elimination method. Multiply Equation (1) by -28 and add.

$$\begin{aligned} -28x - 28y &= -8400 \\ 28x + 40y &= 10,800 \\ \hline 12y &= 2400 \\ y &= 200 \end{aligned}$$

Now substitute 200 for y in Equation (1) and solve for x.

$$x + y = 300$$
$$x + 200 = 300$$
$$x = 100$$

Check. The sum of 100 and 200 is 300. The amount of fungicide in the mixture is $0.28(100) + 0.4(200)$, or $28 + 80$, or 108 L. These numbers check.

State. 100 L of the 28%-fungicide solution and 200 L of the 40%-fungicide solution should be used in the mixture.

25. ***Familiarize***. Let x and y represent the number of gallons of 87-octane gas and 93-octane gas to be blended, respectively. We organize the given information in a table.

Type of gasoline	87-octane	93-octane	91-octane
Amount of gas	x	y	12
Octane rating	87	93	91
Mixture	$87x$	$93y$	$91 \cdot 12$, or 1092

Translate. We get a system of equations from the first and third rows of the table.

$$x + y = 12, \quad (1)$$
$$87x + 93y = 1092 \quad (2)$$

Carry out. We use the elimination method. First we multiply Equation (1) by -87 and add.

$$\begin{aligned} -87x - 87y &= -1044 \\ 87x + 93y &= 1092 \\ \hline 6y &= 48 \\ y &= 8 \end{aligned}$$

Now substitute 8 for y in Equation (1) and solve for x.

$$x + y = 12$$
$$x + 8 = 12$$
$$x = 4$$

Check. The sum of 4 and 8 is 12. The mixture is $87(4) + 93(8)$, or $348 + 744$, or 1092. These numbers check.

State. 4 gal of 87-octane gas and 8 gal of 93-octane gas should be blended.

27. ***Familiarize***. Let x = the number of 1300-word pages that were filled and y = the number of 1850-word pages that were filled. Then the number of words on the x pages that hold 1300 words each is $1300x$ and on the y pages that hold 1850 words each is $1850y$.

Translate. We reword the problem and translate.

$$\underbrace{\text{Total number of pages}}_{\downarrow} \ \text{is} \ 12.$$
$$x + y \quad = \quad 12$$

$$\underbrace{\text{Total number of words}}_{\downarrow} \ \text{is} \ 18,350.$$
$$1300x + 1850y \quad = \quad 18,350$$

The resulting system of equations is

$$x + y = 12, \quad (1)$$
$$1300x + 1850y = 18,350. \quad (2)$$

Carry out. We solve using the elimination method. First we multiply Equation (1) by -1300 and add.

$$\begin{aligned} -1300x - 1300y &= -15,600 \\ 1300x + 1850y &= 18,350 \\ \hline 550y &= 2750 \\ y &= 5 \end{aligned}$$

Substitute 5 for y in Equation (1) and solve for x.

$$x + y = 12$$
$$x + 5 = 12$$
$$x = 7$$

Check. If $x = 7$ and $y = 5$, then $7 + 5$, or 12 pages are filled. The 1300-word pages contain $1300 \cdot 7$, or 9100 words and the 1850-word pages contain $1850 \cdot 5$ or 9250 words. The total number of words is $9100 + 9250$, or 18,350. The numbers check.

State. The typesetter used 7 1300-word pages and 5 1850-word pages.

29. ***Familiarize***. Let f = the number of foul shots made and t = the number of two point shots made. Then Chamberlain scored f points from foul shots and $2t$ points from two-pointers.

Translate. We reword the problem and translate.

$$\underbrace{\text{Total number of shots}}_{\downarrow} \ \text{is} \ 64.$$
$$f + t \quad = \quad 64$$

$$\underbrace{\text{Total number of points}}_{\downarrow} \ \text{is} \ 100.$$
$$f + 2t \quad = \quad 100$$

The resulting system of equations is

$$f + t = 64, \quad (1)$$
$$f + 2t = 100. \quad (2)$$

Carry out. We solve using the elimination method. First we multiply Equation (1) by -1 and add.

$$\begin{aligned} -f - t &= -64 \\ f + 2t &= 100 \\ \hline t &= 36 \end{aligned}$$

Substitute 36 for t in Equation (1) and solve for f.

$$f+t=64$$
$$f+36=64$$
$$f=28$$

Check. If Chamberlain made 28 foul shots and 36 two point shots, he made $28+36$, or 64 shots for a total of $28+2\cdot36$, or $28+72$, or 100 points. The numbers check.

State. Chamberlain made 28 foul shots and 36 two point shots.

31. ***Familiarize.*** Let $x=$ the number of fluid ounces of Kinney's suntan lotion that should be used and $y=$ the number of fluid ounces of Coppertone that should be used.

Translate. We present the information in a table.

	Kinney's	Coppertone	Mixture
spf Rating	15	30	20
Amount	x	y	50
spf Value	$15x$	$30y$	$20\cdot50$, or 1000

The last two rows of the table give us a system of equations.

$$x+y=50, \quad (1)$$
$$15x+30y=1000. \quad (2)$$

Carry out. We solve using the elimination method. First we multiply Equation (1) by -15 and add.

$$-15x-15y=-750$$
$$15x+30y=1000$$
$$15y=250$$
$$y=16\frac{2}{3}$$

Substitute $16\frac{2}{3}$ for y in Equation (1) and solve for x.

$$x+y=50$$
$$x+16\frac{2}{3}=50$$
$$x=33\frac{1}{3}$$

Check. If $x=33\frac{1}{3}$ and $y=16\frac{2}{3}$, then the total amount of suntan lotion is $33\frac{1}{3}+16\frac{1}{3}$, or 50 fluid ounces. The spf value of the mixture is $15\left(33\frac{1}{3}\right)+30\left(16\frac{2}{3}\right)$, or $500+500$, or 1000. The numbers check.

State. The mixture should contain $33\frac{1}{3}$ fluid ounces of Kinney's and $16\frac{2}{3}$ fluid ounces of Coppertone.

33. Observe that 27.5 is midway between 20 and 35. Thus the mixture would contain equal parts of the cereals that get 20% and 35% of their calories from fat. Since a 40-lb mixture is desired, it should contain 20 lb of New England Natural Bakers Muesli and 20 lb of Breadshop Supernatural granola.

35. *Writing Exercise*

37.
$$7-3x<22$$
$$-3x<15$$
$$x>-5$$

The solution set is $\{x|x>-5\}$.

39.
$$x+2\geq6$$
$$x\geq4$$

0 4

41.
$$6<-\frac{1}{2}x+1$$
$$5<-\frac{1}{2}x$$
$$-10>x, \text{ or } x<-10$$

−10 0

43. *Writing Exercise*

45. ***Familiarize.*** Let $k=$ the number of pounds of pure Kona beans that should be added to the Columbian beans and $m=$ the total weight of the final mixture, in pounds.

Translate. We present the information in a table.

	Kona	Columbian	Total
Amount	k	45	m
Percent of Kona	100%	0%	30%
Amount of Kona	$1\cdot k$, or k	$0\cdot45$, or 0	$0.3m$

The table gives us two equations.

$$k+45=m, \quad (1)$$
$$k=0.3m \quad (2)$$

Carry out. We use the substitution method. First substitute $k+45$ for m in Equation (2) and solve for k.

$$k=0.3(x+45)$$
$$k=0.3x+13.5$$
$$0.7k=13.5$$
$$k=\frac{13.5}{0.7}=\frac{135}{7}=19\frac{2}{7}$$

This is the number the problem asks for. We will also find m so that we can check the answer. Substitute $19\frac{2}{7}$ for k in Equation (1) and compute m.

$$k+45=m$$
$$19\frac{2}{7}+45=m$$
$$64\frac{2}{7}=m$$

Check. If a coffee mixture that weighs $64\frac{2}{7}$ lb contains $19\frac{2}{7}$ lb of Kona coffee, then the percent of Kona coffee in the mixture is $\frac{19\frac{2}{7}}{64\frac{2}{7}}$, or 0.3, or 30%. The answer checks.

State. $19\frac{2}{7}$ lb of Kona coffee should be added to 45 lb of Columbian coffee to obtain the desired mixture.

47. ***Familiarize***. In a table we arrange the information regarding the solution *after* some of the 30% solution is drained and replaced with pure antifreeze. We let x represent the amount of the original (30%) solution remaining, and we let y represent the amount of the 30% mixture that is drained and replaced with pure antifreeze.

Type of solution	Original (30%)	Pure anti-freeze	Mixture
Amount of solution	x	y	6.3
Percent of antifreeze	30%	100%	50%
Amount of antifreeze in solution	$0.3x$	$1 \cdot y$, or y	0.5(6.3), or 3.15

Translate. The table gives us two equations.

Amount of solution: $x + y = 6.3$

Amount of antifreeze in solution: $0.3x + y = 3.15$, or $30x + 100y = 315$

The resulting system is

$$x + y = 63, \quad (1)$$
$$30x + 100y = 315. \quad (2)$$

Carry out. We multiply Equation (1) by -30 and then add.

$$\begin{aligned} -30x - 30y &= -189 \\ 30x + 100y &= 315 \\ \hline 70y &= 126 \\ y &= \frac{126}{70} = 1.8 \end{aligned}$$

Then we substitute 1.8 for y in Equation (1) and solve for x.

$$\begin{aligned} x + y &= 6.3 \\ x + 1.8 &= 6.3 \\ x &= 4.5 \end{aligned}$$

Check. When $x = 4.5$ L and $y = 1.8$ L, the total is 6.3 L. The amount of antifreeze in the mixture is $0.3(4.5)+1.8$, or $1.35+1.8$, or 3.15 L. This is 50% of 6.3 L, so the numbers check.

State. 1.8 L of the original mixture should be drained and replaced with pure antifreeze.

49. ***Familiarize***. Let x represent the number of gallons of 91-octane gas to be added to the tank and let y represent the total number of gallons in the tank after the 91-octane gas is added. We organize the given information in a table.

Type of gasoline	85-octane	91-octane	Mixture
Amount of gas	5	x	y
Octane rating	85	91	87
Mixture	$85 \cdot 5$, or 425	$91x$	$87y$

Translate. We get a system of equations from the first and third rows of the table.

$$5 + x = y, \quad (1)$$
$$425 + 91x = 87y \quad (2)$$

Carry out. Substitute $5 + x$ for y in Equation (2) and solve for x.

$$\begin{aligned} 425 + 91x &= 87y \\ 425 + 91x &= 87(5 + x) \\ 425 + 91x &= 435 + 87x \\ 425 + 4x &= 435 \\ 4x &= 10 \\ x &= 2.5 \end{aligned}$$

Although the original problem asks us to find only x, we will find y also in order to check the answer. Substitute 2.5 for x in Equation (1) and compute y.

$$y = 5 + 2.5 = 7.5$$

Check. The mixture is $425 + 91(2.5)$, or 652.5. This is equal to $87(7.5)$, so the answer checks.

State. Kim should add 2.5 gal of 91-octane gas to her tank.

51. ***Familiarize***. Let x = Juanita's regular hourly pay rate and let y = her overtime pay rate. Since $55 - 40 = 15$, she worked 15 hr of overtime. For the first 40 hr she earned $40x$ dollars and for the 15 overtime hours she earned $15y$ dollars.

Translate.

$$\underbrace{\text{Total pay}} \text{ is } \$812.50.$$
$$40x + 15y = 812.50$$

$$\underbrace{\text{Overtime rate}} \text{ is } 1.5 \text{ times } \underbrace{\text{regular rate.}}$$
$$y = 1.5 \cdot x$$

The resulting system is

$$40x + 15y = 812.50, \quad (1)$$
$$y = 1.5x. \quad (2)$$

Carry out. It will be most efficient to use the substitution method. First substitute $1.5x$ for y in Equation (1) and solve for x.

$$\begin{aligned} 40x + 15(1.5x) &= 812.50 \\ 40x + 22.5x &= 812.50 \\ 62.5x &= 812.50 \\ x &= 13 \end{aligned}$$

Since we are asked to find only the regular hourly rate, we could stop here but we will also find the overtime rate so that we can check our work.

Substitute 13 for x in Equation (2) and compute y.

$$y = 1.5x = 1.5(13) = 19.5$$

Check. \$19.50 is one and a half times \$13. Also, Juanita would earn 40(\$13) + 15(\$19.50), or \$520 + \$292.50, or \$812.50. The answer checks.

State. Juanita's regular hourly pay rate is \$13.

53. ***Familiarize***. Let x = the number of liters of skim milk and y = the number of liters of 2% milk.

Translate. We present the information in a table.

Type of milk	4.6%	Skim	2% (Mixture)
Amount of milk	1000 gal	x	y
Percent of butterfat	4.6%	0%	2%
Amount of butterfat in milk	4.6% × 1000, or 46 gal	$0\% \cdot x$, or 0 L	$2\%y$

The first and third rows of the table give us two equations.

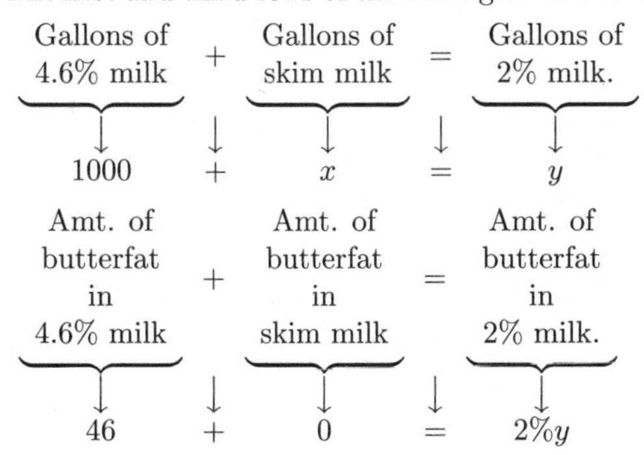

The resulting system is

$$1000 + x = y,$$
$$46 = 2\%y, \text{ or}$$

$$1000 + x = y,$$
$$46 = 0.02y.$$

Carry out. We solve the second equation for y.

$$46 = 0.02y$$
$$\frac{46}{0.02} = y$$
$$2300 = y$$

We substitute 2300 for y in the first equation and solve for x.

$$1000 + x = y$$
$$1000 + x = 2300$$
$$x = 1300$$

Check. We consider x = 1300 gal and y = 2300 gal. The difference between 1300 gal and 2300 gal is 1000 gal. There is no butterfat in the skim milk. There are 46 liters of butterfat in the 1000 liters of the 4.6% milk. Thus there are 46 liters of butterfat in the mixture. This checks because 2% of 2300 is 46.

State. 1300 gal of skim milk should be added.

55. ***Familiarize***. Let x represent the ten's digit and y the one's digit. Then the number is $10x + y$.

Translate.

The number is 6 times the sum of its digits.

$$10x + y = 6 \cdot (x + y)$$

The ten's digit is 1 more than the one's digit.

$$x = 1 + y$$

We simplify the first equation.

$$10x + y = 6(x + y)$$
$$10x + y = 6x + 6y$$
$$4x - 5y = 0$$

The system of equations is

$$4x - 5y = 0, \quad (1)$$
$$x = 1 + y. \quad (2)$$

Carry out. We use the substitution method. We substitute $1 + y$ for x in Equation (1) and solve for y.

$$4(1 + y) - 5y = 0$$
$$4 + 4y - 5y = 0$$
$$4 - y = 0$$
$$4 = y$$

Then we substitute 4 for y in Equation (2) and compute x.

$$x = 1 + y = 1 + 4 = 5$$

Check. We consider the number 54. The number is 6 times the sum of the digits, 9. The ten's digit is 1 more than the one's digit. This number checks.

State. The number is 54.

57. ***Familiarize***. Let x = Tweedledum's weight and y = Tweedledee's weight, in pounds.

Translate.

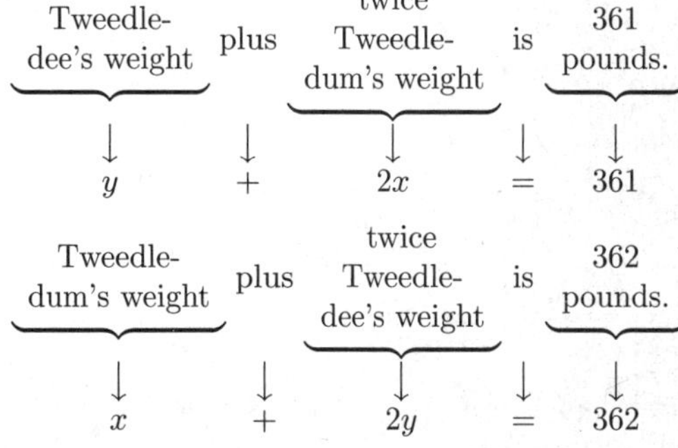

We have a system of equations.

$$y + 2x = 361,$$
$$x + 2y = 362, \text{ or}$$

$$2x + y = 361, \quad (1)$$
$$x + 2y = 362 \quad (2)$$

Carry out. We use elimination. First multiply Equation (2) by -2 and add.

$$\begin{array}{rl} 2x + \;\; y &= 361 \\ -2x - \;\; 4y &= -724 \\ \hline -3y &= -363 \\ y &= 121 \end{array}$$

Now substitute 121 for y in Equation (2) and solve for x.

$$\begin{aligned} x + 2y &= 362 \quad (2) \\ x + 2 \cdot 121 &= 362 \quad \text{Substituting} \\ x + 242 &= 362 \\ x &= 120 \end{aligned}$$

Check. If Tweedledum weighs 120 lb and Tweedledee weighs 121 lb, then the sum of Tweedledee's weight and twice Tweedledum's is $121+2\cdot 120$, or $121+240$, or 361 lb. The sum of Tweedledum's weight and twice Tweedledee's is $120+2\cdot 121$, or $120+242$, or 362 lb. The answer checks.

State. Tweedledum weighs 120 lb, and Tweedledee weighs 121 lb.

Exercise Set 7.5

1. True; see the box on page 466 in the text.

3. True; see the box on page 466 in the text.

5. We use alphabetical order of variables. We replace x by -3 and y by -5.

$$\begin{array}{r|l} \multicolumn{2}{c}{x + 3y < -18} \\ \hline -3 + 3(-5) & -18 \\ -3 - 15 & \\ & \\ \multicolumn{2}{c}{-18 \stackrel{?}{=} -18 \quad \text{FALSE}} \end{array}$$

Since $-18 < -18$ is false, $(-3, -5)$ is not a solution.

7. We use alphabetical order of variables. We replace x by $\frac{7}{8}$ and y by $\frac{1}{2}$.

$$\begin{array}{r|l} \multicolumn{2}{c}{6y + 5x \geq -3} \\ \hline 6 \cdot \frac{1}{2} + 5\left(\frac{7}{8}\right) & -3 \\ 3 + \frac{35}{8} & \\ \multicolumn{2}{c}{\frac{59}{8} \stackrel{?}{=} -3 \quad \text{TRUE}} \end{array}$$

Since $\frac{59}{8} \geq 7$ is true, $\left(\frac{7}{8}, \frac{1}{2}\right)$ is a solution.

9. Graph $y \leq x + 5$.

First graph the line $y = x+5$. The intercepts are $(0, 5)$ and $(-5, 0)$. We draw a solid line since the inequality symbol is $\leq$. Then we pick a test point that is not on the line We try $(0, 0)$.

$$\begin{array}{r|l} \multicolumn{2}{c}{y \leq x + 5} \\ \hline 0 & 0 + 5 \\ \multicolumn{2}{c}{0 \stackrel{?}{=} 5 \quad \text{TRUE}} \end{array}$$

We see that $(0, 0)$ is a solution of the inequality, so we shade the region that contains $(0, 0)$.

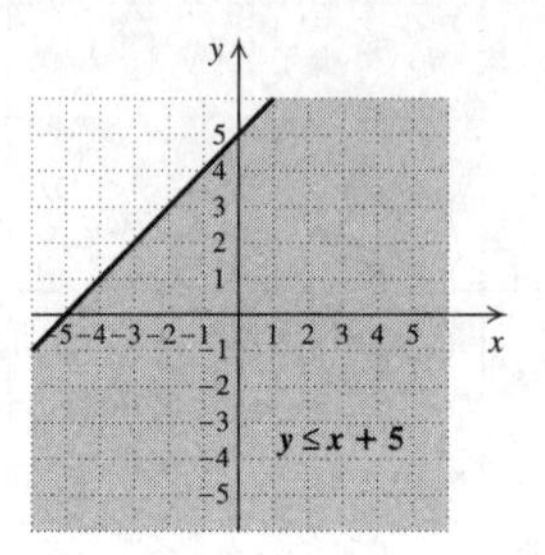

11. Graph $y < x - 2$.

First graph the line $y = x - 2$. The intercepts are $(0, -2)$ and $(2, 0)$. We draw a dashed line since the inequality symbol is $<$. Then we pick a test point that is not on the line. We try $(0, 0)$.

$$\begin{array}{r|l} \multicolumn{2}{c}{y < x - 2} \\ \hline 0 & 0 - 2 \\ \multicolumn{2}{c}{0 \stackrel{?}{=} -2 \quad \text{FALSE}} \end{array}$$

Since $(0, 0)$ is not a solution of the inequality, we shade the region that does not contain $(0, 0)$.

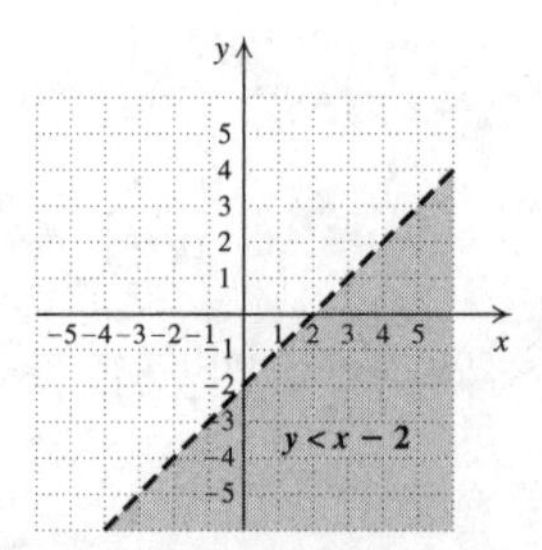

13. Graph $y \geq x - 3$.

First graph the line $y = x - 3$. The intercepts are $(0, -3)$ and $(3, 0)$. We draw a solid line since the inequality symbol is $\geq$. Then we test the point $(0, 0)$.

$$\begin{array}{r|l} \multicolumn{2}{c}{y \geq x - 3} \\ \hline 0 & 0 - 3 \\ \multicolumn{2}{c}{0 \stackrel{?}{=} -3 \quad \text{TRUE}} \end{array}$$

Since $(0, 0)$ is a solution of the inequality, we shade the region containing $(0, 0)$.

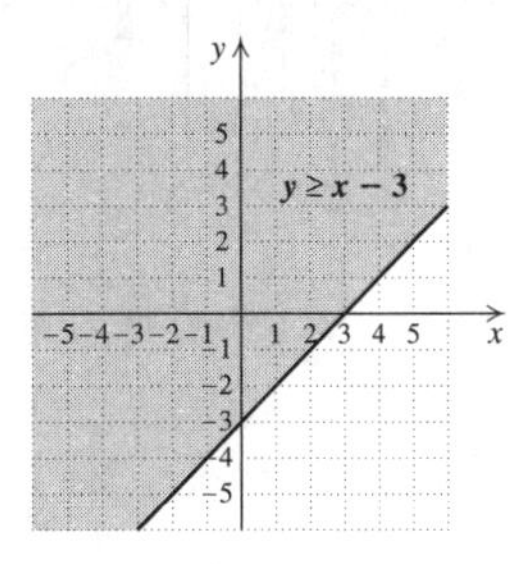

15. Graph $y \leq 2x - 1$.

First graph the line $y = 2x - 1$. The intercepts are $(0, -1)$ and $\left(\frac{1}{2}, 0\right)$. We draw a solid line since the inequality symbol is $\leq$. Then we test the point $(0, 0)$.

$$\begin{array}{c|c} y & \leq 2x - 1 \\ \hline 0 & 2 \cdot 0 - 1 \end{array}$$

$$0 \stackrel{?}{=} -1 \qquad \text{FALSE}$$

Since $(0, 0)$ is not a solution of the inequality, we shade the region that does not contain $(0, 0)$.

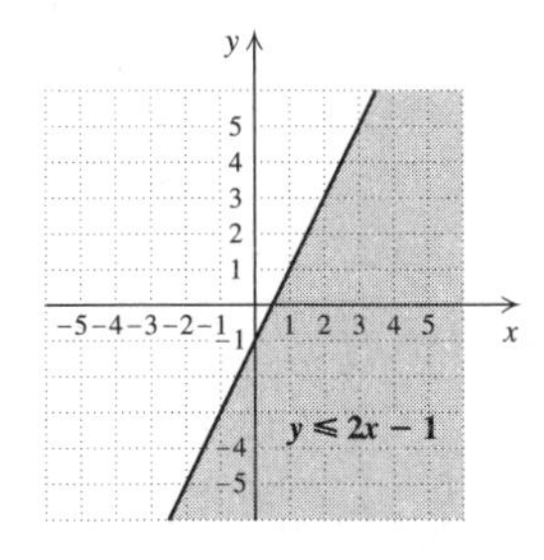

17. Graph $x + y \leq 4$.

First graph the line $x + y = 4$. The intercepts are $(0, 4)$ and $(4, 0)$. We draw a solid line since the inequality symbol is $\leq$. Then we test the point $(0, 0)$.

$$\begin{array}{c|c} x + y & \leq 4 \\ \hline 0 + 0 & 4 \end{array}$$

$$0 \stackrel{?}{=} 4 \quad \text{TRUE}$$

Since $(0, 0)$ is a solution of the inequality, we shade the region that contains $(0, 0)$.

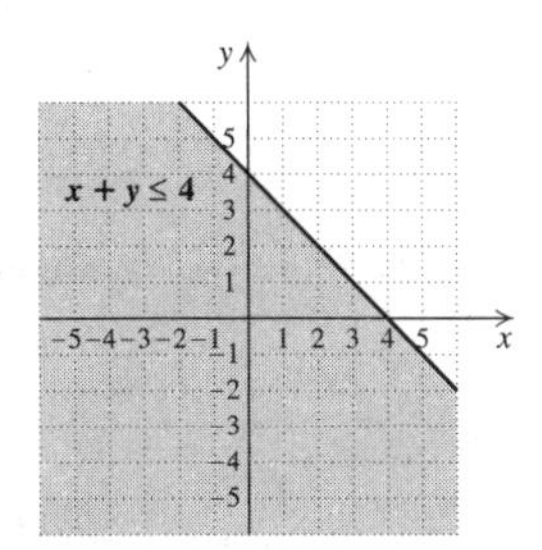

19. Graph $x - y > 7$.

First graph the line $x - y = 7$. The intercepts are $(0, -7)$ and $(7, 0)$. We draw a dashed line since the inequality symbol is $>$. Then we test the point $(0, 0)$.

$$\begin{array}{c|c} x - y & > 7 \\ \hline 0 - 0 & 7 \end{array}$$

$$0 \stackrel{?}{=} 7 \quad \text{FALSE}$$

Since $(0, 0)$ is not a solution of the inequality, we shade the region that does not contain $(0, 0)$.

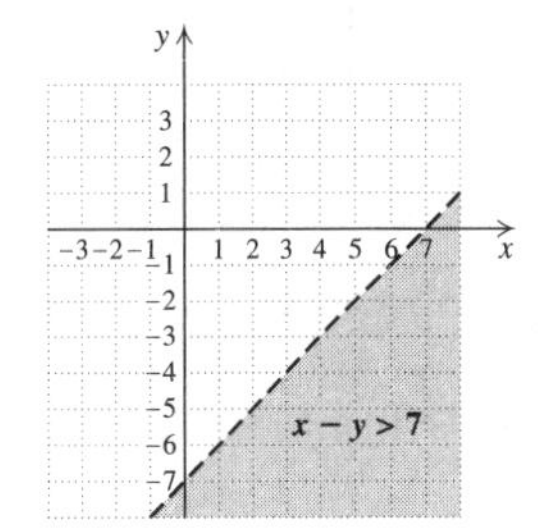

21. Graph $y \geq 1 - 2x$.

First graph the line $y = 1 - 2x$. The intercepts are $(0, 1)$ and $\left(\frac{1}{2}, 0\right)$. We draw a solid line since the inequality symbol is $\geq$. Then we test the point $(0, 0)$.

$$\begin{array}{c|c} y & \geq 1 - 2x \\ \hline 0 & 1 - 2 \cdot 0 \end{array}$$

$$0 \stackrel{?}{=} 1 \qquad \text{FALSE}$$

Since $(0, 0)$ is not a solution of the inequality, we shade the region that does not contain $(0, 0)$.

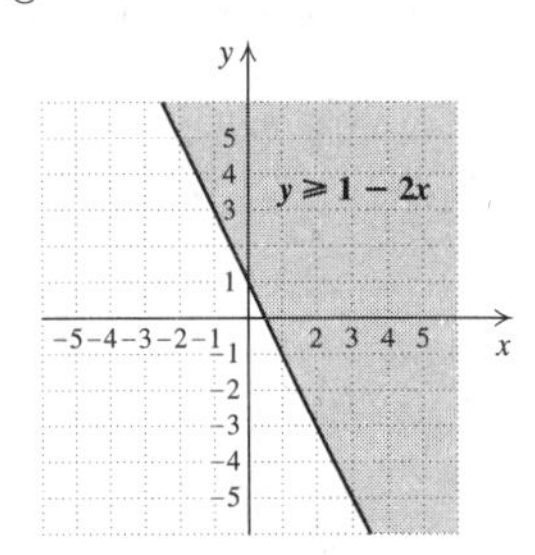

23. Graph $y - 3x > 0$.

First graph the line $y - 3x = 0$, or $y = 3x$. Two points on the line are $(0, 0)$ and $(1, 3)$. We draw a dashed line, since the inequality symbol is $>$. Then we test the point $(2, 1)$, which is not a point on the line.

$$\begin{array}{c|c} y - 3x & > 0 \\ \hline 1 - 3 \cdot 2 & 0 \\ 1 - 6 & \end{array}$$

$$-5 \stackrel{?}{=} 0 \quad \text{FALSE}$$

Since $(2, 1)$ is not a solution of the inequality, we shade the region that does not contain $(2, 1)$.

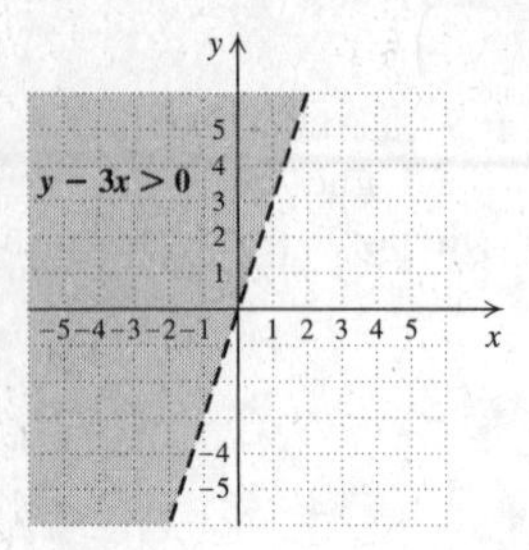

25. Graph $x \ge 4$.

First graph the line $x = 4$ using a solid line since the inequality symbol is $\ge$. Then use $(5, -3)$ as a test point. We can write the inequality as $x + 0y \ge 4$.

$$\begin{array}{c|c} \multicolumn{2}{c}{x + 0y \ge 4} \\ \hline 5 + 0(-3) & 4 \end{array}$$

$$5 \stackrel{?}{=} 4 \quad \text{TRUE}$$

Since $(5, -3)$ is a solution of the inequality, we shade the region containing $(5, -3)$.

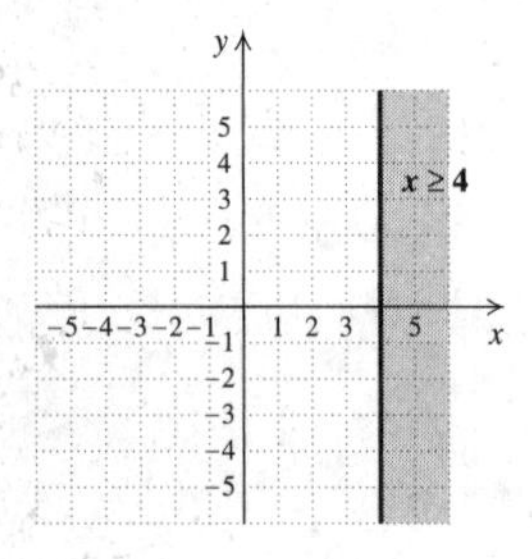

27. Graph $y \le 4$.

Graph the line $y = 4$ using a solid line since the inequality symbol is $\le$. Then use $(1, -2)$ as a test point. We can write the inequality as $0x + y \le 4$.

$$\begin{array}{c|c} \multicolumn{2}{c}{x + 0y \le 4} \\ \hline 0 \cdot 1 + (-2) & 4 \end{array}$$

$$-2 \stackrel{?}{=} 4 \quad \text{TRUE}$$

Since $(1, -2)$ is a solution of the inequality, we shade the region containing $(1, -2)$.

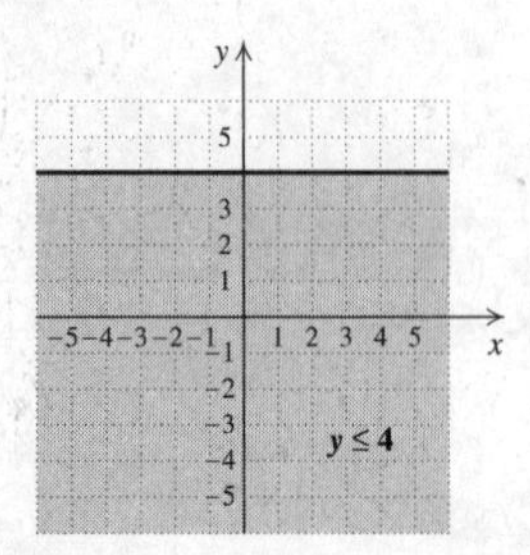

29. Graph $y \ge -5$.

Graph the line $y = -5$ using a solid line since the inequality symbol is $\ge$. Then use $(2, 3)$ as a test point. We can write the inequality as $0x + y \ge -5$.

$$\begin{array}{c|c} \multicolumn{2}{c}{0x + y \ge -5} \\ \hline 0 \cdot 2 + 3 & -5 \end{array}$$

$$3 \stackrel{?}{=} -5 \quad \text{TRUE}$$

Since $(2, 3)$ is a solution of the inequality, we shade the region containing $(2, 3)$.

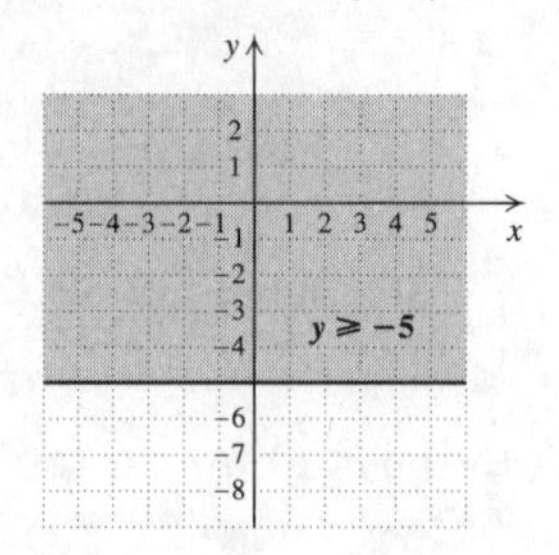

31. Graph $x < 3$.

Graph the line $x = 3$ using a dashed line since the inequality symbol is $<$. Then use $(-1, 2)$ as a test point. We can write the inequality as $x + 0y < 3$.

$$\begin{array}{c|c} \multicolumn{2}{c}{x + 0y < 3} \\ \hline -1 + 0 \cdot 2 & 3 \end{array}$$

$$-1 \stackrel{?}{=} 3 \quad \text{TRUE}$$

Since $(-1, 2)$ is a solution of the inequality, we shade the region containing $(-1, 2)$.

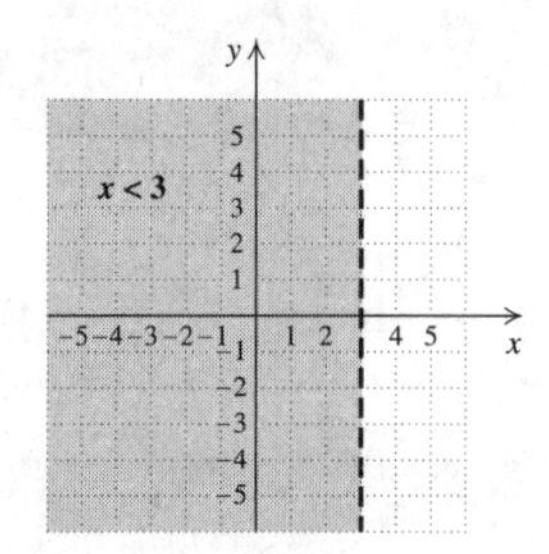

33. Graph $x - y < -10$.

Graph the line $x - y = -10$. The intercepts are $(0, 10)$ and $(-10, 0)$. We draw a dashed line since the inequality symbol is $<$. Then we test the point (0,0).

$$\begin{array}{c|c} \multicolumn{2}{c}{x - y < -10} \\ \hline 0 - 0 & -10 \end{array}$$

$$0 \stackrel{?}{=} -10 \quad \text{FALSE}$$

Since $(0, 0)$ is not a solution of the inequality, we shade the region that does not contain $(0, 0)$.

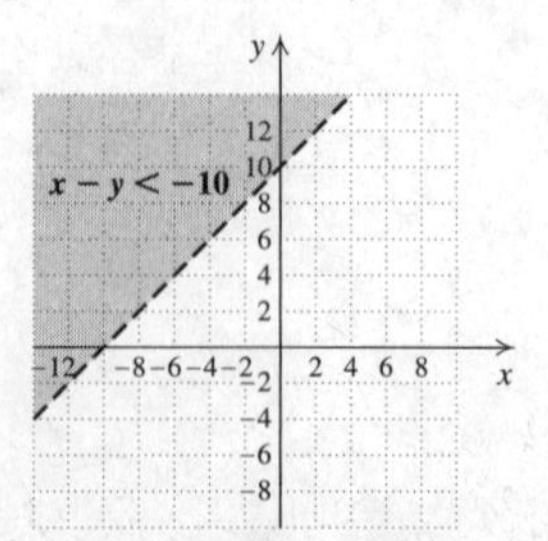

35. Graph $2x + 3y \leq 12$.

First graph the line $2x+3y = 12$. The intercepts are $(0, 4)$ and $(6, 0)$. We draw a solid line since the inequality symbol is $\leq$. Then we test the point $(0, 0)$.

$$\begin{array}{c|c} \multicolumn{2}{c}{2x + 3y \leq 12} \\ \hline 2 \cdot 0 + 3 \cdot 0 & 12 \\ \multicolumn{2}{c}{0 \stackrel{?}{=} 12 \quad \text{TRUE}} \end{array}$$

Since $(0, 0)$ is a solution of the inequality, we shade the region containing $(0, 0)$.

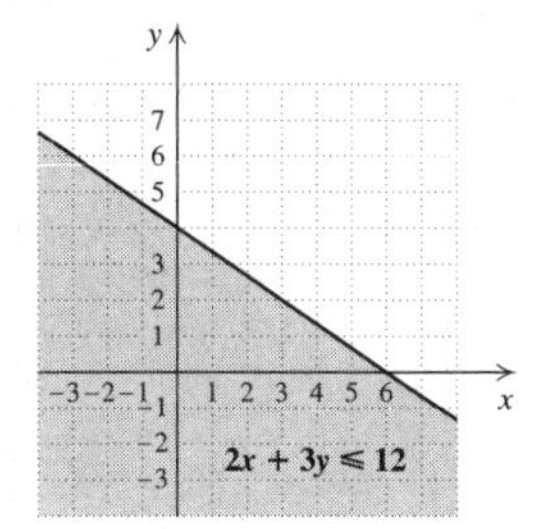

37. *Writing Exercise*

39.
$$\begin{aligned} 3x + 5y &= 3(-2) + 5(4) \\ &= -6 + 20 \\ &= 14 \end{aligned}$$

41. We find June 2002, on the horizontal axis, go up to the line representing corrugated cardboard, then go across to the vertical axis, and read the price there. We see that recyclers were paying about \$100 per ton for corrugated cardboard in June 2002.

43. Find the highest point on the graph representing newspaper and then go directly down to the horizontal axis and read the date there. We see that the value of newspaper peaked in June 2000.

45. Horizontal segments of the graph represent periods during which the price paid remained constant. On the graph representing the price paid for corrugated cardboard, there are horizontal segments from December 2000 to June 2001 and from December 2002 to June 2003.

47. *Writing Exercise*

49. The c children weigh $75c$ lb, and the a adults weigh $150a$ lb. Together, the children and adults weigh $75c + 150a$ lb. When this total is more than 1000 lb the elevator is overloaded, so we have $75c + 150a > 1000$. (Of course, c and a would also have to be nonnegative, so we show only the portion of the graph that is in the first quadrant.)

To graph $75c + 150a > 1000$, we first graph $75c + 150a = 1000$ using a dashed line. (Remember to use alphabetical order of variables.) Then we test the point $(0, 0)$.

$$\begin{array}{c|c} \multicolumn{2}{c}{75c + 150a > 1000} \\ \hline 75 \cdot 0 + 150 \cdot 0 & 1000 \\ \multicolumn{2}{c}{0 \stackrel{?}{=} 1000 \quad \text{FALSE}} \end{array}$$

Since $(0, 0)$ is not a solution of the inequality, we shade the region that does not contain $(0, 0)$.

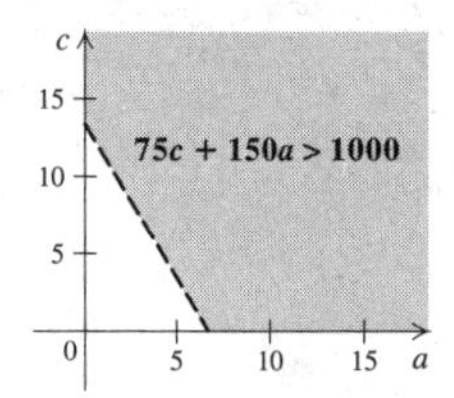

51. The sum of the riser and tread is given by $r + t$. This should not be less than 17 inches, so we write the inequality $r + t \geq 17$. To graph $r + t \geq 17$, we first graph $r + t = 17$ using a solid line. The intercepts are $(0, 17)$ and $(17, 0)$. Then we test the point $(0, 0)$.

$$\begin{array}{c|c} \multicolumn{2}{c}{r + t \geq 17} \\ \hline 0 + 0 & 17 \\ \multicolumn{2}{c}{0 \stackrel{?}{=} 17 \quad \text{FALSE}} \end{array}$$

Since $(0, 0)$ is not a solution of the inequality, we shade the region that does not contain $(0, 0)$.

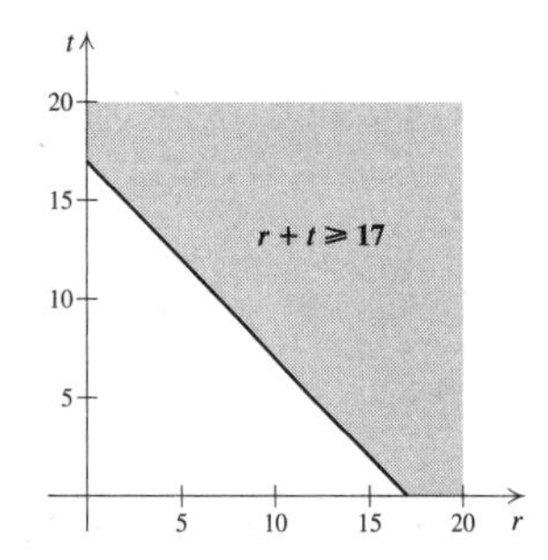

53. First find the equation of the line containing the points $(2, 0)$ and $(0, -2)$. The slope is

$$\frac{-2 - 0}{0 - 2} = \frac{-2}{-2} = 1.$$

We know that the y-intercept is $(0, -2)$, so we write the equation using the slope-intercept form.

$$\begin{aligned} y &= mx + b \\ y &= 1 \cdot x + (-2) \\ y &= x - 2 \end{aligned}$$

Since the line is dashed, the inequality symbol will be $<$ or $>$. To determine which, we substitute the coordinates of a point in the shaded region. We will use $(0, 0)$.

$$\begin{array}{c|c} y & x - 2 \\ \hline 0 & 0 - 2 \\ \multicolumn{2}{c}{0 \stackrel{?}{=} -2} \end{array}$$

Since $0 > -2$ is true, the correct symbol is $>$. The inequality is $y > x - 2$.

55. Graph $xy \leq 0$.

From the principle of zero products, we know that $xy = 0$ when $x = 0$ or $y = 0$. Therefore, the graph contains the lines $x = 0$ and $y = 0$, or the y- and x-axes. Also, $xy < 0$ when x and y have different signs. This is the case for all points in the second quadrant (x is negative and y is positive) and in the fourth quadrant (x is positive and y is negative). Thus, we shade the second and fourth quadrants.

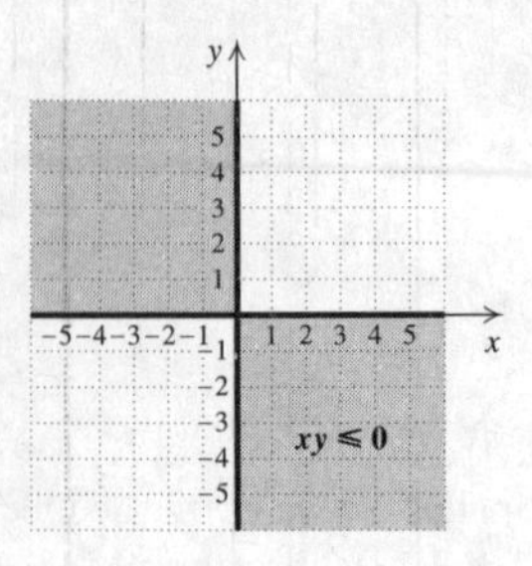

57. First solve the equation for y.

$$y + 3x \leq 4.9$$
$$y \leq -3x + 4.9$$

Then graph the line $y = -3x + 4.9$ and shade below the line.

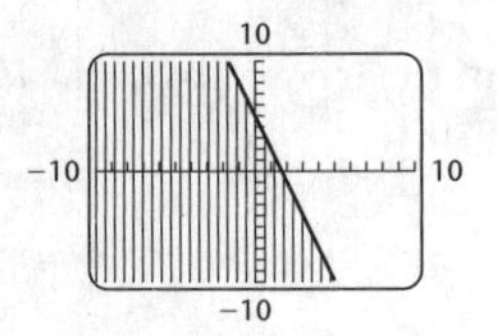

Exercise Set 7.6

1. $x + y \leq 7,$
$x - y \leq 4$

We graph the lines $x + y = 7$ and $x - y = 4$ using solid lines. We indicate the region for each inequality by the arrows at the ends of the lines. We shade the area where the regions overlap.

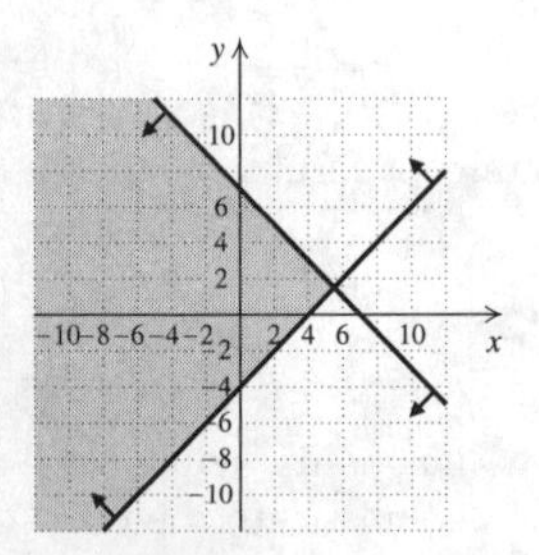

3. $y - 2x > 1,$
$y - 2x < 3$

We graph the lines $y - 2x = 1$ and $y - 2x = 3$ using dashed lines. We indicate the region for each inequality by the arrows at the ends of the lines. We shade the area where the regions overlap.

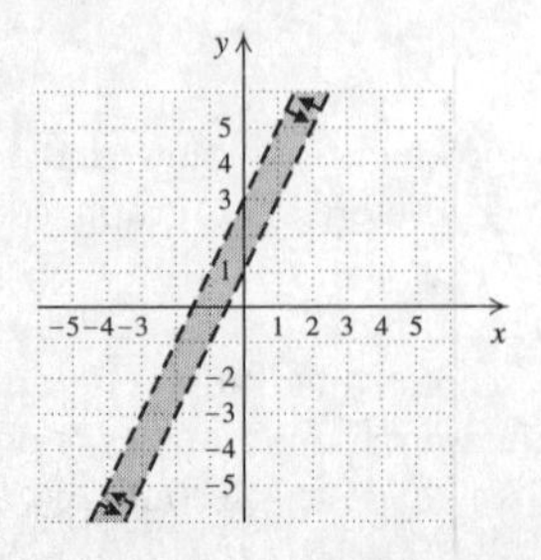

5. $y \geq -1,$
$x > 3 + y$

We graph the line $y = -1$ using a solid line and the line $x = 3 + y$ using a dashed line. We indicate the region for each inequality by the arrows at the ends of the lines. We shade the area where the regions overlap.

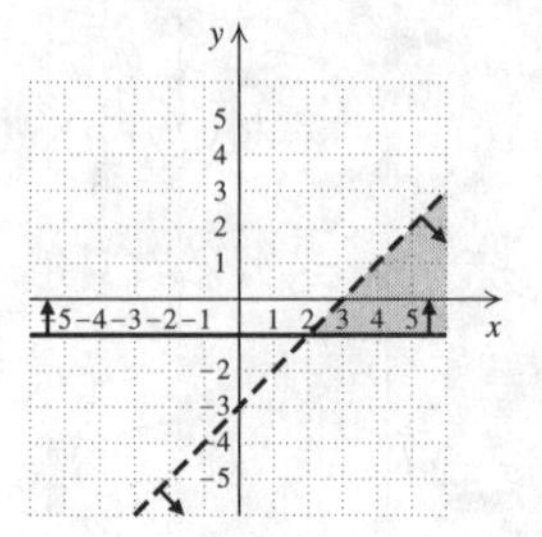

7. $y > 3x - 2,$
$y < -x + 4$

We graph the lines $y = 3x - 2$ and $y = -x + 4$ using dashed lines. We indicate the region for each inequality by the arrows at the ends of the lines. We shade the area where the regions overlap.

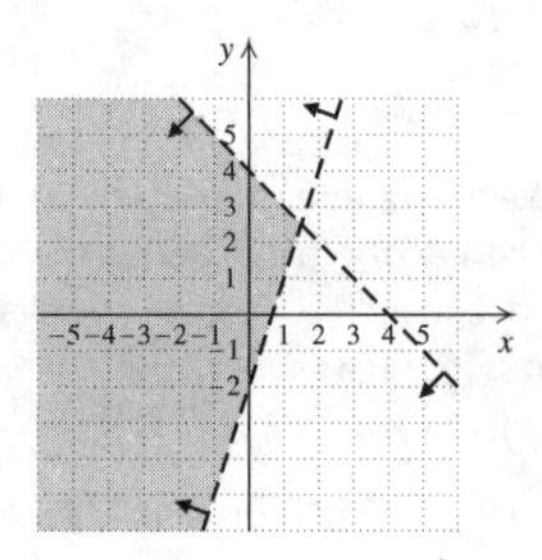

9. $x \leq 4,$
$y \leq 5$

We graph the lines $x = 4$ and $y = 5$ using sold lines. We indicate the region for each inequality by the arrows at the ends of the lines. We shade the area where the regions overlap.

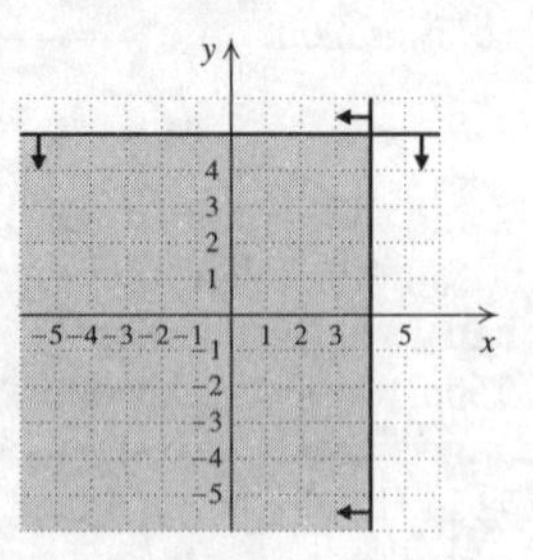

11. $x \leq 0,$
$y \leq 0$

We graph the lines $x = 0$ and $y = 0$ using solid lines. We indicate the region for each inequality by the arrows at the ends of the lines. We shade the area where the regions overlap.

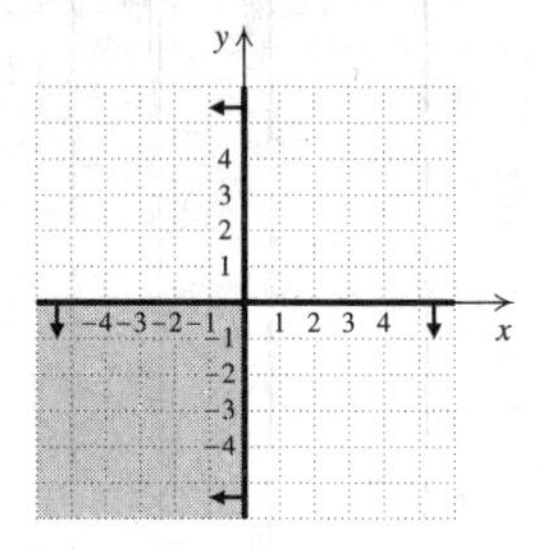

13. $2x - 3y \geq 9,$

$2y + x > 6$

We graph the line $2x - 3y = 9$ using a solid line and the line $2y + x = 6$ using a dashed line. We indicate the region for each inequality by the arrows at the ends of the lines. We shade the area where the regions overlap.

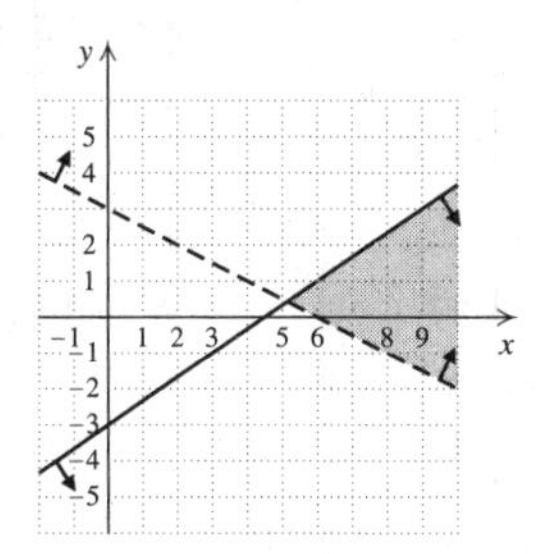

15. $y > 4x - 2,$

$y \leq 1 - x$

We graph the line $y = 4x - 2$ using a dashed line and the line $y = 1 - x$ using a solid line. We indicate the region for each inequality by the arrows at the ends of the lines. We shade the area where the regions overlap.

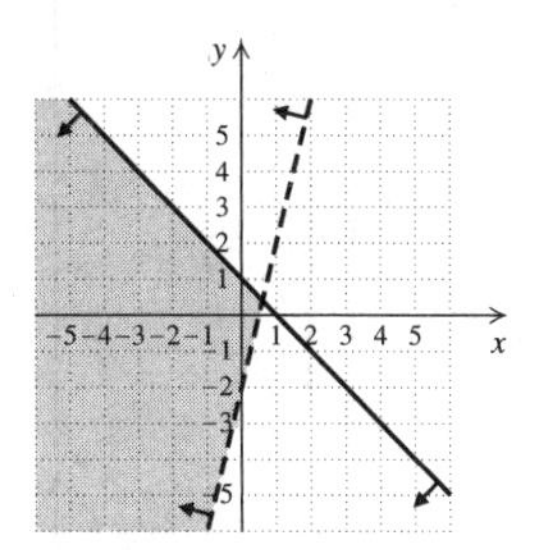

17. $x + y \leq 5,$

$x \geq 0,$

$y \geq 0,$

$y \leq 3$

We graph the lines $x + y = 5$, $x = 0$, $y = 0$, and $y = 3$ using solid lines. We indicate the region for each inequality by the arrows at the ends of the lines. We shade the area where the regions overlap.

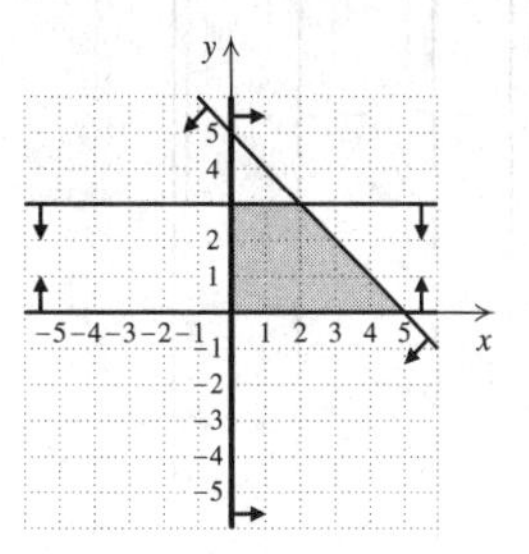

19. $y - x \geq 1,$

$y - x \leq 3,$

$x \leq 5,$

$x \geq 2$

We graph the lines $y - x = 1$, $y - x = 3$, $x = 5$, and $x = 2$ using solid lines. We indicate the region for each inequality by the arrows at the ends of the lines. We shade the area where the regions overlap.

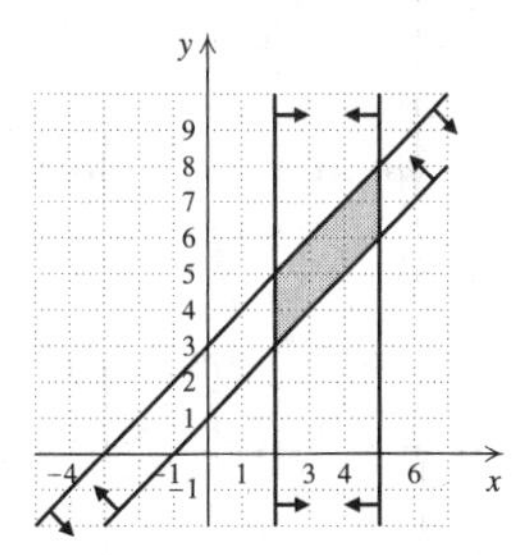

21. $y \leq x,$

$x \geq -2,$

$x \leq -y$

We graph the lines $y = x$, $x = -2$ and $x = -y$ using solid lines. We indicate the region for each inequality by the arrows at the ends of the lines. We shade the area where the regions overlap.

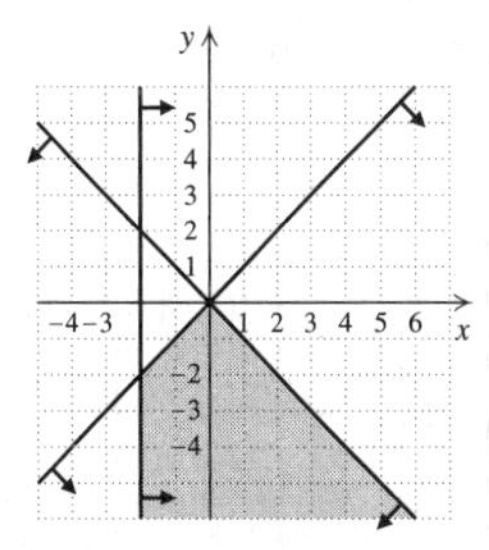

23. *Writing Exercise*

25.
$$7 = \frac{k}{5}$$
$$5 \cdot 7 = 5 \cdot \frac{k}{5}$$
$$35 = k$$

The solution is 35.

27. $18 = k \cdot 3$

$\frac{18}{3} = \frac{k \cdot 3}{3}$

$6 = k$

The solution is 6.

29. $5 = k \cdot 45$

$\frac{5}{45} = \frac{k \cdot 45}{45}$

$\frac{1}{9} = k$

The solution is $\frac{1}{9}$.

31. *Writing Exercise*

33. $3r + 6t \geq 36,$

$2r + 3t \geq 21,$

$3r + 3t \geq 30,$

$t \geq 0,$

$r \geq 0$

We graph the boundary lines, find the region for each inequality, and shade the area where the regions overlap.

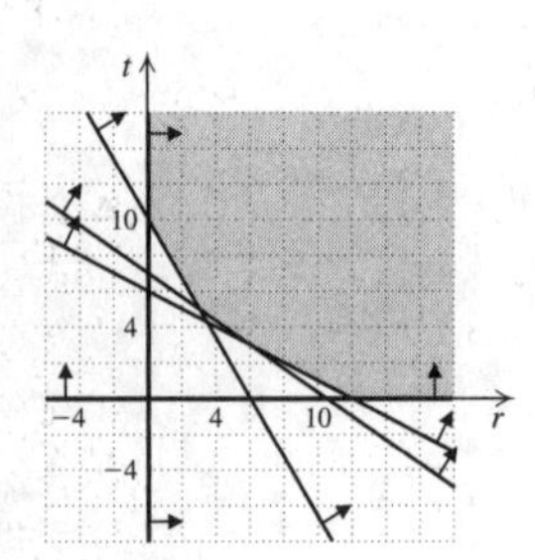

35. $x + 3y \leq 6,$

$2x + y \geq 4,$

$3x + 9y \geq 18$

We graph the boundary lines and find the region for each inequality. We find that the solution set consists of a single point, $\left(\frac{6}{5}, \frac{8}{5}\right)$, or $(1.2, 1.6)$.

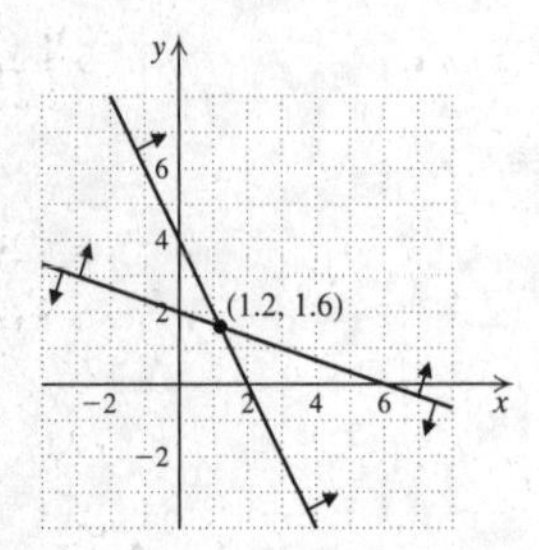

37. Note that the graphs of the related equations for the first two inequalities are parallel lines since $4x+6y = 2(2x+3y)$ and $9 \neq 2 \cdot 1$. Also observe that we would shade below the first line when graphing the solution set of the first inequality and we would shade above the second line when graphing the solution set of the second inequality. Thus there is no region of overlap for these two inequalities and hence the system of inequalities has no solution.

39. *Writing Exercise*

41.

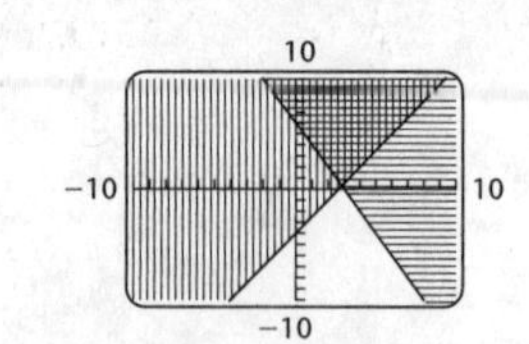

Exercise Set 7.7

1. As the number of copiers increases, the time required to complete the job decreases, so the situation reflects inverse variation.

3. As the number of cabinets increases, the number of flaws increases, so the situation reflects direct variation.

5. As the number of heaters increases, the time required to heat the room decreases, so the situation reflects inverse variation.

7. We substitute to find k.

$y = kx$ y varies directly as x.

$20 = k \cdot 4$ Substituting 20 for y and 4 for x

$\frac{20}{4} = k$

$5 = k$ k is the constant of variation.

The equation of variation is $y = 5x$.

9. We substitute to find k.

$y = kx$ y varies directly as x.

$0.7 = k \cdot 0.4$ Substituting 0.7 for y and 0.4 for x

$\frac{0.7}{0.4} = k$

$\frac{7}{4} = k$, or $k = 1.75$

The equation of variation is $y = \frac{7}{4}x$, or $1.75x$.

11. We substitute to find k.

$y = kx$

$400 = k \cdot 75$ Substituting 400 for y and 75 for x

$\frac{400}{75} = k$

$\frac{16}{3} = k$

The equation of variation is $y = \frac{16}{3}x$.

13. We substitute to find k.

$y = kx$

$200 = k \cdot 300$ Substituting 200 for y and 300 for x

$\frac{200}{300} = k$

$\frac{2}{3} = k$

The equation of variation is $y = \frac{2}{3}x$.

15. We substitute to find k.

$$y = \frac{k}{x}$$

$$45 = \frac{k}{2} \quad \text{Substituting 45 for } y \text{ and 2 for } x$$

$$90 = k$$

The equation of variation is $y = \frac{90}{x}$.

17. We substitute to find k.

$$y = \frac{k}{x}$$

$$7 = \frac{k}{10} \quad \text{Substituting 7 for } y \text{ and 10 for } x$$

$$70 = k$$

The equation of variation is $y = \frac{70}{x}$.

19. We substitute to find k.

$$y = \frac{k}{x}$$

$$6.25 = \frac{k}{25} \quad \text{Substituting 6.25 for } y \text{ and 25 for } x$$

$$156.25 = k$$

The equation of variation is $y = \frac{156.25}{x}$.

21. We substitute to find k.

$$y = \frac{k}{x}$$

$$42 = \frac{k}{5} \quad \text{Substituting 42 for } y \text{ and 5 for } x$$

$$210 = k$$

The equation of variation is $y = \frac{210}{x}$.

23. ***Familiarize and Translate.*** The problem states that we have direct variation between the variables P and H. Thus, an equation $P = kH$ applies.

Carry out. First find an equation of variation.

$$P = kH$$

$$135 = k \cdot 15 \quad \text{Substituting 135 for } P \text{ and 15 for } H$$

$$\frac{135}{15} = k$$

$$9 = k$$

The equation of variation is $P = 9H$. When $H = 23$, we have:

$$P = 9H$$

$$P = 9(23) \quad \text{Substituting 23 for } H$$

$$P = 207$$

Check. This check might be done by repeating the computations. We might also do some reasoning about the answer. The paycheck increased from \$135 to \$207. Similarly, the hours increased from 15 to 23. The ratios 15/135 and 23/207 are the same value: $0.\overline{1}$.

State. For 23 hours work, the paycheck is \$207.

25. ***Familiarize and Translate.*** The problem states that we have inverse variation between V and P. Thus, an equation $V = \frac{k}{P}$ applies.

Carry out. First find an equation of variation.

$$V = \frac{k}{P}$$

$$200 = \frac{k}{32}$$

$$6400 = k$$

The equation of variation is $V = \frac{6400}{P}$. When $P = 20$, we have:

$$V = \frac{6400}{P}$$

$$V = \frac{6400}{20} \quad \text{Substituting 20 for } P$$

$$V = 320$$

Check. A check might be done by repeating the computation. Also note that as the pressure decreased, the volume increased as expected.

State. Under a pressure of 20 kg/cm^2, the volume of the gas is 320 cm^3.

27. ***Familiarize and Translate.*** The problem states that we have direct variation between the variables M and E. Thus, an equation $M = kE$ applies.

Carry out. First find an equation of variation.

$$M = kE$$

$$32 = k \cdot 192 \quad \text{Substituting 32 for } M \text{ and 192 for } E$$

$$\frac{32}{192} = k$$

$$\frac{1}{6} = k$$

The equation of variation is $M = \frac{1}{6}E$.

When $E = 185$,

$$M = \frac{1}{6}E$$

$$M = \frac{1}{6} \cdot 185 \quad \text{Substituting 185 for } E$$

$$M \approx 30.8$$

Check. In addition to repeating the computations we can do some reasoning. The weight on the earth decreased from 192 lb to 185 lb. Similarly, the weight on the moon decreased from 32 lb to about 30.8 lb. The ratios 192/32 and 185/30.8 are about the same value: 6.

State. David Ellenbogen would weigh about 30.8 lb on the moon.

29. ***Familiarize and Translate.*** The problem states that we have inverse variation between the variables P and W. Thus, an equation $P = \frac{k}{W}$ applies.

Carry out. First find an equation of variation.

$$P = \frac{k}{W}$$

$$440 = \frac{k}{2.4} \quad \text{Substituting 440 for } P \text{ and 2.4 for } W$$

$$1056 = k$$

The equation of variation is $P = \dfrac{1056}{W}$. When $P = 660$, we have

$$P = \frac{1056}{W}$$

$$660 = \frac{1056}{W} \quad \text{Substituting 660 for } P$$

$$660W = 1056$$

$$W = \frac{1056}{660} = 1.6$$

Check. A check might be done by repeating the computations. We can also do some reasoning about the answer. Note that as the pitch increases, the wavelength decreases as expected. Also note that 2.4(440) and 1.6(660) are both 1056.

State. The wavelength of a trumpet's E above concert A is 1.6 ft.

31. ***Familiarize and Translate***. The problem states that we have inverse variation between the variables t and r. Thus, an equation $t = \dfrac{k}{r}$ applies.

Carry out. First find an equation of variation.

$$t = \frac{k}{r}$$

$$90 = \frac{k}{1200} \quad \text{Substituting 90 for } t \text{ and 1200 for } r$$

$$108,000 = k$$

The equation of variation is $t = \dfrac{108,000}{r}$. When $r = 2000$, we have

$$t = \frac{108,000}{r}$$

$$t = \frac{108,000}{2000} \quad \text{Substituting 2000 for } r$$

$$t = 54$$

Check. In addition to repeating the computations, we note that the rate increased and the time decreased as we would expect.

State. At a rate of 2000 L/min, it would take 54 min to fill the pool.

33. Observe that \$50 is twice \$25. Since we have inverse variation and the cost doubles, the number of people would be divided by 2, so there would be 30/2, or 15 people going fishing.

35. *Writing Exercise*

37. *Writing Exercise*

39. $(-7)^2 = (-7)(-7) = 49$

41. $13^2 = 13 \cdot 13 = 169$

43. $(3x)^2 = 3^2 \cdot x^2 = 9x^2$

45. $(a^2b)^2 = (a^2)^2(b)^2 = a^4b^2$

47. *Writing Exercise*

49. $P = kv^3$

51. $P^2 = kt$

53. $N = k \cdot \dfrac{T}{P}$

55. $P = kS$

Since an octagon has 8 sides, $k = 8$, and we have $P = 8S$.

57. $A = kr^2$

From the formula for the area of a circle we know that $k = \pi$, and we have $A = \pi r^2$.

Chapter 8

Radical Expressions and Equations

Exercise Set 8.1

1. The name for an expression written under a radical is a radicand, so choice (c) is correct.

3. The name for a number that is real but not rational is an irrational number, so choice (a) is correct.

5. When x represents a negative number, then $3x$ is the product of a positive number and a negative number so the sign of $3x$ is negative. Choice (b) is correct.

7. $8^2 = 64$, $9^2 = 81$, and 67 is between 64 and 81, so it is true that $\sqrt{67}$ is between 8 and 9.

9. $11^2 = 121$, $12^2 = 144$, and 150 is not between 121 and 144, so it is false that $\sqrt{150}$ is between 11 and 12.

11. The square roots of 9 are 3 and -3, since $3^2 = 9$ and $(-3)^2 = 9$.

13. The square roots of 16 are 4 and -4, because $4^2 = 16$ and $(-4)^2 = 16$.

15. The square roots of 49 are 7 and -7, because $7^2 = 49$ and $(-7)^2 = 49$.

17. The square roots of 144 are 12 and -12, because $12^2 = 144$ and $(-12)^2 = 144$.

19. $\sqrt{4} = 2$, taking the principal square root.

21. $\sqrt{1} = 1$, so $-\sqrt{1} = -1$.

23. $\sqrt{0} = 0$

25. $\sqrt{121} = 11$, so $-\sqrt{121} = -11$.

27. $\sqrt{900} = 30$

29. $\sqrt{144} = 12$

31. $\sqrt{625} = 25$, so $-\sqrt{625} = -25$.

33. The radicand is the expression under the radical, $a + 7$.

35. The radicand is the expression under the radical, $t^3 - 2$.

37. The radicand is the expression under the radical, $\dfrac{7}{x+y}$.

39. $\sqrt{100}$ is rational, since 100 is a perfect square.

41. $\sqrt{8}$ is irrational, since 8 is not a perfect square.

43. $\sqrt{32}$ is irrational, since 32 is not a perfect square.

45. $\sqrt{98}$ is irrational, since 98 is not a perfect square.

47. $-\sqrt{4}$ is rational, since 4 is a perfect square.

49. Since the radicand is expressed as a number squared, we know that it is a perfect square and thus the number in the exercise is rational.

51. 2.236

53. 4.123

55. 9.644

57. For any real number A, $\sqrt{A^2} = |A|$, so $\sqrt{x^2} = |x|$.

59. For any real number A, $\sqrt{A^2} = |A|$, so $\sqrt{(x-3)} = |x-3|$.

61. For any real number A, $\sqrt{A^2} = |A|$, so $\sqrt{(8x)^2} = |8x|$, or $8|x|$.

63. For any real number A, $\sqrt{A^2} = |A|$, so $\sqrt{(2x+1)^2} = |2x+1|$.

65. $\sqrt{t^2} = t$ Since t is assumed to be nonnegative

67. $\sqrt{9x^2} = \sqrt{(3x)^2} = 3x$ Since x is assumed to be nonnegative

69. $\sqrt{(7a)^2} = 7a$ Since a is assumed to be nonnegative

71. $\sqrt{(x+1)^2} = x + 1$ Since x is assumed to be nonnegative

73. a) We substitute 36 in the formula.

$N = 2.5\sqrt{36} = 2.5(6) = 15$

For an average of 36 arrivals, 15 spaces are needed.

b) We substitute 29 in the formula. We use a calculator or Table 2 to find an approximation.

$N = 2.5\sqrt{29} \approx 2.5(5.385) \approx 13.463 \approx 14$

For an average of 29 arrivals, 14 spaces are needed.

75. Substitute 40 in the formula.

$T = 0.144\sqrt{40} \approx 0.91$ sec

77. *Writing Exercise*

79. $(7x)^2 = 7^2x^2 = 49x^2$

81. $(4t^7)^2 = 4^2(t^7)^2 = 16t^{7\cdot 2} = 16t^{14}$

83. $3a \cdot 16a^{10} = 3 \cdot 16 \cdot a \cdot a^{10} = 48a^{1+10} = 48a^{11}$

85. *Writing Exercise*

87. We find the inner square root first.

$\sqrt{\sqrt{81}} = \sqrt{9} = 3$

89. $-\sqrt{36} < -\sqrt{33} < -\sqrt{25}$, or $-6 < -\sqrt{33} < -5$, so $-\sqrt{33}$ is between -6 and -5.

91. If $\sqrt{t^2} = 7$, then $t^2 = 7^2$, or 49. Thus, $t = 7$ or $t = -7$. The solutions are 7 and -7.

93. If $-\sqrt{x^2} = -3$, then $\sqrt{x^2} = 3$ and $x^2 = 3^2$, or 9. Thus, $x = 3$ or $x = -3$. The solutions are 3 and -3.

95. The values of x for which $\sqrt{(x+5)^2} = x+5$ is false are the values of x for which $x+5 < 0$, so we have $x < -5$, or $\{x|x < -5\}$.

97. $\sqrt{\frac{144x^8}{36y^6}} = \sqrt{\frac{4x^8}{y^6}} = \sqrt{\left(\frac{2x^4}{y^3}\right)^2} = \frac{2x^4}{y^3}$

99. $\sqrt{\frac{400}{m^{16}}} = \sqrt{\left(\frac{20}{m^8}\right)^2} = \frac{20}{m^8}$

101. a) Locate 3 on the x-axis, move up vertically to the graph, and then move left horizontally to the y-axis to read the approximation.

$\sqrt{3} \approx 1.7$ (Answers may vary.)

b) Locate 5 on the x-axis, move up vertically to the graph, and then move left horizontally to the y-axis to read the approximation.

$\sqrt{5} \approx 2.2$ (Answers may vary.)

c) Locate 7 on the x-axis, move up vertically to the graph, and then move left horizontally to the y-axis to read the approximation.

$\sqrt{7} \approx 2.6$ (Answers may vary.)

103. We substitute 5 in the formula.

$V = \frac{1087\sqrt{273+5}}{16.52}$

$V = \frac{1087\sqrt{278}}{16.52}$

$V \approx \frac{1087(16.673)}{16.52}$

$V \approx 1097.1$ ft/sec

105.

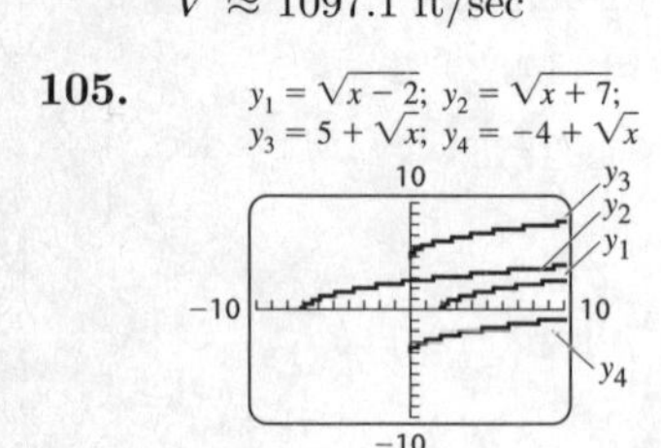

Exercise Set 8.2

1. $\sqrt{3} \cdot \sqrt{7} = \sqrt{3 \cdot 7} = \sqrt{21}$, so choice (h) is correct.

3. $\sqrt{3 \cdot 49} = \sqrt{3} \cdot \sqrt{49} = \sqrt{3} \cdot 7$, so choice (f) is correct.

5. $\sqrt{a^2} = a$, so choice (i) is correct.

7. $\sqrt{a^6} = \sqrt{(a^3)^2}$, so choice (b) is correct.

9. $\sqrt{ab} \cdot \sqrt{bc} = \sqrt{ab \cdot bc} = \sqrt{ab^2c} = \sqrt{b^2ac}$, so choice (c) is correct.

11. $\sqrt{7}\sqrt{3} = \sqrt{7 \cdot 3} = \sqrt{21}$

13. $\sqrt{4}\sqrt{3} = \sqrt{12}$, or

$\sqrt{4}\sqrt{3} = 2\sqrt{3}$ Taking the square root of 4

15. $\sqrt{\frac{2}{5}}\sqrt{\frac{3}{4}} = \sqrt{\frac{2 \cdot 3}{5 \cdot 4}} = \sqrt{\frac{3}{10}}$

17. $\sqrt{14}\sqrt{14} = \sqrt{14 \cdot 14} = \sqrt{196} = 14$

19. $\sqrt{25}\sqrt{3} = \sqrt{75}$, or

$\sqrt{25}\sqrt{3} = 5\sqrt{3}$ Taking the square root of 25

21. $\sqrt{2}\sqrt{x} = \sqrt{2 \cdot x} = \sqrt{2x}$

23. $\sqrt{7}\sqrt{2a} = \sqrt{7 \cdot 2a} = \sqrt{14a}$

25. $\sqrt{5x} \cdot 7 = \sqrt{5x \cdot 7} = \sqrt{35x}$

27. $\sqrt{3a}\sqrt{2c} = \sqrt{3a \cdot 2c} = \sqrt{6ac}$

29. $\sqrt{20} = \sqrt{4 \cdot 5}$ 4 is a perfect square.

$= \sqrt{4}\sqrt{5}$ Factoring into a product of radicals

$= 2\sqrt{5}$

31. $\sqrt{50} = \sqrt{25 \cdot 2}$ 25 is a perfect square.

$= \sqrt{25}\sqrt{2}$ Factoring into a product of radicals

$= 5\sqrt{2}$

33. $\sqrt{700} = \sqrt{100 \cdot 7} = \sqrt{100}\sqrt{7} = 10\sqrt{7}$

35. $\sqrt{4tx} = \sqrt{4 \cdot t} = \sqrt{4}\,\sqrt{t} = 2\sqrt{t}$

37. $\sqrt{75a} = \sqrt{25 \cdot 3a} = \sqrt{25}\sqrt{3a} = 5\sqrt{3a}$

39. $\sqrt{16a} = \sqrt{16 \cdot a} = \sqrt{16}\,\sqrt{a} = 4\sqrt{a}$

41. $\sqrt{64y^2} = \sqrt{64}\,\sqrt{y^2} = 8y$, or

$\sqrt{64y^2} = \sqrt{(8y)^2} = 8y$

43. $\sqrt{13x^2} = \sqrt{13}\,\sqrt{x^2} = \sqrt{13} \cdot x$, or $x\sqrt{13}$

45. $\sqrt{28t^2} = \sqrt{4 \cdot t^2 \cdot 7} = \sqrt{4}\,\sqrt{t^2}\,\sqrt{7} = 2t\sqrt{7}$

47. $\sqrt{b^2 - 4ac} = \sqrt{4^2 - 4 \cdot 2(-3)}$

$= \sqrt{16 - 8(-3)}$

$= \sqrt{16 + 24}$

$= \sqrt{40}$

$= \sqrt{2 \cdot 2 \cdot 2 \cdot 5}$

$= \sqrt{4} \cdot \sqrt{10}$

$= 2\sqrt{10}$

49. $\sqrt{b^2 - 4ac} = \sqrt{1^2 - 4 \cdot 1(-12)}$

$= \sqrt{1 + 48}$

$= \sqrt{49}$

$= 7$

51. $\sqrt{b^2-4ac} = \sqrt{(-6)^2-4\cdot 3(-4)}$
$= \sqrt{36-12(-4)}$
$= \sqrt{36+48}$
$= \sqrt{84}$
$= \sqrt{2\cdot 2\cdot 3\cdot 7}$
$= \sqrt{4}\cdot\sqrt{21}$
$= 2\sqrt{21}$

53. $\sqrt{a^{14}} = \sqrt{(a^7)^2} = a^7$

55. $\sqrt{x^{16}} = \sqrt{(x^8)^2} = x^8$

57. $\sqrt{r^7} = \sqrt{r^6\cdot r}$ One factor is a perfect square
$= \sqrt{r^6}\,\sqrt{r}$
$= \sqrt{(r^3)^2}\,\sqrt{r}$
$= r^3\sqrt{r}$

59. $\sqrt{t^{19}} = \sqrt{t^{18}\cdot t} = \sqrt{t^{18}}\,\sqrt{t} = \sqrt{(t^9)^2}\,\sqrt{t} = t^9\sqrt{t}$

61. $\sqrt{40a^3} = \sqrt{4\cdot a^2\cdot 10\cdot a} = \sqrt{4}\sqrt{a^2}\sqrt{10a} = 2a\sqrt{10a}$

63. $\sqrt{8a^5} = \sqrt{4a^4(2a)} = \sqrt{4(a^2)^2(2a)} = \sqrt{4}\sqrt{(a^2)^2}\sqrt{2a} = 2a^2\sqrt{2a}$

65. $\sqrt{104p^{17}} = \sqrt{4p^{16}(26p)} = \sqrt{4(p^8)^2(26p)} = \sqrt{4}\sqrt{(p^8)^2}\sqrt{26p} = 2p^8\sqrt{26p}$

67. $\sqrt{7}\cdot\sqrt{14} = \sqrt{7\cdot 14}$ Multiplying
$= \sqrt{7\cdot 2\cdot 7}$ Writing the prime factorization
$= \sqrt{2}\sqrt{7^2}$
$= \sqrt{2}\cdot 7$, or
$7\sqrt{2}$

69. $\sqrt{3}\cdot\sqrt{27} = \sqrt{3\cdot 27}$ Multiplying
$= \sqrt{3\cdot 3\cdot 3\cdot 3}$ Writing the prime factorization
$= \sqrt{3^4}$
$= 3^2$
$= 9$

71. $\sqrt{3x}\sqrt{12y} = \sqrt{3x\cdot 12y}$
$= \sqrt{3\cdot x\cdot 2\cdot 2\cdot 3\cdot y}$
$= \sqrt{2^2}\sqrt{3^2}\sqrt{xy}$
$= 2\cdot 3\sqrt{xy}$
$= 6\sqrt{xy}$

73. $\sqrt{17}\sqrt{17x} = \sqrt{17\cdot 17x} = \sqrt{17\cdot 17\cdot x} = \sqrt{17^2}\sqrt{x} = 17\sqrt{x}$

75. $\sqrt{10b}\sqrt{50b} = \sqrt{10b\cdot 50b}$
$= \sqrt{10\cdot b\cdot 5\cdot 10\cdot b}$
$= \sqrt{10^2}\sqrt{b^2}\sqrt{5}$
$= 10b\sqrt{5}$

77. Since the radicands are identical, the product will be the radicand, $8t$. We could also do this problem as follows.
$\sqrt{8t}\cdot\sqrt{8t} = \sqrt{8t\cdot 8t} = \sqrt{(8t)^2} = 8t$

79. $\sqrt{ab}\sqrt{ac} = \sqrt{a^2bc} = \sqrt{a^2}\sqrt{bc} = a\sqrt{bc}$

81. $\sqrt{3x^3}\sqrt{6x} = \sqrt{3x^3\cdot 6x}$
$= \sqrt{3\cdot 2\cdot 3\cdot x^4}$
$= \sqrt{3^2}\sqrt{2}\sqrt{x^4}$
$= 3\cdot\sqrt{2}\cdot x^2$, or
$3x^2\sqrt{2}$

83. $\sqrt{x^2y^3}\sqrt{xy^4}$
$= \sqrt{x^2y^3}\sqrt{x(y^2)^2}$ x^2 and $(y^2)^2$ are perfect squares
$= \sqrt{x^2}\cdot\sqrt{y^3}\cdot\sqrt{x}\cdot\sqrt{(y^2)^2}$
$= x\cdot\sqrt{y^3}\cdot\sqrt{x}\cdot y^2$
$= xy^2\sqrt{y^3\cdot x}$
$= xy^2\sqrt{x\cdot y^2\cdot y}$
$= xy^2\sqrt{y^2}\sqrt{xy}$
$= xy^2\cdot y\cdot\sqrt{xy}$
$= xy^3\sqrt{xy}$

85. $\sqrt{10ab}\sqrt{50a^2b^7} = \sqrt{10ab}\cdot\sqrt{25}\cdot\sqrt{2}\cdot\sqrt{a^2}\cdot\sqrt{(b^3)^2\cdot b}$
$= \sqrt{10ab}\cdot 5\cdot\sqrt{2}\cdot a\cdot b^3\sqrt{b}$
$= 5ab^3\sqrt{10ab\cdot 2\cdot b}$
$= 5ab^3\sqrt{20ab^2}$
$= 5ab^3\sqrt{4\cdot 5\cdot a\cdot b^2}$
$= 5ab^3\cdot 2\cdot\sqrt{5a}\cdot b$
$= 10ab^4\sqrt{5a}$

87. In 2000, $x = 2000-1980 = 20$.
$h = 10.7 + 13.3\sqrt{20} \approx 70.2$ million
In 2010, $x = 2010-1980 = 30$.
$h = 10.7 + 13.3\sqrt{30} \approx 83.5$ million

89. First we substitute 20 for L in the formula:
$r = 2\sqrt{5L} = 2\sqrt{5\cdot 20} = 2\sqrt{100} = 2\cdot 10 = 20$ mph
Then we substitute 150 for L:
$r = 2\sqrt{5\cdot 150} = 2\sqrt{750} = 2\sqrt{25\cdot 30} = 2\sqrt{25}\,\sqrt{30} = 2\cdot 5\sqrt{30} = 10\sqrt{30} \approx 10(5.477) \approx 54.77$ mph, or 54.8 mph (rounded to the nearest tenth)

91. *Writing Exercise*

93. $\dfrac{a^7b^3}{a^2b} = a^{7-2}b^{3-1} = a^5b^2$

95. $\dfrac{3x}{5y}\cdot\dfrac{2x}{7y} = \dfrac{3x\cdot 2x}{5y\cdot 7y} = \dfrac{6x^2}{35y^2}$

97. $\dfrac{2r^3}{7t}\cdot\dfrac{rt}{rt} = \dfrac{2r^3}{7t}\cdot 1 = \dfrac{2r^3}{7t}$

99. *Writing Exercise*

101. $\sqrt{0.01} = \sqrt{(0.1)^2} = 0.1$

103. $\sqrt{0.0625} = \sqrt{(0.25)^2} = 0.25$

105. $4\sqrt{14} = \sqrt{16 \cdot 14} = \sqrt{224}$ and $15 = \sqrt{225}$, so
$4\sqrt{14} < 15$.

107. $3\sqrt{11} = \sqrt{9}\sqrt{11} = \sqrt{99}$ and
$7\sqrt{2} = \sqrt{49}\sqrt{2} = \sqrt{98}$, so
$3\sqrt{11} > 7\sqrt{2}$.

109. $8^2 = 64$
$(\sqrt{15} + \sqrt{17})^2 = 15 + 2\sqrt{255} + 17 = 32 + 2\sqrt{255}$
Now $\sqrt{255} < \sqrt{256}$, or $\sqrt{255} < 16$, so
$2\sqrt{255} < 2 \cdot 16 = 32$. Then $32 + 2\sqrt{255} < 32 + 32$, or
$(\sqrt{15} + \sqrt{17})^2 < 64$. Thus, $8 > \sqrt{15} + \sqrt{17}$.

111.
$$\begin{aligned} &\sqrt{54(x+1)}\,\sqrt{6y(x+1)^2} \\ &= \sqrt{54(x+1) \cdot 6y(x+1)^2} \\ &= \sqrt{9 \cdot 6 \cdot (x+1) \cdot 6 \cdot y \cdot (x+1)^2} \\ &= \sqrt{9 \cdot 6 \cdot 6 \cdot (x+1)^2 \cdot y(x+1)} \\ &= \sqrt{9}\,\sqrt{6 \cdot 6}\,\sqrt{(x+1)^2}\,\sqrt{y(x+1)} \\ &= 3 \cdot 6 \cdot (x+1)\sqrt{y(x+1)} \\ &= 18(x+1)\sqrt{y(x+1)} \end{aligned}$$

113. $\sqrt{x^9}\,\sqrt{2x}\,\sqrt{10x^5} = \sqrt{x^9 \cdot 2x \cdot 10x^5} =$
$\sqrt{x^8 \cdot x \cdot 2 \cdot x \cdot 2 \cdot 5 \cdot x^4 \cdot x} =$
$\sqrt{x^8 \cdot x \cdot x \cdot 2 \cdot 2 \cdot x^4 \cdot 5 \cdot x} =$
$\sqrt{x^8}\,\sqrt{x \cdot x}\,\sqrt{2 \cdot 2}\,\sqrt{x^4}\,\sqrt{5x} =$
$x^4 \cdot x \cdot 2 \cdot x^2\sqrt{5x} = 2x^7\sqrt{5x}$

115. $7x^{14}\sqrt{6x^7} = \sqrt{(7x^{14})^2} \cdot \sqrt{6x^7} = \sqrt{49x^{28}} \cdot \sqrt{6x^7} =$
$\sqrt{49x^{28} \cdot 6x^7} = \sqrt{294x^{35}} = \sqrt{21x^9 \cdot 14x^{26}} =$
$\sqrt{21x^9} \cdot \underline{\sqrt{14x^{26}}}$

117. $\sqrt{x^{16n}} = \sqrt{(x^{8n})^2} = x^{8n}$

119. If n is an odd whole number greater than or equal to 3, then $n = 2k + 1$, where k is a natural number.
$\sqrt{y^n} = \sqrt{y^{2k+1}} = \sqrt{y^{2k} \cdot y} = y^k\sqrt{y} = y^{(1/2)(n-1)}\sqrt{y}$.

Exercise Set 8.3

1. The statement is true by the quotient rule for square roots.

3. The statement is false. See Example 4 and 5.

5. $\dfrac{\sqrt{20}}{\sqrt{5}} = \sqrt{\dfrac{20}{5}} = \sqrt{4} = 2$

7. $\dfrac{\sqrt{75}}{\sqrt{3}} = \sqrt{\dfrac{75}{3}} = \sqrt{25} = 5$

9. $\dfrac{\sqrt{35}}{\sqrt{5}} = \sqrt{\dfrac{35}{5}} = \sqrt{7}$

11. $\dfrac{\sqrt{3}}{\sqrt{75}} = \sqrt{\dfrac{3}{75}} = \sqrt{\dfrac{1}{25}} = \dfrac{1}{5}$

13. $\dfrac{\sqrt{18}}{\sqrt{32}} = \sqrt{\dfrac{18}{32}} = \sqrt{\dfrac{9}{16}} = \dfrac{3}{4}$

15. $\dfrac{\sqrt{8x}}{\sqrt{2x}} = \sqrt{\dfrac{8x}{2x}} = \sqrt{4} = 2$

17. $\dfrac{\sqrt{63y^3}}{\sqrt{7y}} = \sqrt{\dfrac{63y^3}{7y}} = \sqrt{9y^2} = 3y$

19. $\dfrac{\sqrt{27x^5}}{\sqrt{3x}} = \sqrt{\dfrac{27x^5}{3x}} = \sqrt{9x^4} = 3x^2$

21. $\dfrac{\sqrt{21a^9}}{\sqrt{7a^3}} = \sqrt{\dfrac{21a^9}{7a^3}} = \sqrt{3a^6} = a^3\sqrt{3}$

23. $\sqrt{\dfrac{36}{25}} = \dfrac{\sqrt{36}}{\sqrt{25}} = \dfrac{6}{5}$

25. $\sqrt{\dfrac{49}{16}} = \dfrac{\sqrt{49}}{\sqrt{16}} = \dfrac{7}{4}$

27. $-\sqrt{\dfrac{25}{81}} = -\dfrac{\sqrt{25}}{\sqrt{81}} = -\dfrac{5}{9}$

29. $\sqrt{\dfrac{2a^5}{50a}} = \sqrt{\dfrac{a^4}{25}} = \dfrac{\sqrt{a^4}}{\sqrt{25}} = \dfrac{a^2}{5}$

31. $\sqrt{\dfrac{6x^7}{32x}} = \sqrt{\dfrac{3x^6}{16}} = \dfrac{\sqrt{3x^6}}{\sqrt{16}} = \dfrac{x^3\sqrt{3}}{4}$

33. $\sqrt{\dfrac{21t^9}{28t^3}} = \sqrt{\dfrac{3t^6}{4}} = \dfrac{\sqrt{3t^6}}{\sqrt{4}} = \dfrac{t^3\sqrt{3}}{2}$

35. $\dfrac{5}{\sqrt{3}} = \dfrac{5}{\sqrt{3}} \cdot \dfrac{\sqrt{3}}{\sqrt{3}} = \dfrac{5\sqrt{3}}{3}$

37. $\dfrac{\sqrt{3}}{\sqrt{7}} = \dfrac{\sqrt{3}}{\sqrt{7}} \cdot \dfrac{\sqrt{7}}{\sqrt{7}} = \dfrac{\sqrt{21}}{7}$

39. $\dfrac{\sqrt{4}}{\sqrt{27}} = \dfrac{\sqrt{4}}{\sqrt{9}\sqrt{3}} = \dfrac{2}{3\sqrt{3}} = \dfrac{2}{3\sqrt{3}} \cdot \dfrac{\sqrt{3}}{\sqrt{3}} = \dfrac{2\sqrt{3}}{3 \cdot 3} = \dfrac{2\sqrt{3}}{9}$

41. $\dfrac{\sqrt{7}}{\sqrt{3}} = \dfrac{\sqrt{7}}{\sqrt{3}} \cdot \dfrac{\sqrt{3}}{\sqrt{3}} = \dfrac{\sqrt{21}}{3}$

43. $\dfrac{\sqrt{3}}{\sqrt{50}} = \dfrac{\sqrt{3}}{\sqrt{25}\sqrt{2}} = \dfrac{\sqrt{3}}{5\sqrt{2}} = \dfrac{\sqrt{3}}{5\sqrt{2}} \cdot \dfrac{\sqrt{2}}{\sqrt{2}} =$
$\dfrac{\sqrt{6}}{5 \cdot 2} = \dfrac{\sqrt{6}}{10}$

45. $\dfrac{\sqrt{2a}}{\sqrt{45}} = \dfrac{\sqrt{2a}}{\sqrt{9}\sqrt{5}} = \dfrac{\sqrt{2a}}{3\sqrt{5}} = \dfrac{\sqrt{2a}}{3\sqrt{5}} \cdot \dfrac{\sqrt{5}}{\sqrt{5}} =$
$\dfrac{\sqrt{10a}}{3 \cdot 5} = \dfrac{\sqrt{10a}}{15}$

47. $\sqrt{\dfrac{12}{5}} = \dfrac{\sqrt{4}\sqrt{3}}{\sqrt{5}} = \dfrac{2\sqrt{3}}{\sqrt{5}} = \dfrac{2\sqrt{3}}{\sqrt{5}} \cdot \dfrac{\sqrt{5}}{\sqrt{5}} = \dfrac{2\sqrt{15}}{5}$

49. $\sqrt{\dfrac{5}{z}} = \dfrac{\sqrt{5}}{\sqrt{z}} = \dfrac{\sqrt{5}}{\sqrt{z}} \cdot \dfrac{\sqrt{z}}{\sqrt{z}} = \dfrac{\sqrt{5z}}{z}$

51. $\sqrt{\frac{t}{32}} = \frac{\sqrt{t}}{\sqrt{32}} = \frac{\sqrt{t}}{\sqrt{16}\sqrt{2}} = \frac{\sqrt{t}}{4\sqrt{2}} =$

$\frac{\sqrt{t}}{4\sqrt{2}} \cdot \frac{\sqrt{2}}{\sqrt{2}} = \frac{\sqrt{2t}}{4 \cdot 2} = \frac{\sqrt{2t}}{8}$

53. $\sqrt{\frac{x}{90}} = \frac{\sqrt{x}}{\sqrt{90}} = \frac{\sqrt{x}}{\sqrt{9}\sqrt{10}} = \frac{\sqrt{x}}{3\sqrt{10}} = \frac{\sqrt{x}}{3\sqrt{10}} \cdot \frac{\sqrt{10}}{\sqrt{10}} =$

$\frac{\sqrt{10x}}{3 \cdot 10} = \frac{\sqrt{10x}}{30}$

55. Since the denominator, 25, is a perfect square we need only to simplify the expression in order to rationalize the denominator.

$\sqrt{\frac{3a}{25}} = \frac{\sqrt{3a}}{\sqrt{25}} = \frac{\sqrt{3a}}{5}$

57. $\sqrt{\frac{5x^3}{12x}} = \sqrt{\frac{5x^2}{12}} = \frac{\sqrt{5x^2}}{\sqrt{12}} = \frac{\sqrt{5x^2}}{\sqrt{4}\sqrt{3}} = \frac{x\sqrt{5}}{2\sqrt{3}} =$

$\frac{x\sqrt{5}}{2\sqrt{3}} \cdot \frac{\sqrt{3}}{\sqrt{3}} = \frac{x\sqrt{15}}{2 \cdot 3} = \frac{x\sqrt{15}}{6}$

59. $V = \frac{18,500}{\sqrt{t+1.0565}} = \frac{18,500}{\sqrt{2+1.0565}} = \frac{18,500}{\sqrt{3.0565}} \approx \$10,582$

61. $V = \frac{18,500}{\sqrt{t+1.0565}} = \frac{18,500}{\sqrt{7+1.0565}} = \frac{18,500}{\sqrt{8.0565}} \approx \$6518$

63. 32 ft: $T \approx 2(3.14)\sqrt{\frac{32}{32}} \approx 6.28\sqrt{1} \approx 6.28(1) \approx$

6.28 sec

50 ft: $T \approx 2(3.14)\sqrt{\frac{50}{32}} \approx 6.28\sqrt{\frac{25}{16}} \approx 6.28\frac{\sqrt{25}}{\sqrt{16}} \approx$

$6.28\left(\frac{5}{4}\right) \approx 7.85$ sec

65. Substitute $\frac{2}{\pi^2}$ for L in the formula.

$T = 2\pi\sqrt{\frac{L}{32}} = 2\pi\sqrt{\frac{\frac{2}{\pi^2}}{32}} = 2\pi\sqrt{\frac{2}{\pi^2} \cdot \frac{1}{32}} =$

$2\pi\sqrt{\frac{2}{32\pi^2}} = 2\pi\sqrt{\frac{1}{16\pi^2}} = 2\pi \cdot \frac{\sqrt{1}}{\sqrt{16\pi^2}} = 2\pi \cdot \frac{1}{4\pi} = \frac{2\pi}{4\pi} =$
0.5 sec

It takes 0.5 sec to move from one side to the other and back.

67. Substitute 72 for L in the formula.

$T \approx 2(3.14)\sqrt{\frac{72}{32}} \approx 6.28\sqrt{2.25} \approx 9.42$ sec

69. *Writing Exercise*

71. $5x + 9 + 7x + 4 = 5x + 7x + 9 + 4 = 12x + 13$

73. $2a^3 - a^2 - 3a^3 - 7a^2 = 2a^3 - 3a^3 - a^2 - 7a^2 = -a^3 - 8a^2$

75. $9x(2x - 7) = 9x \cdot 2x - 9x \cdot 7 = 18x^2 - 63x$

77. $(3 + 4x)(2 + 5x) = 3 \cdot 2 + 3 \cdot 5x + 4x \cdot 2 + 4x \cdot 5x =$
$6 + 15x + 8x + 20x^2 = 6 + 23x + 20x^2$

79. *Writing Exercise*

81. $\sqrt{\frac{7}{1000}} = \frac{\sqrt{7}}{\sqrt{1000}} = \frac{\sqrt{7}}{\sqrt{100}\sqrt{10}} = \frac{\sqrt{7}}{10\sqrt{10}} =$

$\frac{\sqrt{7}}{10\sqrt{10}} \cdot \frac{\sqrt{10}}{\sqrt{10}} = \frac{\sqrt{70}}{10 \cdot 10} = \frac{\sqrt{70}}{100}$

83. $\sqrt{\frac{3x^2}{8x^7y^3}} = \sqrt{\frac{3}{8x^5y^3}} = \frac{\sqrt{3}}{\sqrt{8x^5y^3}} =$

$\frac{\sqrt{3}}{\sqrt{4x^4y^2}\sqrt{2xy}} = \frac{\sqrt{3}}{2x^2y\sqrt{2xy}} =$

$\frac{\sqrt{3}}{2x^2y\sqrt{2xy}} \cdot \frac{\sqrt{2xy}}{\sqrt{2xy}} = \frac{\sqrt{6xy}}{2x^2y \cdot 2xy} = \frac{\sqrt{6xy}}{4x^3y^2}$

85. $\sqrt{\frac{2a}{5b^3c^9}} = \frac{\sqrt{2a}}{\sqrt{5b^3c^9}} = \frac{\sqrt{2a}}{\sqrt{b^2c^8}\sqrt{5bc}} = \frac{\sqrt{2a}}{bc^4\sqrt{5bc}} =$

$\frac{\sqrt{2a}}{bc^4\sqrt{5bc}} \cdot \frac{\sqrt{5bc}}{\sqrt{5bc}} = \frac{\sqrt{10abc}}{bc^4 \cdot 5bc} = \frac{\sqrt{10abc}}{5b^2c^5}$

87. $\frac{3}{\sqrt{\sqrt{7}}} = \frac{3}{\sqrt{\sqrt{7}}} \cdot \frac{\sqrt{7\sqrt{7}}}{\sqrt{7\sqrt{7}}} = \frac{3\sqrt{7\sqrt{7}}}{\sqrt{\sqrt{7} \cdot 7\sqrt{7}}} = \frac{3\sqrt{7\sqrt{7}}}{\sqrt{7 \cdot 7}} =$
$\frac{3\sqrt{7\sqrt{7}}}{7}$

89. $\sqrt{\frac{1}{x^2} - \frac{2}{xy} + \frac{1}{y^2}}$, LCD is x^2y^2

$= \sqrt{\frac{1}{x^2} \cdot \frac{y^2}{y^2} - \frac{2}{xy} \cdot \frac{xy}{xy} + \frac{1}{y^2} \cdot \frac{x^2}{x^2}}$

$= \sqrt{\frac{y^2 - 2xy + x^2}{x^2y^2}}$

$= \sqrt{\frac{(y-x)^2}{x^2y^2}}$

$= \frac{\sqrt{(y-x)^2}}{\sqrt{x^2y^2}}$

$= \frac{y-x}{xy}$

An alternate method of simplifying this expression is shown below.

$\sqrt{\frac{1}{x^2} - \frac{2}{xy} + \frac{1}{y^2}} = \sqrt{\left(\frac{1}{x} - \frac{1}{y}\right)^2}$

$= \frac{1}{x} - \frac{1}{y}$

The two answers are equivalent.

91. $\sqrt{\dfrac{2x-3}{8}} = \dfrac{5}{2}$

If this equation is true, we have:

$$\frac{2x-3}{8} = \left(\frac{5}{2}\right)^2$$
$$\frac{2x-3}{8} = \frac{25}{4}$$
$$2x-3 = 50 \quad \text{Multiplying by 8}$$
$$2x = 53$$
$$x = \frac{53}{2}$$

The number $\dfrac{53}{2}$ checks in the original equation, so the solution is $\dfrac{53}{2}$.

Exercise Set 8.4

1. (b); see page 505 in the text.

3. (d); see page 505 in the text.

5. $2\sqrt{7}+4\sqrt{7}$
$= (2+4)\sqrt{7}$ Using the distributive law
$= 6\sqrt{7}$

7. $9\sqrt{5}-6\sqrt{5}$
$= (9-6)\sqrt{5}$ Using the distributive law
$= 3\sqrt{5}$

9. $6\sqrt{x}+7\sqrt{x} = (6+7)\sqrt{x} = 13\sqrt{x}$

11. $9\sqrt{x}-11\sqrt{x} = (9-11)\sqrt{x} = -2\sqrt{x}$

13. $5\sqrt{2a}+3\sqrt{2a} = (5+3)\sqrt{2a} = 8\sqrt{2a}$

15. $9\sqrt{10y}-\sqrt{10y} = (9-1)\sqrt{10y} = 8\sqrt{10y}$

17. $6\sqrt{7}+2\sqrt{7}+4\sqrt{7} = (6+2+4)\sqrt{7} = 12\sqrt{7}$

19. $5\sqrt{2}-9\sqrt{2}+8\sqrt{2} = (5-9+8)\sqrt{2} = 4\sqrt{2}$

21. $5\sqrt{3}+\sqrt{8} = 5\sqrt{3}+\sqrt{4\cdot 2}$ Factoring 8
$= 5\sqrt{3}+\sqrt{4}\sqrt{2}$
$= 5\sqrt{3}+2\sqrt{2}$

$5\sqrt{3}+\sqrt{8}$, or $5\sqrt{3}+2\sqrt{2}$, cannot be simplified further.

23. $\sqrt{x}-\sqrt{9x} = \sqrt{x}-\sqrt{9}\sqrt{x}$
$= \sqrt{x}-3\sqrt{x}$
$= (1-3)\sqrt{x}$
$= -2\sqrt{x}$

25. $2\sqrt{3}-4\sqrt{75} = 2\sqrt{3}-4\sqrt{25\cdot 3}$
$= 2\sqrt{3}-4\sqrt{25}\sqrt{3}$
$= 2\sqrt{3}-4\cdot 5\sqrt{3}$
$= 2\sqrt{3}-20\sqrt{3}$
$= -18\sqrt{3}$

27. $6\sqrt{18}+5\sqrt{8} = 6\sqrt{9\cdot 2}+5\sqrt{4\cdot 2}$
$= 6\sqrt{9}\sqrt{2}+5\sqrt{4}\sqrt{2}$
$= 6\cdot 3\sqrt{2}+5\cdot 2\sqrt{2}$
$= 18\sqrt{2}+10\sqrt{2}$
$= 28\sqrt{2}$

29. $\sqrt{72}+\sqrt{98} = \sqrt{36\cdot 2}+\sqrt{49\cdot 2}$
$= 6\sqrt{2}+7\sqrt{2}$
$= 13\sqrt{2}$

31. $9\sqrt{8}+\sqrt{72}-9\sqrt{8}$

Observe that $9\sqrt{8}-9\sqrt{8} = 0$, so we need only to simplify $\sqrt{72}$.

$\sqrt{72} = \sqrt{36\cdot 2} = \sqrt{36}\sqrt{2} = 6\sqrt{2}$

33. $7\sqrt{12}-2\sqrt{27}+\sqrt{75}$
$= 7\sqrt{4\cdot 3}-2\sqrt{9\cdot 3}+\sqrt{25\cdot 3}$
$= 7\cdot 2\sqrt{3}-2\cdot 3\sqrt{3}+5\sqrt{3}$
$= 14\sqrt{3}-6\sqrt{3}+5\sqrt{3}$
$= 13\sqrt{3}$

35. $\sqrt{9x}+\sqrt{49x}-9\sqrt{x} = 3\sqrt{x}+7\sqrt{x}-9\sqrt{x}$
$= (3+7-9)\sqrt{x}$
$= 1\sqrt{x}$
$= \sqrt{x}$

37. $\sqrt{7}(\sqrt{5}+\sqrt{2})$
$= \sqrt{7}\sqrt{5}+\sqrt{7}\sqrt{2}$ Using the distributive law
$= \sqrt{35}+\sqrt{14}$

39. $\sqrt{5}(\sqrt{6}-\sqrt{10}) = \sqrt{5}\sqrt{6}-\sqrt{5}\sqrt{10}$
$= \sqrt{30}-\sqrt{50}$
$= \sqrt{30}-\sqrt{25\cdot 2}$
$= \sqrt{30}-5\sqrt{2}$

41. $(3+\sqrt{2})(4+\sqrt{2})$
$= 3\cdot 4+3\cdot\sqrt{2}+\sqrt{2}\cdot 4+\sqrt{2}\cdot\sqrt{2}$ Using FOIL
$= 12+3\sqrt{2}+4\sqrt{2}+2$
$= 14+7\sqrt{2}$

43. $(\sqrt{7}-2)(\sqrt{7}-5)$
$= \sqrt{7}\cdot\sqrt{7}-\sqrt{7}\cdot 5-2\cdot\sqrt{7}+2\cdot 5$ Using FOIL
$= 7-5\sqrt{7}-2\sqrt{7}+10$
$= 17-7\sqrt{7}$

45. $(\sqrt{5}+4)(\sqrt{5}-4)$
$= (\sqrt{5})^2-4^2$ Using $(A+B)(A-B) = A^2-B^2$
$= 5-16$
$= -11$

47. $(\sqrt{6}-\sqrt{3})(\sqrt{6}+\sqrt{3})$
$= (\sqrt{6})^2-(\sqrt{3})^2$ Using $(A-B)(A+B) = A^2-B^2$
$= 6-3$
$= 3$

49. $(2+3\sqrt{2})(3-\sqrt{2})$

$= 2\cdot 3 - 2\cdot\sqrt{2} + 3\sqrt{2}\cdot 3 - 3\sqrt{2}\cdot\sqrt{2}$ Using FOIL

$= 6 - 2\sqrt{2} + 9\sqrt{2} - 3\cdot 2$

$= 6 - 2\sqrt{2} + 9\sqrt{2} - 6$

$= 7\sqrt{2}$

51. $(7+\sqrt{3})^2$

$= 7^2 + 2\cdot 7\cdot\sqrt{3} + (\sqrt{3})^2$ Using $(A+B)^2 = A^2 + 2AB + B^2$

$= 49 + 14\sqrt{3} + 3$

$= 52 + 14\sqrt{3}$

53. $(1-2\sqrt{3})^2$

$= 1^2 - 2\cdot 1\cdot 2\sqrt{3} + (2\sqrt{3})^2$ Using $(A-B)^2 = A^2 - 2AB + B^2$

$= 1 - 4\sqrt{3} + 4\cdot 3$

$= 1 - 4\sqrt{3} + 12$

$= 13 - 4\sqrt{3}$

55. $(\sqrt{x}-\sqrt{10})^2 = (\sqrt{x})^2 - 2\sqrt{x}\sqrt{10} + (\sqrt{10})^2$

$= x - 2\sqrt{10x} + 10$

57. $\dfrac{3}{2+\sqrt{5}}$

$= \dfrac{3}{2+\sqrt{5}}\cdot\dfrac{2-\sqrt{5}}{2-\sqrt{5}}$ Multiplying by 1

$= \dfrac{3(2-\sqrt{5})}{(2+\sqrt{5})(2-\sqrt{5})}$

$= \dfrac{6-3\sqrt{5}}{2^2-(\sqrt{5})^2}$

$= \dfrac{6-3\sqrt{5}}{4-5}$

$= \dfrac{6-3\sqrt{5}}{-1}$

$= -6+3\sqrt{5}$

59. $\dfrac{6}{2-\sqrt{7}}$

$= \dfrac{6}{2-\sqrt{7}}\cdot\dfrac{2+\sqrt{7}}{2+\sqrt{7}}$

$= \dfrac{6(2+\sqrt{7})}{(2-\sqrt{7})(2+\sqrt{7})}$

$= \dfrac{12+6\sqrt{7}}{2^2-(\sqrt{7})^2}$

$= \dfrac{12+6\sqrt{7}}{4-7}$

$= \dfrac{12+6\sqrt{7}}{-3}$ Since 3 is a common factor, we simplify.

$= \dfrac{\not{3}(4+2\sqrt{7})}{\not{3}(-1)}$ Factoring and removing a factor equal to 1

$= \dfrac{4+2\sqrt{7}}{-1}$

$= -4-2\sqrt{7}$

61. $\dfrac{2}{\sqrt{7}+5}$

$= \dfrac{2}{\sqrt{7}+5}\cdot\dfrac{\sqrt{7}-5}{\sqrt{7}-5}$

$= \dfrac{2(\sqrt{7}-5)}{(\sqrt{7})^2-5^2}$

$= \dfrac{2\sqrt{7}-10}{7-25}$

$= \dfrac{2\sqrt{7}-10}{-18}$ Since 2 is a common factor, we simplify.

$= \dfrac{\not{2}(\sqrt{7}-5)}{\not{2}(-9)}$ Factoring and removing a factor equal to 1

$= \dfrac{\sqrt{7}-5}{-9}$

$= \dfrac{-(\sqrt{7}-5)}{9}$

$= \dfrac{5-\sqrt{7}}{9}$

63. $\dfrac{\sqrt{6}}{\sqrt{6}-5} = \dfrac{\sqrt{6}}{\sqrt{6}-5}\cdot\dfrac{\sqrt{6}+5}{\sqrt{6}+5} = \dfrac{\sqrt{6}\sqrt{6}+\sqrt{6}\cdot 5}{(\sqrt{6})^2-5^2} =$

$\dfrac{6+5\sqrt{6}}{6-25} = \dfrac{6+5\sqrt{6}}{-19} = -\dfrac{6+5\sqrt{6}}{19}$, or $\dfrac{-6-5\sqrt{6}}{19}$

65. $\dfrac{\sqrt{7}}{\sqrt{7}-\sqrt{3}} = \dfrac{\sqrt{7}}{\sqrt{7}-\sqrt{3}}\cdot\dfrac{\sqrt{7}+\sqrt{3}}{\sqrt{7}+\sqrt{3}} = \dfrac{\sqrt{7}\sqrt{7}+\sqrt{7}\sqrt{3}}{(\sqrt{7})^2-(\sqrt{3})^2} =$

$\dfrac{7+\sqrt{21}}{7-3} = \dfrac{7+\sqrt{21}}{4}$

67. $\dfrac{\sqrt{3}}{\sqrt{5}-\sqrt{3}} = \dfrac{\sqrt{3}}{\sqrt{5}-\sqrt{3}}\cdot\dfrac{\sqrt{5}+\sqrt{3}}{\sqrt{5}+\sqrt{3}} =$

$\dfrac{\sqrt{3}\sqrt{5}+\sqrt{3}\sqrt{3}}{(\sqrt{5})^2-(\sqrt{3})^2} = \dfrac{\sqrt{15}+3}{5-3} = \dfrac{\sqrt{15}+3}{2}$

69. $\dfrac{2}{\sqrt{7}+\sqrt{2}} = \dfrac{2}{\sqrt{7}+\sqrt{2}}\cdot\dfrac{\sqrt{7}-\sqrt{2}}{\sqrt{7}-\sqrt{2}} =$

$\dfrac{2\sqrt{7}-2\sqrt{2}}{(\sqrt{7})^2-(\sqrt{2})^2} = \dfrac{2\sqrt{7}-2\sqrt{2}}{7-2} = \dfrac{2\sqrt{7}-2\sqrt{2}}{5}$

71. $\dfrac{\sqrt{6}-\sqrt{x}}{\sqrt{6}+\sqrt{x}} = \dfrac{\sqrt{6}-\sqrt{x}}{\sqrt{6}+\sqrt{x}}\cdot\dfrac{\sqrt{6}-\sqrt{x}}{\sqrt{6}-\sqrt{x}}$

$= \dfrac{(\sqrt{6}-\sqrt{x})^2}{(\sqrt{6}+\sqrt{x})(\sqrt{6}-\sqrt{x})}$

$= \dfrac{(\sqrt{6})^2-2\sqrt{6}\sqrt{x}+(\sqrt{x})^2}{(\sqrt{6})^2-(\sqrt{x})^2}$

$= \dfrac{6-2\sqrt{6x}+x}{6-x}$

73. *Writing Exercise*

75.
$$\begin{aligned} 3x+5+2(x-3) &= 4-6x \\ 3x+5+2x-6 &= 4-6x \\ 5x-1 &= 4-6x \\ 11x-1 &= 4 \\ 11x &= 5 \\ x &= \frac{5}{11} \end{aligned}$$

The solution is $\frac{5}{11}$.

77.
$$\begin{aligned} x^2-5x &= 6 \\ x^2-5x-6 &= 0 \\ (x+1)(x-6) &= 0 \\ x+1=0 \quad &or \quad x-6=0 \\ x=-1 \quad &or \quad x=6 \end{aligned}$$

The solutions are -1 and 6.

79. ***Familiarize***. Let x = the number of liters of Jolly Juice and y = the number of liters of Real Squeeze in the mixture. We organize the given information in a table.

	Jolly Juice	Real Squeeze	Mixture
Amount	x	y	8
Percent real fruit juice	3%	6%	5.4%
Amount of real fruit juice	$0.03x$	$0.06y$	0.054(8), or 0.432

Translate. We get two equation from the first and third rows of the table.

$$\begin{aligned} x + \quad y &= \quad 8, \\ 0.03x + 0.06y &= 0.432 \end{aligned}$$

Clearing decimals gives

$$\begin{aligned} x + \quad y &= \quad 8, \quad (1) \\ 30x + 60y &= 432. \quad (2) \end{aligned}$$

Carry out. We use elimination. Multiply Equation (1) by -30 and add.

$$\begin{aligned} -30x - 30y &= -240 \\ 30x + 60y &= \quad 432 \\ \hline 30y &= \quad 192 \\ y &= \quad 6.4 \end{aligned}$$

Now substitute 6.4 for y in Equation (1) and solve for x.

$$\begin{aligned} x+y &= 8 \\ x+6.4 &= 8 \\ x &= 1.6 \end{aligned}$$

Check. The sum of 1.6 and 6.4 is 8. The amount of real fruit juice in this mixture is $0.03(1.6)+0.06(6.4)$, or $0.048+0.384$, or 0.432 L. The answer checks.

State. 1.6 L of Jolly Juice and 6.4 L of Real Squeeze should be used.

81. *Writing Exercise*

83.
$$\begin{aligned} &\sqrt{\frac{25}{x}} + \frac{\sqrt{x}}{2x} - \frac{5}{\sqrt{2}} \\ &= \frac{5}{\sqrt{x}} + \frac{\sqrt{x}}{2x} - \frac{5}{\sqrt{2}} \\ &= \frac{5}{\sqrt{x}} \cdot \frac{\sqrt{x}}{\sqrt{x}} + \frac{\sqrt{x}}{2x} - \frac{5}{\sqrt{2}} \cdot \frac{\sqrt{2}}{\sqrt{2}} \\ &= \frac{5\sqrt{x}}{x} + \frac{\sqrt{x}}{2x} - \frac{5\sqrt{2}}{2} \\ &= \frac{5\sqrt{x}}{x} \cdot \frac{2}{2} + \frac{\sqrt{x}}{2x} - \frac{5\sqrt{2}}{2} \cdot \frac{x}{x} \\ &= \frac{10\sqrt{x}}{2x} + \frac{\sqrt{x}}{2x} - \frac{5x\sqrt{2}}{2x} \\ &= \frac{11\sqrt{x} - 5x\sqrt{2}}{2x} \end{aligned}$$

85.
$$\begin{aligned} &\sqrt{8x^6y^3} - x\sqrt{2y^7} - \frac{x}{3}\sqrt{18x^2y^9} \\ &= \sqrt{4x^6y^2 \cdot 2y} - x\sqrt{y^6 \cdot 2y} - \frac{x}{3}\sqrt{9x^2y^8 \cdot 2y} \\ &= 2x^3y\sqrt{2y} - xy^3\sqrt{2y} - \frac{x}{3} \cdot 3xy^4\sqrt{2y} \\ &= 2x^3y\sqrt{2y} - xy^3\sqrt{2y} - x^2y^4\sqrt{2y} \\ &= xy\sqrt{2y}(2x^2 - y^2 - xy^3) \end{aligned}$$

87.
$$\begin{aligned} &7x\sqrt{12xy^2} - 9y\sqrt{27x^3} + 5\sqrt{300x^3y^2} \\ &= 7x\sqrt{4y^2 \cdot 3x} - 9y\sqrt{9x^2 \cdot 3x} + 5\sqrt{100x^2y^2 \cdot 3x} \\ &= 7x \cdot 2y\sqrt{3x} - 9y \cdot 3x\sqrt{3x} + 5 \cdot 10xy\sqrt{3x} \\ &= 14xy\sqrt{3x} - 27xy\sqrt{3x} + 50xy\sqrt{3x} \\ &= (14xy - 27xy + 50xy)\sqrt{3x} \\ &= 37xy\sqrt{3x} \end{aligned}$$

89.
$$\begin{aligned} &\sqrt{10} + \sqrt{50} = \sqrt{10} + \sqrt{10}\,\sqrt{5} = \sqrt{10}(1+\sqrt{5}) \\ &\sqrt{10} + \sqrt{50} = \sqrt{10} + \sqrt{25 \cdot 2} = \sqrt{10} + 5\sqrt{2} \\ &\sqrt{10} + \sqrt{50} = \sqrt{2}\,\sqrt{5} + \sqrt{2}\,\sqrt{25} = \\ &\sqrt{2}(\sqrt{5} + \sqrt{25}) = \sqrt{2}(\sqrt{5} + 5), \text{ or } \sqrt{2}(5 + \sqrt{5}) \end{aligned}$$

All three are correct.

Exercise Set 8.5

1. The statement is true by the principle of squaring.

3. The statement is false. See Example 1(b).

5.
$$\begin{aligned} \sqrt{x} &= 7 \\ (\sqrt{x})^2 &= 7^2 \quad \text{Squaring both sides} \\ x &= 49 \quad \text{Simplifying} \end{aligned}$$

Check:
$$\begin{array}{c|c} \multicolumn{2}{c}{\sqrt{x} = 7} \\ \hline \sqrt{49} & 7 \end{array}$$
$$7 \stackrel{?}{=} 7 \quad \text{TRUE}$$

The solution is 49.

7. $\sqrt{x}+4=12$

$$\begin{aligned} \sqrt{x} &= 8 \quad \text{Subtracting 4} \\ (\sqrt{x})^2 &= 8^2 \\ x &= 64 \end{aligned}$$

Check:
$$\begin{array}{c|c} \sqrt{x}+4 & 12 \\ \hline \sqrt{64}+4 & 12 \\ 8+4 & \\ 12 & \stackrel{?}{=} 12 \end{array}$$
TRUE

The solution is 64.

9. $\sqrt{2x+3}=11$

$$\begin{aligned} (\sqrt{2x+3})^2 &= 11^2 \quad \text{Squaring both sides} \\ 2x+3 &= 121 \\ 2x &= 118 \\ x &= 59 \end{aligned}$$

Check:
$$\begin{array}{c|c} \sqrt{2x+3} & 11 \\ \hline \sqrt{2\cdot 59+3} & 11 \\ \sqrt{121} & \\ 11 & \stackrel{?}{=} 11 \end{array}$$
TRUE

The solution is 59.

11. $2+\sqrt{3-y}=9$

$$\begin{aligned} \sqrt{3-y} &= 7 \quad \text{Subtracting 2} \\ (\sqrt{3-y})^2 &= 7^2 \quad \text{Squaring both sides} \\ 3-y &= 49 \\ -y &= 46 \\ y &= -46 \end{aligned}$$

Check:
$$\begin{array}{c|c} 2+\sqrt{3-y} & 9 \\ \hline 2+\sqrt{3-(-46)} & 9 \\ 2+\sqrt{49} & \\ 2+7 & \\ 9 & \stackrel{?}{=} 9 \end{array}$$
TRUE

The solution is -46.

13. $6-3\sqrt{5n}=0$

$$\begin{aligned} 6 &= 3\sqrt{5n} \quad \text{Adding } 3\sqrt{5n} \\ 2 &= \sqrt{5n} \quad \text{Dividing by 3} \\ 2^2 &= (\sqrt{5n})^2 \quad \text{Squaring both sides} \\ 4 &= 5n \\ \frac{4}{5} &= n \end{aligned}$$

Check:
$$\begin{array}{c|c} 6-3\sqrt{5n} & 0 \\ \hline 6-3\sqrt{5\cdot\frac{4}{5}} & 0 \\ 6-3\sqrt{4} & \\ 6-3\cdot 2 & \\ 6-6 & \\ 0 & \stackrel{?}{=} 0 \end{array}$$
TRUE

The solution is $\frac{4}{5}$.

15. $\sqrt{4x-6}=\sqrt{x+9}$

$$\begin{aligned} (\sqrt{4x-6})^2 &= (\sqrt{x+9})^2 \\ 4x-6 &= x+9 \\ 3x-6 &= 9 \\ 3x &= 15 \\ x &= 5 \end{aligned}$$

Check:
$$\begin{array}{c|c} \sqrt{4x-6} & \sqrt{x+9} \\ \hline \sqrt{4\cdot 5-6} & \sqrt{5+9} \\ \sqrt{20-6} & \sqrt{14} \\ \sqrt{14} & \stackrel{?}{=} \sqrt{14} \end{array}$$
TRUE

The solution is 5.

17. $3\sqrt{x}=-7$

$$\sqrt{x} = -\frac{7}{3}$$

Since the principal square root of a number cannot be negative, we see that this equation has no solution.

19. $\sqrt{2t+5}=\sqrt{3t+7}$

$$\begin{aligned} (\sqrt{2t+5})^2 &= (\sqrt{3t+7})^2 \quad \text{Squaring both sides} \\ 2t+5 &= 3t+7 \\ 5 &= t+7 \\ -2 &= t \end{aligned}$$

Check:
$$\begin{array}{c|c} \sqrt{2t+5} & \sqrt{3t+7} \\ \hline \sqrt{2(-2)+5} & \sqrt{3(-2)+7} \\ \sqrt{-4+5} & \sqrt{-6+7} \\ \sqrt{1} & \sqrt{1} \\ 1 & \stackrel{?}{=} 1 \end{array}$$
TRUE

The solution is -2.

21. $\sqrt{3x+1}=x-3$

$$\begin{aligned} (\sqrt{3x+1})^2 &= (x-3)^2 \\ 3x+1 &= x^2-6x+9 \\ 0 &= x^2-9x+8 \\ 0 &= (x-1)(x-8) \end{aligned}$$

$x-1=0$ *or* $x-8=0$

$x=1$ *or* $x=8$

Check:
$$\begin{array}{c|c} \multicolumn{2}{c}{\sqrt{3x+1} = x-3} \\ \hline \sqrt{3\cdot 1+1} & 1-3 \\ \sqrt{4} & -2 \\ \end{array}$$
$$2 \stackrel{?}{=} -2 \qquad \text{FALSE}$$

$$\begin{array}{c|c} \multicolumn{2}{c}{\sqrt{3x+1} = x-3} \\ \hline \sqrt{3\cdot 8+1} & 8-3 \\ \sqrt{25} & 5 \\ \end{array}$$
$$5 \stackrel{?}{=} 5 \qquad \text{TRUE}$$

The number 1 does not check, but 8 does. The solution is 8.

23.
$$\begin{aligned} a-9 &= \sqrt{a-3} \\ (a-9)^2 &= (\sqrt{a-3})^2 \\ a^2-18a+81 &= a-3 \\ a^2-19a+84 &= 0 \\ (a-12)(a-7) &= 0 \end{aligned}$$
$$a-12 = 0 \quad or \quad a-7=0$$
$$a = 12 \quad or \quad a=7$$

Check:
$$\begin{array}{c|c} \multicolumn{2}{c}{a-9 = \sqrt{a-3}} \\ \hline 12-9 & \sqrt{12-3} \\ 3 & \sqrt{9} \\ \end{array}$$
$$3 \stackrel{?}{=} 3 \qquad \text{TRUE}$$

$$\begin{array}{c|c} \multicolumn{2}{c}{a-9 = \sqrt{a-3}} \\ \hline 7-9 & \sqrt{7-3} \\ -2 & \sqrt{4} \\ \end{array}$$
$$-2 \stackrel{?}{=} 2 \qquad \text{FALSE}$$

The number 12 checks, but 7 does not. The solution is 12.

25.
$$\begin{aligned} x+1 &= 6\sqrt{x-7} \\ (x+1)^2 &= (6\sqrt{x-7})^2 \\ x^2+2x+1 &= 36(x-7) \\ x^2+2x+1 &= 36x-252 \\ x^2-34x+253 &= 0 \\ (x-11)(x-23) &= 0 \end{aligned}$$
$$x-11 = 0 \quad or \quad x-23=0$$
$$x = 11 \quad or \quad x=23$$

Check:
$$\begin{array}{c|c} \multicolumn{2}{c}{x+1 = 6\sqrt{x-7}} \\ \hline 11+1 & 6\sqrt{11-7} \\ 12 & 6\sqrt{4} \\ & 6\cdot 2 \\ \end{array}$$
$$12 \stackrel{?}{=} 12 \qquad \text{TRUE}$$

$$\begin{array}{c|c} \multicolumn{2}{c}{x+1 = 6\sqrt{x-7}} \\ \hline 23+1 & 6\sqrt{23-7} \\ 24 & 6\sqrt{16} \\ & 6\cdot 4 \\ \end{array}$$
$$24 \stackrel{?}{=} 24 \qquad \text{TRUE}$$

The solutions are 11 and 23.

27.
$$\begin{aligned} \sqrt{5x+21} &= x+3 \\ (\sqrt{5x+21})^2 &= (x+3)^2 \\ 5x+21 &= x^2+6x+9 \\ 0 &= x^2+x-12 \\ 0 &= (x+4)(x-3) \end{aligned}$$
$$x+4 = 0 \quad or \quad x-3=0$$
$$x = -4 \quad or \quad x=3$$

Check:
$$\begin{array}{c|c} \multicolumn{2}{c}{\sqrt{5x+21} = x+3} \\ \hline \sqrt{5(-4)+21} & -4+3 \\ \sqrt{1} & -1 \\ \end{array}$$
$$1 \stackrel{?}{=} -1 \qquad \text{FALSE}$$

$$\begin{array}{c|c} \multicolumn{2}{c}{\sqrt{5x+21} = x+3} \\ \hline \sqrt{5\cdot 3+21} & 3+3 \\ \sqrt{36} & 6 \\ \end{array}$$
$$6 \stackrel{?}{=} 6 \qquad \text{TRUE}$$

The number 3 checks, but -4 does not. The solution is 3.

29.
$$\begin{aligned} t+4 &= 4\sqrt{t+1} \\ (t+4)^2 &= (4\sqrt{t+1})^2 \\ t^2+8t+16 &= 16(t+1) \\ t^2+8t+16 &= 16t+16 \\ t^2-8t &= 0 \\ t(t-8) &= 0 \end{aligned}$$
$$t = 0 \quad or \quad t-8=0$$
$$t = 0 \quad or \quad t=8$$

Check:
$$\begin{array}{c|c} \multicolumn{2}{c}{t+4 = 4\sqrt{t+1}} \\ \hline 0+4 & 4\sqrt{0+1} \\ 4 & 4\sqrt{1} \\ & 4\cdot 1 \\ \end{array}$$
$$4 \stackrel{?}{=} 4 \qquad \text{TRUE}$$

$$\begin{array}{c|c} \multicolumn{2}{c}{t+4 = 4\sqrt{t+1}} \\ \hline 8+4 & 4\sqrt{8+1} \\ 12 & 4\sqrt{9} \\ & 4\cdot 3 \\ \end{array}$$
$$12 \stackrel{?}{=} 12 \qquad \text{TRUE}$$

The solutions are 0 and 8.

31. $\sqrt{x^2+6}-x+3=0$

$\sqrt{x^2+6}=x-3$ Isolating the radical

$(\sqrt{x^2+6})^2=(x-3)^2$

$x^2+6=x^2-6x+9$

$-3=-6x$ Adding $-x^2$ and -9

$\frac{1}{2}=x$

Check: $\sqrt{x^2+6}-x+3=0$

$\sqrt{\left(\frac{1}{2}\right)^2+6}-\frac{1}{2}+3 \;\big|\; 0$

$\sqrt{\frac{25}{4}}-\frac{1}{2}+3$

$\frac{5}{2}-\frac{1}{2}+3$

$2+3$

$5\overset{?}{=}0$ FALSE

The number $\frac{1}{2}$ does not check. There is no solution.

33. $\sqrt{(p+6)(p+1)}-2=p+1$

$\sqrt{(p+6)(p+1)}=p+3$ Isolating the radical

$\left(\sqrt{(p+6)(p+1)}\right)^2=(p+3)^2$

$(p+6)(p+1)=p^2+6p+9$

$p^2+7p+6=p^2+6p+9$

$p=3$

The number 3 checks. It is the solution.

35. $\sqrt{3-7x}=\sqrt{5-2x}$

$(\sqrt{3-7x})^2=(\sqrt{5-2x})^2$

$3-7x=5-2x$

$3=5+5x$

$-2=5x$

$-\frac{2}{5}=x$

The number $-\frac{2}{5}$ checks. It is the solution.

37. $x-1=\sqrt{(x+1)(x-2)}$

$(x-1)^2=(\sqrt{(x+1)(x-2)})^2$

$x^2-2x+1=(x+1)(x-2)$

$x^2-2x+1=x^2-x-2$

$-2x+1=-x-2$ Adding $-x^2$ on both sides

$-x+1=-2$

$-x=-3$

$x=3$

The number 3 checks. It is the solution.

39. $r=2\sqrt{5L}$

$40=2\sqrt{5L}$ Substituting 40 for r

$20=\sqrt{5L}$

$20^2=(\sqrt{5L})^2$

$400=5L$

$80=L$

The car will skid 80 ft at 40 mph.

$60=2\sqrt{5L}$ Substituting 60 for r

$30=\sqrt{5L}$

$30^2=(\sqrt{5L})^2$

$900=5L$

$180=L$

The car will skid 180 ft at 60 mph.

41. ***Familiarize***. We will use the formula $s=21.9\sqrt{5t+2457}$, where t is in degrees Fahrenheit and s is in feet per second.

Translate. We substitute 1113 for s in the formula.

$1113=21.9\sqrt{5t+2457}$

Carry out. We solve for t.

$1113=21.9\sqrt{5t+2457}$

$\frac{1113}{21.9}=\sqrt{5t+2457}$

$\left(\frac{1113}{21.9}\right)^2=(\sqrt{5t+2457})^2$

$2582.9\approx 5t+2457$

$125.9\approx 5t$

$25.2\approx t$

Check. We can substitute 25.2 for t in the formula.

$21.9\sqrt{5(25.2)+2457}=21.9\sqrt{2583}\approx 1113$

The answer checks.

State. The temperature was about 25.2°F.

43. ***Familiarize and Translate***. We substitute 99.4 for V in equation $V=3.5\sqrt{h}$.

$99.4=3.5\sqrt{h}$

Carry out. We solve the equation.

$99.4=3.5\sqrt{h}$

$28.4=\sqrt{h}$ Dividing by 3.5

$(28.4)^2=(\sqrt{h})^2$

$806.56=h$

Check. We go over the computation.

State. The balloon is 806.56 m high.

45. ***Familiarize and Translate***. We substitute 84 for V in the equation $V=3.5\sqrt{h}$.

$84=3.5\sqrt{h}$

Carry out.

$$84 = 3.5\sqrt{h}$$
$$24 = \sqrt{h} \quad \text{Dividing by 3.5}$$
$$24^2 = (\sqrt{h})^2$$
$$576 = h$$

Check. We go over the computation.

State. The altitude of the scout's eyes is 576 m.

47.
$$T = 2\pi\sqrt{\frac{L}{32}}$$
$$4.4 = 2(3.14)\sqrt{\frac{L}{32}} \quad \text{Substituting 4.4 for } T \text{ and 3.14 for } \pi$$
$$4.4 = 6.28\sqrt{\frac{L}{32}}$$
$$\frac{4.4}{6.28} = \sqrt{\frac{L}{32}}$$
$$\left(\frac{4.4}{6.28}\right)^2 = \left(\sqrt{\frac{L}{32}}\right)^2$$
$$0.4909 \approx \frac{L}{32}$$
$$15.71 \approx L$$

The pendulum is about 15.71 ft long.

49. *Writing Exercise*

51. ***Familiarize.*** Let a and b represent the number of questions of type A and type B answered correctly, respectively. Then Amy scores $10a$ points on the type A items and $15b$ points on the type B items.

Translate. Sixteen questions were answered correctly, so we can write one equation:

$$a + b = 16$$

Amy scored a total of 180 points, so we can write a second equation:

$$10a + 15b = 180$$

The resulting system is

$$a + b = 16, \quad (1)$$
$$10a + 15b = 180. \quad (2)$$

Carry out. We first multiply Equation (1) by -10 and add.

$$-10a - 10b = -160$$
$$\underline{10a + 15b = 180}$$
$$5b = 20$$
$$b = 4$$

Now substitute 4 for b in Equation (1) and solve for a.

$$a + b = 16$$
$$a + 4 = 16$$
$$a = 12$$

Check. If Amy correctly answers 12 questions of type A and 4 questions of type B, she answers a total of $12 + 4$, or 16 questions. She scores $10 \cdot 12 + 15 \cdot 4$, or $120 + 60$, 180 points. The answer checks.

State. Amy answered 12 questions of type A and 4 questions of type B correctly.

53. ***Familiarize.*** We present the information in a table.

$d = r \cdot t$

	Distance	Speed	Time
First car	d	56	t
Second car	d	84	$t - 1$

Translate. From the rows of the table we get two equations:

$$d = 56t,$$
$$d = 84(t - 1).$$

Carry out. We use the substitution method.

$$56t = 84(t - 1) \quad \text{Substituting } 56t \text{ for } d$$
$$56t = 84t - 84$$
$$-28t = -84$$
$$t = 3$$

The problem asks how far from Parkton the second car will overtake the first, so we need to find d. Substitute 3 for t in the first equation.

$$d = 56t$$
$$d = 56 \cdot 3$$
$$d = 168$$

Check. If $t = 3$, then the first car travels $56 \cdot 3$, or 168 km, and the second car travels $84(3 - 1)$, or $84 \cdot 2$, or 168 km. Since the distances are the same, the answer checks.

State. The second car overtakes the first 168 km from Parkton.

55. *Writing Exercise*

57. ***Familiarize.*** Let x represent the number. Then three times its square root is $3\sqrt{x}$ and the opposite of three times its square root is $-3\sqrt{x}$.

Translate. We reword the problem.

$$\underbrace{\text{The opposite of three times the square root of a number}}_{-3\sqrt{x}} \ \underset{=}{\text{is}} \ \underset{-33}{-33.}$$

Carry out. We solve the equation.

$$-3\sqrt{x} = -33$$
$$\sqrt{x} = 11 \quad \text{Dividing by } -3$$
$$(\sqrt{x})^2 = 11^2$$
$$x = 121$$

Check. $\sqrt{121} = 11$ and $-3 \cdot 11 = -33$. The answer checks.

State. The number is 121.

59.
$$1+\sqrt{x}=\sqrt{x+9}$$
$$(1+\sqrt{x})^2=(\sqrt{x+9})^2$$
$$1+2\sqrt{x}+x=x+9$$
$$1+2\sqrt{x}=9 \quad \text{Adding } -x$$
$$2\sqrt{x}=8$$
$$\sqrt{x}=4$$
$$(\sqrt{x})^2=4^2$$
$$x=16$$

The number 16 checks. It is the solution.

61.
$$\sqrt{t+4}=1-\sqrt{3t+1}$$
$$(\sqrt{t+4})^2=(1-\sqrt{3t+1})^2$$
$$t+4=1-2\sqrt{3t+1}+3t+1$$
$$t+4=2-2\sqrt{3t+1}+3t$$
$$-2t+2=-2\sqrt{3t+1}$$
Isolating the radical
$$t-1=\sqrt{3t+1}$$
Multiplying by $-\frac{1}{2}$
$$(t-1)^2=(\sqrt{3t+1})^2$$
$$t^2-2t+1=3t+1$$
$$t^2-5t=0$$
$$t(t-5)=0$$
$$t=0 \quad or \quad t-5=0$$
$$t=0 \quad or \quad t=5$$

Check: $\sqrt{t+4}=1-\sqrt{3t+1}$

$\sqrt{0+4}$	$1-\sqrt{3\cdot 0+1}$
$\sqrt{4}$	$1-\sqrt{1}$
2	$1-1$

$2\stackrel{?}{=}0$ FALSE

$\sqrt{t+4}=1-\sqrt{3t+1}$

$\sqrt{5+4}$	$1-\sqrt{3\cdot 5+1}$
$\sqrt{9}$	$1-\sqrt{16}$
3	$1-4$

$3\stackrel{?}{=}-3$ FALSE

Neither number checks. There is no solution.

63.
$$\sqrt{y+1}-\sqrt{y-2}=\sqrt{2y-5}$$
$$(\sqrt{y+1}-\sqrt{y-2})^2=(\sqrt{2y-5})^2$$
$$y+1-2\sqrt{(y+1)(y-2)}+y-2=2y-5$$
$$2y-1-2\sqrt{(y+1)(y-2)}=2y-5$$
$$-2\sqrt{(y+1)(y-2)}=-4 \quad \text{Adding } -2y \text{ and } 1$$
$$\sqrt{(y+1)(y-2)}=2 \quad \text{Dividing by } -2$$
$$\left(\sqrt{(y+1)(y-2)}\right)^2=2^2$$
$$(y+1)(y-2)=4$$
$$y^2-y-2=4$$
$$y^2-y-6=0$$
$$(y-3)(y+2)=0$$
$$y-3=0 \quad or \quad y+2=0$$
$$y=3 \quad or \quad y=-2$$

The number 3 checks, but -2 does not. The solution is 3.

65.
$$2\sqrt{x-1}-\sqrt{x-9}=\sqrt{3x-5}$$
$$(2\sqrt{x-1}-\sqrt{x-9})^2=(\sqrt{3x-5})^2$$
$$4(x-1)-4\sqrt{(x-1)(x-9)}+x-9=3x-5$$
$$4x-4-4\sqrt{x^2-10x+9}+x-9=3x-5$$
$$5x-13-4\sqrt{x^2-10x+9}=3x-5$$
$$-4\sqrt{x^2-10x+9}=-2x+8$$
$$2\sqrt{x^2-10x+9}=x-4$$
Multiplying by $-\frac{1}{2}$
$$(2\sqrt{x^2-10x+9})^2=(x-4)^2$$
$$4(x^2-10x+9)=x^2-8x+16$$
$$4x^2-40x+36=x^2-8x+16$$
$$3x^2-32x+20=0$$
$$(3x-2)(x-10)=0$$
$$3x-2=0 \quad or \quad x-10=0$$
$$3x=2 \quad or \quad x=10$$
$$x=\frac{2}{3} \quad or \quad x=10$$

The number 10 checks, but $\frac{2}{3}$ does not. The solution is 10.

67. ***Familiarize***. We will use the formula $V=3.5\sqrt{h}$. We present the information in a table.

	Height	Distance to the horizon
First sighting	h	V
Second sighting	$h+100$	$V+20$

Translate. The rows of the table give us two equations.

$$V=3.5\sqrt{h}, \quad (1)$$
$$V+20=3.5\sqrt{h+100} \quad (2)$$

Carry out. We substitute $3.5\sqrt{h}$ for V in Equation (2) and solve for h.

$$3.5\sqrt{h}+20 = 3.5\sqrt{h+100}$$
$$(3.5\sqrt{h}+20)^2 = (3.5\sqrt{h+100})^2$$
$$12.25h+140\sqrt{h}+400 = 12.25(h+100)$$
$$12.25h+140\sqrt{h}+400 = 12.25h+1225$$
$$140\sqrt{h} = 825$$
$$28\sqrt{h} = 165 \quad \text{Multiplying by } \frac{1}{5}$$
$$(28\sqrt{h})^2 = (165)^2$$
$$784h = 27,225$$
$$h = 34\frac{569}{784}, \text{ or}$$
$$h \approx 34.726$$

Check. When $h \approx 34.726$, then $V \approx 3.5\sqrt{34.726} \approx 20.625$ km. When $h \approx 100 + 34.726$, or 134.726, then $V \approx 3.5\sqrt{134.726} \approx 40.625$ km. This is 20 km more than 20.625. The answer checks.

State. The climber was at a height of $34\frac{569}{784}$ m, or about 34.726 m when the first computation was made.

69. Graph $y = \sqrt{x}$.

We make a table of values. Note that we must choose nonnegative values of x in order to have a nonnegative radicand.

x	y
0	0
1	1
2	1.414
4	2
5	2.236

We plot these points and connect them with a smooth curve.

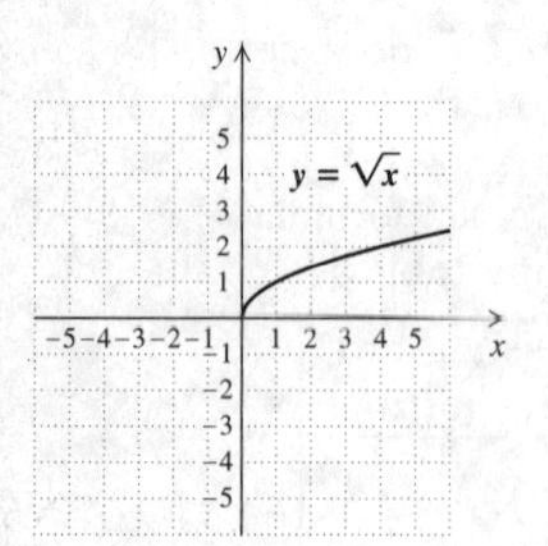

71. Graph $y = \sqrt{x-4}$.

We make a table of values. Note that we must choose values for x that are greater than or equal to 3 in order to have a nonnegative radicand.

x	y
4	0
5	1
6	1.414
7	1.732
8	2

We plot these points and connect them with a smooth curve.

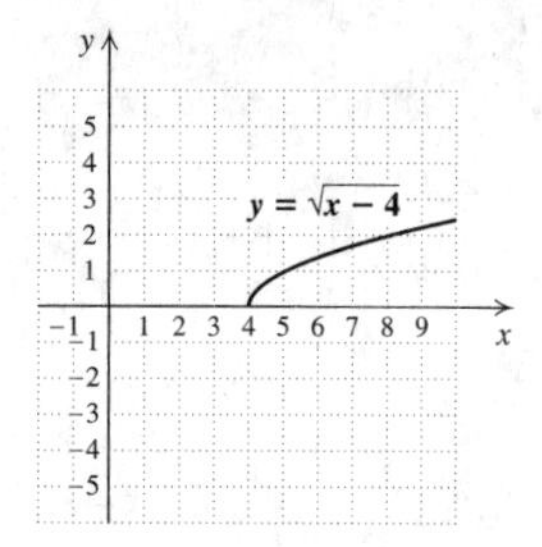

73. Given a car that is 15 ft long and a speed r, in mph, one car length per 10 mph of speed is represented by $15 \cdot \frac{r}{10}$, or $\frac{3r}{2}$. Substitute $\frac{3r}{2}$ for L in the formula and solve for r.

$$r = 2\sqrt{5L}$$
$$r = 2\sqrt{5 \cdot \frac{3r}{2}}$$
$$r = 2\sqrt{\frac{15r}{2}}$$
$$r^2 = \left(2\sqrt{\frac{15r}{2}}\right)^2$$
$$r^2 = 4 \cdot \frac{15r}{2}$$
$$r^2 = 30r$$
$$r^2 - 30r = 0$$
$$r(r-30) = 0$$
$$r = 0 \quad or \quad r - 30 = 0$$
$$r = 0 \quad or \quad r = 30$$

The number 0 has no meaning in this problem. The number 30 checks, so the answer is 30 mph.

75. First we make a table of values for each equation.

For $y = 1 + \sqrt{x}$:

x	y
0	1
1	2
4	3
9	4
16	5

For $y = \sqrt{x+9}$:

x	y
−9	0
−8	1
−5	2
0	3
7	4

We graph the equations.

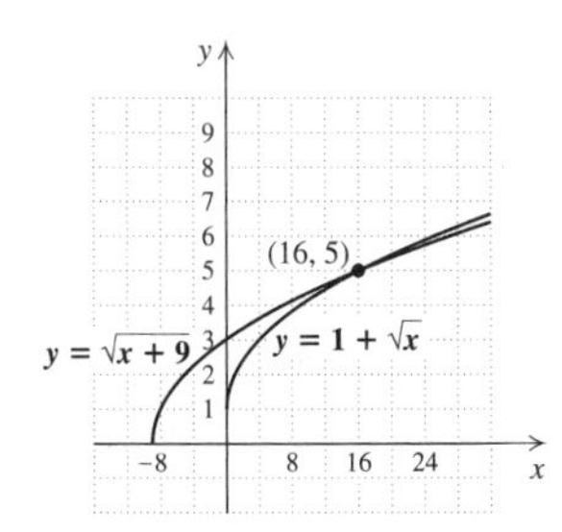

The graphs intersect at $(16, 5)$, so the solution of $1 + \sqrt{x} = \sqrt{x+9}$ is 16.

77. Graph $y_1 = -\sqrt{x+3}$ and $y_2 = 2x - 1$ and then find the first coordinate(s) of the point(s) of intersection. The solution is about -0.32.

Exercise Set 8.6

1. hypotenuse; see page 515 in the text.

3. approximation; see Examples 1-4.

5.
$$a^2 + b^2 = c^2$$
$$8^2 + 15^2 = x^2 \quad \text{Substituting}$$
$$64 + 225 = x^2$$
$$289 = x^2$$
$$\sqrt{289} = x$$
$$17 = x$$

7.
$$a^2 + b^2 = c^2$$
$$6^2 + 6^2 = x^2 \quad \text{Substituting}$$
$$36 + 36 = x^2$$
$$72 = x^2$$
$$\sqrt{72} = x \quad \text{Exact answer}$$
$$8.485 \approx x \quad \text{Approximation}$$

9.
$$a^2 + b^2 = c^2$$
$$5^2 + x^2 = 13^2$$
$$25 + x^2 = 169$$
$$x^2 = 144$$
$$x = 12$$

11.
$$a^2 + b^2 = c^2$$
$$(\sqrt{110})^2 + x^2 = 12^2$$
$$110 + x^2 = 144$$
$$x^2 = 34$$
$$x = \sqrt{34}$$
$$x \approx 5.831$$

13.
$$a^2 + b^2 = c^2$$
$$12^2 + 5^2 = c^2$$
$$144 + 25 = c^2$$
$$169 = c^2$$
$$13 = c$$

15.
$$a^2 + b^2 = c^2$$
$$9^2 + b^2 = 15^2$$
$$81 + b^2 = 225$$
$$b^2 = 144$$
$$b = 12$$

17.
$$a^2 + b^2 = c^2$$
$$a^2 + 1^2 = (\sqrt{5})^2$$
$$a^2 + 1 = 5$$
$$a^2 = 4$$
$$a = 2$$

19.
$$a^2 + b^2 = c^2$$
$$1^2 + b^2 = (\sqrt{3})^2$$
$$1 + b^2 = 3$$
$$b^2 = 2$$
$$b = \sqrt{2} \quad \text{Exact answer}$$
$$b \approx 1.414 \quad \text{Approximation}$$

21.
$$a^2 + b^2 = c^2$$
$$a^2 + (5\sqrt{3})^2 = 10^2$$
$$a^2 + 25 \cdot 3 = 100$$
$$a^2 + 75 = 100$$
$$a^2 = 25$$
$$a = 5$$

23. ***Familiarize***. Referring to the drawing in the text, let l represent the length of the string of lights.

Translate. We use the Pythagorean theorem, substituting 8 for a, 15 for b, and l for c.

$$8^2 + 15^2 = l^2.$$

Carry out. We solve the equation.

$$8^2 + 15^2 = l^2$$
$$64 + 225 = l^2$$
$$289 = l^2$$
$$17 = l$$

Check. We check by substituting 8, 15, and 17 in the Pythagorean theorem.

$a^2+b^2=c^2$	
8^2+15^2	17^2
$64+225$	289

$289 \stackrel{?}{=} 289$ TRUE

State. The string of lights needs to be 17 ft long.

25. ***Familiarize.*** Let h = the height of the back of the jump, in inches.

Translate. We use the Pythagorean theorem, substituting 30 for a, h for b, and 33 for c.

$$30^2+h^2=33^2$$

Carry out. We solve the equation.

$$30^2+h^2=33^2$$
$$900+h^2=1089$$
$$h^2=189$$
$$h=\sqrt{189}=\sqrt{9\cdot 21}$$
$$h=3\sqrt{21} \quad \text{Exact answer}$$
$$h\approx 13.748$$

Check. We check by substituting 30, $\sqrt{189}$, and 33 in the Pythagorean theorem.

$a^2+b^2=c^2$	
$30^2+(\sqrt{189})^2$	33^2
$900+189$	1089

$1089 \stackrel{?}{=} 1089$ TRUE

State. The back of the jump should be $3\sqrt{21}$ in., or about 13.748 in. high.

27. ***Familiarize.*** We first make a drawing. We label the unknown length w.

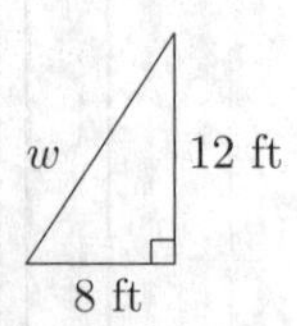

Translate. We use the Pythagorean theorem, substituting 8 for a, 12 for b, and w for c.

$$8^2+12^2=w^2$$

Carry out. We solve the equation.

$$8^2+12^2=w^2$$
$$64+144=w^2$$
$$208=w^2$$
$$\sqrt{208}=w$$
$$\sqrt{16\cdot 13}=w$$
$$4\sqrt{13}=w \quad \text{Exact answer}$$
$$14.422\approx w \quad \text{Approximation}$$

Check. We check by substituting 8, 12, and $\sqrt{208}$ into the Pythagorean theorem:

$a^2+b^2=c^2$	
8^2+12^2	$(\sqrt{208})^2$
$64+144$	208

$208 \stackrel{?}{=} 208$ TRUE

State. The pipe should be $4\sqrt{13}$ feet or about 14.422 feet long.

29. ***Familiarize.*** Let d = the distance from first base to third base, in feet.

Translate. We use the Pythagorean theorem, substituting 90 for a, 90 for b, and d for c.

$$90^2+90^2=d^2$$

Carry out. We solve the equation.

$$90^2+90^2=d^2$$
$$8100+8100=d^2$$
$$16,200=d^2$$
$$\sqrt{16,200}=d$$
$$\sqrt{8100\cdot 2}=d$$
$$90\sqrt{2}=d \quad \text{Exact answer}$$
$$127.279\approx d$$

Check. We check by substituting 90, 90, and $\sqrt{16,200}$ into the Pythagorean theorem.

$a^2+b^2=c^2$	
90^2+90^2	$(\sqrt{16,200})^2$
$8100+8100$	$16,200$

$16,200 \stackrel{?}{=} 16,200$ TRUE

State. It is $90\sqrt{2}$ ft, or about 127.279 ft, from first base to third base.

31. ***Familiarize.*** We first make a drawing. We label the diagonal d.

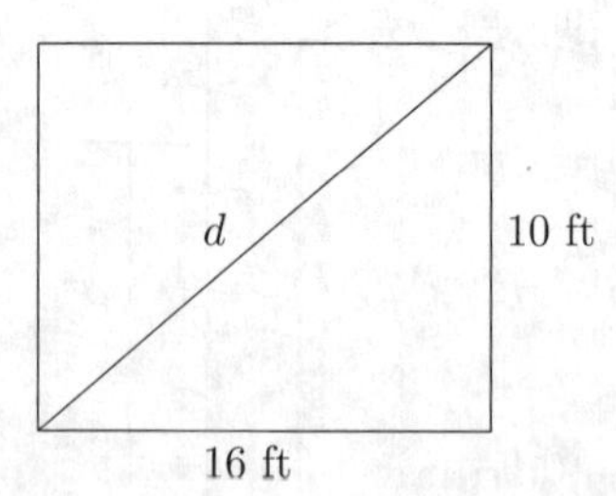

Translate.We use the Pythagorean theorem, substituting 16 for a, 10 for b, and d for c.

$$a^2+b^2=c^2$$
$$16^2+10^2=d^2$$

Carry out. We solve the equation.

$$16^2+10^2=d^2$$
$$256+100=d^2$$
$$356=d^2$$
$$\sqrt{356}=d$$
$$18.868\approx d$$

Check. We check by substituting 16, 10, and $\sqrt{356}$ in the Pythagorean equation.

$a^2 + b^2$	$= c^2$
$16^2 + 10^2$	$(\sqrt{356})^2$
$256 + 100$	356

$356 \stackrel{?}{=} 356$ TRUE

State. The wire needs to be $\sqrt{356}$ ft, or about 18.868 ft long.

33. ***Familiarize***. We make a drawing. We label the diagonal d.

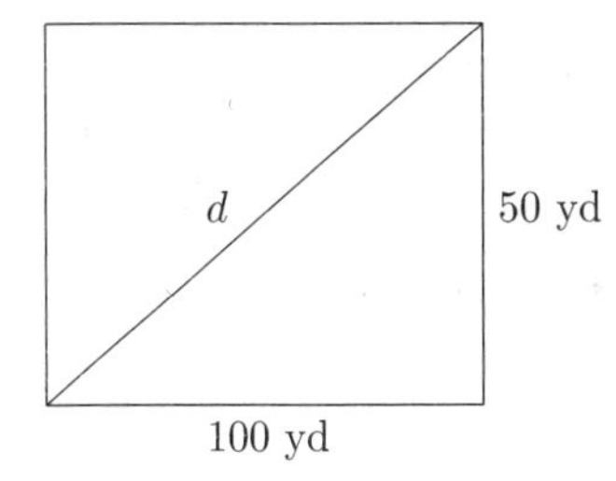

Translate. We use the Pythagorean theorem, substituting 50 for a, 100 for b, and d for c.

$$50^2 + 100^2 = d^2$$

Carry out. We solve the equation.

$$50^2 + 100^2 = d^2$$
$$2500 + 10,000 = d^2$$
$$12,500 = d^2$$
$$\sqrt{12,500} = d$$
$$111.803 \approx d$$

Check. We check by substituting 50 for a, 100 for b, and $\sqrt{12,500}$ for c in the Pythagorean theorem.

$a^2 + b^2$	$= c^2$
$50^2 + 100^2$	$(\sqrt{12,500})^2$
$2500 + 10,000$	$12,500$

$12,500 \stackrel{?}{=} 12,500$ TRUE

State. The length of a diagonal is $\sqrt{12,500}$ yd, or about 111.803 yd.

35. ***Familiarize***. We make a drawing. We let h = the height the hose can reach.

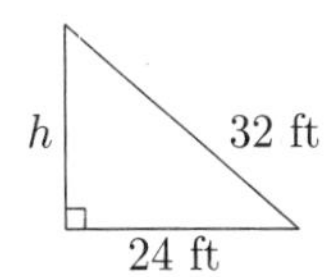

Translate. We use the Pythagorean theorem, substituting h for a, 24 for b, and 32 for c.

$$h^2 + 24^2 = 32^2$$

Carry out. We solve the equation.

$$h^2 + 24^2 = 32^2$$
$$h^2 + 576 = 1024$$
$$h^2 = 448$$
$$h = \sqrt{448} \quad \text{Exact answer}$$
$$h \approx 21.166$$

Check. We check by substituting $\sqrt{448}$ for a, 24 for b, and 32 for c in the Pythagorean equation.

$a^2 + b^2$	$= c^2$
$(\sqrt{448})^2 + 24^2$	32^2
$448 + 576$	1024

$1024 \stackrel{?}{=} 1024$ TRUE

State. The hose can reach $\sqrt{448}$ ft, or about 21.166 ft, up the far corner of the house.

37. *Writing Exercise*

39. Rational numbers can be expressed in the form $\frac{a}{b}$, where a and b are integers and $b \neq 0$. Decimal notation for rational numbers either terminates or repeats. The rational numbers in the given list are -45, -9.7, 0, $\frac{2}{7}$, 5.09, and 19.

41. Each of the numbers in the list is either rational or irrational, so all of them are real numbers.

43. $(-2)^5 = (-2)(-2)(-2)(-2)(-2) = -32$

45. $(2a)^4 = 2^4a^4 = 16a^4$

47. $x^2 - 10x + 25 = x^2 - 2 \cdot x \cdot 5 + 5^2 = (x - 5)^2$

49. *Writing Exercise*

51. ***Familiarize***. First we find the length s of each strand of lights. Then we will find the total length of the 10 strands and finally the cost of each type of strand.

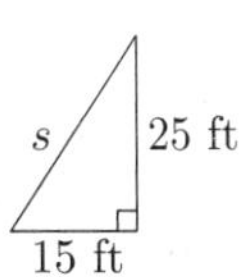

Translate. We use the Pythagorean theorem, substituting 15 for a 25 for b, and s for c.

$$15^2 + 25^2 = s^2$$

Carry out. We solve the equation.

$$15^2 + 25^2 = s^2$$
$$225 + 625 = s^2$$
$$850 = s^2$$
$$\sqrt{850} = s^2$$
$$29.155 \approx s$$

The 25-ft strands cannot be used because they will not reach 29.155 ft from the top of the tree to the ground. If 35-ft strands are used it will take 10 of them at $6.99

each for a total cost of 10($6.99), or $69.99. Since 68 ft is more than twice 29.155 ft, each 68-ft strand could be used to make 2 strands from the tree to the ground. Thus the decorations would require 5 68-ft strands at a cost of $11.99 each for a total cost of 5($11.99), or $59.95. We see that the most economical purchase will be 5 68-ft strands.

Check. We repeat the calculations.

State. Julia should purchase 5 68-ft strands of lights for $59.95.

53. ***Familiarize.*** First we find d, the shortest distance, in feet, from the pool to the corner of the building. This is the hypotenuse of a right triangle with legs of 600 ft and 400 ft. Then we find l, the shortest distance, in feet, from the pool to Virginia's apartment. This is the hypotenuse of a right triangle with legs of 180 ft and d.

Translate. To find d we use the Pythagorean theorem, substituting 600 for a, 400 for b, and d for c.

$$600^2 + 400^2 = d^2$$

After we find d we will use the Pythagorean theorem again, this time substituting 180 for a, d for b, and l for c. (Since we will use d^2 in this equation we need only solve the equation above for d^2.)

$$180^2 + d^2 = l^2$$

Carry out. First find d^2.

$$\begin{aligned} 600^2 + 400^2 &= d^2 \\ 360,000 + 160,000 &= d^2 \\ 552,400 &= d^2 \end{aligned}$$

Now we substitute 552,400 for d^2 in the second equation in the Translate step.

$$\begin{aligned} 180^2 + 552,400 &= l^2 \\ 32,400 + 552,400 &= l^2 \\ 584,800 &= l^2 \\ \sqrt{584,800} &= l \\ 765 &\approx l \end{aligned}$$

Check. We repeat the calculations.

State. The distance from the pool to Virginia's apartment is about 765 ft. Since this distance is less than 1000 ft, Virginia can use the phone at the pool.

55. ***Familiarize.*** Let s = the length of a side of the square, in feet. We make a drawing.

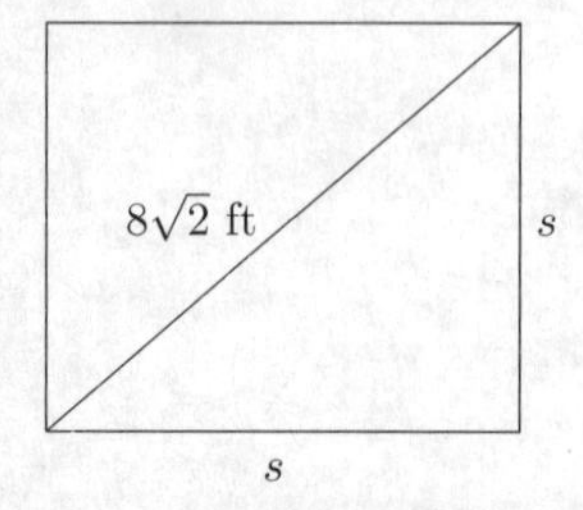

Translate. We use the Pythagorean theorem, substituting s for a, s for b, and $8\sqrt{2}$ for c.

$$s^2 + s^2 = (8\sqrt{2})^2$$

Carry out. We solve the equation.

$$\begin{aligned} s^2 + s^2 &= (8\sqrt{2})^2 \\ 2s^2 &= 64 \cdot 2 \\ s^2 &= 64 \\ s &= 8 \end{aligned}$$

Check. We check by substituting 8 for a, 8 for b, and $8\sqrt{2}$ for c in the Pythagorean theorem.

$$\begin{array}{c|c} \multicolumn{2}{c}{a^2 + b^2 = c^2} \\ \hline 8^2 + 8^2 & (8\sqrt{2})^2 \\ 64 + 64 & 64 \cdot 2 \\ \end{array}$$

$$128 \stackrel{?}{=} 128 \quad \text{TRUE}$$

State. The length of a side of the square is 8 ft.

57. ***Familiarize.*** Let x, $x+2$, and $x+4$ represent the lengths of the sides of the triangle.

Translate. We use the Pythagorean theorem.

$$x^2 + (x+2)^2 = (x+4)^2$$

Carry out. We solve the equation.

$$\begin{aligned} x^2 + (x+2)^2 &= (x+4)^2 \\ x^2 + x^2 + 4x + 4 &= x^2 + 8x + 16 \\ 2x^2 + 4x + 4 &= x^2 + 8x + 16 \\ x^2 - 4x - 12 &= 0 \\ (x+2)(x-6) &= 0 \end{aligned}$$

$$x + 2 = 0 \quad or \quad x - 6 = 0$$
$$x = -2 \quad or \quad x = 6$$

Check. The length of a side cannot be negative, so we check only 6. If $x = 6$, then $x + 2 = 6 + 2 = 8$ and $x + 4 = 6 + 4 = 10$. We substitute 6 for a, 8 for b, and 10 for c in the Pythagorean theorem.

$$\begin{array}{c|c} \multicolumn{2}{c}{a^2 + b^2 = c^2} \\ \hline 6^2 + 8^2 & 10^2 \\ 36 + 64 & 100 \\ \end{array}$$

$$100 \stackrel{?}{=} 100 \quad \text{TRUE}$$

State. The lengths of the sides of the triangle are 6, 8, and 10.

59.

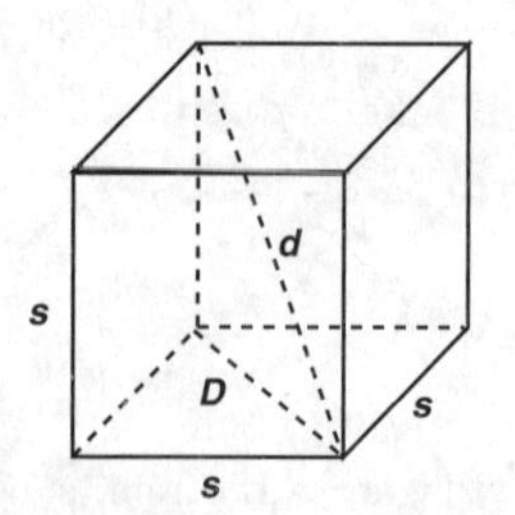

From the drawing we see that the diagonal d of the cube is the hypotenuse of a right triangle with one leg of length s, where s is the length of a side of the cube, and the other leg of length D, where D is the length of the diagonal of the base of the cube. First we find D:

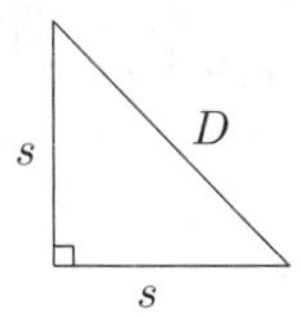

Using the Pythagorean theorem we have:

$$s^2+s^2=D^2$$
$$2s^2=D^2$$
$$\sqrt{2s^2}=D$$
$$s\sqrt{2}=D$$

Then we find d:

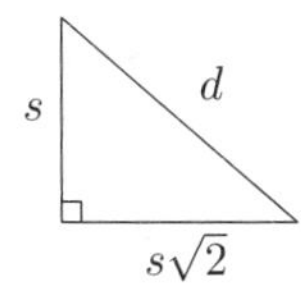

Using the Pythagorean theorem again we have:

$$s^2+(s\sqrt{2})^2=d^2$$
$$s^2+2s^2=d^2$$
$$3s^2=d^2$$
$$\sqrt{3s^2}=d$$
$$s\sqrt{3}=d$$

61. Using the Pythagorean theorem we get:

$$h^2+\left(\frac{a}{2}\right)^2=a^2$$
$$h^2+\frac{a^2}{4}=a^2$$
$$h^2=a^2-\frac{a^2}{4}$$
$$h^2=\frac{3a^2}{4}$$
$$h=\sqrt{\frac{3a^2}{4}}$$
$$h=\frac{a\sqrt{3}}{2}$$

63. After one-half hour, the car traveling east has gone $\frac{1}{2}\cdot 50$, or 25 mi, and the car traveling south has gone $\frac{1}{2}\cdot 60$, or 30 mi. We make a drawing. We label the distance between the cars, d, where d is in miles.

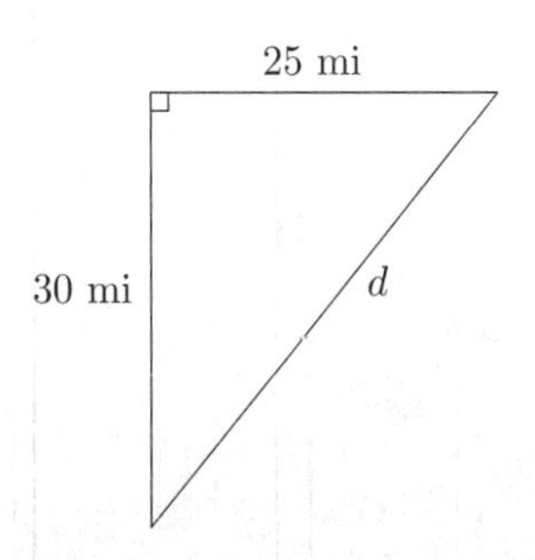

We use the Pythagorean theorem, substituting 25 for a, 30 for b, and d for c.

$$25^2+30^2=d^2$$

We get $d=\sqrt{1525}$ mi, or $d\approx 39.051$ mi.

After one half hour, the cars are $\sqrt{1525}$ mi ≈ 39.051 mi apart.

65. The perimeter of the smaller square plot is 2 mi, so each side is $\frac{1}{4}\cdot 2$, or $\frac{1}{2}$ mi, and the area is $\left(\frac{1}{2}\text{ mi}\right)^2$, or $\frac{1}{4}$ mi^2.

This tells us that $\frac{1}{4}$ mi^2 is equivalent to 160 acres. The perimeter of the larger square plot is 4 mi, so each side is $\frac{1}{4}\cdot 4$, or 1 mi, and the area is $(1\text{ mi})^2$, or 1 mi^2. Since 1 mi$^2 = 4\cdot\frac{1}{4}$ mi^2, then 1 mi^2 is equivalent to $4\cdot 160$, or 640 acres. Thus, 4 mi of fencing will enclose a square whose area is 640 acres.

Exercise Set 8.7

1. True; see page 526 in the text.

3. True; see page 526 in the text.

5. $\sqrt[3]{-8}=-2 \qquad (-2)^3=(-2)(-2)(-2)=-8$

7. $\sqrt[3]{-125}=-5 \qquad (-5)^3=(-5)(-5)(-5)=-125$

9. $\sqrt[3]{1000}=10 \qquad 10^3=10\cdot 10\cdot 10=1000$

11. $-\sqrt[3]{216}=-6 \qquad \sqrt[3]{216}=6$, so $-\sqrt[3]{216}=-6$.

13. $\sqrt[4]{625}=5 \qquad 5^4=5\cdot 5\cdot 5\cdot 5=625$

15. $\sqrt[5]{0}=0 \qquad 0^5=0\cdot 0\cdot 0\cdot 0\cdot 0=0$

17. $\sqrt[5]{-1}=-1 \quad (-1)^5=(-1)(-1)(-1)(-1)(-1)=-1$

19. $\sqrt[4]{-81}$ is not a real number, because it is an even root of a negative number.

21. $\sqrt[4]{10,000}=10 \quad 10^4=10\cdot 10\cdot 10\cdot 10=10,000$

We might also observe that $10,000=10^4$, so we have $\sqrt[4]{10^4}=10$.

23. $\sqrt[3]{8^3}=8 \qquad 8^3=8\cdot 8\cdot 8$

25. $\sqrt[6]{64}=2 \qquad 2^6=2\cdot 2\cdot 2\cdot 2\cdot 2\cdot 2=64$

27. $\sqrt[9]{a^9}=a \qquad a^9=a\cdot a\cdot a\cdot a\cdot a\cdot a\cdot a\cdot a\cdot a$

29. $\sqrt[3]{54}=\sqrt[3]{27\cdot 2}=\sqrt[3]{27}\sqrt[3]{2}=3\sqrt[3]{2}$

31. $\sqrt[4]{48}=\sqrt[4]{16\cdot 3}=\sqrt[4]{16}\sqrt[4]{3}=2\sqrt[4]{3}$

33. $\sqrt[3]{\frac{64}{125}}=\frac{\sqrt[3]{64}}{\sqrt[3]{125}}=\frac{4}{5}$

35. $\sqrt[5]{\frac{32}{243}}=\frac{\sqrt[5]{32}}{\sqrt[5]{243}}=\frac{2}{3}$

37. $\sqrt[3]{\frac{7}{8}}=\frac{\sqrt[3]{7}}{\sqrt[3]{8}}=\frac{\sqrt[3]{7}}{2}$

39. $\sqrt[4]{\frac{14}{81}} = \frac{\sqrt[4]{14}}{\sqrt[4]{81}} = \frac{\sqrt[4]{14}}{3}$

41. $16^{1/2} = \sqrt{16} = 4$

43. $125^{1/3} = \sqrt[3]{125} = 5$

45. $32^{1/5} = \sqrt[5]{32} = 2$

47. $16^{3/4} = (16^{1/4})^3 = (\sqrt[4]{16})^3 = 2^3 = 8$

49. $9^{5/2} = (9^{1/2})^5 = (\sqrt{9})^5 = 3^5 = 243$

51. $64^{2/3} = (64^{1/3})^2 = (\sqrt[3]{64})^2 = 4^2 = 16$

53. $8^{5/3} = (8^{1/3})^5 = (\sqrt[3]{8})^5 = 2^5 = 32$

55. $4^{5/2} = (4^{1/2})^5 = (\sqrt{4})^5 = 2^5 = 32$

57. $25^{-1/2} = \frac{1}{25^{1/2}} = \frac{1}{\sqrt{25}} = \frac{1}{5}$

59. $256^{-1/4} = \frac{1}{256^{1/4}} = \frac{1}{\sqrt[4]{256}} = \frac{1}{4}$

61. $16^{-3/4} = \frac{1}{16^{3/4}} = \frac{1}{(\sqrt[4]{16})^3} = \frac{1}{2^3} = \frac{1}{8}$

63. $81^{-5/4} = \frac{1}{81^{5/4}} = \frac{1}{(\sqrt[4]{81})^5} = \frac{1}{3^5} = \frac{1}{243}$

65. $8^{-2/3} = \frac{1}{8^{2/3}} = \frac{1}{(\sqrt[3]{8})^2} = \frac{1}{2^2} = \frac{1}{4}$

67. *Writing Exercise*

69.
$$x^2 - 5x - 6 = 0$$
$$(x-6)(x+1) = 0$$
$$x - 6 = 0 \quad or \quad x + 1 = 0$$
$$x = 6 \quad or \quad x = -1$$

The solutions are 6 and -1.

71.
$$4t^2 - 9 = 0$$
$$(2t+3)(2t-3) = 0$$
$$2t + 3 = 0 \quad or \quad 2t - 3 = 0$$
$$2t = -3 \quad or \quad 2t = 3$$
$$t = -\frac{3}{2} \quad or \quad t = \frac{3}{2}$$

The solutions are $-\frac{3}{2}$ and $\frac{3}{2}$.

73.
$$3x^2 + 8x + 4 = 0$$
$$(3x+2)(x+2) = 0$$
$$3x + 2 = 0 \quad or \quad x + 2 = 0$$
$$3x = -2 \quad or \quad x = -2$$
$$x = -\frac{2}{3} \quad or \quad x = -2$$

The solutions are $-\frac{2}{3}$ and -2.

75. *Writing Exercise*

77. Enter 10, press the power key, enter 0.8 (or $(4 \div 5)$), and then press [=].

$$10^{4/5} \approx 6.310$$

(Some calculators have a 10^x key which might have to be accessed using the [SHIFT] key.)

79. Enter 36, press the power key, enter 0.375 (or $(3 \div 8)$), and then press [=].

$$36^{3/8} \approx 3.834$$

81. $a^{1/4}a^{3/2} = a^{1/4+6/4} = a^{7/4}$

83. $m^{-2/3}m^{1/4}m^{3/2} = m^{-2/3+1/4+3/2} = m^{-8/12+3/12+18/12} = m^{13/12}$

85. Graph $y = \sqrt[3]{x}$

We make a table of values.

x	y
-8	-2
-1	-1
0	0
1	1
8	2

We plot these points and connect them with a smooth curve.

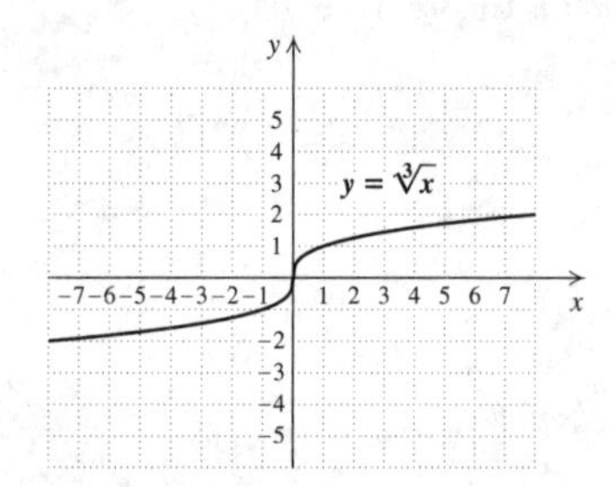

87.

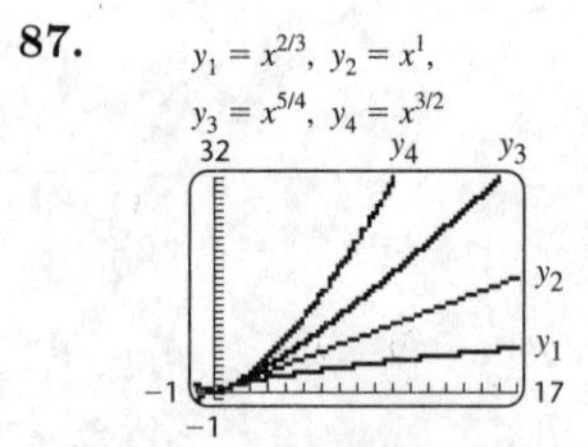

Chapter 9

Quadratic Equations

Exercise Set 9.1

1. True; see page 534 in the text.

3. False; see Examples 2, 3(b), and 4(b).

5. $x^2 = 36$

$x = \sqrt{36}$ *or* $x = -\sqrt{36}$ Using the principle of square roots

$x = 6$ *or* $x = -6$

We can check mentally that $6^2 = 36$ and $(-6)^2 = 36$. The solutions are 6 and -6.

7. $a^2 = 25$

$a = \sqrt{25}$ *or* $a = -\sqrt{25}$ Using the principle of square roots

$a = 5$ *or* $a = -5$

We can check mentally that $5^2 = 25$ and $(-5)^2 = 25$. The solutions are 5 and -5.

9. $t^2 = 17$

$t = \sqrt{17}$ *or* $t = -\sqrt{17}$ Using the principle of square roots

Check: For $\sqrt{17}$:

$t^2 = 17$	
$(\sqrt{17})^2$	17
$17 \stackrel{?}{=} 17$ TRUE	

For $-\sqrt{17}$:

$t^2 = 17$	
$(-\sqrt{17})^2$	17
$17 \stackrel{?}{=} 17$ TRUE	

The solutions are $\sqrt{17}$ and $-\sqrt{17}$.

11. $3x^2 = 27$

$x^2 = 9$ Dividing by 3

$x = \sqrt{9}$ *or* $x = -\sqrt{9}$ Using the principle of square roots

$x = 3$ *or* $x = -3$

Both numbers check. The solutions are 3 and -3.

13. $8t^2 = 0$

Observe that t^2 must be 0, so $t = 0$. The solution is 0.

15. $4 - 9x^2 = 0$

$4 = 9x^2$

$\frac{4}{9} = x^2$

$x = \sqrt{\frac{4}{9}}$ *or* $x = -\sqrt{\frac{4}{9}}$

$x = \frac{2}{3}$ *or* $x = -\frac{2}{3}$

Both numbers check. The solutions are $\frac{2}{3}$ and $-\frac{2}{3}$.

17. $49y^2 - 5 = 15$

$49y^2 = 20$

$y^2 = \frac{20}{49}$

$y = \sqrt{\frac{20}{49}}$ *or* $y = -\sqrt{\frac{20}{49}}$

$y = \frac{\sqrt{20}}{7}$ *or* $y = -\frac{\sqrt{20}}{7}$

$y = \frac{\sqrt{4 \cdot 5}}{7}$ *or* $y = -\frac{\sqrt{4 \cdot 5}}{7}$

$y = \frac{2\sqrt{5}}{7}$ *or* $y = -\frac{2\sqrt{5}}{7}$

The solutions are $\frac{2\sqrt{5}}{7}$ and $-\frac{2\sqrt{5}}{7}$.

19. $8x^2 - 28 = 0$

$8x^2 = 28$

$x^2 = \frac{7}{2}$

$x = \sqrt{\frac{7}{2}}$ *or* $x = -\sqrt{\frac{7}{2}}$ Using the principle of square roots

$x = \frac{\sqrt{7}}{\sqrt{2}}$ *or* $x = -\frac{\sqrt{7}}{\sqrt{2}}$

$x = \frac{\sqrt{7}}{\sqrt{2}} \cdot \frac{\sqrt{2}}{\sqrt{2}}$ *or* $x = -\frac{\sqrt{7}}{\sqrt{2}} \cdot \frac{\sqrt{2}}{\sqrt{2}}$

$x = \frac{\sqrt{14}}{2}$ *or* $x = -\frac{\sqrt{14}}{2}$

The solutions are $\frac{\sqrt{14}}{2}$ and $-\frac{\sqrt{14}}{2}$.

21. $(x - 1)^2 = 49$

$x - 1 = 7$ *or* $x - 1 = -7$ Using the principle of square roots

$x = 8$ *or* $x = -6$

The solutions are 8 and -6.

23. $(x + 4)^2 = 81$

$x + 4 = 9$ *or* $x + 4 = -9$ Using the principle of square roots

$x = 5$ *or* $x = -13$

The solutions are 5 and -13.

25. $(m + 3)^2 = 6$

$m + 3 = \sqrt{6}$ *or* $m + 3 = -\sqrt{6}$

$m = -3 + \sqrt{6}$ *or* $m = -3 - \sqrt{6}$

The solutions are $-3 + \sqrt{6}$ and $-3 - \sqrt{6}$, or $-3 \pm \sqrt{6}$.

27. $(a-7)^2 = 0$

Observe that $a-7$ must be 0, so $a-7=0$, or $a=7$. The solution is 7.

29.
$$(5-x)^2 = 14$$
$$5-x=\sqrt{14} \quad or \quad 5-x=-\sqrt{14}$$
$$-x=-5+\sqrt{14} \quad or \quad -x=-5-\sqrt{14}$$
$$x=5-\sqrt{14} \quad or \quad x=5+\sqrt{14}$$

The solutions are $5-\sqrt{14}$ and $5+\sqrt{14}$, or $5\pm\sqrt{14}$.

31.
$$(t+1)^2=1$$
$$t+1=1 \quad or \quad t+1=-1$$
$$t=0 \quad or \quad t=-2$$

The solutions are 0 and -2.

33.
$$\left(y-\frac{3}{4}\right)^2=\frac{17}{16}$$
$$y-\frac{3}{4}=\sqrt{\frac{17}{16}} \quad or \quad y-\frac{3}{4}=-\sqrt{\frac{17}{16}}$$
$$y-\frac{3}{4}=\frac{\sqrt{17}}{4} \quad or \quad y-\frac{3}{4}=-\frac{\sqrt{17}}{4}$$
$$y=\frac{3}{4}+\frac{\sqrt{17}}{4} \quad or \quad y=\frac{3}{4}-\frac{\sqrt{17}}{4}$$

The solutions are $\frac{3}{4}+\frac{\sqrt{17}}{4}$ and $\frac{3}{4}-\frac{\sqrt{17}}{4}$, or $\frac{3}{4}\pm\frac{\sqrt{17}}{4}$.

35.
$$x^2-10x+25=100$$
$$(x-5)^2=100$$
$$x-5=10 \quad or \quad x-5=-10$$
$$x=15 \quad or \quad x=-5$$

The solutions are 15 and -5.

37.
$$p^2+8p+16=1$$
$$(p+4)^2=1$$
$$p+4=1 \quad or \quad p+4=-1$$
$$p=-3 \quad or \quad p=-5$$

The solutions are -3 and -5.

39.
$$t^2-6t+9=13$$
$$(t-3)^2=13$$
$$t-3=\sqrt{13} \quad or \quad t-3=-\sqrt{13}$$
$$t=3+\sqrt{13} \quad or \quad t=3-\sqrt{13}$$

The solutions are $3+\sqrt{13}$ and $3-\sqrt{13}$, or $3\pm\sqrt{13}$.

41.
$$x^2+12x+36=18$$
$$(x+6)^2=18$$
$$x+6=\sqrt{18} \quad or \quad x+6=-\sqrt{18}$$
$$x+6=3\sqrt{2} \quad or \quad x+6=-3\sqrt{2}$$
$$x=-6+3\sqrt{2} \quad or \quad x=-6-3\sqrt{2}$$

The solutions are $-6+3\sqrt{2}$ and $-6-3\sqrt{2}$, or $-6\pm3\sqrt{2}$.

43. *Writing Exercise*

45. $3x^2+12x+3=3(x^2+4x+1)$

47. $t^2+16t+64=t^2+2\cdot t\cdot 8+8^2=(t+8)^2$

49. $x^2-10x+25=x^2-2\cdot x\cdot 5+5^2=(x-5)^2$

51. *Writing Exercise*

53.
$$x^2+\frac{7}{3}x+\frac{49}{36}=\frac{7}{36}$$
$$\left(x+\frac{7}{6}\right)^2=\frac{7}{36}$$
$$x+\frac{7}{6}=\frac{\sqrt{7}}{6} \quad or \quad x+\frac{7}{6}=-\frac{\sqrt{7}}{6}$$
$$x=-\frac{7}{6}+\frac{\sqrt{7}}{6} \quad or \quad x=-\frac{7}{6}-\frac{\sqrt{7}}{6}$$

The solutions are $-\frac{7}{6}+\frac{\sqrt{7}}{6}$ and $-\frac{7}{6}-\frac{\sqrt{7}}{6}$, or $-\frac{7}{6}\pm\frac{\sqrt{7}}{6}$.

55.
$$m^2-\frac{3}{2}m+\frac{9}{16}=\frac{17}{16}$$
$$\left(m-\frac{3}{4}\right)^2=\frac{17}{16}$$
$$m-\frac{3}{4}=\frac{\sqrt{17}}{4} \quad or \quad m-\frac{3}{4}=-\frac{\sqrt{17}}{4}$$
$$m=\frac{3}{4}+\frac{\sqrt{17}}{4} \quad or \quad m=\frac{3}{4}-\frac{\sqrt{17}}{4}$$

The solutions are $\frac{3}{4}\pm\frac{\sqrt{17}}{4}$.

57.
$$x^2+2.5x+1.5625=9.61$$
$$(x+1.25)^2=9.61$$
$$x+1.25=3.1 \quad or \quad x+1.25=-3.1$$
$$x=1.85 \quad or \quad x=-4.35$$

The solutions are 1.85 and -4.35.

59. From the graph we see that when $y=1$, then $x=-4$ or $x=-2$. Thus, the solutions of $(x+3)^2=1$ are -4 and -2.

61. From the graph we see that when $y=9$, then $x=-6$ or $x=0$. Thus, the solutions of $(x+3)^2=9$ are -6 and 0.

63.
$$f=\frac{kMm}{d^2}$$
$$d^2f=kMm \quad \text{Multiplying by } d^2$$
$$d^2=\frac{kMm}{f} \quad \text{Dividing by } f$$
$$d=\sqrt{\frac{kMm}{f}} \quad \text{Taking the principal square root}$$
$$d=\frac{\sqrt{kMmf}}{f} \quad \text{Rationalizing the denominator}$$

Exercise Set 9.2

1. First complete the square for $x^2 + 6x$:

$$\left(\frac{6}{2}\right)^2 = 3^2 = 9$$

Then we have:

$$\begin{aligned} x^2 + 6x &= 2 \\ x^2 + 6x + 9 &= 2 + 9 \\ x^2 + 6x + 9 &= 11 \end{aligned}$$

Choice (c) is correct.

3. Factoring on the left side, we have $(x + 3)^2 = 10$. Choice (a) is correct.

5. First complete the square for $x^2 + 8x$:

$$\left(\frac{8}{2}\right)^2 = 4^2 = 16$$

Then we have:

$$\begin{aligned} x^2 + 8x &= 2 \\ x^2 + 8x + 16 &= 2 + 16 \\ (x + 4)^2 &= 18 \end{aligned}$$

Choice (d) is correct.

7. To complete the square for $x^2 + 8x$, we take half the coefficient of x and square it:

$$\left(\frac{8}{2}\right)^2 = 4^2 = 16$$

The trinomial $x^2 + 8x + 16$ is the square of $x + 4$.

Check: $(x + 4)^2 = x^2 + 8x + 16$.

9. To complete the square for $x^2 - 12x$, we take half the coefficient of x and square it:

$$\left(\frac{-12}{2}\right)^2 = (-6)^2 = 36$$

The trinomial $x^2 - 12x + 36$ is the square of $x - 6$.

Check: $(x - 6)^2 = x^2 - 12x + 36$.

11. To complete the square for $x^2 - 3x$, we take half the coefficient of x and square it:

$$\left(\frac{-3}{2}\right)^2 = \frac{9}{4}$$

The trinomial $x^2 - 3x + \frac{9}{4}$ is the square of $x - \frac{3}{2}$.

Check: $\left(x - \frac{3}{2}\right)^2 = x^2 - 3x + \frac{9}{4}$.

13. To complete the square for $t^2 + t$, we take half the coefficient of t and square it:

$$\left(\frac{1}{2}\right)^2 = \frac{1}{4}$$

The trinomial $t^2 + t + \frac{1}{4}$ is the square of $t + \frac{1}{2}$.

Check: $\left(t + \frac{1}{2}\right)^2 = t^2 + t + \frac{1}{4}$.

15. To complete the square for $x^2 + \frac{5}{4}x$, we take half the coefficient of x and square it:

$$\left(\frac{1}{2} \cdot \frac{5}{4}\right)^2 = \left(\frac{5}{8}\right)^2 = \frac{25}{64}$$

The trinomial $x^2 + \frac{5}{4}x + \frac{25}{64}$ is the square of $x + \frac{5}{8}$.

Check: $\left(x + \frac{5}{8}\right)^2 = x^2 + \frac{5}{4}x + \frac{25}{64}$.

17. To complete the square for $m^2 - \frac{9}{2}m$, we take half the coefficient of m and square it:

$$\left[\frac{1}{2}\left(-\frac{9}{2}\right)\right]^2 = \left(-\frac{9}{4}\right)^2 = \frac{81}{16}$$

The trinomial $m^2 - \frac{9}{2}m + \frac{81}{16}$ is the square of $m - \frac{9}{4}$.

Check: $\left(x - \frac{9}{4}\right)^2 = m^2 - \frac{9}{2}m + \frac{81}{16}$.

19.

$$\begin{aligned} x^2 + 8x + 12 &= 0 \\ x^2 + 8x \qquad &= -12 \qquad \text{Subtracting 12} \\ x^2 + 8x + 16 &= -12 + 16 \quad \text{Adding 16: } \left(\frac{8}{2}\right)^2 = 4^2 = 16 \\ (x + 4)^2 &= 4 \end{aligned}$$

$x + 4 = 2$ *or* $x + 4 = -2$ Principle of square roots

$x = -2$ *or* $x = -6$

The solutions are -2 and -6.

21.

$$\begin{aligned} x^2 - 24x + 21 &= 0 \\ x^2 - 24x \qquad &= -21 \qquad \text{Subtracting 21} \\ x^2 - 24x + 144 &= -21 + 144 \quad \text{Adding 144: } \left(\frac{-24}{2}\right)^2 = (-12)^2 = 144 \\ (x - 12)^2 &= 123 \end{aligned}$$

$x - 12 = \sqrt{123}$ *or* $x - 12 = -\sqrt{123}$

Principle of square roots

$x = 12 + \sqrt{123}$ *or* $x = 12 - \sqrt{123}$

The solutions are $12 + \sqrt{123}$ and $12 - \sqrt{123}$, or $12 \pm \sqrt{123}$.

23.

$$\begin{aligned} 3x^2 - 6x - 15 &= 0 \\ \frac{1}{3}(3x^2 - 6x - 15) &= \frac{1}{3} \cdot 0 \\ x^2 - 2x - 5 &= 0 \\ x^2 - 2x \qquad &= 5 \\ x^2 - 2x + 1 &= 5 + 1 \quad \text{Adding 1: } \left(\frac{-2}{2}\right)^2 = (-1)^2 = 1 \\ (x - 1)^2 &= 6 \end{aligned}$$

$x - 1 = \sqrt{6}$ *or* $x - 1 = -\sqrt{6}$

$x = 1 + \sqrt{6}$ *or* $x = 1 - \sqrt{6}$

The solutions are $1 \pm \sqrt{6}$.

25.
$$\begin{aligned} x^2 - 22x + 102 &= 0 \\ x^2 - 22x &= -102 \\ x^2 - 22x + 121 &= -102 + 121 \quad \text{Adding 121: } \left(\frac{-22}{2}\right)^2 = (-11)^2 = 121 \\ (x-11)^2 &= 19 \end{aligned}$$

$$x - 11 = \sqrt{19} \quad \textit{or} \quad x - 11 = -\sqrt{19}$$
$$x = 11 + \sqrt{19} \quad \textit{or} \quad x = 11 - \sqrt{19}$$

The solutions are $11 \pm \sqrt{19}$.

27.
$$\begin{aligned} x^2 + 3x - 3 &= 0 \\ x^2 + 3x &= 3 \\ x^2 + 3x + \frac{9}{4} &= 3 + \frac{9}{4} \quad \text{Adding } \frac{9}{4}\text{: } \left(\frac{3}{2}\right)^2 = \frac{9}{4} \\ \left(x + \frac{3}{2}\right)^2 &= \frac{21}{4} \end{aligned}$$

$$x + \frac{3}{2} = \sqrt{\frac{21}{4}} \quad \textit{or} \quad x + \frac{3}{2} = -\sqrt{\frac{21}{4}}$$
$$x + \frac{3}{2} = \frac{\sqrt{21}}{2} \quad \textit{or} \quad x + \frac{3}{2} = -\frac{\sqrt{21}}{2}$$
$$x = -\frac{3}{2} + \frac{\sqrt{21}}{2} \quad \textit{or} \quad x = -\frac{3}{2} - \frac{\sqrt{21}}{2}$$

The solutions are $-\frac{3}{2} \pm \frac{\sqrt{21}}{2}$, or $\frac{-3 \pm \sqrt{21}}{2}$.

29.
$$\begin{aligned} 2x^2 + 6x - 56 &= 0 \\ \frac{1}{2}\left(2x^2 + 6x - 56\right) &= \frac{1}{2} \cdot 0 \\ x^2 + 3x - 28 &= 0 \\ x^2 + 3x &= 28 \\ x^2 + 3x + \frac{9}{4} &= 28 + \frac{9}{4} \quad \text{Adding } \frac{9}{4}\text{: } \left(\frac{3}{2}\right)^2 = \frac{9}{4} \\ \left(x + \frac{3}{2}\right)^2 &= \frac{121}{4} \end{aligned}$$

$$x + \frac{3}{2} = \frac{11}{2} \quad \textit{or} \quad x + \frac{3}{2} = -\frac{11}{2}$$
$$x = \frac{8}{2} \quad \textit{or} \quad x = -\frac{14}{2}$$
$$x = 4 \quad \textit{or} \quad x = -7$$

The solutions are 4 and -7.

31.
$$\begin{aligned} x^2 - \frac{3}{2}x - 2 &= 0 \\ x^2 - \frac{3}{2}x &= 2 \\ x^2 - \frac{3}{2}x + \frac{9}{16} &= 2 + \frac{9}{16} \quad \text{Adding } \frac{9}{16}\text{: } \left[\frac{1}{2}\left(-\frac{3}{2}\right)\right]^2 = \left(-\frac{3}{4}\right)^2 = \frac{9}{16} \\ \left(x - \frac{3}{4}\right)^2 &= \frac{32}{16} + \frac{9}{16} = \frac{41}{16} \end{aligned}$$

$$x - \frac{3}{4} = \frac{\sqrt{41}}{4} \quad \textit{or} \quad x - \frac{3}{4} = -\frac{\sqrt{41}}{4}$$
$$x = \frac{3}{4} + \frac{\sqrt{41}}{4} \quad \textit{or} \quad x = \frac{3}{4} - \frac{\sqrt{41}}{4}$$

The solutions are $\frac{3}{4} \pm \frac{\sqrt{41}}{4}$, or $\frac{3 \pm \sqrt{41}}{4}$.

33.
$$\begin{aligned} 2t^2 - 3t - 8 &= 0 \\ \frac{1}{2}\left(2t^2 - 3t - 8\right) &= \frac{1}{2} \cdot 0 \\ t^2 - \frac{3}{2}t - 4 &= 0 \\ t^2 - \frac{3}{2}t &= 4 \\ t^2 - \frac{3}{2}t + \frac{9}{16} &= 4 + \frac{9}{16} \quad \text{Adding } \frac{9}{16}\text{: } \left[\frac{1}{2}\left(-\frac{3}{2}\right)\right]^2 = \left(-\frac{3}{4}\right)^2 = \frac{9}{16} \\ \left(t - \frac{3}{4}\right)^2 &= \frac{64}{16} + \frac{9}{16} = \frac{73}{16} \end{aligned}$$

$$t - \frac{3}{4} = \frac{\sqrt{73}}{4} \quad \textit{or} \quad t - \frac{3}{4} = -\frac{\sqrt{73}}{4}$$
$$t = \frac{3}{4} + \frac{\sqrt{73}}{4} \quad \textit{or} \quad t = \frac{3}{4} - \frac{\sqrt{73}}{4}$$

The solutions are $\frac{3}{4} \pm \frac{\sqrt{73}}{4}$, or $\frac{3 \pm \sqrt{73}}{4}$.

35.
$$\begin{aligned} 3x^2 - 4x - 3 &= 0 \\ \frac{1}{3}\left(3x^2 - 4x - 3\right) &= \frac{1}{3} \cdot 0 \\ x^2 - \frac{4}{3}x - 1 &= 0 \\ x^2 - \frac{4}{3}x &= 1 \\ x^2 - \frac{4}{3}x + \frac{4}{9} &= 1 + \frac{4}{9} \quad \text{Adding } \frac{4}{9}\text{: } \left[\frac{1}{2}\left(-\frac{4}{3}\right)\right]^2 = \left(-\frac{2}{3}\right)^2 = \frac{4}{9} \\ \left(x - \frac{2}{3}\right)^2 &= \frac{9}{9} + \frac{4}{9} = \frac{13}{9} \end{aligned}$$

$$x - \frac{2}{3} = \frac{\sqrt{13}}{3} \quad \textit{or} \quad x - \frac{2}{3} = -\frac{\sqrt{13}}{3}$$
$$x = \frac{2}{3} + \frac{\sqrt{13}}{3} \quad \textit{or} \quad x = \frac{2}{3} - \frac{\sqrt{13}}{3}$$

The solutions are $\frac{2}{3} \pm \frac{\sqrt{13}}{3}$, or $\frac{2 \pm \sqrt{13}}{3}$.

37.
$$\begin{aligned} 2x^2 &= 5 + 9x \\ 2x^2 - 9x - 5 &= 0 \\ \frac{1}{2}\left(2x^2 - 9x - 5\right) &= \frac{1}{2}\cdot 0 \\ x^2 - \frac{9}{2}x - \frac{5}{2} &= 0 \\ x^2 - \frac{9}{2}x &= \frac{5}{2} \\ x^2 - \frac{9}{2}x + \frac{81}{16} &= \frac{5}{2} + \frac{81}{16} \\ \left(x - \frac{9}{4}\right)^2 &= \frac{121}{16} \end{aligned}$$

$$x - \frac{9}{4} = \frac{11}{4} \quad or \quad x - \frac{9}{4} = -\frac{11}{4}$$
$$x = 5 \quad or \quad x = -\frac{1}{2}$$

The solutions are 5 and $-\frac{1}{2}$.

39.
$$\begin{aligned} 6x^2 + 11x &= 10 \\ \frac{1}{6}\left(6x^2 + 11x\right) &= \frac{1}{6}\cdot 10 \\ x^2 + \frac{11}{6}x &= \frac{5}{3} \\ x^2 + \frac{11}{6}x + \frac{121}{144} &= \frac{5}{3} + \frac{121}{144} \\ \left(x + \frac{11}{12}\right)^2 &= \frac{361}{144} \end{aligned}$$

$$x + \frac{11}{12} = \frac{19}{12} \quad or \quad x + \frac{11}{12} = -\frac{19}{12}$$
$$x = \frac{2}{3} \quad or \quad x = -\frac{5}{2}$$

The solutions are $\frac{2}{3}$ and $-\frac{5}{2}$.

41. *Writing Exercise*

43. $\dfrac{3+6x}{3} = \dfrac{\not{3}(1+2x)}{\not{3}\cdot 1} = 1 + 2x$

45. $\dfrac{15-10x}{5} = \dfrac{\not{5}(3-2x)}{\not{5}\cdot 1} = 3 - 2x$

47. $\dfrac{24-3\sqrt{5}}{9} = \dfrac{\not{3}(8-\sqrt{5})}{\not{3}\cdot 3} = \dfrac{8-\sqrt{5}}{3}$

49. *Writing Exercise*

51. $x^2 + bx + 36$

The trinomial is a square if the square of one-half the x-coefficient is equal to 36. Thus, we have:

$$\begin{aligned} \left(\frac{b}{2}\right)^2 &= 36 \\ \frac{b^2}{4} &= 36 \\ b^2 &= 144 \end{aligned}$$

$b = 12$ *or* $b = -12$

53. $x^2 + bx + 45$

The trinomial is a square if the square of one-half the x-coefficient is equal to 45. Thus, we have:

$$\begin{aligned} \left(\frac{b}{2}\right)^2 &= 45 \\ \frac{b^2}{4} &= 45 \\ b^2 &= 180 \end{aligned}$$

$b = \sqrt{180}$ *or* $b = -\sqrt{180}$

$b = 6\sqrt{5}$ *or* $b = -6\sqrt{5}$

55. $4x^2 + bx + 16$

The trinomial is a square if the square of one-half the x-coefficient is equal to 16. Thus, we have:

$$4\left(x^2 + \frac{b}{4}x + 4\right)$$

$$\begin{aligned} \left(\frac{b/4}{2}\right)^2 &= 4 \\ \left(\frac{b}{8}\right)^2 &= 4 \\ \frac{b^2}{64} &= 4 \\ b^2 &= 256 \end{aligned}$$

$b = 16$ *or* $b = -16$

57. -0.39, -7.61

59. 23.09, 0.91

61. 3.71, -1.21

63. *Writing Exercise*

Exercise Set 9.3

1. Since $x^2 + 5x + 6$ is easily factored, choice (c) is correct.

3. Since the equation is not easily solved using method (a), (c), or (d), choice (b) is correct.

5.
$$\begin{aligned} x^2 + 7x &= 18 \\ x^2 + 7x - 18 &= 0 \quad \text{Standard form} \end{aligned}$$

We can factor.

$$\begin{aligned} x^2 + 7x - 18 &= 0 \\ (x+9)(x-2) &= 0 \end{aligned}$$

$$x + 9 = 0 \quad or \quad x - 2 = 0$$
$$x = -9 \quad or \quad x = 2$$

The solutions are -9 and 2.

7.
$$\begin{aligned} x^2 &= 8x - 16 \\ x^2 - 8x + 16 &= 0 \quad \text{Standard form} \end{aligned}$$

We can factor.

$$\begin{aligned} x^2 - 8x + 16 &= 0 \\ (x-4)(x-4) &= 0 \end{aligned}$$

$$x - 4 = 0 \quad or \quad x - 4 = 0$$
$$x = 4 \quad or \quad x = 4$$

The solution is 4.

9. $3y^2+7y+4=0$

We can factor.

$$3y^2+7y+4=0$$
$$(3y+4)(y+1)=0$$

$$3y+4=0 \quad or \quad y+1=0$$
$$3y=-4 \quad or \quad y=-1$$
$$y=-\frac{4}{3} \quad or \quad y=-1$$

The solutions are $-\frac{4}{3}$ and -1.

11. $4x^2-12x=7$

$$4x^2-12x-7=0$$

We can factor.

$$4x^2-12x-7=0$$
$$(2x+1)(2x-7)=0$$

$$2x+1=0 \quad or \quad 2x-7=0$$
$$2x=-1 \quad or \quad 2x=7$$
$$x=-\frac{1}{2} \quad or \quad x=\frac{7}{2}$$

The solutions are $-\frac{1}{2}$ and $\frac{7}{2}$.

13. $t^2=64$

$t=8$ *or* $t=-8$ Principle of square roots

The solutions are 8 and -8.

15. $x^2+4x-7=0$

We use the quadratic formula.

$a=1,\ b=4,\ c=-7$

$$x=\frac{-b\pm\sqrt{b^2-4ac}}{2a}$$
$$x=\frac{-4\pm\sqrt{4^2-4\cdot1\cdot(-7)}}{2\cdot1}$$
$$x=\frac{-4\pm\sqrt{16+28}}{2}$$
$$x=\frac{-4\pm\sqrt{44}}{2}=\frac{-4\pm\sqrt{4\cdot11}}{2}$$
$$x=\frac{-4\pm2\sqrt{11}}{2}=\frac{2(-2\pm\sqrt{11})}{2\cdot1}$$
$$x=-2\pm\sqrt{11}$$

The solutions are $-2+\sqrt{11}$ and $-2-\sqrt{11}$, or $-2\pm\sqrt{11}$.

17. $y^2-10y+19=0$

We use the quadratic formula.

$a=1,\ b=-10,\ c=19$

$$y=\frac{-b\pm\sqrt{b^2-4ac}}{2a}$$
$$y=\frac{-(-10)\pm\sqrt{(-10)^2-4\cdot1\cdot19}}{2\cdot1}$$
$$y=\frac{10\pm\sqrt{100-76}}{2}$$
$$y=\frac{10\pm\sqrt{24}}{2}=\frac{10\pm\sqrt{4\cdot6}}{2}$$
$$y=\frac{10\pm2\sqrt{6}}{2}=\frac{2(5\pm\sqrt{6})}{2\cdot1}$$
$$y=5\pm\sqrt{6}$$

The solutions are $5+\sqrt{6}$ and $5-\sqrt{6}$, or $5\pm\sqrt{6}$.

19. $x^2+2x+1=7$

Observe that x^2+2x+1 is a perfect-square trinomial. Then we can use the principle of square roots.

$$x^2+2x+1=7$$
$$(x+1)^2=7$$

$$x+1=\sqrt{7} \quad or \quad x+1=-\sqrt{7}$$
$$x=-1+\sqrt{7} \quad or \quad x=-1-\sqrt{7}$$

The solutions are $-1+\sqrt{7}$ and $-1-\sqrt{7}$, or $-1\pm\sqrt{7}$.

21. $3t^2+8t+2=0$

We use the quadratic formula.

$a=3,\ b=8,\ c=2$

$$t=\frac{-b\pm\sqrt{b^2-4ac}}{2a}$$
$$t=\frac{-8\pm\sqrt{8^2-2\cdot3\cdot2}}{2\cdot3}$$
$$t=\frac{-8\pm\sqrt{64-24}}{6}=\frac{-8\pm\sqrt{40}}{6}$$
$$t=\frac{-8\pm\sqrt{4\cdot10}}{6}=\frac{-8\pm2\sqrt{10}}{6}$$
$$t=\frac{2(-4\pm\sqrt{10})}{2\cdot3}=\frac{-4\pm\sqrt{10}}{3}$$

The solutions are $\frac{-4+\sqrt{10}}{3}$ and $\frac{-4-\sqrt{10}}{3}$, or $\frac{-4\pm\sqrt{10}}{3}$, or $-\frac{4}{3}\pm\frac{\sqrt{10}}{3}$.

23. $2x^2-5x=1$

$2x^2-5x-1=0$ Standard form

We use the quadratic formula.

$a=2,\ b=-5,\ c=-1$

$$x=\frac{-b\pm\sqrt{b^2-4ac}}{2a}$$
$$x=\frac{-(-5)\pm\sqrt{(-5)^2-4\cdot2\cdot(-1)}}{2\cdot2}$$
$$x=\frac{5\pm\sqrt{25+8}}{4}=\frac{5\pm\sqrt{33}}{4}$$

The solutions are $\frac{5+\sqrt{33}}{4}$ and $\frac{5-\sqrt{33}}{4}$, or $\frac{5\pm\sqrt{33}}{4}$, or $\frac{5}{4}\pm\frac{\sqrt{33}}{4}$.

25. $4y^2 + 2y - 3 = 0$

We use the quadratic formula.

$a = 4$, $b = 2$, $c = -3$

$$y = \frac{-b \pm \sqrt{b^2 - 4ac}}{2a}$$

$$y = \frac{-2 \pm \sqrt{2^2 - 4 \cdot 4 \cdot (-3)}}{2 \cdot 4}$$

$$y = \frac{-2 \pm \sqrt{4 + 48}}{8} = \frac{-2 \pm \sqrt{52}}{8}$$

$$y = \frac{-2 \pm \sqrt{4 \cdot 13}}{8} = \frac{-2 \pm 2\sqrt{13}}{8}$$

$$y = \frac{2(-1 \pm \sqrt{13})}{2 \cdot 4} = \frac{-1 \pm \sqrt{13}}{4}$$

The solutions are $\frac{-1 + \sqrt{13}}{4}$ and $\frac{-1 - \sqrt{13}}{4}$, or $\frac{-1 \pm \sqrt{13}}{4}$, or $-\frac{1}{4} \pm \frac{\sqrt{13}}{4}$.

27. $2t^2 - 3t + 2 = 0$

We use the quadratic formula.

$a = 2$, $b = -3$, $c = 2$

$$t = \frac{-b \pm \sqrt{b^2 - 4ac}}{2a}$$

$$t = \frac{-(-3) \pm \sqrt{(-3)^2 - 4 \cdot 2 \cdot 2}}{2 \cdot 2}$$

$$t = \frac{3 \pm \sqrt{9 - 16}}{4} = \frac{3 \pm \sqrt{-7}}{4}$$

Since the radicand, -7, is negative, there are no real-number solutions.

29.

$$3x^2 - 5x = 4$$

$$3x^2 - 5x - 4 = 0$$

We use the quadratic formula.

$a = 3$, $b = -5$, $c = -4$

$$x = \frac{-b \pm \sqrt{b^2 - 4ac}}{2a}$$

$$x = \frac{-(-5) \pm \sqrt{(-5)^2 - 4 \cdot 3 \cdot (-4)}}{2 \cdot 3}$$

$$x = \frac{5 \pm \sqrt{25 + 48}}{6} = \frac{5 \pm \sqrt{73}}{6}$$

The solutions are $\frac{5 + \sqrt{73}}{6}$ and $\frac{5 - \sqrt{73}}{6}$, or $\frac{5 \pm \sqrt{73}}{6}$, or $\frac{5}{6} \pm \frac{\sqrt{73}}{6}$.

31.

$$2y^2 - 6y = 10$$

$$2y^2 - 6y - 10 = 0$$

$$y^2 - 3y - 5 = 0 \quad \text{Multiplying by } \frac{1}{2}$$

We use the quadratic formula.

$a = 1$, $b = -3$, $c = -5$

$$y = \frac{-b \pm \sqrt{b^2 - 4ac}}{2a}$$

$$y = \frac{-(-3) \pm \sqrt{(-3)^2 - 4 \cdot 1 \cdot (-5)}}{2 \cdot 1}$$

$$y = \frac{3 \pm \sqrt{9 + 20}}{2} = \frac{3 \pm \sqrt{29}}{2}$$

The solutions are $\frac{3 + \sqrt{29}}{2}$ and $\frac{3 - \sqrt{29}}{2}$, or $\frac{3 \pm \sqrt{29}}{2}$, or $\frac{3}{2} \pm \frac{\sqrt{29}}{2}$.

33.

$$6t^2 + 26t = 20$$

$$6t^2 + 26t - 20 = 0$$

$$2(3t^2 + 13t - 10) = 0$$

$$2(3t - 2)(t + 5) = 0$$

$$3t - 2 = 0 \quad or \quad t + 5 = 0$$

$$3t = 2 \quad or \quad t = -5$$

$$t = \frac{2}{3} \quad or \quad t = -5$$

The solutions are $\frac{2}{3}$ and -5.

35.

$$5t^2 - 7t = -4$$

$$5t^2 - 7t + 4 = 0 \quad \text{Standard form}$$

We use the quadratic formula.

$a = 5$, $b = -7$, $c = 4$

$$t = \frac{-b \pm \sqrt{b^2 - 4ac}}{2a}$$

$$t = \frac{-(-7) \pm \sqrt{(-7)^2 - 4 \cdot 5 \cdot 4}}{2 \cdot 5}$$

$$t = \frac{7 \pm \sqrt{49 - 80}}{10} = \frac{7 \pm \sqrt{-31}}{10}$$

Since the radicand, -31, is negative, there are no real-number solutions.

37. $9y^2 = 162$

$$y^2 = 18 \quad \text{Dividing by 9}$$

$$y = \sqrt{18} \quad or \quad y = -\sqrt{18} \quad \text{Principle of square roots}$$

$$y = 3\sqrt{2} \quad or \quad y = -3\sqrt{2}$$

The solutions are $3\sqrt{2}$ and $-3\sqrt{2}$, or $\pm 3\sqrt{2}$.

39. $x^2 - 4x - 7 = 0$

$a = 1$, $b = -4$, $c = -7$

$$x = \frac{-(-4) \pm \sqrt{(-4)^2 - 4 \cdot 1 \cdot (-7)}}{2 \cdot 1}$$

$$x = \frac{4 \pm \sqrt{16 + 28}}{2} = \frac{4 \pm \sqrt{44}}{2}$$

$$x = \frac{4 \pm \sqrt{4 \cdot 11}}{2} = \frac{4 \pm 2\sqrt{11}}{2}$$

$$x = \frac{2(2 \pm \sqrt{11})}{2} = 2 \pm \sqrt{11}$$

Using a calculator or Table 2, we see that $\sqrt{11} \approx 3.317$:

$$2+\sqrt{11} \approx 2+3.317 \quad or \quad 2-\sqrt{11} \approx 2-3.317$$
$$\approx 5.317 \quad or \quad \approx -1.317$$

The approximate solutions, to the nearest thousandth, are 5.317 and -1.317.

41. $y^2-5y-1=0$

$a=1,\ b=-5,\ c=-1$

$$y=\frac{-b\pm\sqrt{b^2-4ac}}{2a}$$
$$y=\frac{-(-5)\pm\sqrt{(-5)^2-4\cdot 1\cdot(-1)}}{2\cdot 1}$$
$$y=\frac{5\pm\sqrt{25+4}}{2}=\frac{5\pm\sqrt{29}}{2}$$

Using a calculator or Table 2, we see that $\sqrt{29}\approx 5.385$:

$$\frac{5+\sqrt{29}}{2}\approx\frac{5+5.385}{2} \quad or \quad \frac{5-\sqrt{29}}{2}\approx\frac{5-5.385}{2}$$
$$\approx 5.193 \quad or \quad \approx -0.193$$

The approximate solutions, to the nearest thousandth, are 5.193 and -0.193.

43. $4x^2+4x=1$

$4x^2+4x-1=0$ Standard form

$a=4,\ b=4,\ c=-1$

$$x=\frac{-4\pm\sqrt{4^2-4\cdot 4\cdot(-1)}}{2\cdot 4}$$
$$x=\frac{-4\pm\sqrt{16+16}}{8}=\frac{-4\pm\sqrt{32}}{8}$$
$$x=\frac{-4\pm\sqrt{16\cdot 2}}{8}=\frac{-4\pm 4\sqrt{2}}{8}$$
$$x=\frac{4(-1\pm\sqrt{2})}{4\cdot 2}=\frac{-1\pm\sqrt{2}}{2}$$

Using a calculator or Table 2, we see that $\sqrt{2}\approx 1.414$:

$$\frac{-1+\sqrt{2}}{2}\approx\frac{-1+1.414}{2} \quad or \quad \frac{-1-\sqrt{2}}{2}\approx\frac{-1-1.414}{2}$$
$$\approx\frac{0.414}{2} \quad or \quad \approx\frac{-2.414}{2}$$
$$\approx 0.207 \quad or \quad \approx -1.207$$

The approximate solutions, to the nearest thousandth, are 0.207 and -1.207.

45. ***Familiarize***. We will use the formula

$$d=\frac{n^2-3n}{2},$$

where d is the number of diagonals and n is the number of sides.

Translate. We substitute 35 for d.

$$35=\frac{n^2-3n}{2}$$

Carry out. We solve the equation.

$$\frac{n^2-3n}{2}=35$$
$$n^2-3n=70 \quad \text{Multiplying by 2}$$
$$n^2-3n-70=0$$
$$(n-10)(n+7)=0$$
$$n-10=0 \quad or \quad n+7=0$$
$$n=10 \quad or \quad n=-7$$

Check. Since the number of sides cannot be negative, -7 cannot be a solution. To check 10, we substitute 10 for n in the original formula and determine if this yields $d=35$. This is left to the student.

State. The polygon has 10 sides.

47. ***Familiarize***. We will use the formula $s=16t^2$.

Translate. We substitute 1482 for s.

$$1482=16t^2$$

Carry out. We solve the equation.

$$1482=16t^2$$
$$\frac{1482}{16}=t^2$$
$$\sqrt{\frac{1482}{16}}=t \quad or \quad -\sqrt{\frac{1482}{16}}=t \quad \text{Principle of square roots}$$
$$9.62\approx t \quad or \quad -9.62\approx t$$

Check. The number -9.62 cannot be a solution, because time cannot be negative in this situation. We substitute 9.62 in the original equation:

$$s=16(9.62)^2=16(92.5444)\approx 1481.$$

This is close to 1482. Remember that we approximated the solution. Thus, we have a check.

State. It would take about 9.62 sec for an object to fall to the ground from the top of the Petronas Towers.

49. ***Familiarize***. We will use the formula $s=16t^2$.

Translate. We substitute 29.6 for s.

$$29.6=16t^2$$

Carry out. We solve the equation.

$$29.6=16t^2$$
$$\frac{29.6}{16}=t^2$$
$$\sqrt{\frac{29.6}{16}}=t \quad or \quad -\sqrt{\frac{29.6}{16}}=t \quad \text{Principle of square roots}$$
$$1.36\approx t \quad or \quad -1.36\approx t$$

Check. The number -1.36 cannot be a solution, because time cannot be negative in this situation. We substitute 1.36 in the original equation:

$$s=16(1.36)^2=16(1.8496)=29.5936.$$

This is close to 29.6. Remember that we approximated the solution. Thus, we have a check.

State. The free-fall portion of the jump lasted about 1.36 sec.

51. ***Familiarize***. From the drawing in the text we have $s=$ the length of the shorter leg and $s+17=$ the length of the longer leg, in feet.

Translate. We use the Pythagorean theorem.

$$x^2+(x+17)^2=25^2$$

Carry out. We solve the equation.

$$x^2 + x^2 + 34x + 289 = 625$$
$$2x^2 + 34x - 336 = 0$$
$$x^2 + 17x - 168 = 0 \quad \text{Multiplying by } \frac{1}{2}$$
$$(x - 7)(x + 24) = 0$$
$$x - 7 = 0 \quad or \quad x + 24 = 0$$
$$x = 7 \quad or \quad x = -24$$

Check. Since the length of a leg cannot be negative, -24 does not check. But 7 does check. If the smaller leg is 7, the other leg is $7+17$, or 24. Then, $7^2+24^2 = 49+576 = 625$, and $\sqrt{625} = 25$, the length of the hypotenuse.

State. The legs measure 7 ft and 24 ft.

53. ***Familiarize***. From the drawing in the text, we see that w represents the width of the rectangle and $w + 4$ represents the length, in centimeters.

Translate. The area is length $\times$ width. Thus, we have two expressions for the area of the rectangle: $(w+4)w$ and 60. This gives us a translation.

$$(w + 4)w = 60$$

Carry out. We solve the equation.

$$w^2 + 4w = 60$$
$$w^2 + 4w - 60 = 0$$
$$(w + 10)(w - 6) = 0$$
$$w + 10 = 0 \quad or \quad w - 6 = 0$$
$$w = -10 \quad or \quad w = 6$$

Check. Since the length of a side cannot be negative, -10 does not check. But 6 does check. If the width is 6, then the length is $6 + 4$, or 10. The area is 10×6, or 60. This checks.

State. The length is 10 cm, and the width is 6 cm.

55. ***Familiarize***. We make a drawing. We let $w =$ the width of the yard. Then $w + 6 =$ the length, in meters.

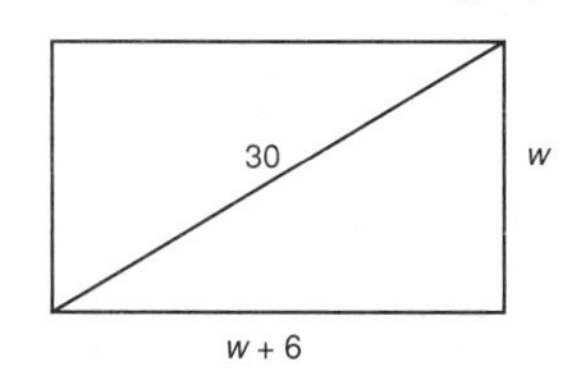

Translate. We use the Pythagorean theorem.

$$w^2 + (w + 6)^2 = 30^2$$

Carry out. We solve the equation.

$$w^2 + w^2 + 12w + 36 = 900$$
$$2w^2 + 12w - 864 = 0$$
$$w^2 + 6w - 432 = 0$$
$$(w + 24)(w - 18) = 0$$
$$w + 24 = 0 \quad or \quad w - 18 = 0$$
$$w = -24 \quad or \quad w = 18$$

Check. Since the width cannot be negative, -24 does not check. But 18 does check. If the width is 18, then the length is $18+6$, or 24, and $18^2+24^2 = 324+576 = 900 = 30^2$.

State. The yard is 18 m by 24 m.

57. ***Familiarize***. We make a drawing. Let $x =$ the length of the shorter leg of the right triangle. Then $x + 2.4 =$ the length of the longer leg, in meters.

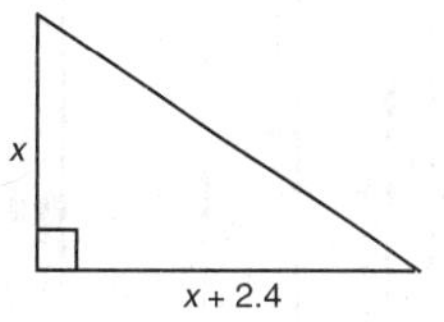

Translate. Using the formula $A = \frac{1}{2}bh$, we substitute 31 for A, $x + 2.4$ for b, and x for h.

$$31 = \frac{1}{2}(x + 2.4)(x)$$

Carry out. We solve the equation.

$$31 = \frac{1}{2}(x + 2.4)(x)$$
$$62 = (x + 2.4)(x) \quad \text{Multiplying by 2}$$
$$62 = x^2 + 2.4x$$
$$0 = x^2 + 2.4x - 62$$
$$0 = 10x^2 + 24x - 620 \quad \text{Multiplying by 10 to clear the decimal}$$
$$0 = 2(5x^2 + 12x - 310)$$

We use the quadratic formula.

$a = 5$, $b = 12$, $b = -310$

$$x = \frac{-12 \pm \sqrt{12^2 - 4 \cdot 5 \cdot (-310)}}{2 \cdot 5}$$
$$x = \frac{-12 \pm \sqrt{6344}}{10}$$
$$x = \frac{-12 + \sqrt{6344}}{10} \quad or \quad x = \frac{-12 - \sqrt{6344}}{10}$$
$$x \approx 6.76 \quad or \quad x \approx -9.16$$

Check. Since the length of a leg cannot be negative, -9.16 does not check. But 6.76 does. If the shorter leg is 6.76 m, then the longer leg is $6.76 + 2.4$, or 9.16 m, and $A = \frac{1}{2}(9.16)(6.76) \approx 31 \approx 30$.

State. The lengths of the legs are about 6.76 m and 9.16 m.

59. ***Familiarize***. We first make a drawing. We let x represent the width and $x + 3$ the length, in inches.

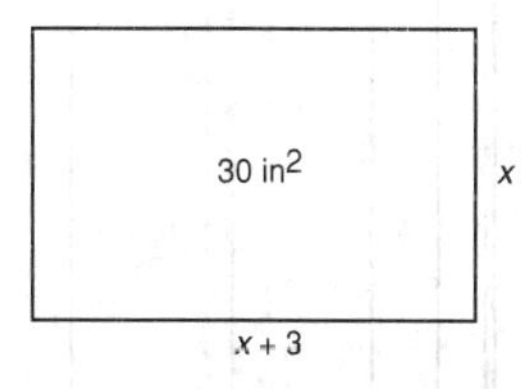

Translate. The area is length $\times$ width. We have two expressions for the area of the rectangle: $(x + 3)x$ and 30. This gives us a translation.

$$(x + 3)x = 30$$

Carry out. We solve the equation.

$$x^2 + 3x = 30$$
$$x^2 + 3x - 30 = 0$$
$$a = 1,\ b = 3,\ c = -30$$
$$x = \frac{-3 \pm \sqrt{3^2 - 4 \cdot 1 \cdot (-30)}}{2 \cdot 1}$$
$$x = \frac{-3 \pm \sqrt{129}}{2}$$
$$x = \frac{-3 + \sqrt{129}}{2} \quad or \quad x = \frac{-3 - \sqrt{129}}{2}$$
$$x \approx 4.18 \quad or \quad x \approx -7.18$$

Check. Since the width cannot be negative, -7.18 does not check. But 4.18 does check. If the width is 4.18 in., then the length is $4.18 + 3$, or 7.18 in., and the area is $7.18(4.18) \approx 30$ in^2.

State. The length is about 7.18 in., and the width is about 4.18 in.

61. ***Familiarize.*** We first make a drawing. We let x represent the width and $2x$ the length, in meters.

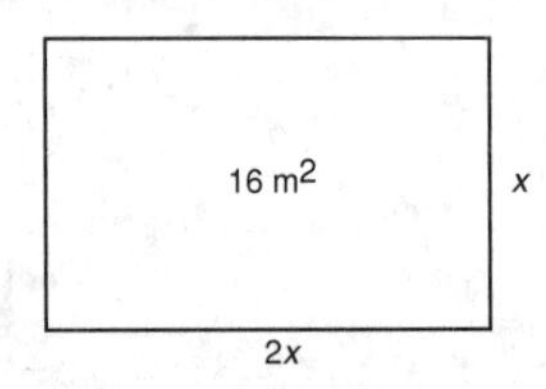

Translate. The area is length $\times$ width. We have two expressions for the area of the rectangle: $2x \cdot x$ and 16. This gives us a translation.

$$2x \cdot x = 16$$

Carry out. We solve the equation.

$$2x^2 = 16$$
$$x^2 = 8$$
$$x = \sqrt{8} \quad or \quad x = -\sqrt{8}$$
$$x = 2.83 \quad or \quad x \approx -2.83 \qquad \text{Using a calculator or Table 2}$$

Check. Since the length cannot be negative, -2.83 does not check. But 2.83 does check. If the width is $\sqrt{8}$ m, then the length is $(2\sqrt{8})$ or 5.66 m. The area is $(5.66)(2.83)$, or $16.0178 \approx 16$.

State. The length is about 5.66 m, and the width is about 2.83 m.

63. ***Familiarize.*** We will use the formula $A = P(1+r)^t$.

Translate. We substitute 2000 for P, 2880 for A, and 2 for t.

$$2880 = 2000(1+r)^2$$

Carry out. We solve the equation.

$$2880 = 2000(1+r)^2$$
$$1.44 = (1+r)^2$$
$$\sqrt{1.44} = 1 + r \quad or \quad -\sqrt{1.44} = 1 + r \qquad \text{Principle of square roots}$$
$$1.2 = 1 + r \quad or \quad -1.2 = 1 + r$$
$$0.2 = r \quad or \quad -2.2 = r$$

Check. Since the interest rate cannot be negative, we check only 0.2, or 20%. We substitute in the formula: $2000(1 + 0.2)^2 = 2000(1.2)^2 = 2000(1.44) = 2880$.

The answer checks.

State. The interest rate is 20%.

65. ***Familiarize.*** We will use the formula $A = P(1+r)^t$.

Translate. We substitute 6000 for P, 6615 for A, and 2 for t.

$$6615 = 6000(1+r)^2$$

Carry out. We solve the equation.

$$6615 = 6000(1+r)^2$$
$$1.1025 = (1+r)^2$$
$$\sqrt{1.1025} = 1 + r \quad or \quad -\sqrt{1.1025} = 1 + r \qquad \text{Principle of square roots}$$
$$1.05 = 1 + r \quad or \quad -1.05 = 1 + r$$
$$0.05 = r \quad or \quad -2.05 = r$$

Check. Since the interest rate cannot be negative, we check only 0.05, or 5%. We substitute in the formula: $6000(1 + 0.05)^2 = 6000(1.05)^2 = 6615$.

The answer checks.

State. The interest rate is 5%.

67. ***Familiarize.*** Let d = the diameter (or width) of the oil slick, in km. Then $d/2$ = the radius. We will use the formula for the area of the circle, $A = \pi r^2$.

Translate. We substitute 20 for A, 3.14 for π, and $d/2$ for r in the formula.

$$20 = 3.14\left(\frac{d}{2}\right)^2$$
$$20 = 3.14\left(\frac{d^2}{4}\right)$$

Carry out. We solve the equation.

$$20 = 3.14\left(\frac{d^2}{4}\right)$$
$$20 = 0.785d^2 \qquad \left(\frac{3.14}{4} = 0.785\right)$$
$$\frac{20}{0.785} = d^2$$
$$d = \sqrt{\frac{20}{0.785}} \quad or \quad d = -\sqrt{\frac{20}{0.785}}$$
$$d \approx 5.05 \quad or \quad d \approx -5.05$$

Check. Since the diameter cannot be negative, -5.05 cannot be a solution. If $d = 5.05$, then $A = 3.14\left(\frac{5.05}{2}\right)^2 \approx 20$. The answer checks.

State. The oil slick was about 5.05 km wide.

69. *Writing Exercise*

71.
$$2x - 7 = 43$$
$$2x = 50 \qquad \text{Adding 7}$$
$$x = 25 \qquad \text{Dividing by 2}$$

The solution is 25.

73. $\frac{3}{5}t + 6 = 15$

$\frac{3}{5}t = 9$ Subtracting 9

$\frac{5}{3} \cdot \frac{3}{5}t = \frac{5}{3} \cdot 9$

$t = 15$

The solution is 15.

75. $\sqrt{4x} - 3 = 5$

$\sqrt{4x} = 8$ Adding 3

$(\sqrt{4x})^2 = 8^2$ Squaring both sides

$4x = 64$

$x = 16$ Dividing by 4

The solution is 16.

77. *Writing Exercise*

79.

$$5x = -x(x-7)$$
$$5x = -x^2 + 7x$$
$$5x + x^2 - 7x = 0$$
$$x^2 - 2x = 0$$
$$x(x-2) = 0$$

$x = 0$ *or* $x - 2 = 0$

$x = 0$ *or* $x = 2$

The solutions are 0 and 2.

81. $3 - x(x-3) = 4$

$$3 - x^2 + 3x = 4$$
$$0 = x^2 - 3x + 1$$

$a = 1$, $b = -3$, $c = 1$

$$x = \frac{-(-3) \pm \sqrt{(-3)^2 - 4 \cdot 1 \cdot 1}}{2 \cdot 1}$$
$$x = \frac{3 \pm \sqrt{5}}{2}$$

The solutions are $\frac{3+\sqrt{5}}{2}$ and $\frac{3-\sqrt{5}}{2}$, or $\frac{3 \pm \sqrt{5}}{2}$, or $\frac{3}{2} \pm \frac{\sqrt{5}}{2}$

83. $(y+4)(y+3) = 15$

$$y^2 + 7y + 12 = 15$$
$$y^2 + 7y - 3 = 0$$

$a = 1$, $b = 7$, $c = -3$

$$y = \frac{-7 \pm \sqrt{7^2 - 4 \cdot 1 \cdot (-3)}}{2 \cdot 1}$$
$$y = \frac{-7 \pm \sqrt{61}}{2}$$

The solutions are $\frac{-7+\sqrt{61}}{2}$ and $\frac{-7-\sqrt{61}}{2}$, or $\frac{-7 \pm \sqrt{61}}{2}$, or $-\frac{7}{2} \pm \frac{\sqrt{61}}{2}$.

85.

$$\frac{x^2}{x+3} - \frac{5}{x+3} = 0, \quad \text{LCM is } x+3$$
$$(x+3)\left(\frac{x^2}{x+3} - \frac{5}{x+3}\right) = (x+3) \cdot 0$$
$$x^2 - 5 = 0$$
$$x^2 = 5$$

$x = \sqrt{5}$ or $x = -\sqrt{5}$ Principle of square roots

Both numbers check. The solutions are $\sqrt{5}$ and $-\sqrt{5}$, or $\pm\sqrt{5}$.

87.

$$\frac{1}{x} + \frac{1}{x+1} = \frac{1}{3}, \quad \text{LCM is } 3x(x+1)$$
$$3x(x+1)\left(\frac{1}{x} + \frac{1}{x+1}\right) = 3x(x+1) \cdot \frac{1}{3}$$
$$3(x+1) + 3x = x(x+1)$$
$$3x + 3 + 3x = x^2 + x$$
$$6x + 3 = x^2 + x$$
$$0 = x^2 - 5x - 3$$
$$x = \frac{-(-5) \pm \sqrt{(-5)^2 - 4 \cdot 1 \cdot (-3)}}{2 \cdot 1}$$
$$x = \frac{5 \pm \sqrt{37}}{2}$$

The solutions are $\frac{5+\sqrt{37}}{2}$ and $\frac{5-\sqrt{37}}{2}$, or $\frac{5 \pm \sqrt{37}}{2}$, or $\frac{5}{2} \pm \frac{\sqrt{37}}{2}$.

89. ***Familiarize.*** From the drawing in the text, we see that we have a right triangle where r = the length of each leg and $r + 2$ = the length of the hypotenuse, in centimeters.

Translate. We use the Pythagorean theorem.

$$r^2 + r^2 = (r+2)^2.$$

Carry out. We solve the equation.

$$2r^2 = r^2 + 4r + 4$$
$$r^2 - 4r - 4 = 0$$

$a = 1$, $b = -4$, $c = -4$

$$r = \frac{-(-4) \pm \sqrt{(-4)^2 - 4 \cdot 1 \cdot (-4)}}{2 \cdot 1}$$
$$r = \frac{4 \pm \sqrt{16+16}}{2} = \frac{4 \pm \sqrt{32}}{2}$$
$$r = \frac{4 \pm \sqrt{16 \cdot 2}}{2} = \frac{4 \pm 4\sqrt{2}}{2}$$
$$r = \frac{2(2 \pm 2\sqrt{2})}{2 \cdot 1} = 2 \pm 2\sqrt{2}$$

$x = 2 - 2\sqrt{2}$ *or* $x = 2 + 2\sqrt{2}$

$x \approx 2 - 2.828$ *or* $x \approx 2 + 2.828$

$x \approx -0.828$ *or* $x \approx 4.828$

$x \approx -0.83$ *or* $x \approx 4.83$ Rounding to the nearest hundredth

Check. Since the length of a leg cannot be negative, -0.83 cannot be a solution of the original equation. When $x \approx 4.83$, then $x + 2 \approx 6.83$ and $(4.83)^2 + (4.83)^2 = 23.3289 + 23.3289 = 46.6578 \approx (5.83)^2$. This checks.

State. In the figure, $r = 2 + 2\sqrt{2}$ cm ≈ 4.83 cm.

91. ***Familiarize.*** Let w = the width of the rectangle, in meters. Then $1.6w$ = the length. Recall that the formula for the area of a rectangle is $A = l \times w$.

Translate. We substitute 9000 for A, $1.6w$ for w, and w for w in the formula.

$$9000 = 1.6w(w)$$
$$9000 = 1.6w^2$$

Carry out. We solve the equation.

$$9000 = 1.6w^2$$
$$\frac{9000}{1.6} = w^2$$
$$w = \sqrt{\frac{9000}{1.6}} \quad or \quad w = -\sqrt{\frac{9000}{1.6}}$$
$$w = 75 \quad or \quad w = -75$$

Check. Since the width of the rectangle cannot be negative we will not check -75. If $w = 75$, then the length is $1.6(75)$, or 120, and the area is $120(75) = 9000$ m^2. The answer checks.

State. The length of the rectangle is 120 m, and the width is 75 m.

93. ***Familiarize.*** We will use the formula $A = P(1+r)^2$ twice. The amount in the account for the \$4000 invested for 2 yr is given by $4000(1+r)^2$. The amount for the \$2000 invested for 1 yr is $2000(1+r)$.

Translate. The total amount in the account at the end of 2 yr is the sum of the amounts above.

$$6510 = 4000(1+r)^2 + 2000(1+r)$$

Carry out. We solve the equation. Begin by letting $x = 1 + r$.

$$6510 = 4000x^2 + 2000x$$
$$0 = 4000x^2 + 2000x - 6510$$
$$0 = 400x^2 + 200x - 651 \quad \text{Dividing by 10}$$

We will use the quadratic formula.

$$a = 400,\ b = 200,\ c = -651$$
$$x = \frac{-200 \pm \sqrt{200^2 - 4 \cdot 400(-651)}}{2 \cdot 400}$$
$$x = \frac{-200 \pm \sqrt{1,081,600}}{800} = \frac{-200 \pm 1040}{800}$$
$$x = \frac{-200 + 1040}{800} \quad or \quad x = \frac{-200 - 1040}{800}$$
$$x = 1.05 \quad or \quad x = -1.55$$
$$1 + r = 1.05 \quad or \quad 1 + r = -1.55$$
$$r = 0.05 \quad or \quad r = -2.55$$

Check. Since the interest rate cannot be negative, we check only 0.05 or 5%. At the end of 2 yr, \$4000 invested at 5% interest has grown to $\$4000(1+0.05)^2$, or \$4410. At the end of 1 yr, \$2000 invested at 5% interest has grown to $\$2000(1 + 0.05)$, or \$2100. Then the total amount in the account is $\$4410 + \$2100$, or \$6510. The answer checks.

State. The interest rate is 5%.

95. ***Familiarize.*** The area of the actual strike zone is $15(40)$, so the area of the enlarged zone is $15(40) + 0.4(15)(40)$, or $1.4(15)(40)$. From the drawing in the text we see that the dimensions of the enlarged strike zone are $15 + 2x$ by $40 + 2x$.

Translate. Using the formula $A = lw$, we write an equation for the area of the enlarged strike zone.

$$1.4(15)(40) = (15 + 2x)(40 + 2x)$$

Carry out. We solve the equation.

$$1.4(15)(40) = (15 + 2x)(40 + 2x)$$
$$840 = 600 + 110x + 4x^2 \quad \text{Multiplying on both sides}$$
$$0 = 4x^2 + 110x - 240$$
$$0 = 2x^2 + 55x - 120 \quad \text{Dividing by 2}$$
$$a = 2,\ b = 55,\ c = -120$$
$$x = \frac{-55 \pm \sqrt{55^2 - 4 \cdot 2 \cdot (-120)}}{2 \cdot 2}$$
$$x = \frac{-55 \pm \sqrt{3985}}{4}$$
$$x \approx 2.03 \quad or \quad x \approx -29.53$$

Check. Since the measurement cannot be negative, -29.53 cannot be a solution. If $x = 2.03$, then the dimensions of the enlarged strike zone are $15 + 2(2.03)$, or 19.06, by $40 + 2(2.03)$, or 44.06, and the area is $19.06(44.06) = 839.7836 \approx 840$. The answer checks.

State. The dimensions of the enlarged strike zone are 19.06 in. by 44.06 in.

Exercise Set 9.4

1. $3x + 8 = 7x + 4$

$$4 = 4x \quad \text{Adding } -3x - 4$$
$$1 = x \quad \text{Dividing by 4}$$

The solution is 1.

3.
$$x^2 = 5x - 6$$
$$x^2 - 5x + 6 = 0 \quad \text{Adding } -5x + 6$$
$$(x - 2)(x - 3) = 0$$
$$x - 2 = 0 \quad or \quad x - 3 = 0$$
$$x = 2 \quad or \quad x = 3$$

The solutions are 2 and 3.

5. $3 + \sqrt{x} = 8$

$$\sqrt{x} = 5 \quad \text{Subtracting 3}$$
$$(\sqrt{x})^2 = 5^2 \quad \text{Squaring both sides}$$
$$x = 25$$

The number 25 checks. It is the solution.

7. $$\frac{5}{x}+\frac{3}{4}=2 \quad \text{Note that } x\neq 0;\ \text{LCD} = 4x.$$

$$4x\left(\frac{5}{x}+\frac{3}{4}\right)=4x\cdot 2 \quad \text{Multiplying by the LCD}$$

$$4x\cdot\frac{5}{x}+4x\cdot\frac{3}{4}=8x$$

$$20+3x=8x$$

$$20=5x$$

$$4=x$$

The number 4 checks. It is the solution.

9. $$4x-7=2(5x-3)$$

$$4x-7=10x-6$$

$$-1=6x \quad \text{Adding } -4x+6$$

$$-\frac{1}{6}=x$$

The solution is $-\frac{1}{6}$.

11. $$\frac{2}{9x}-\frac{5}{6}=1 \quad \text{Note that } x\neq 0;\ \text{LCD} = 18x.$$

$$18x\left(\frac{2}{9x}-\frac{5}{6}\right)=18x\cdot 1$$

$$18x\cdot\frac{2}{9x}-18x\cdot\frac{5}{6}=18x$$

$$4-15x=18x$$

$$4=33x$$

$$\frac{4}{33}=x$$

This number checks. The solution is $\frac{4}{33}$.

13. $$3t^2-7t+2=0$$

$$(3t-1)(t-2)=0$$

$$3t-1=0 \quad \text{or} \quad t-2=0$$

$$3t=1 \quad \text{or} \quad t=2$$

$$t=\frac{1}{3} \quad \text{or} \quad t=2$$

The solutions are $\frac{1}{3}$ and 2.

15. $$11=4\sqrt{3t}-5$$

$$16=4\sqrt{3t}$$

$$4=\sqrt{3t} \quad \text{Dividing by 4}$$

$$4^2=(\sqrt{3t})^2$$

$$16=3t$$

$$\frac{16}{3}=t$$

The number $\frac{16}{3}$ checks. It is the solution.

17. $$\frac{3}{10t}+\frac{t}{5}=1 \quad \text{Note that } t\neq 0;\ \text{LCD} = 10t.$$

$$10t\left(\frac{3}{10t}+\frac{t}{5}\right)=10t\cdot 1$$

$$10t\cdot\frac{3}{10t}+10t\cdot\frac{t}{5}=10t$$

$$3+2t^2=10t$$

$$2t^2-10t+3=0$$

$$a=2,\ b=-10,\ c=3$$

$$t=\frac{-b\pm\sqrt{b^2-4ac}}{2a}$$

$$t=\frac{-(-10)\pm\sqrt{(-10)^2-4\cdot 2\cdot 3}}{2\cdot 2}$$

$$t=\frac{10\pm\sqrt{76}}{4}=\frac{10\pm\sqrt{4\cdot 19}}{4}$$

$$t=\frac{10\pm 2\sqrt{19}}{4}=\frac{2(5\pm\sqrt{19})}{2\cdot 2}$$

$$t=\frac{5\pm\sqrt{19}}{2}$$

Both numbers check. The solutions are $\frac{5\pm\sqrt{19}}{2}$, or $\frac{5}{2}\pm\frac{\sqrt{19}}{2}$.

19. $$7t-1=2(3-t)+5$$

$$7t-1=6-2t+5$$

$$7t-1=11-2t$$

$$9t=12 \quad \text{Adding } 2t+1$$

$$t=\frac{4}{3}$$

The solution is $\frac{4}{3}$.

21. $$3\sqrt{2t-1}=2\sqrt{5t+3}$$

$$(3\sqrt{2t-1})^2=(2\sqrt{5t+3})^2$$

$$9(2t-1)=4(5t+3)$$

$$18t-9=20t+12$$

$$-21=2t \quad \text{Adding } -18t-12$$

$$-\frac{21}{2}=t$$

Check:

$3\sqrt{2t-1}$	$=\ 2\sqrt{5t+3}$
$3\sqrt{2\left(-\frac{21}{2}\right)-1}$	$2\sqrt{5\left(-\frac{21}{2}\right)+3}$
$3\sqrt{-21-1}$	$2\sqrt{-\frac{105}{2}+3}$
$3\sqrt{-22}$	$\stackrel{?}{=}\ 2\sqrt{-\frac{99}{2}}$ UNDEFINED

The equation has no real-number solutions.

23. $2n - 1 = 3n^2$

$0 = 3n^2 - 2n + 1$

$a = 3,\ b = -2,\ c = 1$

$$n = \frac{-b \pm \sqrt{b^2 - 4ac}}{2a}$$

$$n = \frac{-(-2) \pm \sqrt{(-2)^2 - 4 \cdot 3 \cdot 1}}{2 \cdot 3}$$

$$n = \frac{2 \pm \sqrt{-8}}{6}$$

Since the radicand, -8, is negative, there are no real-number solutions.

25. $1 - \dfrac{3}{7n} = \dfrac{5}{14}$ Note that $n \neq 0$; LCD $= 14n$.

$$14n\left(1 - \frac{3}{7n}\right) = 14n \cdot \frac{5}{14}$$

$$14n \cdot 1 - 14n \cdot \frac{3}{7n} = 5n$$

$$14n - 6 = 5n$$

$$-6 = -9n$$

$$\frac{2}{3} = n$$

The solution is $\dfrac{2}{3}$.

27. $s = \dfrac{1}{2}gt^2$

$2s = gt^2$ Multiplying by 2

$\dfrac{2s}{t^2} = g$ Dividing by t^2

29. $A = P(1 + rt)$

$A = P + Prt$

$A - P = Prt$

$\dfrac{A - P}{Pr} = t$, or

$\dfrac{A}{Pr} - \dfrac{1}{r} = t$

31. $d = c\sqrt{h}$

$d^2 = (c\sqrt{h})^2$ Squaring both sides

$d^2 = c^2 \cdot h$

$\dfrac{d^2}{c^2} = h$ Dividing by c^2

33. $\dfrac{1}{R} = \dfrac{1}{r_1} + \dfrac{1}{r_2}$

$Rr_1r_2 \cdot \dfrac{1}{R} = Rr_1r_2\left(\dfrac{1}{r_1} + \dfrac{1}{r_2}\right)$ Multiplying by the LCD, Rr_1r_2, to clear fractions

$r_1r_2 = Rr_1r_2 \cdot \dfrac{1}{r_1} + Rr_1r_2 \cdot \dfrac{1}{r_2}$

$r_1r_2 = Rr_2 + Rr_1$

$r_1r_2 = R(r_2 + r_1)$

$\dfrac{r_1r_2}{r_2 + r_1} = R$

35. $ax^2 + bx + c = 0$

Observe that this is standard form for a quadratic equation. The solution is given by the quadratic formula:

$$x = \frac{-b \pm \sqrt{b^2 - 4ac}}{2a}.$$

37. $\dfrac{m}{n} = p - q$

$m = n(p - q)$ Multiplying by n

$\dfrac{m}{p - q} = n$ Dividing by $p - q$

39. $Ax + By = C$

$By = C - Ax$

$y = \dfrac{C - Ax}{B}$

41. $S = 2\pi r(r + h)$

$\dfrac{S}{2\pi r} = r + h$ Dividing by $2\pi r$

$\dfrac{S}{2\pi r} - r = h$, or

$\dfrac{S - 2\pi r^2}{2\pi r} = h$

43. $\dfrac{s}{h} = \dfrac{h}{t}$ LCD $= ht$

$ht \cdot \dfrac{s}{h} = ht \cdot \dfrac{h}{t}$

$st = h^2$

$\pm\sqrt{st} = h$

45. $mt^2 + nt - p = 0$

$a = m,\ b = n,\ c = -p$

$$t = \frac{-n \pm \sqrt{n^2 - 4 \cdot m \cdot (-p)}}{2 \cdot m}$$

$$t = \frac{-n \pm \sqrt{n^2 + 4mp}}{2m}$$

47. $\dfrac{m}{n} = r$

$m = nr$ Multiplying by n

$\dfrac{m}{r} = n$ Dividing by r

49. $m + t = \dfrac{n}{m}$

$m(m + t) = m \cdot \dfrac{n}{m}$

$m^2 + mt = n$

$m^2 + mt - n = 0$

$a = 1,\ b = t,\ c = -n$

$$m = \frac{-t \pm \sqrt{t^2 - 4 \cdot 1 \cdot (-n)}}{2 \cdot 1}$$

$$m = \frac{-t \pm \sqrt{t^2 + 4n}}{2}$$

51.
$$\begin{aligned} n &= p - 3\sqrt{t+c} \\ n - p &= -3\sqrt{t+c} \\ (n-p)^2 &= (-3\sqrt{t+c})^2 \\ n^2 - 2np + p^2 &= 9(t+c) \\ n^2 - 2np + p^2 &= 9t + 9c \\ n^2 - 2np + p^2 - 9c &= 9t \\ \frac{n^2 - 2np + p^2 - 9c}{9} &= t \end{aligned}$$

53.
$$\begin{aligned} \sqrt{m-n} &= \sqrt{3t} \\ (\sqrt{m-n})^2 &= (\sqrt{3t})^2 \\ m - n &= 3t \\ m &= n + 3t \\ m - 3t &= n \end{aligned}$$

55. *Writing Exercise*

57. $\sqrt{6} \cdot \sqrt{10} = \sqrt{6 \cdot 10} = \sqrt{2 \cdot 3 \cdot 2 \cdot 5} = \sqrt{2 \cdot 2}\sqrt{3 \cdot 5} = 2\sqrt{15}$

59. $\sqrt{150} = \sqrt{25 \cdot 6} = \sqrt{25}\sqrt{6} = 5\sqrt{6}$

61. $\sqrt{4a^7b^4} = \sqrt{4 \cdot a^6 \cdot b^4 \cdot a} = \sqrt{4}\sqrt{a^6}\sqrt{b^4}\sqrt{a} = 2a^3b^2\sqrt{a}$

63. *Writing Exercise*

65. Substitute 8 for c and 224 for d and solve for a.
$$\begin{aligned} c &= \frac{a}{a+12} \cdot d \\ 8 &= \frac{a}{a+12} \cdot 24 \\ \frac{1}{3} &= \frac{a}{a+12} \quad \text{Dividing by 24} \\ 3(a+12) \cdot \frac{1}{3} &= 3(a+12) \cdot \frac{a}{a+12} \\ a + 12 &= 3a \\ 12 &= 2a \\ 6 &= a \end{aligned}$$

The child is 6 years old.

67.
$$\begin{aligned} fm &= \frac{gm - t}{m} \\ fm^2 &= gm - t \quad \text{Multiplying by } m \\ fm^2 - gm + t &= 0 \end{aligned}$$
$a = f,\ b = -g,\ c = t$
$$m = \frac{-(-g) \pm \sqrt{(-g)^2 - 4 \cdot f \cdot t}}{2 \cdot f}$$
$$m = \frac{g \pm \sqrt{g^2 - 4ft}}{2f}$$

69.
$$\begin{aligned} V &= \frac{k}{\sqrt{at+b}} + c \\ V - c &= \frac{k}{\sqrt{at+b}} \\ (V-c)^2 &= \left(\frac{k}{\sqrt{at+b}}\right)^2 \\ (V-c)^2 &= \frac{k^2}{at+b} \\ at + b &= \frac{k^2}{(V-c)^2} \quad \text{Multiplying by } \frac{at+b}{(V-c)^2} \\ at &= -b + \frac{k^2}{(V-c)^2} \\ t &= -\frac{b}{a} + \frac{k^2}{a(V-c)^2} \quad \text{Dividing by } a \end{aligned}$$

71. When $C = F$, we have
$$\begin{aligned} C &= \frac{5}{9}(C - 32) \\ 9C &= 5(C - 32) \\ 9C &= 5C - 160 \\ 4C &= -160 \\ C &= -40 \end{aligned}$$

At $-40°$ the Fahrenheit and Celsius readings are the same.

Exercise Set 9.5

1. False; a complex number $a + bi$ with $b \neq 0$ is not a real number.

3. True; complex numbers $a + bi$ with $b \neq 0$ are imaginary numbers.

5. $\sqrt{-1} = i$

7. $\sqrt{-81} = \sqrt{-1 \cdot 81} = \sqrt{-1} \cdot \sqrt{81} = i \cdot 9 = 9i$

9. $\sqrt{-50} = \sqrt{-1 \cdot 25 \cdot 2} = \sqrt{-1} \cdot \sqrt{25}\sqrt{2} = i \cdot 5\sqrt{2} = 5i\sqrt{2}$, or $5\sqrt{2}i$

11. $-\sqrt{-45} = -\sqrt{-1 \cdot 9 \cdot 5} = -\sqrt{-1} \cdot \sqrt{9} \cdot \sqrt{5} = -i \cdot 3\sqrt{5} = -3i\sqrt{5}$, or $-3\sqrt{5}i$

13. $-\sqrt{-18} = -\sqrt{-1 \cdot 9 \cdot 2} = -\sqrt{-1} \cdot \sqrt{9} \cdot \sqrt{2} = -i \cdot 3\sqrt{2} = -3i\sqrt{2}$, or $-3\sqrt{2}i$

15. $4 + \sqrt{-49} = 4 + \sqrt{-1 \cdot 49} = 4 + \sqrt{-1} \cdot \sqrt{49} = 4 + i \cdot 7 = 4 + 7i$

17. $3 + \sqrt{-9} = 3 + \sqrt{-1 \cdot 9} = 3 + \sqrt{-1} \cdot \sqrt{9} = 3 + i \cdot 3 = 3 + 3i$

19. $5 - \sqrt{-98} = 5 - \sqrt{-1 \cdot 98} = 5 - \sqrt{-1} \cdot \sqrt{98} = 5 - i \cdot 7\sqrt{2} = 5 - 7i\sqrt{2}$

21. $x^2+9=0$

$x^2=-9$

$x=\sqrt{-9}$ *or* $x=-\sqrt{-9}$

$x=\sqrt{-1}\sqrt{9}$ *or* $x=-\sqrt{-1}\sqrt{9}$

$x=3i$ *or* $x=-3i$ Principle of square roots

The solutions are $3i$ and $-3i$, or $\pm 3i$.

23. $x^2=-28$

$x=\sqrt{-28}$ *or* $x=-\sqrt{-28}$ Principle of square roots

$x=\sqrt{-1\cdot 4\cdot 7}$ *or* $x=-\sqrt{-1\cdot 4\cdot 7}$

$x=i\cdot 2\sqrt{7}$ *or* $x=-i\cdot 2\sqrt{7}$

$x=2i\sqrt{7}$ *or* $x=-2i\sqrt{7}$

The solutions are $2i\sqrt{7}$ and $-2i\sqrt{7}$, or $\pm 2i\sqrt{7}$.

25. $t^2+4t+5=0$

$a=1,\ b=4,\ c=5$

$$t=\frac{-b\pm\sqrt{b^2-4ac}}{2a}$$

$$t=\frac{-4\pm\sqrt{4^2-4\cdot1\cdot5}}{2\cdot1}$$

$$t=\frac{-4\pm\sqrt{-4}}{2}=\frac{-4\pm\sqrt{-1}\sqrt{4}}{2}$$

$$t=\frac{-4\pm 2i}{2}$$

$t=-2\pm i$ Writing in the form $a+bi$

The solutions are $-2\pm i$.

27. $(t+3)^2=-16$

$t+3=\pm\sqrt{-16}$ Principle of square roots

$t+3=\pm\sqrt{-1}\sqrt{16}$

$t+3=\pm 4i$

$t=-3\pm 4i$

The solutions are $-3\pm 4i$.

29. $x^2+5=2x$

$x^2-2x+5=0$

$a=1,\ b=-2,\ c=5$

$$x=\frac{-b\pm\sqrt{b^2-4ac}}{2a}$$

$$x=\frac{-(-2)\pm\sqrt{(-2)^2-4\cdot1\cdot5}}{2\cdot1}$$

$$x=\frac{2\pm\sqrt{-16}}{2}=\frac{2\pm i\sqrt{16}}{2}$$

$$x=\frac{2\pm 4i}{2}=\frac{2}{2}\pm\frac{4i}{2}$$

$x=1\pm 2i$

The solutions are $1\pm 2i$.

31. $t^2+7-4t=0$

$t^2-4t+7=0$ Standard form

$a=1,\ b=-4,\ c=7$

$$t=\frac{-b\pm\sqrt{b^2-4ac}}{2a}$$

$$t=\frac{-(-4)\pm\sqrt{(-4)^2-4\cdot1\cdot7}}{2\cdot1}$$

$$t=\frac{4\pm\sqrt{-12}}{2}=\frac{4\pm i\sqrt{12}}{2}$$

$$t=\frac{4\pm 2i\sqrt{3}}{2}=\frac{4}{2}\pm\frac{2\sqrt{3}}{2}i=2\pm\sqrt{3}i$$

The solutions are $2\pm\sqrt{3}i$.

33. $2t^2+6t+5=0$

$a=2,\ b=6,\ t=5$

$$t=\frac{-b\pm\sqrt{b^2-4ac}}{2a}$$

$$t=\frac{-6\pm\sqrt{6^2-4\cdot2\cdot5}}{2\cdot2}$$

$$t=\frac{-6\pm\sqrt{-4}}{4}=\frac{-6\pm i\sqrt{4}}{4}=\frac{-6\pm 2i}{4}$$

$$t=\frac{-6}{4}\pm\frac{2}{4}i=-\frac{3}{2}\pm\frac{1}{2}i$$

The solutions are $-\frac{3}{2}\pm\frac{1}{2}i$.

35. $1+2m+3m^2=0$

$3m^2+2m+1=0$ Standard form

$a=3,\ b=2,\ c=1$

$$m=\frac{-b\pm\sqrt{b^2-4ac}}{2a}$$

$$m=\frac{-2\pm\sqrt{2^2-4\cdot3\cdot1}}{2\cdot3}$$

$$m=\frac{-2\pm\sqrt{-8}}{6}=\frac{-2\pm i\sqrt{8}}{6}=\frac{-2\pm 2i\sqrt{2}}{6}$$

$$m=\frac{-2}{6}\pm\frac{2\sqrt{2}}{6}i=-\frac{1}{3}\pm\frac{\sqrt{2}}{3}i$$

The solutions are $-\frac{1}{3}\pm\frac{\sqrt{2}}{3}i$.

37. *Writing Exercise*

39. Graph $y=\frac{3}{5}x$.

x	y	(x,y)
-5	-3	$(-5,-3)$
0	0	$(0,0)$
5	3	$(5,3)$

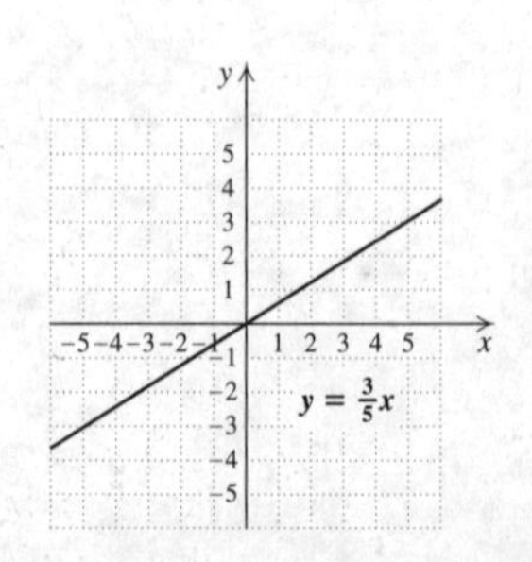

41. Graph $y = -4x$.

x	y	(x, y)
-1	4	$(-1, 4)$
0	0	$(0, 0)$
1	-4	$(1, -4)$

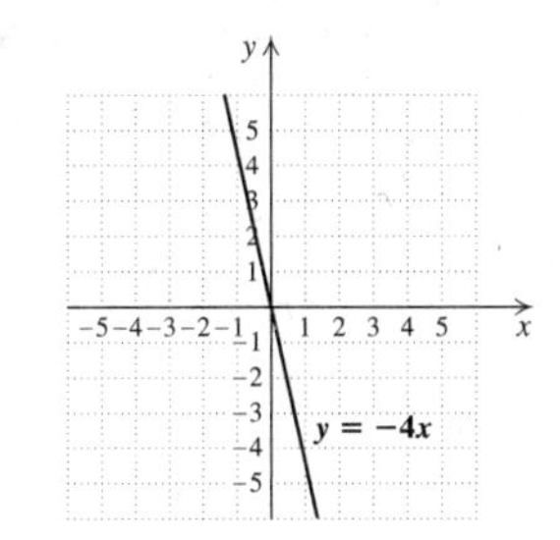

43. $(-17)^2 - (8+9)^2$

Observe that $8 + 9 = 17$, so we have $(-17)^2 - 17^2$. Since $(-17)^2 = 17^2$, then the difference is 0.

45. *Writing Exercise*

47.
$$(x+1)^2 + (x+3)^2 = 0$$
$$x^2 + 2x + 1 + x^2 + 6x + 9 = 0$$
$$2x^2 + 8x + 10 = 0$$
$$x^2 + 4x + 5 = 0 \quad \text{Dividing by 2}$$

$a = 1$, $b = 4$, $c = 5$

$$x = \frac{-b \pm \sqrt{b^2 - 4ac}}{2a}$$
$$x = \frac{-4 \pm \sqrt{4^2 - 4 \cdot 1 \cdot 5}}{2 \cdot 1}$$
$$x = \frac{-4 \pm \sqrt{16 - 20}}{2} = \frac{-4 \pm \sqrt{-4}}{2}$$
$$x = \frac{-4 \pm 2i}{2} = \frac{2(-2 \pm i)}{2 \cdot 1}$$
$$x = -2 \pm i$$

The solutions are $-2 \pm i$.

49. $\dfrac{2x-1}{5} - \dfrac{2}{x} = \dfrac{x}{2}$

We multiply by $10x$, the LCD.

$$10x\left(\frac{2x-1}{5} - \frac{2}{x}\right) = 10x \cdot \frac{x}{2}$$
$$2x(2x-1) - 10 \cdot 2 = 5x \cdot x$$
$$4x^2 - 2x - 20 = 5x^2$$
$$0 = x^2 + 2x + 20$$

$a = 1$, $b = 2$, $c = 20$

$$x = \frac{-b \pm \sqrt{b^2 - 4ac}}{2a}$$
$$x = \frac{-2 \pm \sqrt{2^2 - 4 \cdot 1 \cdot 20}}{2 \cdot 1}$$
$$x = \frac{-2 \pm \sqrt{-76}}{2} = \frac{-2 \pm i\sqrt{76}}{2} = \frac{-2 \pm 2i\sqrt{19}}{2}$$
$$x = \frac{-2}{2} \pm \frac{2\sqrt{19}}{2}i = -1 \pm \sqrt{19}i$$

The solutions are $-1 \pm \sqrt{19}i$.

51. Example 2(a):

Graph $y = x^2 + 3x + 4$.

There are no x-intercepts, so the equation $x^2 + 3x + 4 = 0$ has no real-number solutions.

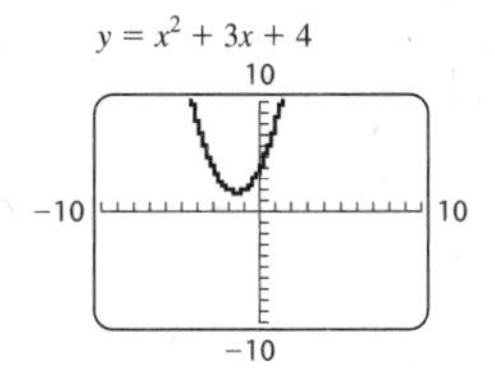

Example 2(b):

Graph $y_1 = x^2 + 2$ and $y_2 = 2x$.

The graphs do not intersect, so the equation $x^2 + 2 = 2x$ has no real-number solutions.

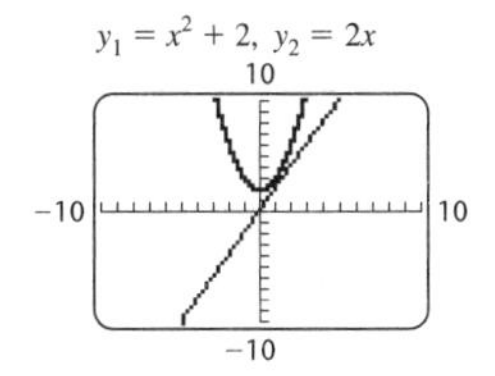

Exercise Set 9.6

1. $y = x^2 + 3$

We first find the vertex. The x-coordinate is

$$-\frac{b}{2a} = -\frac{0}{2 \cdot 1} = 0.$$

We substitute into the equation to find the second coordinate of the vertex.

$$x^2 + 3 = 0^2 + 3 = 3$$

The vertex is $(0, 3)$. The line of symmetry is $x = 0$, the y-axis.

We choose some x-values on both sides of the vertex and graph the parabola.

When $x = 1$, $y = 1^2 + 3 = 1 + 3 = 4$.

When $x = -1$, $y = (-1)^2 + 3 = 1 + 3 = 4$.

When $x = 2$, $y = 2^2 + 3 = 4 + 3 = 7$.

When $x = -2$, $y = (-2)^2 + 3 = 4 + 3 = 7$.

x	y
0	3
1	4
-1	4
2	7
-2	7

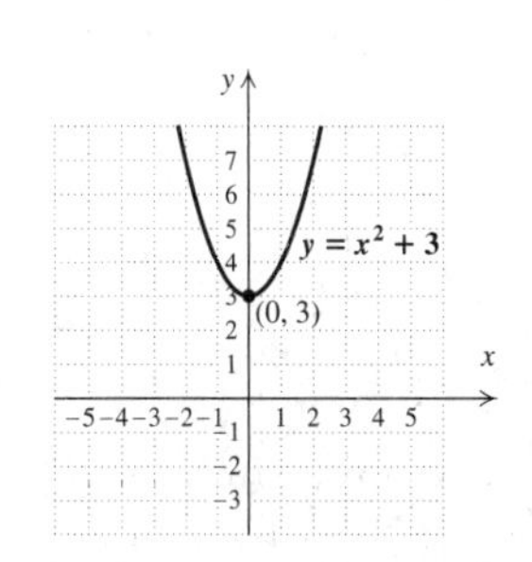

3. $y = -2x^2$

Find the vertex. The x-coordinate is

$$-\frac{b}{2a} = -\frac{0}{2(-2)} = 0.$$

The y-coordinate is

$$-2x^2 = -2 \cdot 0^2 = 0.$$

The vertex is $(0, 0)$. The line of symmetry is $x = 0$, the y-axis.

Choose some x-values on both sides of the vertex and graph the parabola.

When $x = -2$, $y = -2(-2)^2 = -2 \cdot 4 = -8$.
When $x = -1$, $y = -2(-1)^2 = -2 \cdot 1 = -2$.
When $x = 1$, $y = -2 \cdot 1^2 = -2 \cdot 1 = -2$.
When $x = 2$, $y = -2 \cdot 2^2 = -2 \cdot 4 = -8$.

x	y
0	0
−2	−8
−1	−2
1	−2
2	−8

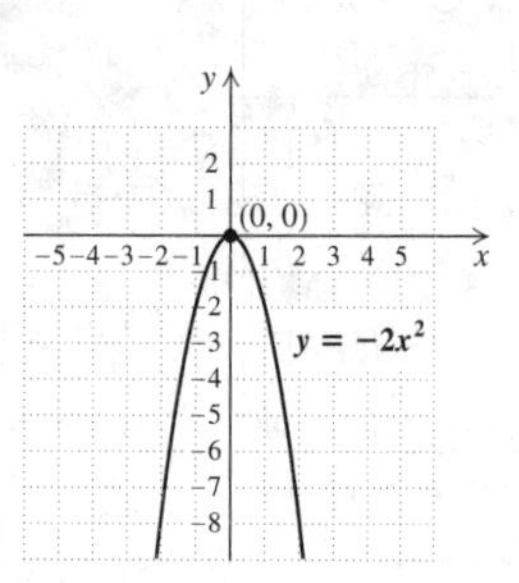

5. $y = x^2 + x - 6$

Find the vertex. The x-coordinate is

$$-\frac{b}{2a} = -\frac{1}{2 \cdot 1} = -\frac{1}{2}.$$

The y-coordinate is

$$x^2 + x - 6 = \left(-\frac{1}{2}\right)^2 + \left(-\frac{1}{2}\right) - 6 = \frac{1}{4} - \frac{1}{2} - 6 = -\frac{25}{4}.$$

The vertex is $\left(-\frac{1}{2}, -\frac{25}{4}\right)$.

We choose some x-values on both sides of the vertex and graph the parabola. We make sure we find y when $x = 0$. This gives us the y-intercept.

x	y
−4	6
−3	0
−1	−6
0	−6
1	−4
3	6

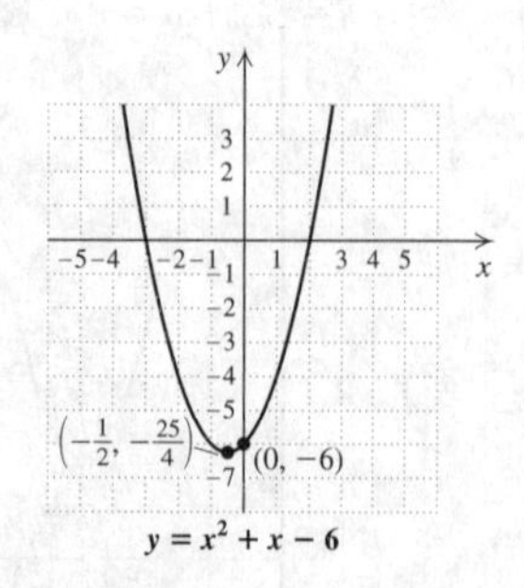

7. $y = x^2 + 2x + 1$

Find the vertex. The x-coordinate is

$$-\frac{b}{2a} = -\frac{2}{2 \cdot 1} = -\frac{2}{2} = -1.$$

The y-coordinate is

$$x^2 + 2x + 1 = (-1)^2 + 2(-1) + 1 = 1 - 2 + 1 = 0.$$

The vertex is $(-1, 0)$.

We choose some x-values on both sides of the vertex and graph the parabola. We make sure we find y when $x = 0$. This gives us the y-intercept.

x	y
−3	4
−2	1
0	1
1	4
2	9

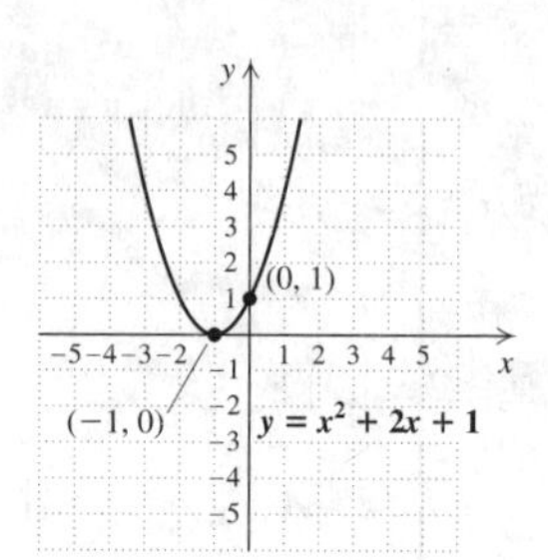

9. $y = 2x^2 - 12x + 13$

Find the vertex. The x-coordinate is

$$-\frac{b}{2a} = -\frac{-12}{2 \cdot 2} = \frac{12}{4} = 3.$$

The y-coordinate is

$$2x^2 - 12x + 13 = 2 \cdot 3^2 - 12 \cdot 3 + 13 = 18 - 36 + 13 = -5.$$

The vertex is $(3, -5)$.

We choose some x-values on both sides of the vertex and graph the parabola. We make sure we find y when $x = 0$. This gives us the y-intercept.

x	y
−1	27
0	13
2	−3
4	−3
5	3

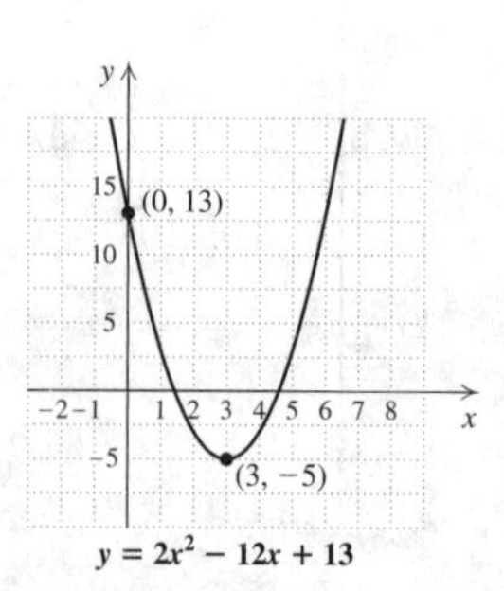

11. $y = -3x^2 - 2x + 8$

Find the vertex. The x-coordinate is

$$-\frac{b}{2a} = -\frac{-2}{2(-3)} = -\frac{1}{3}.$$

The y-coordinate is

$$-3x^2 - 2x + 8 = -3\left(-\frac{1}{3}\right)^2 - 2\left(-\frac{1}{3}\right) + 8 =$$

$$-\frac{1}{3} + \frac{2}{3} + 8 = \frac{25}{3}.$$

We choose some x-values on both sides of the vertex and graph the parabola. We make sure we find y when $x = 0$. This gives us the y-intercept.

x	y
-3	-13
-2	0
0	8
1	3
2	-8

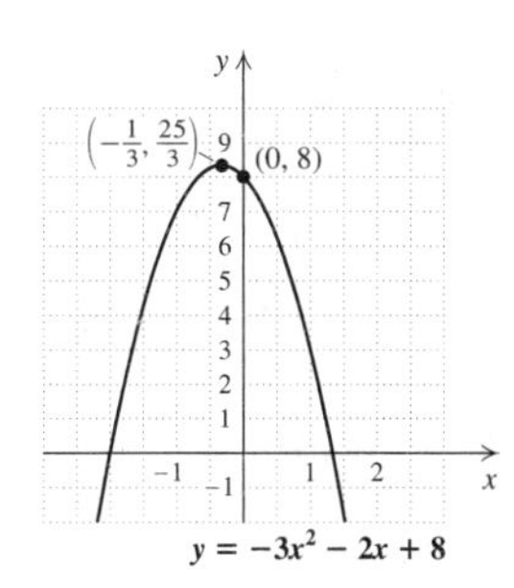

13. $y = -\frac{1}{3}x^2$

Find the vertex. The x-coordinate is

$$-\frac{b}{2a} = -\frac{0}{2\left(-\frac{1}{3}\right)} = 0.$$

The y-coordinate is

$$-\frac{1}{3} \cdot 0^2 = 0.$$

The vertex is $(0, 0)$.

We choose some x-values on both sides of the vertex and graph the parabola.

x	y
-3	-3
-1	$-\frac{1}{3}$
2	$-\frac{4}{3}$
3	3

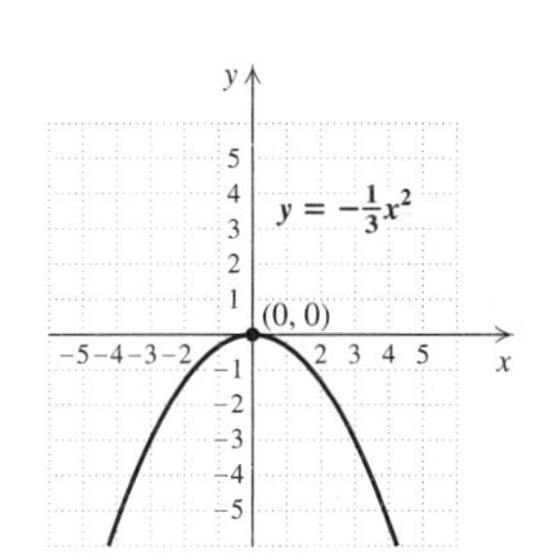

15. $y = -\frac{1}{2}x^2 + 5$

Find the vertex. The x-coordinate is

$$-\frac{b}{2a} = -\frac{0}{2\left(-\frac{1}{2}\right)} = 0.$$

The y-coordinate is

$$-\frac{1}{2} \cdot 0^2 + 5 = 0 + 5 = 5.$$

The vertex is $(0, 5)$.

We choose some x-values on both sides of the vertex and graph the parabola.

x	y
-4	-3
-2	3
2	3
4	-3

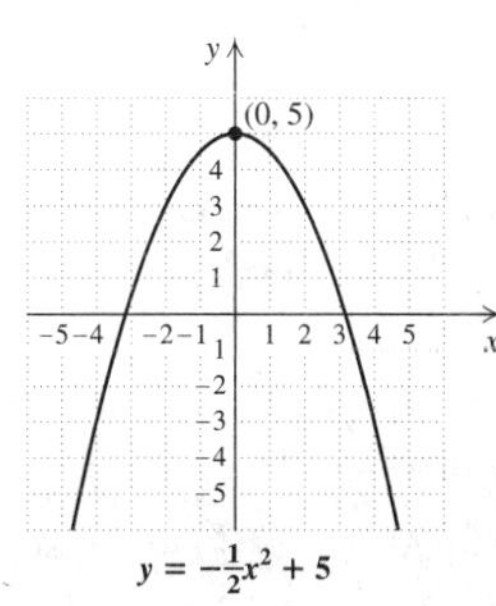

17. $y = x^2 - 3x$

Find the vertex. The x-coordinate is

$$-\frac{b}{2a} = -\frac{-3}{2 \cdot 1} = \frac{3}{2}.$$

The y-coordinate is

$$\left(\frac{3}{2}\right)^2 - 3 \cdot \frac{3}{2} = \frac{9}{4} - \frac{9}{2} = -\frac{9}{4}.$$

The vertex is $\left(\frac{3}{2}, -\frac{9}{4}\right)$.

We choose some x-values on both sides of the vertex and graph the parabola.

x	y
0	0
1	-2
2	-2
3	0

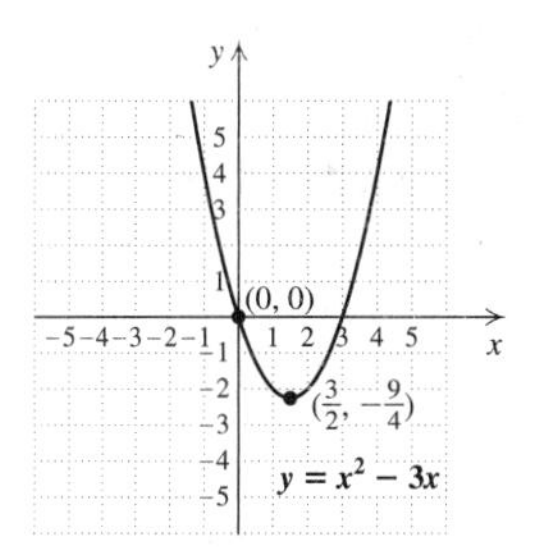

19. $y = x^2 + 2x - 8$

Find the vertex. The x-coordinate is

$$-\frac{b}{2a} = -\frac{2}{2 \cdot 1} = -1.$$

The y-coordinate is

$$(-1)^2 + 2(-1) - 8 = 1 - 2 - 8 = -9.$$

To find the y-intercept we replace x with 0 and compute y:

$$y = 0^2 + 2 \cdot 0 - 8 = 0 + 0 - 8 = -8.$$

The y-intercept is $(0, -8)$.

To find the x-intercepts we replace y with 0 and solve for x.

$$0 = x^2 + 2x - 8$$
$$0 = (x + 4)(x - 2)$$
$$x + 4 = 0 \quad or \quad x - 2 = 0$$
$$x = -4 \quad or \quad x = 2$$

The x-intercepts are $(-4, 0)$ and $(2, 0)$.

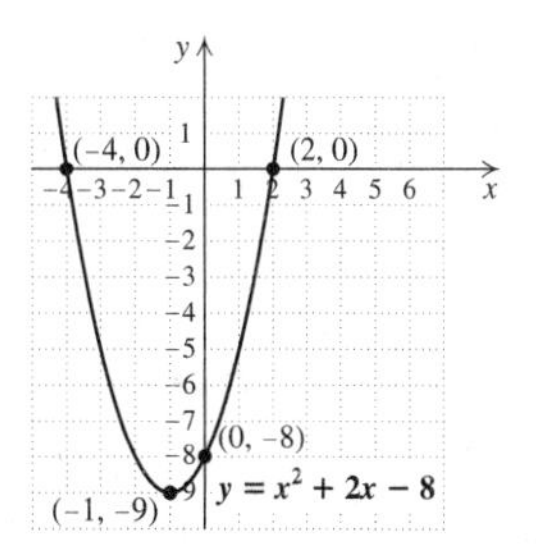

21. $y = 2x^2 - 6x$

Find the vertex. The x-coordinate is

$$-\frac{b}{2a} = -\frac{-6}{2\cdot 2} = \frac{3}{2}.$$

The y-coordinate is

$$2\left(\frac{3}{2}\right)^2 - 6\cdot\frac{3}{2} = \frac{9}{2} - 9 = -\frac{9}{2}.$$

The vertex is $\left(\frac{3}{2}, -\frac{9}{2}\right)$.

To find the y-intercept we replace x with 0 and compute y:

$$y = 2\cdot 0^2 - 6\cdot 0 = 0 - 0 = 0.$$

The y-intercept is $(0,0)$.

To find the x-intercepts we replace y with 0 and solve for x.

$$0 = 2x^2 - 6x$$
$$0 = 2x(x-3)$$
$$2x = 0 \quad or \quad x - 3 = 0$$
$$x = 0 \quad or \quad x = 3$$

The x-intercepts are $(0,0)$ and $(3,0)$.

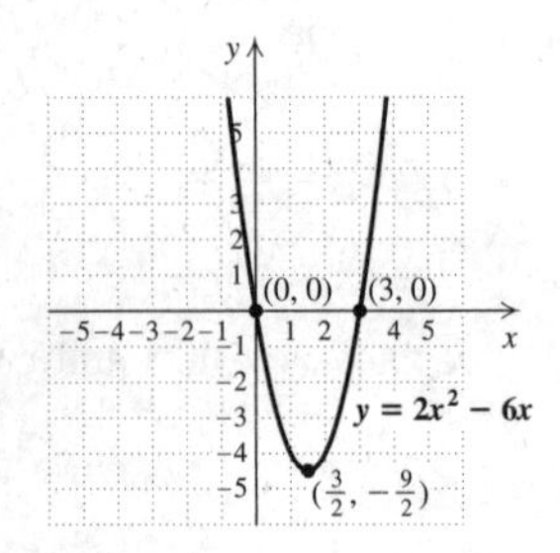

23. $y = -x^2 - x + 12$

Find the vertex. The x-coordinate is

$$-\frac{b}{2a} = -\frac{-1}{2(-1)} = -\frac{1}{2}.$$

The y-coordinate is

$$-\left(-\frac{1}{2}\right)^2 - \left(-\frac{1}{2}\right) + 12 = -\frac{1}{4} + \frac{1}{2} + 12 = \frac{49}{4}.$$

The vertex is $\left(-\frac{1}{2}, \frac{49}{4}\right)$.

To find the y-intercept we replace x with 0 and compute y:

$$y = -0^2 - 0 + 12 = -0 - 0 + 12 = 12.$$

The y-intercept is $(0,12)$.

To find the x-intercepts we replace y with 0 and solve for x.

$$0 = -x^2 - x + 12$$
$$0 = x^2 + x - 12 \qquad \text{Multiplying by } -1$$
$$0 = (x+4)(x-3)$$
$$x + 4 = 0 \quad or \quad x - 3 = 0$$
$$x = -4 \quad or \quad x = 3$$

The x-intercepts are $(-4,0)$ and $(3,0)$.

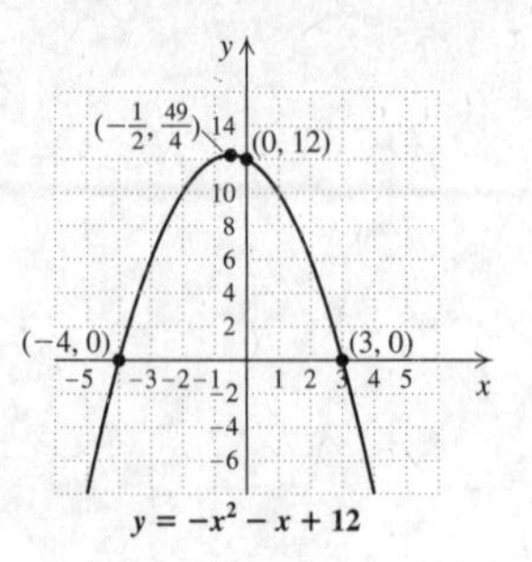

25. $y = 3x^2 - 6x + 1$

Find the vertex. The x-coordinate is

$$-\frac{b}{2a} = -\frac{-6}{2\cdot 3} = 1.$$

The y-coordinate is

$$3\cdot 1^2 - 6\cdot 1 + 1 = 3 - 6 + 1 = -2.$$

The vertex is $(1,-2)$.

To find the y-intercept we replace x with 0 and compute y:

$$y = 3\cdot 0^2 - 6\cdot 0 + 1 = 0 - 0 + 1 = 1.$$

The y-intercept is $(0,1)$.

To find the x-intercepts we replace y with 0 and solve for x.

$$0 = 3x^2 - 6x + 1$$
$$x = \frac{-b \pm \sqrt{b^2 - 4ac}}{2a}$$
$$x = \frac{-(-6) \pm \sqrt{(-6)^2 - 4\cdot 3\cdot 1}}{2\cdot 3}$$
$$x = \frac{6 \pm \sqrt{36-12}}{6} = \frac{6 \pm \sqrt{24}}{6}$$
$$x = \frac{6 \pm 2\sqrt{6}}{6} = \frac{2(3\pm\sqrt{6})}{2\cdot 3}$$
$$x = \frac{3\pm\sqrt{6}}{3}$$

The x-intercepts are $\left(\frac{3-\sqrt{6}}{3}, 0\right)$ and $\left(\frac{3+\sqrt{6}}{3}, 0\right)$, or about $(0.184, 0)$ and $(1.816, 0)$.

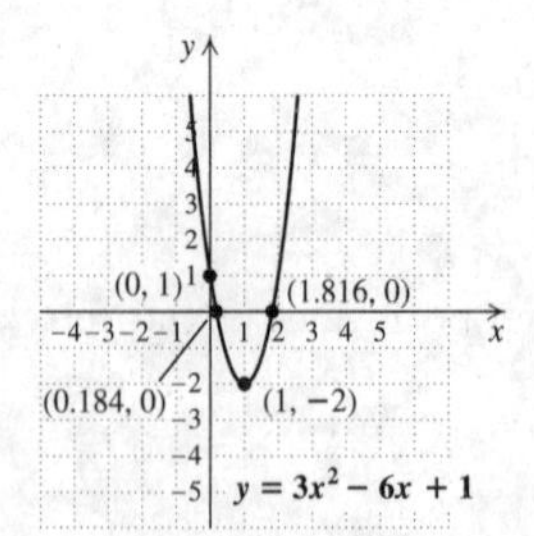

27. $y = x^2 + 2x + 3$

Find the vertex. The x-coordinate is

$$-\frac{b}{2a} = -\frac{2}{2\cdot 1} = -1.$$

The y-coordinate is

$$y = (-1)^2 + 2(-1) + 3 = 1 - 2 + 3 = 2.$$

The vertex is $(-1, 2)$.

To find the y-intercept we replace x with 0 and compute y:

$$y = 0^2 + 2 \cdot 0 + 3 = 0 + 0 + 3 = 3.$$

The y-intercept is $(0, 3)$.

To find the x-intercepts we replace y with 0 and solve for x.

$$0 = x^2 + 2x + 3$$
$$x = \frac{-b \pm \sqrt{b^2 - 4ac}}{2a}$$
$$x = \frac{-2 \pm \sqrt{2^2 - 4 \cdot 1 \cdot 3}}{2 \cdot 1}$$
$$x = \frac{-2 \pm \sqrt{4 - 12}}{2} = \frac{-2 \pm \sqrt{-8}}{2}$$

Because the radicand, -8, is negative the equation has no real-number solutions. Thus, there are no x-intercepts.

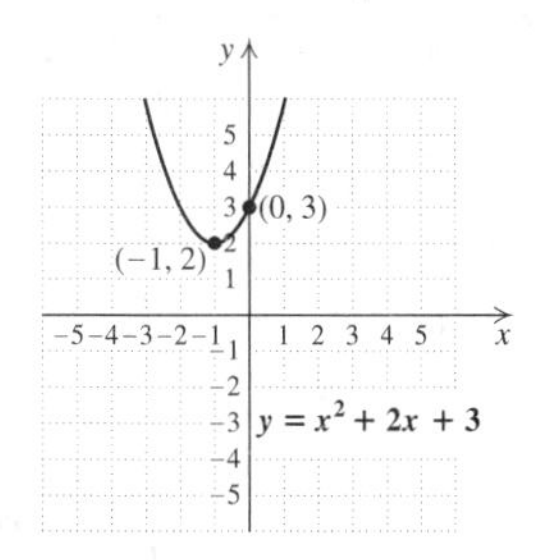

29. $y = 3 - 4x - 2x^2$, or $y = -2x^2 - 4x + 3$

Find the vertex. The x-coordinate is

$$-\frac{b}{2a} = -\frac{-4}{2(-2)} = -1.$$

The y-coordinate is

$$y = 3 - 4(-1) - 2(-1)^2 = 3 + 4 - 2 = 5.$$

The vertex is $(-1, 5)$.

To find the y-intercept we replace x with 0 and compute y:

$$y = 3 - 4 \cdot 0 - 2 \cdot 0^2 = 3 - 0 - 0 = 3.$$

The y-intercept is $(0, 3)$.

To find the x-intercepts we replace y with 0 and solve for x.

$$0 = -2x^2 - 4x + 3$$
$$x = \frac{-b \pm \sqrt{b^2 - 4ac}}{2a}$$
$$x = \frac{-(-4) \pm \sqrt{(-4)^2 - 4(-2)(3)}}{2(-2)}$$
$$x = \frac{4 \pm \sqrt{16 + 24}}{-4} = \frac{4 \pm \sqrt{40}}{-4}$$
$$x = \frac{4 \pm 2\sqrt{10}}{-4} = -\frac{2 \pm \sqrt{10}}{2}$$

The x-intercepts are $\left(\frac{2 - \sqrt{10}}{2}, 0\right)$ and $\left(\frac{2 + \sqrt{10}}{2}, 0\right)$, or about $(0.581, 0)$ and $(-2.581, 0)$.

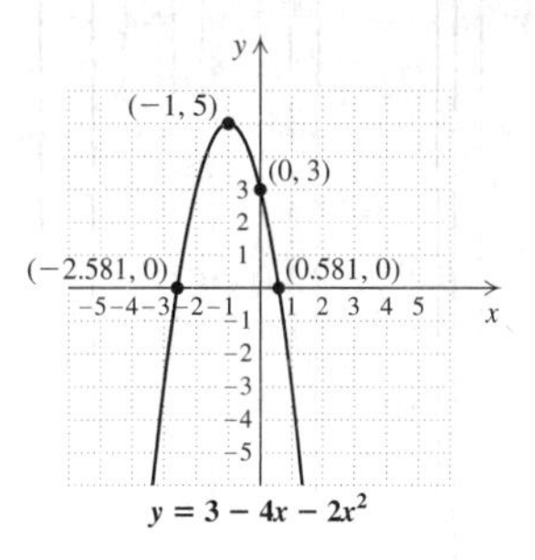

31. *Writing Exercise*

33. $3a^2 - 5a = 3(-1)^2 - 5(-1)$
$= 3 \cdot 1 - 5(-1)$
$= 3 + 5$
$= 8$

35. $5x^3 - 2x = 5(-1)^3 - 2(-1)$
$= 5(-1) - 2(-1)$
$= -5 + 2$
$= -3$

37. $-9.8x^5 + 3.2x = -9.8(1)^5 + 3.2(1)$
$= -9.8(1) + 3.2(1)$
$= -9.8 + 3.2$
$= -6.6$

39. *Writing Exercise*

41. a) We substitute 128 for H and solve for t:

$$128 = -16t^2 + 96t$$
$$16t^2 - 96t + 128 = 0$$
$$16(t^2 - 6t + 8) = 0$$
$$16(t - 2)(t - 4) = 0$$

$t - 2 = 0$ *or* $t - 4 = 0$
$t = 2$ *or* $t = 4$

The projectile is 128 ft from the ground 2 sec after launch and again 4 sec after launch. The graph confirms this.

b) We find the first coordinate of the vertex of the function $H = -16t^2 + 96t$:

$$-\frac{b}{2a} = -\frac{96}{2(-16)} = -\frac{96}{-32} = -(-3) = 3$$

The projectile reaches its maximum height 3 sec after launch. The graph confirms this.

c) Since it takes 3 sec for the projectile to reach its maximum height, it will take another 3 sec for it to return to the ground. Thus, the projectile returns to the ground 6 sec after launch.

We could also do this exercise as shown below.

We substitute 0 for H and solve for t:

$$0 = -16t^2 + 96t$$
$$0 = -16t(t - 6)$$

$-16t = 0$ *or* $t - 6 = 0$
$t = 0$ *or* $t = 6$

At $t = 0$ sec the projectile has not yet been launched. Thus, we use $t = 6$. The projectile returns to the ground 6 sec after launch. The graph confirms this.

43. a) For $r = 25$, $d = 25 + 0.05(25)^2 = 56.25$ ft

For $r = 40$, $d = 40 + 0.05(40)^2 = 120$ ft

For $r = 55$, $d = 55 + 0.05(55)^2 = 206.25$ ft

For $r = 65$, $d = 65 + 0.05(65)^2 = 276.25$ ft

For $r = 75$, $d = 75 + 0.05(75)^2 = 356.25$ ft

For $r = 100$, $d = 100 + 0.05(100)^2 = 600$ ft

b)

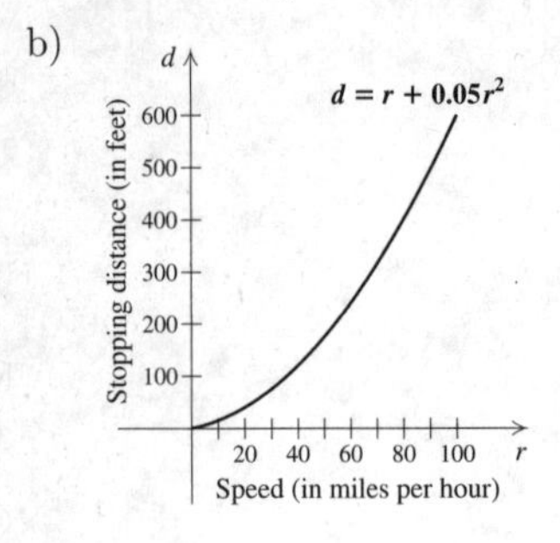

45. *Writing Exercise*

47. $D = (p - 6)^2$

p	D
0	36
1	25
2	16
3	9
4	4
5	1
6	0

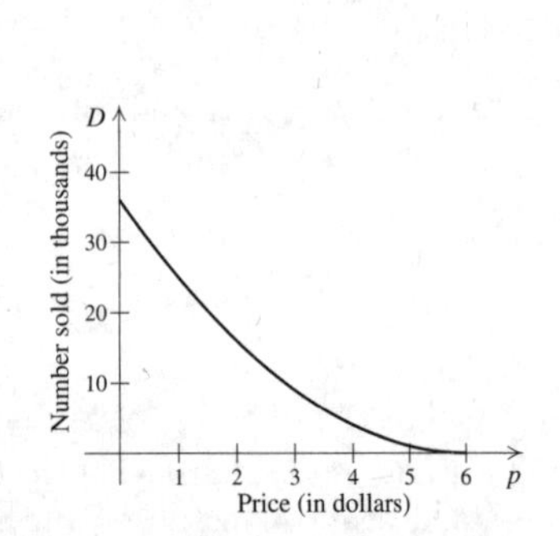

49. Graph $y = x^2 - 5$ and find the first coordinate of the right-hand x-intercept. We find that $\sqrt{5} \approx 2.2361$.

Exercise Set 9.7

1. True; see page 575 in the text.

3. False; see page 578 in the text.

5. Yes; each member of the domain is matched to only one member of the range.

7. Yes; each member of the domain is matched to only one member of the range.

9. No; a member of the domain is matched to more than one member of the range. In fact, each member of the domain is matched to 3 members of the range.

11. Yes; each member of the domain is matched to only one member of the range.

13. $f(x) = x + 5$

$f(4) = 4 + 5 = 9$

$f(7) = 7 + 5 = 12$

$f(-2) = -2 + 5 = 3$

15. $h(p) = 3p$

$h(-7) = 3(-7) = -21$

$h(5) = 3 \cdot 5 = 15$

$h(10) = 3 \cdot 10 = 30$

17. $g(s) = 3s + 4$

$g(1) = 3 \cdot 1 + 4 = 3 + 4 = 7$

$g(-5) = 3(-5) + 4 = -15 + 4 = -11$

$g(6.7) = 3(6.7) + 4 = 20.1 + 4 = 24.1$

19. $F(x) = 2x^2 - 3x$

$F(0) = 2 \cdot 0^2 - 3 \cdot 0 = 0 - 0 = 0$

$F(-1) = 2(-1)^2 - 3(-1) = 2 + 3 = 5$

$F(2) = 2 \cdot 2^2 - 3 \cdot 2 = 8 - 6 = 2$

21. $f(t) = (t + 1)^2$

$f(-5) = (-5 + 1)^2 = (-4)^2 = 16$

$f(0) = (0 + 1)^2 = 1^2 = 1$

$f\left(-\frac{9}{4}\right) = \left(-\frac{9}{4} + 1\right)^2 = \left(-\frac{5}{4}\right)^2 = \frac{25}{16}$

23. $g(t) = t^3 + 3$

$g(1) = 1^3 + 3 = 1 + 3 = 4$

$g(-5) = (-5)^3 + 3 = -125 + 3 = -122$

$g(0) = 0^3 + 3 = 0 + 3 = 3$

25. $F(x) = 2.75x + 71.48$

a) $F(32) = 2.75(32) + 71.48$

$= 88 + 71.48$

$= 159.48$ cm

b) $F(30) = 2.75(30) + 71.48$

$= 82.5 + 71.48$

$= 153.98$ cm

27. $T(5) = 10 \cdot 5 + 20 = 50 + 20 = 70°$ C

$T(20) = 10 \cdot 20 + 20 = 200 + 20 = 220°$ C

$T(1000) = 10 \cdot 1000 + 20 = 10,000 + 20 = 10,020°$ C

29. $P(d) = 1 + \frac{d}{33}$

$P(20) = 1 + \frac{20}{33} = 1\frac{20}{33}$ atm, or $1.\overline{60}$ atm

$P(30) = 1 + \frac{30}{33} = 1\frac{10}{11}$ atm, or $1.\overline{90}$ atm

$P(100) = 1 + \frac{100}{33} = 1 + 3\frac{1}{33} = 4\frac{1}{33}$ atm, or $4.\overline{03}$ atm

31. Graph $f(x) = 2x - 3$

Make a list of function values in a table.

When $x = -1$, $f(-1) = 2(-1) - 3 = -2 - 3 = -5$.

When $x = 0$, $f(0) = 2 \cdot 0 - 3 = 0 - 3 = -3$.

When $x = 2$, $f(2) = 2 \cdot 2 - 3 = 4 - 3 = 1$.

x	$f(x)$
-1	-5
0	-3
2	1

Plot these points and connect them.

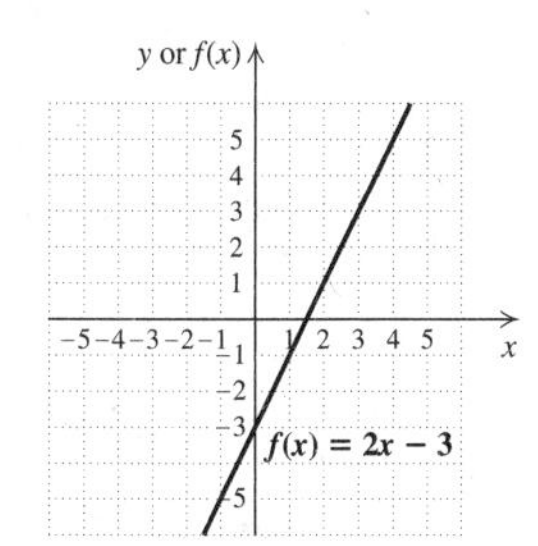

33. Graph $g(x) = -x + 4$

Make a list of function values in a table.

When $x = -1$, $g(-1) = -(-1) + 4 = 1 + 4 = 5$.

When $x = 0$, $g(0) = -0 + 4 = 4$.

When $x = 3$, $g(3) = -3 + 4 = 1$.

x	$g(x)$
-1	5
0	4
3	1

Plot these points and connect them.

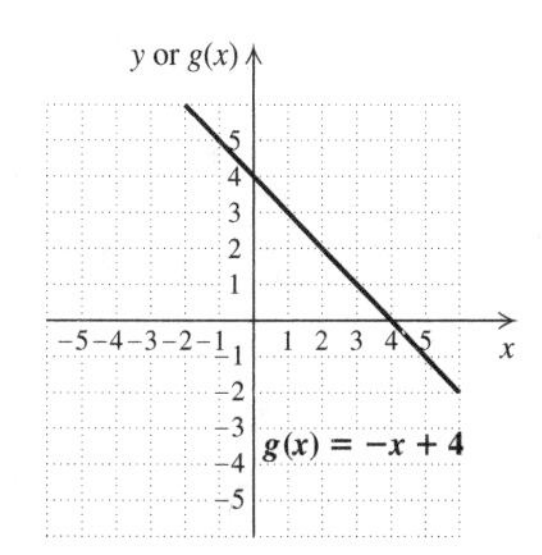

35. Graph $f(x) = \frac{1}{2}x + 1$.

Make a list of function values in a table.

When $x = -2$, $f(-2) = \frac{1}{2}(-2) + 1 = -1 + 1 = 0$.

When $x = 0$, $f(0) = \frac{1}{2} \cdot 0 + 1 = 0 + 1 = 1$.

When $x = 4$, $f(4) = \frac{1}{2} \cdot 4 + 1 = 2 + 1 = 3$.

x	$f(x)$
-2	0
0	1
4	3

Plot these points and connect them.

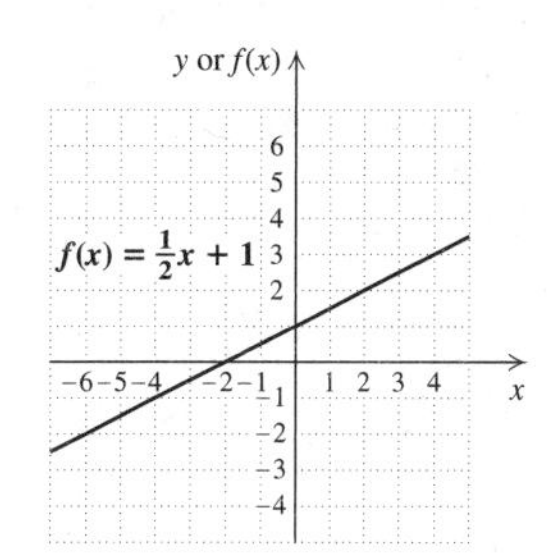

37. Graph $g(x) = 2|x|$.

Make a list of function values in a table.

When $x = -3$, $g(-3) = 2|-3| = 2 \cdot 3 = 6$.

When $x = -1$, $g(-1) = 2|-1| = 2 \cdot 1 = 2$.

When $x = 0$, $g(0) = 2|0| = 2 \cdot 0 = 0$.

When $x = 1$, $g(1) = 2|1| = 2 \cdot 1 = 2$.

When $x = 3$, $g(3) = 2|3| = 2 \cdot 3 = 6$.

x	$g(x)$
-3	6
-1	2
0	0
1	2
3	6

Plot these points and connect them.

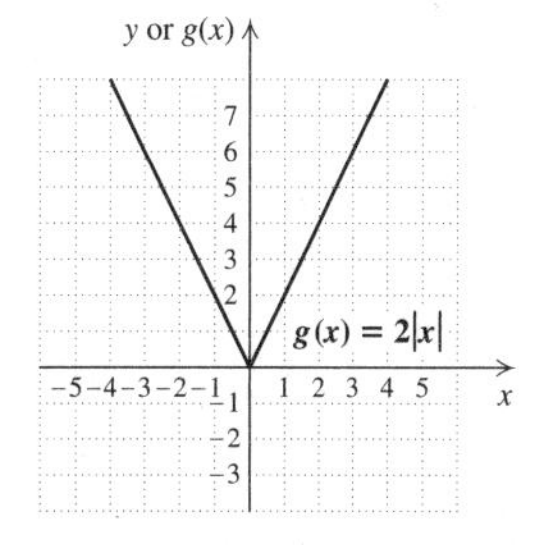

39. Graph $g(x) = x^2$.

Recall from Section 10.5 that the graph is a parabola. Make a list of function values in a table.

When $x = -2$, $g(-2) = (-2)^2 = 4$.

When $x = -1$, $g(-1) = (-1)^2 = 1$.

When $x = 0$, $g(0) = 0^2 = 0$.

When $x = 1$, $g(1) = 1^2 = 1$.

When $x = 2$, $g(2) = 2^2 = 4$.

x	$g(x)$
-2	4
-1	1
0	0
1	1
2	4

Plot these points and connect them.

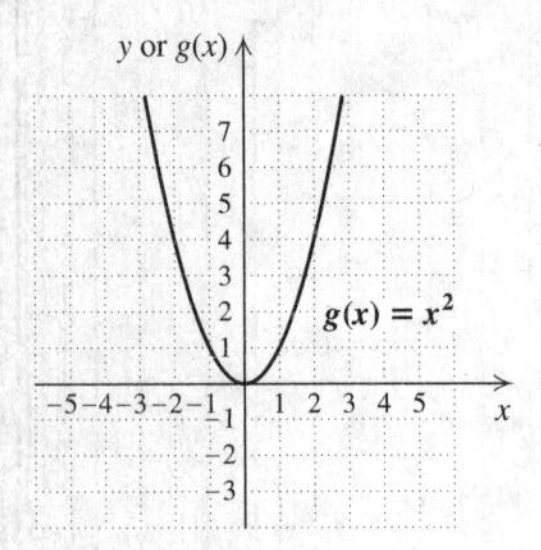

41. Graph $f(x) = x^2 - x - 2$.

Recall from Section 10.5 that the graph is a parabola. Make a list of function values in a table.

When $x = -1$, $f(-1) = (-1)^2 - (-1) - 2 = 1 + 1 - 2 = 0$.

When $x = 0$, $f(0) = 0^2 - 0 - 2 = -2$.

When $x = 1$, $f(1) = 1^2 - 1 - 2 = 1 - 1 - 2 = -2$.

When $x = 2$, $f(2) = 2^2 - 2 - 2 = 4 - 2 - 2 = 0$.

x	$f(x)$
-1	0
0	-2
1	-2
2	0

Plot these points and connect them.

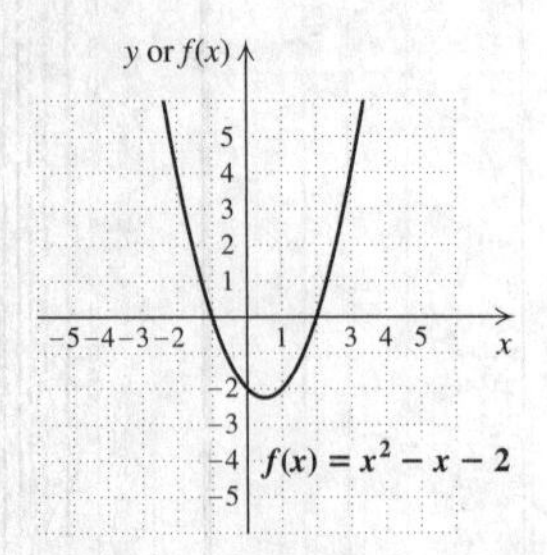

43. The graph is that of a function because no vertical line can cross the graph at more than one point.

45. The graph is not that of a function because a vertical line can cross the graph at more than one point.

47. The graph is that of a function because no vertical line can cross the graph at more than one point.

49. *Writing Exercise*

51. The first equation is in slope-intercept form:

$$y = \frac{3}{4}x - 7,\ m = \frac{3}{4}$$

We write the second equation in slope-intercept form.

$$3x + 4y = 7$$
$$4y = -3x + 7$$
$$y = -\frac{3}{4}x + \frac{7}{4},\ m = -\frac{3}{4}$$

Since the slopes are different, the equations do not represent parallel lines.

53. We write the equations in slope-intercept form.

$$2x = 3y \qquad 4x = 6y - 1$$
$$\frac{2}{3}x = y \qquad 4x + 1 = 6y$$
$$\frac{2}{3}x + \frac{1}{6} = y$$

Since the slopes are the same $\left(m = \frac{2}{3}\right)$ and the y-intercepts are different, the equations represent parallel lines.

55.
$$x - 3y = 2,\quad (1)$$
$$3x - 9y = 6\quad (2)$$

Solve Equation (1) for x.

$$x - 3y = 2 \qquad (1)$$
$$x = 3y + 2 \quad \text{Adding } 3y$$

Substitute $3y + 2$ for x in Equation (2) and solve for y.

$$3x - 9y = 6 \quad (2)$$
$$3(3y + 2) - 9y = 6$$
$$9y + 6 - 9y = 6$$
$$6 = 6$$

We get an equation that is true for all values of y, so there are an infinite number of solutions.

57. *Writing Exercise*

59. Graph $g(x) = x^3$.

Make a list of function values in a table. Then plot the points and connect them.

x	$g(x)$
-2	-8
-1	-1
0	0
1	1
2	8

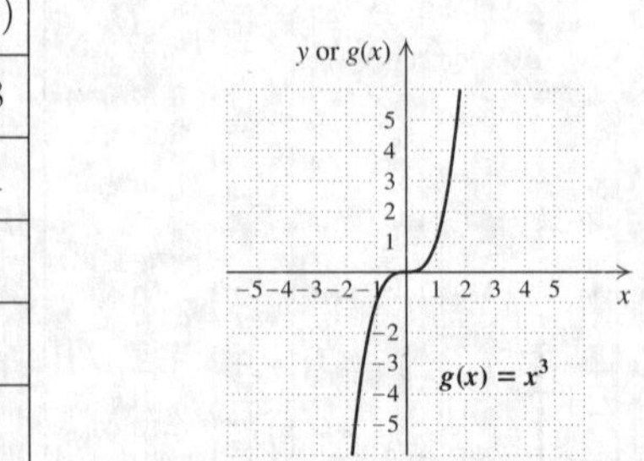

61. Graph $g(x) = |x| + x$.

Make a list of function values in a table. Then plot the points and connect them.

x	$f(x)$
-3	0
-2	0
-1	0
0	0
1	2
2	4
3	6

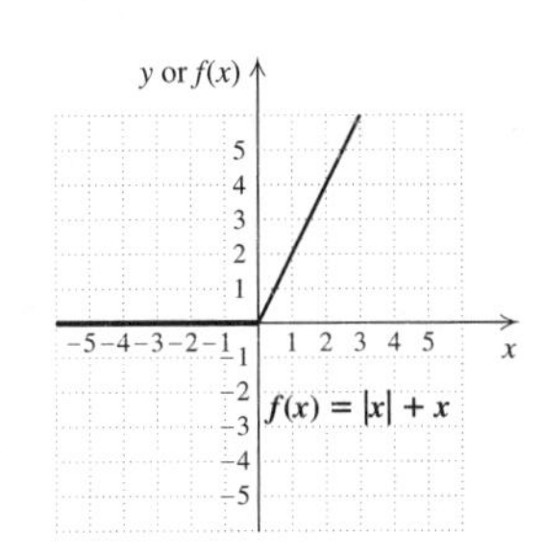

63. Answers may vary.

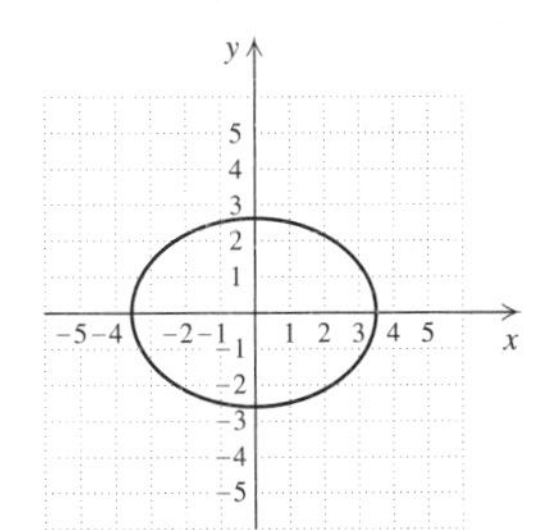

65. $g(x) = ax^2 + bx + c$

$$-4 = a \cdot 0^2 + b \cdot 0 + c, \text{ or } -4 = c \quad (1)$$
$$0 = a(-2)^2 + b(-2) + c, \text{ or } 0 = 4a - 2b + c \quad (2)$$
$$0 = a \cdot 2^2 + b \cdot 2 + c, \text{ or } 0 = 4a + 2b + c \quad (3)$$

Substitute -4 for c in Equations (2) and (3).

$$0 = 4a - 2b - 4, \text{ or } 4 = 4a - 2b \quad (5)$$
$$0 = 4a + 2b - 4, \text{ or } 4 = 4a + 2b \quad (6)$$

Add Equations (5) and (6).

$$8 = 8a$$
$$1 = a$$

Substitute 1 for a in Equation (6).

$$4 = 4 \cdot 1 + 2b$$
$$0 = 2b$$
$$0 = b$$

We have $a = 1$, $b = 0$, $c = -4$, so $g(x) = x^2 - 4$.

67. $g(t) = t^2 - 5$

The domain is the set $\{-3, -2, -1, 0, 1\}$.

$g(-3) = (-3)^2 - 5 = 9 - 5 = 4$

$g(-2) = (-2)^2 - 5 = 4 - 5 = -1$

$g(-1) = (-1)^2 - 5 = 1 - 5 = -4$

$g(0) = 0^2 - 5 = 0 - 5 = -5$

$g(1) = 1^2 - 5 = 1 - 5 = -4$

The range is the set $\{-5, -4, -1, 4\}$.

69. $f(m) = m^3 + 1$

The domain is the set $\{-2, -1, 0, 1, 2\}$.

$f(-2) = (-2)^3 + 1 = -8 + 1 = -7$

$f(-1) = (-1)^3 + 1 = -1 + 1 = 0$

$f(0) = 0^3 + 1 = 0 + 1 = 1$

$f(1) = 1^3 + 1 = 1 + 1 = 2$

$f(2) = 2^3 + 1 = 8 + 1 = 9$

The range is the set $\{-7, 0, 1, 2, 9\}$.

Answers for Exercises in the Appendixes

Exercise Set A

1. $(t+2)(t^2-2t+4)$ **2.** $(p+3)(p^2-3p+9)$

3. $(a-4)(a^2+4a+16)$ **4.** $(w-1)(w^2+w+1)$

5. $(z+5)(z^2-5z+25)$ **6.** $(x+1)(x^2-x+1)$

7. $(2a-1)(4a^2+2a+1)$ **8.** $(3x-1)(9x^2+3x+1)$

9. $(y-3)(y^2+3y+9)$ **10.** $(p-2)(p^2+2p+4)$

11. $(4+5x)(16-20x+25x^2)$

12. $(2+3b)(4-6b+9b^2)$ **13.** $(5p-1)(25p^2+5p+1)$

14. $(4w-1)(16w^2+4w+1)$

15. $(3m+4)(9m^2-12m+16)$

16. $(2t+3)(4t^2-6t+9)$ **17.** $(p-q)(p^2+pq+q^2)$

18. $(a+b)(a^2-ab+b^2)$ **19.** $\left(x+\frac{1}{2}\right)\left(x^2-\frac{1}{2}x+\frac{1}{4}\right)$

20. $\left(y+\frac{1}{3}\right)\left(y^2-\frac{1}{3}y+\frac{1}{9}\right)$ **21.** $2(y-4)(y^2+4y+16)$

22. $3(z-1)(z^2+z+1)$ **23.** $3(2a+1)(4a^2-2a+1)$

24. $2(3x+1)(9x^2-3x+1)$ **25.** $r(s-4)(s^2+4s+16)$

26. $a(b+5)(b^2-5b+25)$

27. $5(x+2z)(x^2-2xz+4z^2)$

28. $2(y-3z)(y^2+3yz+9z^2)$

29. $(x+0.1)(x^2-0.1x+0.01)$

30. $(y+0.5)(y^2-0.5y+0.25)$ **31.** Demonstrate that the product $(a-b)(a^2+b^2) \neq a^3-b^3$. **32.** The number c is not a perfect cube. Otherwise, x^3+c could be factored as $(x+\sqrt[3]{c})(x^2-\sqrt[3]{c}x+(\sqrt[3]{c})^2)$.

33. $(5c^2+2d^2)(25c^4-10c^2d^2+4d^4)$

34. $8(2x^2+t^2)(4x^4-2x^2t^2+t^4)$

35. $3(x^a-2y^b)(x^{2a}+2x^ay^b+4y^{2b})$

36. $\left(\frac{2}{3}x-\frac{1}{4}y\right)\left(\frac{4}{9}x^2++\frac{1}{6}xy+\frac{1}{16}y^2\right)$

37. $\frac{1}{3}\left(\frac{1}{2}xy+z\right)\left(\frac{1}{4}x^2y^2-\frac{1}{2}xyz+z^2\right)$

38. $\frac{1}{2}\left(\frac{1}{2}x^a+y^{2a}z^{3b}\right)\left(\frac{1}{4}x^{2a}-\frac{1}{2}x^ay^{2a}z^{3b}+y^{4a}z^{6b}\right)$

Exercise Set B

1. Mean: 21; median: 18.5; mode: 29 **2.** Mean: 84; median: 85; mode: none **3.** Mean: 21; median: 20; modes: 5, 20 **4.** Mean: 22; median: 25; mode: 13 **5.** Mean: 5.2; median: 5.7; mode: 7.4 **6.** Mean: 20.575; median: 13.4; mode: 13.4 **7.** Mean: 239.5; median: 234; mode: 234 **8.** Mean: \$29.91; median: \$29.95; mode: \$29.95 **9.** Average: $23.\overline{8}$; media: 15; mode: 1 **10.** Average: \$6.51; median: \$6.79; modes: \$5.99, \$6.79 **11.** Mean: 897.2; median: 798; mode: none **12.** Mean: 87.5; median: 87.5; mode: none

13. Average: \$8.19; median: \$8.49; mode: \$6.99

14. Average: 222.25; median: 224; mode: 224

15. 10 home runs **16.** 263 days **17.** $a=30$, $b=58$

18. 177.5 cm

Exercise Set C

1. $\{3,4,5,6,7\}$ **2.** $\{82,83,84,85,86,87,88,89\}$

3. $\{41,43,45,47,49\}$ **4.** $\{15,20,25,30,35\}$

5. $\{-3,3\}$ **6.** $\{0.008\}$ **7.** False **8.** True **9.** True

10. True **11.** True **12.** True **13.** True **14.** True

15. True **16.** False **17.** False **18.** True

19. $\{c,\ d,\ e\}$ **20.** $\{u,\ i\}$ **21.** $\{1,10\}$ **22.** $\{0,1\}$

23. $\emptyset$ **24.** $\emptyset$ **25.** $\{a,\ e,\ i,\ o,\ u,\ q,\ c,\ k\}$

26. $\{a,\ b,\ c,\ d,\ e,\ f,\ g\}$ **27.** $\{0,1,2,5,7,10\}$

28. $\{1,2,5,10,0,7\}$

29. $\{a,\ e,\ i,\ o,\ u,\ m,\ n,\ f,\ g,\ h\}$

30. $\{1,2,5,10,\ a,\ b\}$ **31.** Set-builder notation allows us to name a very large set compactly. **32.** Roster notation allows us to see all the members of a set. It is also useful when a set cannot be named using a general rule or statement. **33.** The set of integers **34.** $\emptyset$

35. The set of real numbers **36.** The set of positive even integers **37.** $\emptyset$ **38.** The set of integers **39.** **(a)** A; **(b)** A; **(c)** A; **(d)** $\emptyset$ **40.** **(a)** Yes; **(b)** no; **(c)** no; **(d)** yes; **(e)** yes; **(f)** no **41.** True